C0-BVE-886

Ecological Studies.

Analysis and Synthesis

Edited by

W.D. Billings, Durham (USA) F. Golley, Athens (USA)

O.L. Lange, Würzburg (FRG) J.S. Olson, Oak Ridge (USA)

H. Remmert, Marburg (FRG)

Volume 63

Ecological Studies

Paul A. Delcourt   Hazel R. Delcourt

Long-Term Forest Dynamics of the Temperate Zone

A Case Study of Late-Quaternary Forests
in Eastern North America

With 90 Illustrations in 131 Parts and 333 Maps

Springer-Verlag
New York Berlin Heidelberg
London Paris Tokyo

PAUL A. DELCOURT
Program for Quaternary Studies
  of the Southeastern United States
Graduate Program in Ecology
Department of Geological Sciences
University of Tennessee
Knoxville, TN 37996, USA

HAZEL R. DELCOURT
Program for Quaternary Studies
  of the Southeastern United States
Graduate Program in Ecology
Department of Botany
University of Tennessee
Knoxville, TN 37996, USA

Library of Congress Cataloging-in-Publication Data
Delcourt, Paul A.
  Long-term forest dynamics of the Temperate Zone.
  (Ecological studies ; v. 63)
  Bibliography: p.
  Includes index.
  1. Pollen, Fossil. 2. Paleoecology—North America.
3. Paleobotany—Quaternary. I. Delcourt, Hazel R.
II. Title. III. Series.
QE993.2.D45 1987   561'13   87-12859

Typeset by David Seham Associates, Inc., Metuchen, New Jersey.
Printed and bound by Quinn-Woodbine Inc., Woodbine, New Jersey.
Printed in the United States of America.

9 8 7 6 5 4 3 2 1

ISBN 0-387-96495-9 Springer-Verlag New York Berlin Heidelberg
ISBN 3-540-96495-9 Springer-Verlag Berlin Heidelberg New York

To our Paleoecological Mentors

Ronald O. Kapp
Herbert E. Wright, Jr.

Preface

The synthesis presented in this volume is a direct outgrowth of our ten-year FORMAP Project (Forest Mapping Across Eastern North America from 20,000 yr B.P. to the Present). Many previous research efforts in paleoecology have used plant-fossil evidence as proxy information for primarily geologic or climatic reconstructions or as a bio-stratigraphic basis for correlation of regional events. In contrast, in this book, we deal with ecological questions that require a holistic perspective that integrates the interactions of biota with their dynamically changing environments over time scales up to tens of thousands of years.

In the FORMAP Project, our major research objective has been to use late-Quaternary plant-ecological data sets to evaluate long-term patterns and processes in forest development. In order to accomplish this objective, we have prepared subcontinent-scale calibrations that quantitatively relate the production and dispersal of arboreal pollen to dominance in the vegetation for the major tree types of eastern North America. Quantification of pollen-vegetation relationships provides a basis for developing quantitative plant-ecological data sets that allow further ecological analysis of both individual taxa and forest communities through time. Application of these calibrations to fossil-pollen records for interpreting forest history thus represents a fundamental step beyond traditional summaries based upon pollen percentages.

The scope of this book includes an introduction to the nature of long-term vegetational change, dealing with different kinds of environmental and biotic changes that result in dynamic patterns and processes in forest communities. Quantification of modern pollen-vegetation relationships permits the development of techniques for reconstruction of past forests, when applied to plant-fossil sequences in, for example, a case study of eastern North America. We use these pollen-vegetation calibrations to reconstruct forest history based upon radiocarbon-dated, late-Quaternary fossil-pollen sequences of 162 sites distributed across the eastern half of North America. The resulting FORMAP time series of maps depict the changing distribution and dominance patterns for populations of major temperate and boreal tree taxa. Areographic analysis of the map patterns provides new insights into migrational strategies of trees during times of environmental change. The role of competition in structuring past forest communities is examined using multiple-taxa population-growth models in conjunction with empirical paleoecological data. Ordination of the paleoecological data using Detrended Correspondence Analysis gives information concerning changes in position and steepness of ecotones. This study represents the first use of DCA (DECORANA) in direct gradient analysis of changing ecoclines through both space and time on a scale relevant to the dynamic development of temperate forest communities. We also develop perspectives in Quaternary landscape ecology, integrating climatic and geomorphic processes with changes in vegetational patterns through time. We compare late-Quaternary pattern and process in development of eastern North American forests with forest dynamics of the Temperate Zone of Europe.

The Appendix to this volume includes a listing with supplementary information for all fossil-pollen sites used in the FORMAP project. The complete data bases for modern forest composition and pollen spectra, as well as all fossil-pollen data and forest reconstructions, are on file at the Program for Quaternary Studies of the Southeastern United States, University of Tennessee, Knoxville.

We acknowledge financial support from the Ecology Program of the United States National Science Foundation for the FORMAP Project (grants DEB-80-04168 and BSR-83-06915) and for related research (grants BSR-83-00345 and BSR-84-15652). We are grateful to P. Wright and D. Broach for their assistance in computer programming. We express our appreciation to the many palynologists who have graciously made their late-Quaternary pollen data available to us for this purpose. We collaborated with T. Webb III in the compilation and exchange of both modern and fossil pollen data for eastern North America. We thank I.C. Prentice for his advice concerning appropriate statistical techniques for evaluating paleoecological data. We thank F.B. Golley, S.A. Hall, and P.S. White for their constructive reviews of earlier drafts of this manuscript. This volume constitutes Contribution No. 43 of the Program for Quaternary Studies of the Southeastern United States, University of Tennessee, Knoxville. This book also represents a United States Contribution to the International Geological Correlation Programme (I.G.C.P.) Project 158B, dealing with late-Quaternary biotic and environmental changes in the Temperate Zone.

Knoxville, Tennessee, 1987   Paul A. Delcourt
                                                              Hazel R. Delcourt

Contents

plant communities (Delcourt and Delcourt 1985a; Jacobson and Grimm 1986; Prentice 1986). From a series of paleoecological sites distributed in appropriate geographic locations, maps can be compiled for populations of individual taxa that show either their distribution and dominance at a given time or changes in their populations over time (Huntley and Birks 1983). From such maps, strategies of migration for plant taxa that are analogous to life-history or plant-successional strategies may be evaluated (Rapoport 1982).

The recent focus on questions of pollen representation has brought Quaternary palynologists closer to attaining goals relating to quantification of the paleoecological record and of increased sophistication in ecological and biogeographical interpretation. It is therefore timely to consider the nature of long-term forest dynamics in the Temperate Zone (*sensu* Walter 1979), with an emphasis on examples from the geographic regions that have been most intensively studied by Quaternary plant ecologists. In this volume, we summarize much of the available literature on late-Quaternary palynology in eastern North America in the form of calibrated, mapped summaries of changes in tree populations through the past 20,000 years, as well as in the form of population trajectories at key sites and across broad geographic regions. We then compare the history of temperate forests in eastern North America with that of Europe (as recently summarized by Huntley and Birks 1983). This paleoecological information forms the basis for integration of patterns and processes in vegetation, climate, and geomorphology as a holistic view of Quaternary landscape ecology for the northern Temperate Zone.

Spatial and temporal scale

As integrating and organizing themes, hierarchies of scale in space and time have emerged as paradigms in ecology (Allen and Starr 1982; Delcourt *et al.* 1982). Not only do fundamental environmental forcing functions and types of biotic responses differ at different spatial and temporal scales (Fig. 1.1), but so do the vegetational patterns we perceive as a consequence of the window through which we view vegetational change.

The scale paradigm

For convenience, we organize space and time into a series of "domains" within which different environmental forcing functions result in specific biotic responses perceptible as discrete vegetational patterns (Fig. 1.1). In each space–time domain, certain specific dynamics of vegetation can be recorded within a given vegetational pattern. Both temporal and spatial scales contain some overlap in patterns perceived. For example, both individual stands of trees and forest subtypes are resolvable on a scale of 10^2 m^2 and 10^2 years (Fig. 1.1). Other scales have no overlap between length of time and area over which changes in vegetation can be measured. For instance, daily measurements of physiological changes within the xylem and phloem of an individual tree may

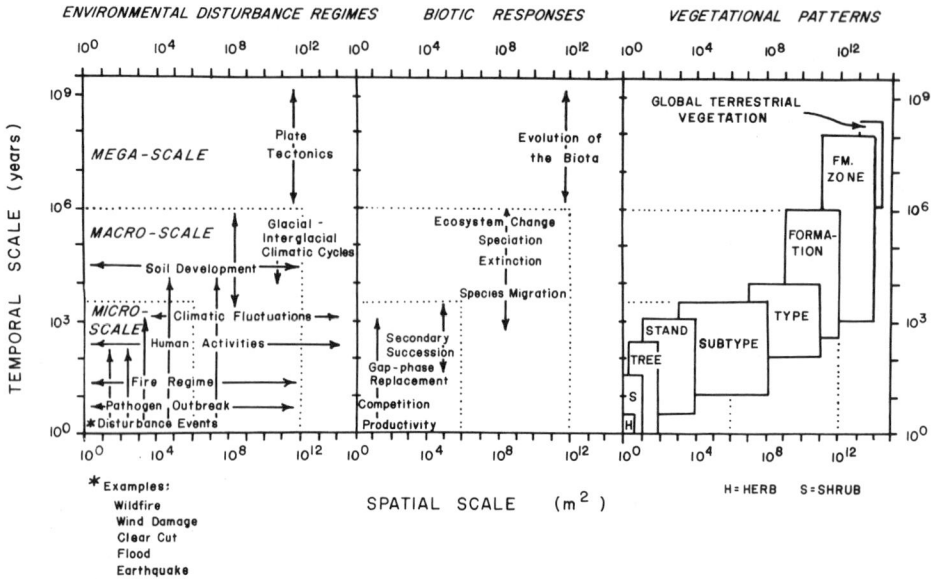

Figure 1.1. Environmental disturbance regimes, biotic responses, and vegetational patterns viewed in the context of space–time domains in which the scale for each process or pattern reflects the sampling intervals required to observe it. The time scale for the vegetational patterns is the time interval required to record their dynamics. The vegetational units are graphed as a nested series of vegetational patterns (from Delcourt *et al.* 1982, © Pergamon Press, Ltd, reprinted with permission).

be required in order to determine cold-hardiness adaptations that allow it to survive annual cycles of weather (Burke *et al.* 1975). Short-term demographic studies within a local population may suffice to explain the relationship between seed-set, establishment, and mortality of individuals (Harper 1977; White 1985). However, understanding the development of large forest types or of vegetational formations requires a perspective that can only be gained from thousands of years of empirical observation of secular climatic change along with changes in dominance and shifts in population centers of entire tree species' populations (Delcourt and Delcourt 1981, 1984).

The mega-scale domain

The mega-scale spatial–temporal domain (Fig. 1.1) encompasses the entire area of global terrestrial vegetation, $1.5 \cdot 10^{14} \text{ m}^2$ (Leith and Whittaker 1975), and it extends back in time to the origin of terrestrial vascular-plant life (approximately $4 \cdot 10^8$ years ago; Taylor 1981). On time scales greater than 10^6 years and over areas greater than 10^{12} m^2, the most significant environmental changes have been conditioned primarily by plate tectonics. Continents have changed orientation with respect to the equator and the poles; they have rifted apart and rejoined in different configurations (Hamilton 1983). Mountain chains have developed along collision zones between continental plates, creating topographic diversity across the land masses (Hamilton 1983). The sizes and positions of oceans have changed dramatically through time, fluctuating in the extent of shoreline surrounding continental land masses. Global climate (Fig. 1.2) has oscillated greatly through time (Frakes 1979). Over geologic time, major changes have occurred in the biosphere in response to alteration of the geosphere, hydrosphere, and atmosphere (Cloud 1972). In response to changes in paleogeography, evolution of the terrestrial biota has resulted in the development of terrestrial vegetation on a global scale (Niklas *et al.* 1985). Within the global terrestrial vegetation, major formation zones or biomes constitute the predominant resolvable pattern, representing fundamental global ecological regions (Walter 1979).

Development of temperate vegetation as we know it today followed the diversification of the angiosperms as dominant terrestrial vascular plants in the

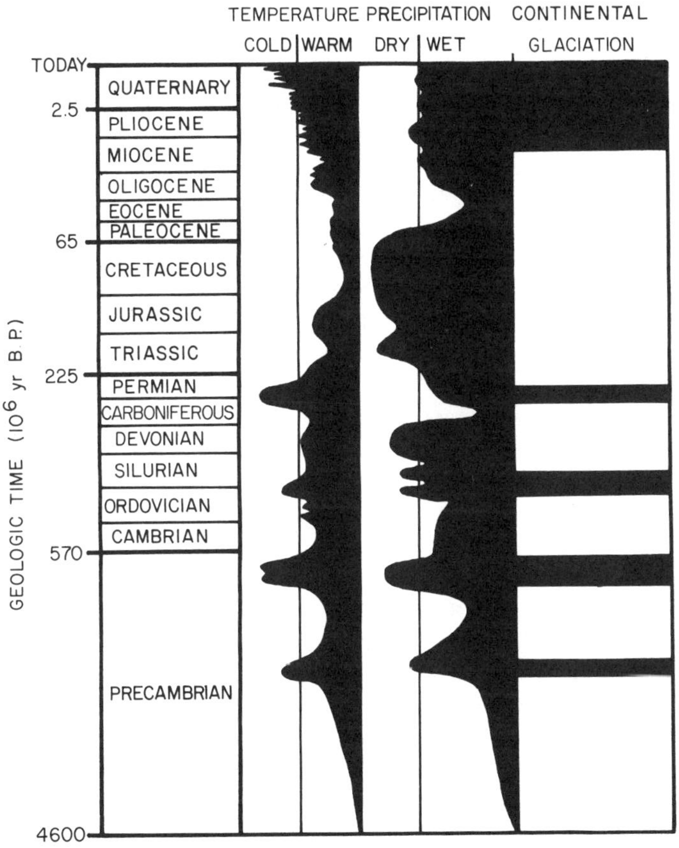

Figure 1.2. Generalized temperature and precipitation history of the Earth. Trends are dashed where data are very sparse. The curves are drawn to represent postulated departures from present global means, but only relative values are indicated. The timescale is progressively expanded in younger time units (modified from Frakes 1979).

early Cretaceous Period (Hughes 1976), about 140 million years ago. In North America, temperate tree genera are known from pollen and macrofossil evidence as early in the Tertiary Period as 50 to 60 million years ago (Graham 1972; Muller 1981; Tiffney 1985).

Extensive temperate forests first developed in the Paleocene to Eocene Epochs in mountainous regions ranging from northern North America to Greenland (Axelrod 1983; Tiffney 1985). With climatic cooling in the late Tertiary, temperate forest taxa spread into lowland basins, both along coastal areas and in continental interiors. By the Miocene Epoch, most temperate woody taxa were similar in morphology to extant species (Axelrod 1983). Adaptive success of temperate angiosperm trees in occupying extensive areas of the continents during the Tertiary Period was in part related to their pollination biology (Whitehead 1985). Today, wind pollination is most frequent in plants living at higher elevations and more northern latitudes (Regal 1982). In angiosperm trees, the evolution of floral structures adapted for wind pollination may be related to both seasonality of climates in temperate regions and spatial clumping of individuals of the same species, itself a consequence of the lower species diversity within temperate forests relative to tropical ecosystems (Whitehead 1985). Speciation within modern North American genera of angiosperm trees, such as in the oaks (*Quercus*), occurred during the Miocene (Axelrod 1983) in response to gradual climatic cooling and drying and increasing topographic diversity that accompanied tectonic uplift of the Coast Ranges and the Cascades (Graham 1972).

Mega-scale climatic and geologic processes have resulted in the fragmentation of temperate forests into disjunct remnants that have persisted to varying degrees throughout the Quaternary in eastern Asia, western and south-central Europe, and North America (Taylor 1981; Tiffney 1985). Mega-scale forcing functions thus have resulted in floristic patterns observable today. The relationship of existing floristic assemblages to plant-community structure and composition, patterns of diversity, and distribution of species along environmental gradients is, however, not explainable by mega-scale processes (White 1983). Tertiary-age communities were not strict templates for modern temperate forests, because temperate vegetation has been greatly affected by macro-scale and micro-scale environmental changes of the Quaternary Period (Davis 1983).

Another factor affecting the development of major floristic regions on the mega-scale is the establishment of intercontinental connections via land bridges (Tiffney 1985). The closure of the Isthmus of Panama about 3 million years ago resulted in the joining of North and South America, two continents with distinct floras and faunas. After the land bridge closed, migrations occurred in both directions, followed by extinctions of species that resulted from competitive exclusion, as predicted for smaller islands by MacArthur-Wilson species equilibrium theory (Marshall *et al.* 1982).

The macro-scale domain

Whereas the processes operative at the mega-scale of resolution have been responsible for evolution of the biota, observations made at the macro-scale are most relevant for examining regional and subcontinental vegetational changes during most of the Quaternary Period. The macro-scale domain (Fig. 1.1) encompasses phenomena occurring over $5 \cdot 10^3$ to 10^6 years and from 10^6 to 10^{12} m^2.

During the Quaternary (approximately the last 2.5 million years), climatic change has been the predominant environmental forcing function for vegetational change in temperate regions. The magnitude of climatic change through glacial–interglacial cycles has resulted in major changes in temperate ecosystems, with limited speciation, but with widespread extinctions of both tree species (Van der Hammen et al. 1971) and megafauna (Kurten and Anderson 1980; Martin and Klein 1984). Extinctions of temperate tree species have been differential across the Temperate Zone and were primarily related to the existence and geographical orientation of migrational barriers and corridors (Davis 1983). Changing climates and composition of plant communities have influenced long-term soil development (Van der Hammen et al. 1971; Stockmarr 1975). In the Holocene (the past 10,000 to 12,500 years), human activities have become an increasingly important factor influencing vegetation composition and cover (Behre 1981; Barker 1985; Delcourt et al. 1986a). On the macro-scale, vegetation patterns that are primarily influenced by these processes are those units mapped as vegetation types and formations (Fig. 1.1).

As the quantity of paleoecological data has increased, and particularly as pollen records have been produced for which chronologies have been established through radiocarbon dating, Quaternary paleoecologists have attempted to summarize the information in the form of paleovegetation maps. These mapped summaries represent generalizations drawn from analogies between modern pollen assemblages collected within known vegetation types (Davis and Webb 1975). Using a widely dispersed grid of sample points, and with pollen-assemblage data generally resolvable at 500- to 1000-year intervals, such maps depict an interpretation of broad physiognomic and compositional patterns based on structural dominants at specified times of greatest interest, e.g., the full-glacial interval (Delcourt and Delcourt 1979, 1981).

Maps of late-Quaternary vegetational patterns have been produced for several regions of the Temperate Zone, including Japan (Tsukada 1985), the U.S.S.R (Grichuk 1984; Khotinskiy 1984), Europe (Van der Hammen et al. 1971; Huntley and Birks 1983), and North America (Martin and Mehringer 1965; Whitehead 1973; Ritchie 1976, 1984; Delcourt and Delcourt 1979, 1981; Richard 1985; Davis and Jacobson 1985).

Maps produced for the time-scale of the last glacial–interglacial cycle (20,000 years Before Present or ''yr B.P.'' to the present) illustrate that major changes have occurred in both the distribution and the composition of vegetation types. Southward displacements of vegetation during times of climatic cooling were not strictly zonal (Whitehead 1973), although broadly defined map units such as boreal coniferous forest and temperate southeastern evergreen forest in many cases remained largely intact, changing primarily in area and location (Delcourt and Delcourt, 1981). Use of quantitative measures of dissimilarity between full-glacial and modern pollen spectra demonstrate that, with an adequate data base of modern pollen samples, close modern analogues can be found for certain full-glacial assemblages, such as the boreal forest of eastern North America (Delcourt and Delcourt 1985a). Because tree species respond individualistically to climatic changes (Davis 1981a), considerable variability in community composition is expectable across biomes that span broad latitudinal belts, especially during times of major climatic change such as the late-glacial to early-Holocene transition (16,500 to 10,000 years ago). During times of climatic change, changes in vegetation, in terms of both composition and physiognomy, represent a continuum in time (Davis and Jacobson 1985) analogous to continua observed across climatic gradients in space at a given time (Whittaker 1956).

The micro-scale domain

The micro-scale of resolution (from 1 year to $5 \cdot 10^3$ years, 1 m^2 to 10^6 m^2) is primarily the domain of traditional plant-ecological studies (Fig. 1.1) but also overlaps substantially with Quaternary paleoecological research. On the micro-scale, local disturbance events, such as outbreaks of wildfire, wind damage during storms, floods, earthquakes, and extensive cutting of forests by humans, influence plant succession within individual forest stands. Short-term climatic fluctuations affect the productivity of tree species, as well as influencing seed-set, germination, and survivorship of genets (Harper 1977). Competition for light, water, and nutrients on microsites results in a landscape patchwork of plant succession, with resulting patterns determined by prevailing disturbance regime (Forman and Godron 1986). Certain kinds of disturbance, such as fire regime and pathogen outbreaks (Davis 1981b), extend beyond the bounds of the micro-scale and are effective in determining more long-term and broad-scale patterns of vegetation.

Pollen records from lakes, ponds, and pools within woodland hollows represent vegetation on a series of nested spatial scales that vary in radius from several kilometers to as fine grained as 20 meters from the coring site (Jacobson and Bradshaw 1981). Pairs of relatively small paleoecological sites, similar in size but located on different site types, can be compared in order to investigate vegetation processes linked directly to site position on topographic gradients, different slope aspects, or soil types (Brubaker 1975; Jacobson 1979). Successional relationships of deciduous tree species such as lime (*Tilia cordata*) in woodlands of southern England have been examined through study of small (<20 meters in diameter) woodland hollows on sites with contrasting histories of disturbance (Bradshaw 1981a,b). In the Great Smoky Mountains National Park of Tennessee, Davidson (1983) used pollen and plant-macrofossil assemblages to reconstruct the history of changes in canopy closure and areal extent of the wetland over the past 6600 years from a woodland hollow located beneath a modern forest canopy of sweetgum (*Liquidambar styraciflua*).

When vegetation histories are documented from several nearby sites that represent a suite of spatial scales, insights can be gained in distinguishing vegetation changes that are local in nature from those that are characteristic of the larger region. In a study in central Wisconsin, Heide (1984) compared regional and local vegetation changes through the late Holocene by contrasting the records from a medium-sized (25 ha) lake and a small (<0.5 ha) hollow beneath a closed canopy of mixed conifer–northern hardwoods forest. The pollen record from the lake recorded broad-scale phenomena such as the immigration of hemlock (*Tsuga canadensis*), whereas the woodland hollow exemplified the local interactions of late-successional taxa such as sugar maple (*Acer saccharum*) that tend to be sparsely represented in the regional pollen rain (Heide 1984).

Quaternary environmental changes as forcing functions

The Quaternary Period has extended over the last several million years. The Tertiary–Quaternary boundary is variously dated as starting between 2.5 and 1.6 million years ago (Bowen 1978; Boellstorff 1978). During the Quaternary, the Earth experienced a progressive cooling, with a series of at least 23 glacial–interglacial cycles (van Donk 1976). The Quaternary is not unique as the time of the Great Ice Age (Geikie 1874); rather, it is the last of a number of major episodes of widespread continental glaciation. Based upon oxygen isotope and other geologic evidence (Fig. 1.2), the broad patterns of changing temperature and precipitation have been reconstructed for the globe's terrestrial and oceanic surfaces for the past 4 billion years (Frakes 1979).

Episodes of continental glaciation in Earth history

Throughout geologic time, several recurring sets of conditions have led to continental glaciation. The spatial distribution of continents has been a critical factor for meridional transport of moisture from the equatorial belt to high latitudes (Frakes 1979). Prior to each episode of glaciation, warm and humid climates resulted in a progressive increase of atmospheric water vapor. Evaporation occurred in low-latitude oceans, and moisture was transported poleward to the continents. Regionally, shallow marine embayments and epeiric seas were an additional moisture source. Continents positioned at or near the geographic poles provided platforms within colder climates that accumulated snow and ice over long periods of time. This combination of conditions has occurred five times since the Earth's beginning 4.6 billion years ago (Fig. 1.2). Major continental glaciations occurred in the middle Precambrian (circa 2300 million years ago), the late Precambrian (from 950 to 615 million years ago), the late Ordovician and early Silurian (from about 440 to 430 million years ago), the late Carboniferous (Pennsylvanian) and Permian (lasting from 330 to 240 million years ago), and the late-Cenozoic interval (beginning 22.5 million years ago in the early Miocene Epoch of the late Tertiary Period and lasting through the Quaternary Period) (Frakes 1979; Crowell 1982).

Plate-tectonic changes thus set the stage for the late-Cenozoic Ice Ages. During the middle to late Tertiary Period, the northward drifting of Australia away from Antarctica reduced the continuity of the Tethys equatorial current system and separated the Indian and Pacific oceans. The westward-flowing

Pacific equatorial current was deflected southward along eastern Australia into a warm subtropical gyre of circulation; this gyre provided a high-latitude source for evaporation and brought snowfall to the Antarctic continent (Berggren 1982). The Antarctic continent then shifted over the South Pole. This polar platform was initially glaciated during the Oligocene; however, buildup of a major polar ice cap did not occur until about 14 million years ago in the middle Miocene (Shackleton and Kennett 1975). The Tethys Seaway, formerly trending east–west from the Pacific Ocean to the Atlantic Ocean, became the Mediterranean Sea. The western portion of the Mediterranean was closed with the collision of the European and African continental plates. The Asian continent moved northward, constricting the Arctic basin. This resulted in continental land masses predominantly located at very high latitudes (Menard 1971).

During the late Pliocene Epoch, the Isthmus of Panama was closed by volcanic activity. By 3 million years ago, the Isthmus was a continuous land bridge between North and South America, eliminating the exchange of equatorial water between the Pacific and Atlantic Oceans (Berggren and Hollister 1974). Thereafter, the Gulf Stream strengthened in intensity and flowed northeastward out of the Gulf of Mexico, along the Atlantic Coastal Plain, and across the northern Atlantic Ocean.

Between 3.0 and 2.5 million years ago, mountain glaciation developed in the high latitudes of the Northern Hemisphere. The source of moisture necessary for the long-term accumulation of glacial ice in Greenland and northeastern Canada was the warm, high-salinity water introduced by the Gulf Stream into the North Atlantic (Cronin *et al.* 1981; Berggren 1982). The cooling trend that began in the Southern Hemisphere in the mid-Miocene Epoch culminated in widespread continental glaciation in the Northern Hemisphere in the late Pliocene and Quaternary.

Thus, with the development of the modern plate-tectonic configuration 2.5 million years ago, climates at mid- and high latitudes became sensitive to relatively modest seasonal variations in the incoming solar radiation intercepted by the Earth's atmosphere.

Pacemaker of the late-Cenozoic Ice Ages

Climatic changes during the late Pliocene and the Quaternary have been characterized by both a steepening in the global cooling trend and an increase in the magnitude and frequency of climatic oscillations between glacial and interglacial extremes (Fig. 1.2). Climatic change can be measured as a change in (1) a specific parameter, such as mean annual temperature; (2) the distributional pattern of climatic regimes, involving the spatial and temporal dynamics of elements within atmospheric and oceanic circulation systems; or (3) changing modes for interaction of dominant components of the climate system, including their time constants and their response times (lags). Different kinds of climatic changes occur on a variety of spatial and temporal scales (Kutzbach 1976; Budyko 1977).

Sediment cores from deep-ocean basins provide a detailed proxy-paleoclimatic record that extends over hundreds of thousands or millions of years. Oscillations in climate as primarily reflected by changes in volume of glacial ice can be determined by examining the systematic variation of oxygen isotopes (the ratio of ^{16}O to ^{18}O) preserved in calcite or silicate shells of marine fossils (Emiliani 1966). Estimates of changes in sea-surface temperatures can be made from changing composition of planktonic foraminiferal assemblages that lived and responded to water temperature near the ocean surface (Imbrie and Kipp 1971). Time-series analysis of oxygen-isotopic data has identified periodic variation in sediment cores of Quaternary age. These fluctuations in the paleoclimatic record can be explained by a dominant climatic cycle of 100,000 years, and by less prominent cycles of 43,000 years, 24,000 years, and 19,000 years (CLIMAP 1976, 1981, 1984; Hays *et al.* 1976). This evidence provides a powerful test of the astronomical hypothesis, proposed by James Croll in A.D. 1864 (Imbrie and Imbrie 1979) and subsequently refined by Milutin Milankovich (Milankovitch 1941). The Milankovitch hypothesis proposed that changes in the Earth's orbit around the sun generate cyclic changes in solar radiation received by the Earth, producing climatic changes that alternately favor the growth of glaciers, then their disintegration. Hays *et al.* (1976) concluded that systematic and predictable variations in the Earth's orbit about the sun serve as a "pacemaker" to trigger the glacial–interglacial cycles of the Quaternary.

The glacial–interglacial cycle responds to a 3.5% change in the solar radiation over each 100,000-year Milankovitch cycle. This cycle is caused by the variation in the eccentricity of the Earth's elliptical path around the Sun. At its closest approach, the distance between the Sun and the Earth is about 147.1 million km and at the farthest part of the orbit it is 152.1 million km. The other two Milankovitch cycles influence the seasonal contrast in the total radiation received at different latitudes. The 41,000-year Milankovitch cycle is produced by the Earth's wobble about its axis of rotation; the current angle of tilt is 23.5° with respect to the plane of the Earth's orbit. The tilt varies between 22° and 24.5°. The higher degree of the tilt accentuates the seasonal contrast between the amount of solar radiation in winter and summer. The 21,000-year Milankovitch cycle (the average of two periodicities, 23,000 and 19,000 years) represents variation in solar radiation due to precession of the equinoxes. Because of precession of the Earth's rotational axis, the positions of equinox (March 20 and September 22) and of solstice (June 21 and December 21) shift progressively with each annual journey of the Earth around the sun (Imbrie and Imbrie 1979). The 41,000-year cycle primarily influences latitudes above 65°N, the critical region for the initiation or elimination of continental ice sheets. The 21,000-year cycle plays a key role in amplifying climate changes at lower latitudes between 65°S and 65°N (Ruddiman and McIntyre 1981a,b). These two Milankovitch cycles trigger minor glacial advances (stadials) and minor retreats (interstadials) superimposed upon the longer 100,000-year glacial–interglacial cycle.

The 100,000-year Milankovitch cycle has become dominant in the last 900,000 years (Pisias and Moore 1981). Broecker and van Donk (1970) described the oxygen-isotopic curve as sawtoothed, however, not sinusoidal as predicted by

Milankovitch's calculations (Berger 1978). Broecker and van Donk (1970) observed that the buildup of glacial ice on continents occurs gradually, over intervals averaging 90,000 years. In contrast, termination of the glacial mode and the onset of rapid deglaciation during a interglacial interval typically lasts only 10,000 years.

Changes in quantity and seasonal distribution of solar radiation are not sufficient to cause glacial expansion directly during Quaternary Ice Ages. Rather, during a glacial–interglacial cycle, the Milankovitch cycles trigger self-amplifying feedbacks within the Earth's climate system that perpetuate glacial climatic regimes until they are abruptly replaced by a warm interglacial mode. Such a glacial–interglacial cycle has three phases: the initiation of glaciation, feedback of the climate system, and termination of glaciation (Denton and Hughes 1983).

In the Northern Hemisphere, glaciers are initiated on continental mountain ranges and plateaus situated generally between 45°N and 75°N latitude. The growth of continental glaciers is strongly dependent upon the radiation regime of the summer season when the glacial ice is susceptible to melting. In mid- and high latitudes, glaciers grow only when the solar input is low in summer. Glacial expansion can be rapid even under "equable" climatic conditions, however, if solar radiation and seasonal contrasts are minimized, producing relatively warm winters (with mean temperatures still below freezing) and cool summers (Ives *et al.* 1975; Denton and Hughes 1983). Heavy snow accumulation may persist in snowbanks, particularly in the high Ungava, Baffin Island, and Keewatin Plateaus of northern Canada. Increased snow cover increases the reflectance of solar radiation (albedo) of the landscape, leading to progressively colder microclimates favoring the year-round persistence of snow patches (Williams 1978). Coalescence of permanent snow patches across wide areas could take as little as 100 to several thousand years. Decreased summer temperatures and diminished melting of snow would favor the rapid growth of glaciers with an immediate onset of a new Ice Age (Ives *et al.* 1975). With such an albedo-feedback mechanism, a large, continent-based glacier such as the Laurentide Ice Sheet centered over northern North America might form in only 10,000 years (Andrews and Mahaffy 1976).

Once established, the major ice sheets in the Northern Hemisphere represent planetary heat sinks because of their reflective surfaces. The topography of the continent-based ice masses deflects storm tracks along their southern margins (Andrews 1982) and thus concentrates moisture there by steering the flow of airmasses (Kutzbach and Wright 1985). Increases in both area and thickness of ice domes on land cause drops in global sea level. Ice fronts extend to the ocean by lateral flow, developing thick ice shelves grounded on offshore knolls and buttressing the interior dome. Thus, during 90,000 years of ice buildup in each glacial–interglacial cycle, continent-based glaciers expand into adjacent oceans with the development of extensive marine-based ice shelves.

During glacial maxima, the global system of ice sheets is highly vulnerable to deterioration (Denton and Hughes 1983). Glacial growth is terminated over a 10,000-year interglacial interval when the 100,000-year Milankovitch cycle is reinforced by the peak of the 21,000-year cycle (Imbrie and Imbrie 1979). Increased solar radiation and maximum seasonal contrast (very cold winters and very hot summers) in middle latitudes facilitates melting of the principal ice sheets in the Northern Hemisphere (Denton and Hughes 1983).

During the transition from glacial to interglacial conditions, glacial meltwater returned to the oceans by major rivers produces a rise in sea level that submerses coastal plains (Leventer *et al.*, 1982). Rising sea levels float ice shelves off coastal shoals. Along the marine ice margin, streams of ice flow from the interior of the continental glacier to the sea, which provides a heat source sufficient to melt the large volumes of ice rapidly (Denton and Hughes 1983). Calving and melting of ice bergs initially chills the North Atlantic and North Pacific Oceans and produces a freshwater lid over the denser sea water. This cold freshwater layer provides little moisture to replenish the adjacent continental glaciers. Its evaporation is minimal in summer and, in winter, the freshwater layer freezes into a thick layer of sea ice. During this phase of glacial disintegration, winter snowfall is reduced on the continents and the glaciers starve (Ruddiman and McIntyre 1981a,b).

Rapid collapse of continental ice sheets during a glacial–interglacial transition is a response first to climatic influences and then to ice–ocean interactions. Late-glacial climatic warming first melts the margins of mid-latitudinal continental glaciers. Then, the rise in sea level removes the buttressing effect of ice shelves. Ice streams sap the remaining ice domes by transporting much of the glacial ice to the heat reservoir of the ocean. The remnants of glacial ice that remain stranded in the continental interior are beyond the marine influence, but they still require many thousands of years to melt within a warm, interglacial climatic regime (Denton and Hughes 1981, 1983).

Leads and lags in response of environment and biota to climatic change

In the Quaternary, the development of major ice sheets at mid- and high latitudes substantially increased the latitudinal gradient in temperature, intensifying the circulation in both the atmosphere (Williams *et al.* 1974; Johnson 1977; Barry 1983) and oceans (CLIMAP 1981). During glacial times, the primary boundaries (major frontal positions) separating polar and subtropical airmasses were displaced by as much as 5° of latitude toward the equator from their positions during interglacial intervals (Johnson 1977; Delcourt and Delcourt 1984). Oscillation between glacial and interglacial climatic regimes has repercussions on both physical and biological systems, with each component of the natural system responding at a different rate (Wright 1984).

Expansion of ice sheets into middle latitudes displaced periglacial zones both adjacent to the ice margin and situated at higher elevations along montane crests in mid- and low latitudes (Pewe 1983). Within the periglacial zone of discontinuous permafrost, the combination of seasonal frost-churning of soils and poorly drained, summer-thaw layers favored the establishment of both herbaceous and woody pioneer species in tundra communities to the exclusion of arboreal taxa.

Advancing glaciers diverted the paths of rivers fed by meltwater from the ice itself. For example, during the last or Wisconsinan Continental Glaciation in North America, the Laurentide Ice Sheet supplied meltwater and glacially-scoured rock debris to the drainage network of the Mississippi River. As a consequence, from about 80,000 to about 9000 years ago, sandy sediments were deposited in braided streams that occupied the entire Lower Mississippi Alluvial Valley, a riverine corridor 100 km wide and extending 1000 km south from Cairo, Illinois, to the Gulf of Mexico (Saucier 1974, 1978). The seasonal flooding and erosion within the braided-streams channels constituted a phytogeographic barrier for plant migrations as upland forests were separated by the Valley into eastern and western sectors of the northern Gulf Coastal Plain (Delcourt *et al.* 1980).

The ice boundary was dynamic, fluctuating between glacial surges and short-lived episodes of ice stagnation and melting. Along the glacial margin, ice-cored moraines were produced, mantled by till layers of unsorted rock debris. These terminal and recessional moraines were earthen dams that blocked meltwater outlets and generated ephemeral proglacial lakes between the glacial ice and the moraines. For example, Glacial Lake Agassiz periodically inundated as much as 950,000 km^2 in the region of central Canada, extending east from the Alberta Rocky Mountains to the Lake Superior basin and north to Hudson Bay during the interval from 11,700 to 7500 years ago (Teller *et al.* 1983). From the full-glacial to the early-Holocene interval, this network of proglacial lakes continually changed location in response to isostatic uplift of newly degaciated terrain and to new drainage outlets breaching through ice-cored moraines or the retreating glacial ice (Prest 1970). Expansions and contractions of proglacial lakes displaced tundra and boreal plant communities and posed migrational barriers to terrestrial plant and animal species (Davis and Jacobson 1985). The proglacial lakes adjacent to the glacier may have hastened the retreat of the ice front; the glacial ice margin, buoyed up by the lake water, may have broken along crevasse cracks. Calving of icebergs into proglacial lakes would have increased the surface area of ice exposed to the melting action of the relatively warmer lake water. Within moraines in freshly deglaciated terrain, stagnant ice blocks remained buried, capped by surficial till layers. Such ice blocks required from hundreds to as much as 9000 years to melt (Wright 1980). Kettle-shaped lakes formed as water filled the depressions formerly occupied by the ice blocks. The unstable deglaciated landscape underwent "topographic reversal" as buried, stagnant ice lenses and blocks melted out at different rates and as the surficial mantle of rock debris on topographic ridges was reworked laterally into nearby depressions. Downslope slumping of rock debris exposed ice along the ridge crests and accelerated the localized melting of ridges, subsequently producing lake depressions.

As the glacial ice thinned, then disappeared, the landscape stabilized and pioneer plants colonized the land surface, initiating primary succession (Wright 1980). The establishment of plant cover facilitated soil development, the accumulation of an organic layer, and the increased production of humic acids. A progressive decrease in soil pH reflected the gradual dissolution of particulate grains of carbonate minerals and the increased acidic leaching and differentiation of the upper soil layers (Jacobson and Birks 1980). Progressive acidification of surface and ground waters resulted in long-term acidification of lakes in deglaciated terrain (Davis 1987). Initially following deglaciation, soils rich in carbonate minerals favored the growth of early-successional, calcicolous plants such as mosses (for example, *Campylium stellatum*, *Ditrichum flexicaule*, and *Drepanocladus revolvens* var. *intermedius*) and certain vascular plants, including *Saxifraga aizoides* and *Oxyria digyna* (Miller 1973a, 1980). The subsequent restriction of these species in mid- and late interglacial landscapes may have resulted from increased competition for space by plant species that arrived later (Miller 1980). Habitats for pioneer species also may have been lost during the Holocene on such landscapes because of podsolization of the soils. Quantities of soil nitrogen increased following the establishment of alders *(Alnus)*, *Dryas*, and legumes such as *Hedysarum*. These arctic and boreal plants are all associated with nitrogen-fixing bacteria (Ugolini 1968; Jacobson and Birks 1980; Oechel and Lawrence 1985). The accelerated productivity of both terrestrial and aquatic systems may be tied to the migrational histories of nitrogen-fixing plants that increase limiting quantities of key nutrients.

During peak glacial times, mean sea level was between 100 and 130 m lower than today (Bloom 1977, 1983). As a result, broad expanses of coastal plain were exposed for occupation by biotic communities. In coastal areas such as the central Atlantic Seaboard and the northern Gulf Coastal Plain, the underlying bedrock consisted of quartz-rich sandstones, siltstones, and shales of Tertiary and Quaternary age. The groundwater table was perched by colloidal, humate layers and impermeable shale layers (Thom 1967); sufficient soil moisture was accessible for the continuous growth of forests throughout the late Quaternary (Delcourt 1980; Whitehead 1981). However, the impact of lowered sea levels upon both the hydrology and the biotic communities was dramatically different in carbonate regions such as the Yucatan Peninsula, the Florida Peninsula, and the broad platforms of the Bahama Banks. In terrain underlain by porous, permeable limestones and dolomites, the carbonate rock is honeycombed by subterranean mazes of caverns where the rock has been dissolved by the downward percolation of rainwater. In the central and southern portion of the Florida Peninsula, the groundwater table dropped by about 20 m during the last full-glacial interval (Watts 1980a; Kutzbach and Wright 1985). The result was an open, full-glacial landscape dominated by sparse, herbaceous and scrubby vegetation and interspersed with patches of active sand dunes. Populations of drought-tolerant trees were probably more prominent in more suitable sites near the coast; in the interior of the peninsula, trees were probably restricted to valleys of the few major rivers and around the basin peripheries of deep, water-filled cenotes (Watts 1975, 1980a). Subsequent late-glacial and Holocene expansion of forest communities in coastal carbonate areas was directly linked to increases in soil moisture. Enhanced soil moisture reflected increases in both precipitation and groundwater available to the tree root systems, where the position of the groundwater table was in turn tied to the rising position of sea level (Watts 1980a; Delcourt 1985a). Thus, changes in the hydrologic setting of coastal environments have directly influenced predominant life form and species composition in biotic communities. Position of the groundwater table dictates soil-moisture levels, immediate nutrient availability to the plants, long-term rates of soil development, and prevalent disturbance regimes, such as aeolian action and fire frequencies, modulated by accumulation of organic fuel as soil litter.

Over the time scale of glacial–interglacial cycles, there have been changes in the area available for occupation by plants and animals, the quality of habitats determined by amounts of moisture and nutrients accessible to plants, and in the suite of geomorphic processes shaping the environment and serving as physical disturbances initiating episodes of plant succession. During glacial times of each 100,000-year cycle, expansion of continental ice sheets and steepening of climatic gradients restricted the distribution of the biota to a narrower latitudinal belt than is occupied today. Coevolutionary relationships between plants and animals may have developed during the Quaternary in response to glacial-age conditions very different from the interglacial conditions of today.

Anthropogenic influences

In eastern North America, humans have played an important role in shaping vegetational development, particularly in the mid- and late-Holocene intervals (Delcourt 1987). Since the arrival of nomadic tribes approximately 12,000 yr B.P., both structure and composition of forests in temperate latitudes have been influenced by exploitation by human populations. In certain regions, Native Americans were effective over the long term in modifying and fragmenting portions of the forested landscape, a result of many factors, including (1) the expanding human-population size and geographic distribution; (2) the use of fire; (3) technological advances and their cultural implementation; (4) prehistoric and historic introduction of exotic plant species for cultivation; and (5) historic introduction of domesticated grazing animals (Delcourt 1987).

Conservative archaeological evidence for the arrival of the first Native Americans documents their initial colonization of central and eastern North America between 12,000 and 11,000 yr B.P. (West 1983). Human activity consisted of exploration of the landscape south of the glacial-ice limit during the Paleoindian cultural period, from about 12,000 until 10,000 yr B.P. (Morse and Morse 1983). During this time interval of major climatic, geomorphic, and vegetational change, hunting pressure by the nomadic Indian tribes may have contributed to the extinction of late-Pleistocene megafauna (Martin 1973; Martin and Wright 1967; Martin and Klein 1984). With the virtual elimination of big game, Paleoindians shifted the specialized focus of their quarry to the more generalized hunting of smaller game animals.

During the Archaic cultural period, from about 10,000 to about 2800 yr B.P., Native Americans occupied virtually the entire forested landscape of the eastern half of the United States and of Canada. Although the total aboriginal population size was relatively small and dispersed, temporary camps for foraging and hunting were shifted seasonally in order to utilize the rich spectrum of native plant and animal species available as a food resource. Between 7000 and 4000 yr B.P., exotics such as squash *(Cucurbita pepo)* and bottle gourd *(Lagenaria siceraria)* were introduced across eastern North America from Mexico or southeast Texas (Asch and Asch 1985). Used for food and storage containers, squash and gourd were grown on annually flooded bottomlands in small garden plots near Archaic settlements. Early horticultural experimentation with native herbaceous taxa such as sumpweed *(Iva annua)*, sunflower *(Helianthus annuus)*, maygrass *(Phalaris caroliniana)*, and goosefoot *(Chenopodium berlandieri)* may have led to development of many independent centers for plant domestication (Smith 1986).

Intensification of horticultural activity occurred in the Woodland cultural period, between 2800 and 1000 yr B.P. The preferential clearing of bottomland forests for cultivation during that time interval may have reflected the expanding size of the human population and the increasing need for both a greater quantity and a more predictable source of foodstuffs (King 1985). The area of cultivated fields increased along the floodplain and low terraces of major river systems of the temperate forest region (Bareis and Porter 1984; Smith 1986; Delcourt *et al.* 1986a). The long-term labor investment necessary to plant and maintain crop fields may have provided the incentive to establish more permanent settlements with broadly defined limits of territoriality. Technological advances during this cultural period involved the invention of pottery, used for food preparation and storage.

The Mississippian cultural period lasted from about 1000 to 500 yr B.P. Within the sphere of influence of Mississippian cultures, hierarchical societies were controlled within a structured network of politics, religion, and trade across a large geographic region (Fig. 1.3) (Smith 1978; Williams 1982). Warfare and territorial demarcation of tribal chiefdoms is reflected in the archaeological record by the construction of large palisaded cities and towns, situated in easily defended positions near rich agricultural lands. Utilization of the exotic cultigens maize *(Zea mays)* and beans *(Phaseolus vulgaris)* by Mississippian-age peoples was facilitated by the technical advance of shell or limestone tempering of pottery. This permitted increase in both size and strength of food-storage jars and improved the palatibility of native and introduced plant foods when boiled in shell-tempered pots (Morse and Morse 1983).

Ethnobotanical and paleoecolgical evidence document that, by Late-Mississippian times, the vegetation along substantial portions of major river valleys had been altered by human activities. For example, interdisciplinary research in the Little Tennessee River Valley, East Tennessee, provides a case study of progressive aboriginal impact on native ecosystems over the past 10,000 years (Delcourt *et al.* 1986a). Within this riverine corridor, native plant materials were used for food, firewood, construction, and tools throughout the Holocene. However, within this valley, the shift to an agricultural subsistence base accelerated the rate of forest clearance during the Late Woodland and Mississippian periods. This long-term transformation from natural to progressively managed landscape reflected the human impact over ever-expanding areas of

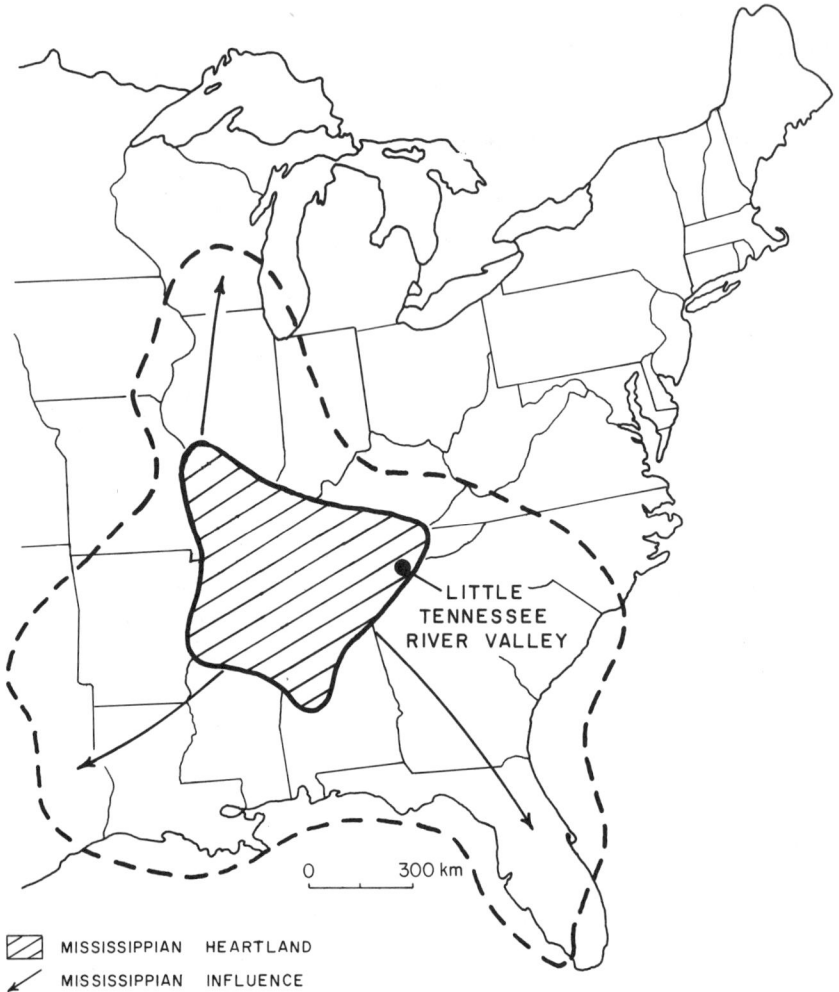

Figure 1.3. Location map of the Mississippian cultural heartland and peripheral zone of influence, ca. A.D. 900 to A.D. 1400 (modified after Williams 1982).

permanent settlements and their adjacent field systems. By Late-Mississippian times, a swath of land extending laterally several kilometers from the Little Tennessee River had been cleared of forests by human use of fire and simple tools of stone and iron. This zone of aboriginal impact was focused along the alluvial corridor and became less severe with distance into the adjacent upland interfluves (Chapman 1985; Delcourt *et al.* 1986a).

During the last 500 years of the Historic cultural period, eastern North America has been colonized by a new wave of settlers from the Old World. These pioneers, traveling initially from European and African countries, established permanent settlements along the Atlantic and Gulf coasts and then penetrated the continental interior via the principal navigatible river systems (Marschner 1959).

The westward spread of the American Frontier between A.D. 1790 and 1880 was accompanied by widespread deforestation and conversion of forest land into agricultural land that resulted in major losses of both aboveground and belowground reserves of organic carbon (Delcourt and Harris 1980) as well as progressive fragmentation of once nearly contiguous forests into smaller, discontinuous woodlots (Burgess and Sharpe 1981). Forest management practices in the 20th Century have resulted in the biomass distributions and in many cases the compositional makeup of forests occurring today in temperate latitudes of eastern North America (Delcourt *et al.* 1981, 1983a).

Models of stability and dynamics of forest communities

In the macro-scale domain, climatic change is an important environmental forcing function for vegetational change. The specific effects of climatic changes, however, may be differential across broad geographic regions and latitudinal zones. For example, during postglacial times, areas that were glaciated would be freshly exposed to colonization by plants and might be characterized by continual vegetational changes through the present interglacial or Holocene interval. Observations based upon paleoecological investigations of lake sites that formed in glaciated terrain tend to emphasize the extent of changes that involve the individualistic nature of species migrations and consequent replacement of species populations over time (*e.g.*, Davis 1981a). However, in geographic areas remote from the direct influence of continental glaciers, macro-scale climatic change, and hence vegetational responses, may have been much more subtle. The nature of those changes depends upon the extent of changes in frequencies of predominant airmasses, emigration patterns (or local extinction rates) of species, and competition among established and immigrating species (Delcourt and Delcourt 1983). In addition, lags may exist between specific changes in climate and vegetational changes that result. Macro-scale models of stability and dynamics of natural forest communities must take into account the differential nature, intensity, and rapidity of climatic changes across regions as large as North America, as well as the compositional mix within which species interact during times of either climatic stability or climatic change.

A north-to-south transect through the last 20,000 years

Based upon Quaternary paleoecological studies, we have proposed that three fundamental types of vegetational response to climatic change have occurred in different geographic regions of eastern North America since the last full-glacial period (Fig. 1.4): (1) continual vegetational stability and dynamic equilibrium in the Gulf Coastal Plain; (2) periods of vegetational stability and dynamic equilibrium alternating with periods of instability and disequilibrium in mid-latitudes of the Temperate Zone; and (3) continual vegetational instability and

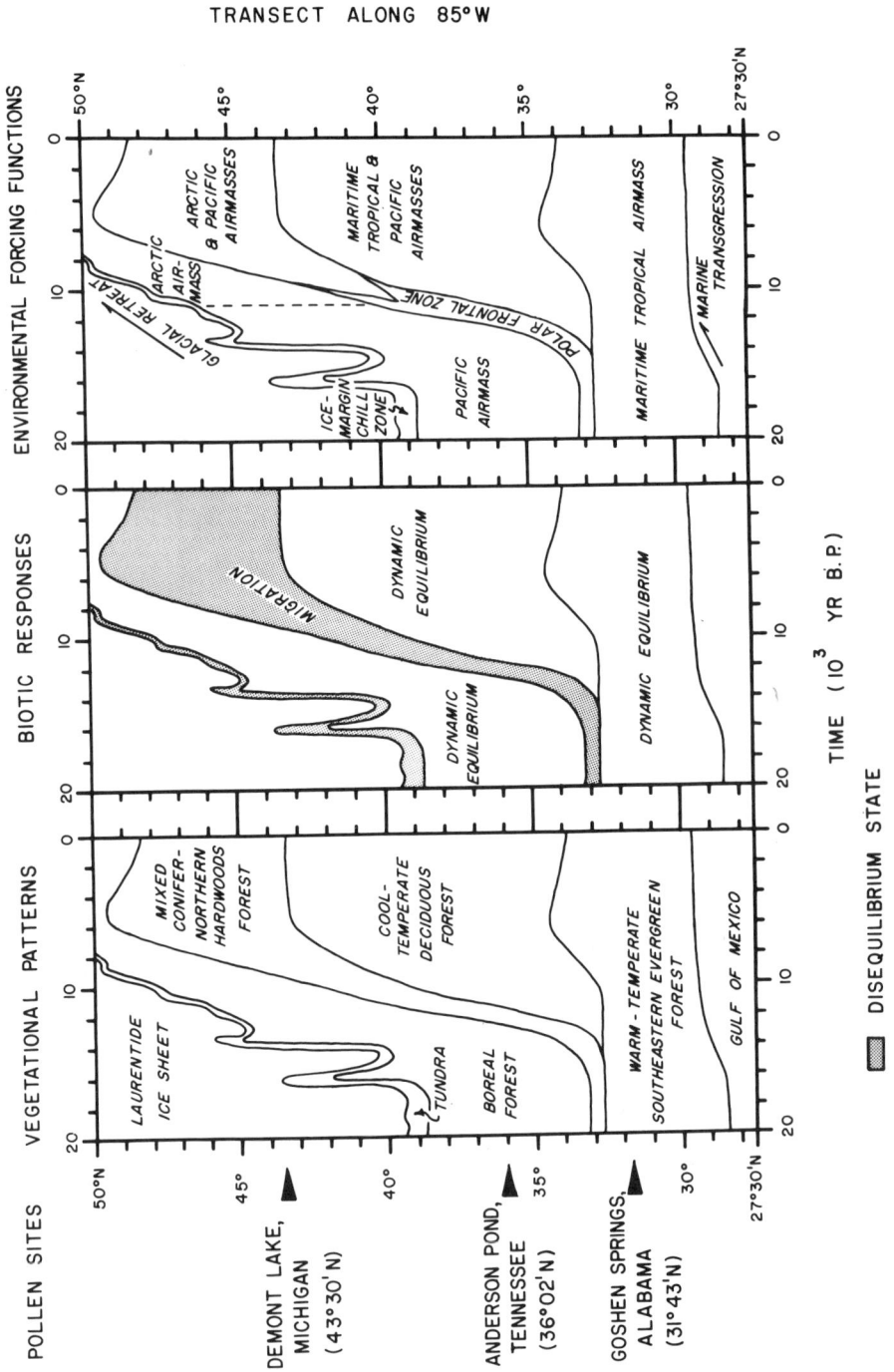

TRANSECT ALONG 85°W

disequilibrium generally north of the former continental ice margin (Delcourt and Delcourt 1983; for a quantitative test of this model, see Chapter 8). Vegetational stability is defined as the tendency for vegetation to return to an equilibrium or steady state after a major disturbance or perturbation (Botkin and Sobel 1975; Bormann and Likens 1979). Stability in vegetation can also be considered to represent a dynamic equilibrium on a landscape resulting from regeneration of a patchwork of successional stages following disturbances such as wind throw or recurrent fires (Pickett 1976; Sprugel and Bormann 1981; Romme and Knight 1982; Pickett and White 1985). Such a dynamic equilibrium depends on a relatively constant environment and predominant disturbance regime. Minor changes in moisture availability, for example, may shift the balance of dominance among taxa along a given slope, and hence on a given landscape their population sizes may change or shift in aspect to adjacent slopes. If the environmental changes are within the biological tolerances of the species present, and if the floristic assemblage does not change either through extinction of populations of species in the assemblage or by immigration of new species into the region, then a dynamic equilibrium can be maintained even with minor variations in temperature and precipitation. However, if the region under consideration is positioned close to a major ecotone across which the flora and vegetation change greatly within short distances, then even a relatively minor climatic change can tip the competitive balance among potential community dominants.

Major climatic change along the ecotone can eliminate one set of plant communities and allow a new assemblage of immigrants to occupy the space. In the extreme situation of continual, major, long-term climatic change, particularly across landscapes that have become ice-free for the first time in tens of thousands of years, sequential plant invasions following primary succession would result in continual vegetational disequilibrium and species turnover.

The zones in eastern North America characterized by each of these situations in the late Quaternary are mappable (Fig. 1.4), following the treatment of paleovegetation and paleoclimate maps in Delcourt and Delcourt (1981, 1984). Discrete boundaries are drawn between vegetation units; however, some forest

◁————————————————————————————————————

Figure 1.4. Schematic diagram showing changes in vegetational patterns, inferred biotic responses, and environmental forcing functions (*i.e.*, dominant airmass boundaries defining climatic regions *sensu* Bryson 1966; Bryson and Wendland 1967; Bryson and Hare 1974) on a transect along 85°W longitude in eastern North America over the last 20,000 years. The shaded pattern for biotic responses denotes intervals of dynamic vegetational disequilibrium. The Polar Frontal Zone represents the mean position of the polar front in winter. The full-glacial and late-glacial region of boreal forest is coincident with the climatic region of the modified Pacific Airmass (Bryson and Wendland 1967). On the panel for environmental forcing functions, the vertical dashed line at 11,000 yr B.P. represents the removal of the Laurentide Ice Sheet as a significant topographic barrier to southward flow of Arctic air (Bryson and Wendland 1967); the northern and southern boundaries of the modern boreal forest coincide with the mean summer and winter position of the Arctic Airmass, respectively (Bryson 1966) (modified from Delcourt and Delcourt 1983).

types have changed more dramatically in species composition than others (Delcourt and Delcourt 1981).

For example, in the Gulf Coast Region, represented by the plant-fossil record from Goshen Springs, Alabama (Delcourt 1980), warm-temperate southeastern evergreen forest has persisted for at least approximately the past 60,000 years. The pollen record at Goshen Springs is characterized by continued presence of warm-temperate elements such as magnolia (*Magnolia*), holly (*Ilex*), sweetgum (*Liquidambar styraciflua*), tupelo gum (*Nyssa*), and southern Diploxylon pine (*Pinus*) throughout the record. Fluctuations in the relative abundances of oak (*Quercus*), sweetgum, and pine occur throughout the pollen record, with the greatest change occurring at about 5000 yr B.P., with a change in dominance from oak to pine. Records of sea-surface temperature in the nearby Gulf of Mexico (Brunner 1982) indicate that the mean annual temperature of surface waters has changed less than 2°C over the last 125,000 years (the time since the last interglacial interval) at that latitude. Goshen Springs thus records only minor environmental fluctuations through the late Quaternary and indicates that the vegetation has been in dynamic equilibrium over long time periods, in the sense that no major invasions or extirpations of species have occurred, and the same forest-community types (*e.g.*, coastal swamp, magnolia–beech, pine, oak) have persisted throughout the late Quaternary. The mosaic of community states, that is, the proportion of area occupied by the different forest communities, has changed through time, as evidenced in the mid-Holocene interval by the shift in relative dominance of southern pine and upland oaks.

The cause of the changeover in forest dominance from oak to southern pine on the Gulf Coastal Plain in the mid-Holocene interval is not yet resolved. The expansion in pine populations could be a result of increased fire frequency in the coastal plain due to hunting practices of Archaic-period Native Americans. During Historic times, with another change in anthropogenic disturbance regime that included logging of the original pine forests and fire suppression, the dominant vegetation of the region has been shifting from southern pinelands through secondary succession to southern mixed hardwoods forest (Quarterman and Keever 1962; Delcourt and Delcourt 1977a). Yet the region as a whole may be mapped as remaining within a southeastern evergreen forest type throughout at least the last glacial–interglacial cycle.

Farther north, between about 33°N and 39°N latitude (Fig. 1.4), relatively stable full-glacial boreal forest was replaced by cool-temperate deciduous forest in the Holocene, after an interval of compositional instability and successive migrations during the late-glacial and early-Holocene intervals. During the full-glacial interval, the zone of transition between boreal and temperate forests was sharp and narrow relative to today's equivalent ecological "tension zone" in the Midwestern United States and eastern Canada (Delcourt and Delcourt 1979, 1981). Jack pine (*Pinus banksiana*) and spruce (*Picea glauca, P. mariana,* and *P. rubens*) dominated large areas of the middle latitudes of eastern North America. Anderson Pond, Tennessee (Fig. 1.4) is one site that documents the nature of the late-glacial transition from jack pine-dominated forest, through mixed northern hardwoods forest with increasing contributions of oak, hickory

(*Carya*), beech (*Fagus grandifolia*), and sugar maple (*Acer saccharum*), to mixed deciduous forest dominated by oak and hickory (Delcourt 1979). At Anderson Pond (Fig. 1.4), jack pine and spruce populations persisted through the full-glacial interval until 16,500 yr B.P. Thereafter, jack pine declined steadily, probably because of lengthening growing season to which deciduous species were better adapted. Oak, ash (*Fraxinus*), hornbeam (*Ostrya/Carpinus* pollen type), and other deciduous taxa increased throughout the late-glacial and early-Holocene intervals, from 16,500 to about 9000 yr B.P. During the last 9000 years, temperate deciduous forest predominated in Middle Tennessee (Delcourt, 1979). Additional sites in the central Appalachian Mountains have been characterized by Watts (1979) as following a pattern of full-glacial vegetational stability (tundra or boreal forest); late-glacial conditions of disequilibrium, instability, and species migrations/invasions (transition to mixed northern hardwoods forest); and a return to vegetational equilibrium (deciduous forest) in the Holocene.

Climatic changes during the late-glacial interval can be envisioned as a dynamic interplay of prevalent airmasses over middle latitudes (Delcourt and Delcourt 1984). The full-glacial position of the Polar Frontal Zone, separating boreal from warm-temperate climatic and vegetational regions in unglaciated North America, can be traced eastward from Cape Hatteras across the North Atlantic Ocean to central Spain (CLIMAP 1981). The full-glacial and late-glacial regions of boreal forest coincided with a climatic region characterized by zonal flow and probably dominated by a modified Pacific Airmass (Bryson and Wendland 1967). The full-glacial dome of the Laurentide Ice Sheet served as a topographic barrier to southward incursions of the frigid Arctic Airmass, which was consequently restricted to high latitudes (Wright 1984; Kutzbach and Wright 1985). Late-glacial melting of an ice-free corridor between the western Laurentide Ice Sheet and the Cordilleran Ice Sheet permitted the southward flow of the Arctic Airmass into central North America by 11,000 yr B.P. Return of the Arctic Airmass south of the Laurentide Ice Sheet resulted in greater seasonal extremes in temperature during the Holocene throughout the regions of both boreal forest and northern mixed hardwoods forest (Delcourt and Delcourt 1984).

As the glacial ice retreated, new areas were opened for vegetation. Sequential species migrations became the dominant vegetation process through the Holocene north of 43°N latitude in eastern North America, as seen for example at Demont Lake (Fig. 1.4), located on the modern ecological tension zone of central lower Michigan (Kapp 1977a), or at Rogers Lake, Connecticut (Davis 1969). Because different tree taxa migrated from different refuge areas at different rates, continued vegetational disequilibrium through the Holocene at these sites was in part due to differential lags in response of tree species populations to late-glacial and early-Holocene climatic changes (Davis 1981a).

The role of changing disturbance regimes

Under an assumption of constant climatic conditions and vegetational equilibrium, plant succession in temperate forested regions is typically thought to proceed from the pioneer-herb/woody-seedling stage, through a shrub/sapling stage, toward mature mixed-age, mixed-composition forest stands dominated by relatively few taxa (Bormann and Likens 1979). In such an idealized successional sequence, accumulating biomass begins as a small amount, increases logistically with time, and plateaus out in a typical S-shaped curve to an asymptotic value conditioned by site type and maximum age and size of typical individuals of the dominant species (Bormann and Likens 1972).

It is now widely recognized (Loucks 1970; Pickett and White 1985) that a variety of natural and anthropogenic disturbances periodically "reset" the clock of temperate forest succession such that late-successional, mature forests may seldom be realized. In an elegant application of time-series analysis to paleo-ecological data, Green (1981; Walker 1982) demonstrated the response times of tree taxa characteristic of mixed conifer–northern hardwoods communities to recurrent fire. By analyzing fluctuations in frequencies of charcoal particles in a radiocarbon-dated sediment sequence from Everitt Lake, Nova Scotia, Green (1981) determined the recurrence interval for major fires on the watershed to be approximately 350 years. Cross-correlograms between charcoal and pollen frequencies showed that peak response of early-successional taxa such as pine, birch *(Betula)*, and spruce occurred within 50 years of major charcoal peaks, but that a much longer time lag was evident for recovery of populations of late-successional taxa such as beech and sugar maple (Green 1981; Walker 1982). Prevailing disturbance regimes would have been greatly affected by late-Quaternary climatic changes (Botkin and Sobel 1975), thereby altering the patterns of plant succession occurring on the landscape such as the watershed surrounding Everitt Lake, as well as on individual micro-scale sites.

Under a constant climate, continuous availability of the same set of plant species for colonization, and a uniform disturbance regime, vegetational equilibrium may be envisioned as a trajectory that is relatively stable and that varies little through time (Fig. 1.5). Under these assumptions, individual cycles of

▷

Figure 1.5. Trajectories of vegetational change (percent composition in macro-sites) and in plant-successional stage (structure in micro-sites) as influenced by climate and disturbance regimes. Trajectory A illustrates minimal overall compositional change (*i.e.,* < 10%) over an interval of 1000 years. Trajectory A corresponds to overall compositional change within a vegetational mosaic on the macro-site spatial scale; this vegetational equilibrium in plant composition reflects stable conditions in climate, species availability, and a disturbance regime with a typical recurrence interval for disturbances. The saw-toothed curve denotes the sequence of plant succession and the corresponding structural change in vegetation for a particular micro-site, following a specified series of disturbances of known magnitude. Trajectories B and C indicate vegetational disequilibrium with > 10% change in vegetation composition on a macro-site over a period of 1000 years. Trajectory B reflects disequilibrium vegetational response to a climatic sequence of rapid change from time 0 to 400 years, stable climate and corresponding equilibrium in vegetational composition from 400 years to 600 years, and a return to changing climates and vegetation in the next 400 years. Trajectory C records a dynamic state of vegetational disequilibrium accompanying a rapid climatic change of great magnitude (from Delcourt *et al.* 1982, © Pergamon Press, Ltd, reprinted with permission).

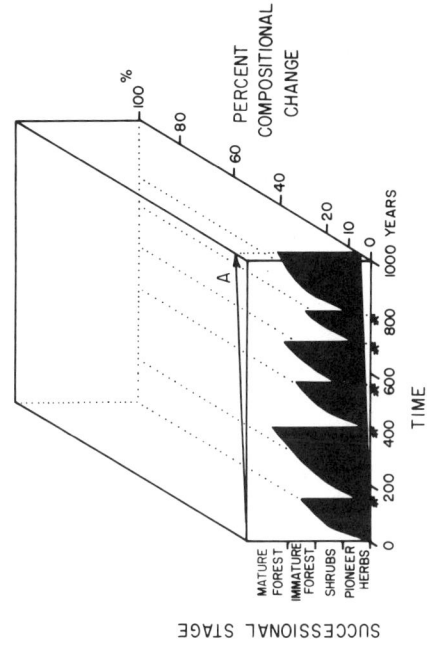

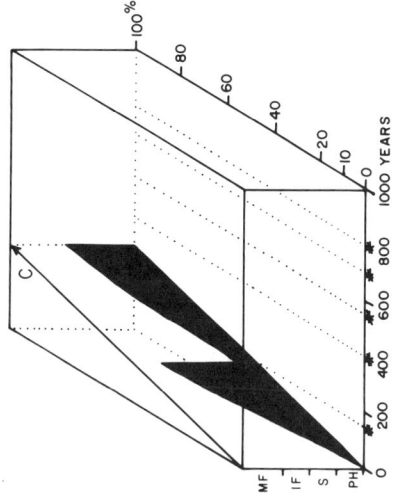

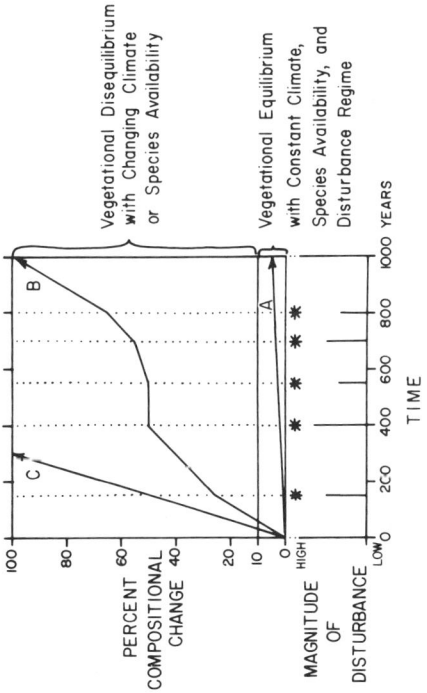

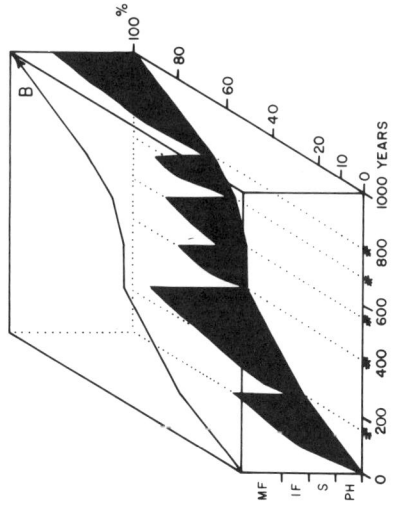

plant succession would proceed toward late-successional, mature forest, in-terrupted at relatively infrequent intervals by disturbance events such as fire that would set the system back to an early-successional stage. Compositional change of forest stands would be restricted to those changes occurring primarily as a function of life-history strategy and relative tolerance to shade. For example, in northern mixed hardwoods communities, pin cherry (*Prunus pennsylvanica*) typically germinates in light gaps but is later eliminated from the stand because it is intolerant of the denser shade provided by the canopy of longer-lived hard-woods such as yellow birch (*Betula alleghaniensis*) and sugar maple (Marks 1974). If the same mix of plant species were available for invasion of recently disturbed forest plots, then the overall composition of the community as rep-resented by all successional stages would remain constant (Trajectory A in Fig. 1.5).

On the other hand, vegetational disequilibrium can result from (1) changing frequency, intensity, or kind of prevalent disturbance; (2) gradual and unidi-rectional climatic change (warming, cooling, a change in seasonality, an increase or decrease in precipitation, or some combination governed by complex seasonal shifts in prevailing airmasses); or (3) a change in species availability (itself pos-sibly a response to changing climate or disturbance regimes). The possible tra-jectories for vegetational change under these circumstances are several. For example, Trajectory B in Fig. 1.5 represents the disequilibrium vegetational response to climatic changes that are rapid from year 0 to year 400, then stable climate and equilibration of vegetation for the next 200 years, then changing climate for the following 400 years. Over such a thousand-year time interval, substantial changes can occur in total community composition. The rate of change in the composition of plant communities is in part conditioned by the lag time for response of constituent taxa (for quantification of rates of com-positional change during the late Quaternary, see discussion of beta-diversity gradients in Chapter 8). This model represents a segment of late-Quaternary time that is realistic in terms of the extent and duration of Holocene climatic and vegetational changes in temperate regions.

In a series of experiments using a forest-stand simulation model of gap-phase dynamics for the mixed conifer–northern hardwoods region, Davis and Botkin (1985) demonstrated that in order to effect a compositional change substantial enough to be recorded in pollen records, thresholds of tolerance of key species must be exceeded in both the magnitude and the duration of climatic change. Using a simulation model of forest dynamics on a mesic site with no physical disturbances such as fire (Fig. 1.6), the duration of climatic cooling was varied, increasing from 25 years to 200 years. Short-term climatic fluctuations (less than 100 years in duration) did not greatly affect the overall composition of these cool-temperate forest communities, partly because individual mature trees are broadly tolerant of year-to-year changes. In the absence of major distur-bances, long-lived, shade-tolerant species such as sugar maple tend to persist even without substantial regeneration within the understory (Davis and Botkin 1985). With progressive cooling that lasted for a simulated span of 200 years, sugar maple was replaced on mesic sites by red spruce (Fig. 1.6). Northern mixed conifer–hardwoods communities were less sensitive to climatic change

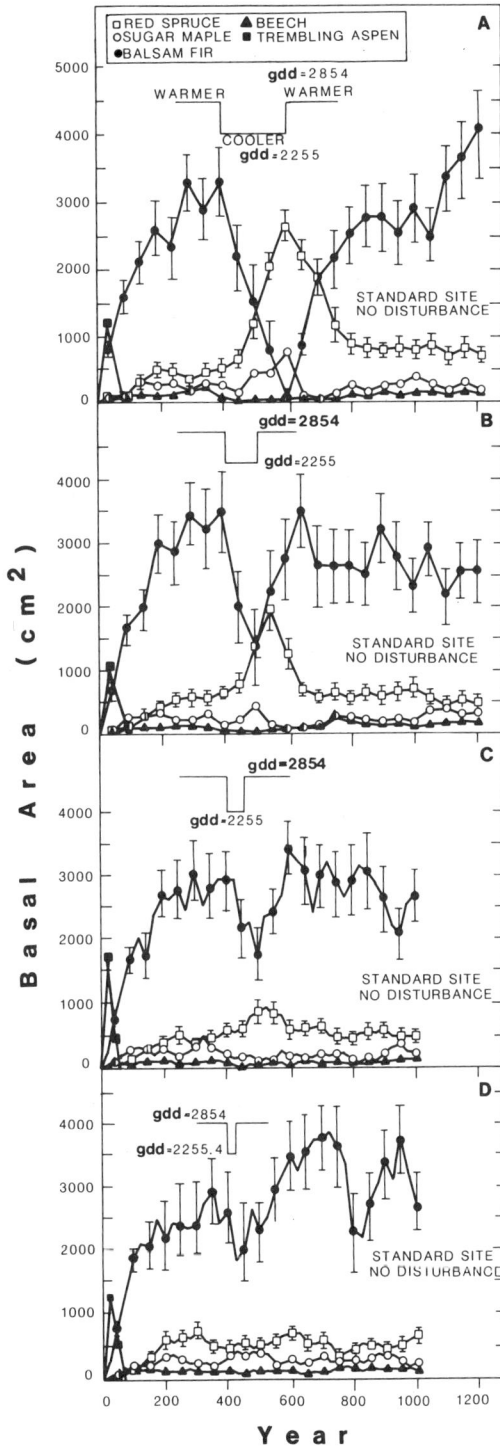

Figure 1.6. Basal area for dominant species of trees plotted against time in a 1200-year simulation of forest growth on good soils. Climate changed from 2854 to 2255 growing-degree days (gdd) starting in Year 400 for A, 200-year interval; B, 100-year interval; C, 50-year interval; and D, 25-year interval (from Davis and Botkin 1985).

on sites with nutrient-poor soils, but both the observed changes in community composition and the heterogeneity of the successional patchwork were greater in simulations that included disturbances that opened the canopy to more rapid regeneration (Davis and Botkin 1985).

A third possibility is that posed in the 20th and 21st Centuries A.D. (Fig 1.5, Trajectory C). Very large and very rapid climatic changes (such as projected warming up to 10°C in the next century; Kellogg 1983) may result in rapid extinctions of species populations and turnover in plant communities on a time scale of 100 years (Delcourt *et al.* 1982). Changes in community composition that would occur with presumed increases in global temperatures due to increases in atmospheric carbon dioxide (Kellogg 1983) may be accelerated in response time if the communities involved are also stressed by atmospheric pollution or other anthropogenic disturbances. The spruce–fir ecosystem of high elevation sites in the southern Appalachian Mountains is an example of a forest community that has decreased in area by 50% because of human activities since the time of EuroAmerican settlement (White 1984). Current trends in dieback of both spruce and fir attributed to pathogen attacks and acid deposition would be accelerated by climatic warming. Without immediate regeneration of seedlings of the coniferous species, elimination of the canopy may result in local extinction of red spruce populations and total extinction of the species Fraser fir (*Abies fraseri*) within several decades (White 1984; White *et al.* 1985).

Conclusions

1. Fundamental environmental forcing functions and types of biotic responses differ across a spectrum of spatial and temporal scales; thus, the vegetational patterns that we perceive are in part a consequence of the window through which we view vegetational change.
2. Within the Temperate Zone, patterns of species richness and forest composition can be interpreted directly as biotic responses to Quaternary climatic, environmental, and anthropogenic changes.
3. Ecological analysis of contemporary forest communities characterizes vegetational response to short-term environmental changes and human impacts. However, an understanding of vegetational dynamics requires recognition that relatively infrequent disturbances recurring over hundreds to thousands of years may shape long-term trajectories of compositional change.
4. Different kinds of vegetational dynamics have occurred within the Temperate Zone during the late Quaternary, according to both the geographic location of biotic regions and the historical context of both rate and magnitude of environmental change to which they have been exposed. Kinds of vegetational dynamics may vary from long-term equilibrium, to alternating periods of equilibrium and disequilibrium, or to continual disequilibrium.

2. Modern Pollen–Vegetation Relationships

Production and dispersal of pollen grains

The establishment of the field of Quaternary paleoecology is commonly considered to be Lennart von Post's presentation before the 16th Convention of Scandinavian Naturalists held in Oslo, Norway, in 1916. In his lecture entitled "Forest tree pollen in south Swedish peat bog deposits", von Post described his pioneering efforts to examine fossil-pollen assemblages preserved within late-Quaternary peat deposits of Scandinavian bogs. By counting a number of grains of different pollen types at each of many levels within the peat deposits, von Post noted the succession of changing pollen assemblages that he interpreted in terms of the postglacial development of plant communities. He identified the replacement of mixed oak forest by beech–spruce forest and interpreted this change in terms of changing climate, immigrations of tree species, and competition at both the species and community level. Von Post was the first to recognize the need for documentation of modern pollen–vegetation relationships as a prerequisite for interpreting changes in forest communities through time. According to von Post (1916, translated by Davis and Faegri 1967), "As long as we have no indices to express the relative pollen productivity of the various trees, nor to express the different degrees to which their pollen is dispersed, we have no right to seek in the [pollen] percentage figures an adequate expression of the composition of the [past] forest communities" (p. 390, words in brackets added for clarification).

The representation of pollen in Quaternary sediments is a function not only of the abundance and distribution of plants, but also of their pollen productivity; the timing and mode of pollen dispersal; the nature of processes associated with deposition, incorporation, and preservation of pollen within sediments; and the source area from which the pollen assemblages are ultimately derived.

Pollen production and mechanisms for dispersal

Certain plants are rarely represented by pollen grains in the fossil record. These include aquatic plants such as *Ceratophyllum* that are pollinated within the water column. These aquatic plants have pollen grains that typically lack or have only a thin exine (the protective outer wall of the pollen grain) that is easily decomposed. Other plants are self-fertilizing and/or cleistogamous, producing relatively little pollen that is released into the atmosphere. With such an effective mode of pollination, only small quantities of pollen grains are dispersed beyond the flower and, consequently, few grains are potentially preserved within lakes, bogs, or swamps. A third group of plants has pollen grains dispersed by animals (zoophily) or, in the more specific case, by insects (entomophily) (Faegri and Iversen 1975). In this situation, pollen grains are carried from the anther of one flower to the stigma of another by an insect, a bird, or a bat. The pollen grains are commonly large (greater than 40 μm and up to 250 μm in diameter), with an ornate sculpture and a sticky coating of oil on their outer exine surface (Whitehead 1969). The flowers may produce surplus pollen or nectar that serves as food and lures insects to the anthers. The dispersal units may consist of clusters of four pollen grains (tetrads), or the tetrads may break apart into smaller clusters of two grains (dyads) or occur as single pollen grains (monads). The more specialized and effective examples of zoophilous pollination are found associated with plants that produce a relatively small number of pollen dispersal units. As a consequence, the pollen of zoophilous plants is commonly found in forest soils (Andersen 1984) or on bog surfaces (Janssen 1984), accumulating near the plants that produced the pollen grains. Types of zoophilous pollen are rarely dispersed far distances and subsequently are only infrequently found preserved in lake sediments (Faegri and Iversen 1975; Janssen 1966, 1973, 1984).

Pollen grains most consistently and abundantly found within Quaternary-age sediments are from anemophilous or wind-pollinated plant species. Individual plants produce relatively large amounts of pollen. For example, a male plant of the herbaceous weed dock *(Rumex acetosa)* can generate approximately 400,000,000 grains in a single growing season (Pohl 1937). Great quantities of pollen grains are liberated by anemophilous species into the air and are scattered across the landscape as "pollen rain" (Fig. 2.1, Hebda 1985; Janssen 1984). Within areas of temperate vegetation, wind-pollinated plants emit between 0.01 and 10 metric tons of pollen annually across each square kilometer of land (Solomon 1979). Anemophilous plants typically produce small pollen grains that range in diameter from 20 μm to 40 μm and are readily transported by wind currents (Whitehead 1969). These pollen grains separate from the tetrad and

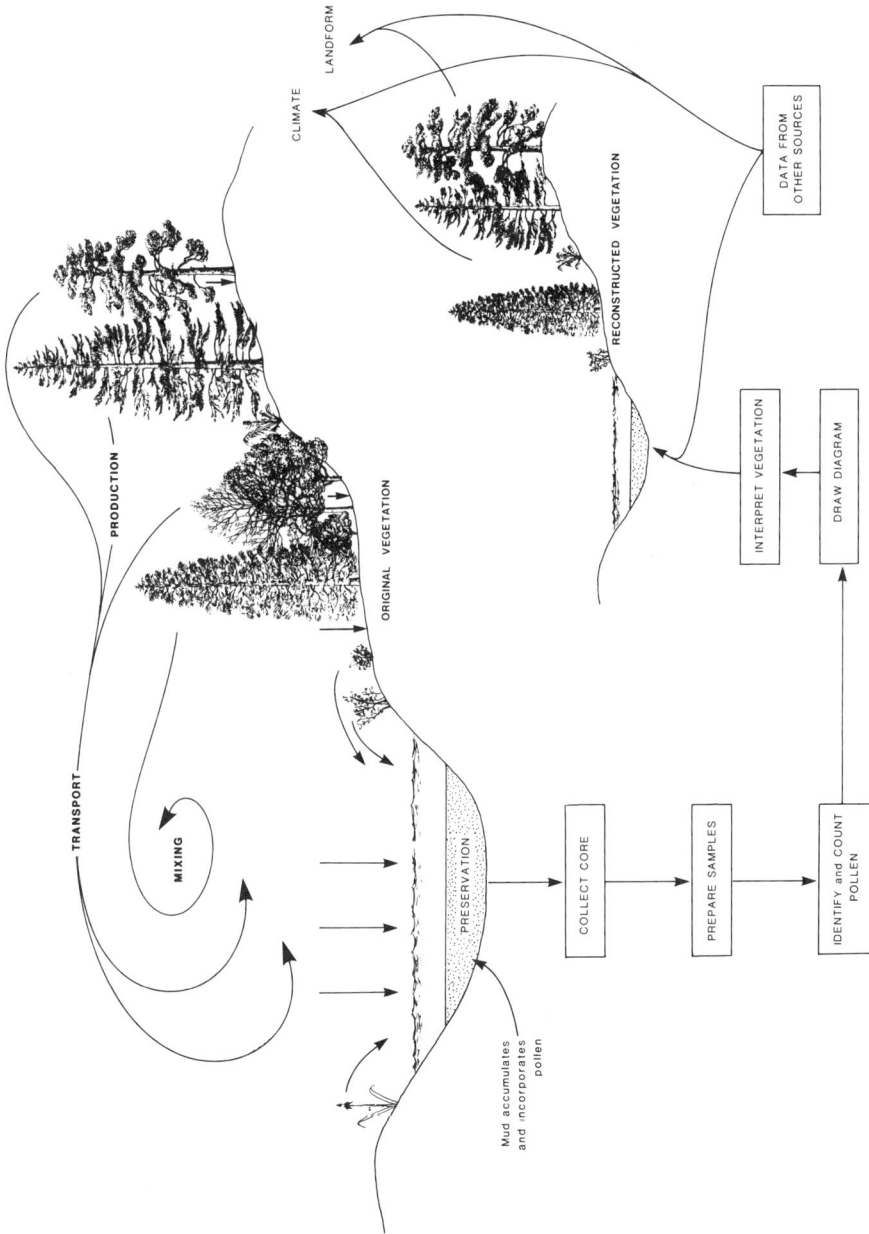

Figure 2.1. Production, dispersal, deposition, and preservation of a pollen assemblage and steps toward recovery, analysis, and interpretation of its vegetational and environmental significance (from Hebda 1985).

are dispersed as four individual grains. The exterior of the grains is dry with no surface coating of oil. The exine surface is usually smooth (psilate) or with sculpturing patterns of relatively low relief. Pollen grains of anemophilous plants characteristically have few or no openings such as elongate furrows or circular pores through the exine. The fewer the apertures, the less water is lost from the interior of the grain. Several morphological characteristics lower the overall density and increase the buoyancy of the pollen grain during atmospheric transport. Airborne pollen types often possess thin, light exines. Some have internal cavae (air-filled cavities) as found in genera of Compositae such as *Ambrosia* (ragweed). Others have external, air-filled wings common in conifers such as *Picea* (spruce), *Abies* (fir), and *Pinus* (pine). Flowers of wind-pollinated species often have reduced perianths; both anthers and structurally complex stigmas may extend beyond the corolla, where they are directly exposed to wind currents (Whitehead 1969, 1985). Intricate, feathery stigmas redirect the aerodynamic flow of air currents moving around them and increase the likelihood of pollen capture on their large receiving surfaces (Faegri and van der Pijl 1979). Inflorescences typically occur on the upper portion of the plant, insuring maximal access for entrainment of pollen by wind currents (Whitehead 1985).

Temporal patterns of pollen dispersal

Temporal variability on diurnal, seasonal, and even biennial cycles can be detected in the release of pollen by wind-pollinated plants. The physical process of pollen discharge from the flower is tied to the phenology of the plant in response to environmental cues. Concentration of airborne pollen correlates with plant response to diurnal changes in temperature and to local weather conditions (Sheldon and Hewson 1960; Solomon 1979). During a typical 24-hr period, peak concentrations of airborne pollen generally occur in the morning within 4 hr following sunrise (Ogden *et al.* 1974, p. 165). Under fair-weather conditions, low air humidity and increasing temperatures during the morning trigger the opening of the flowers and the extension of anthers beyond the corolla. As the anther tissues dry and then split apart, pollen grains are shed from the opened anthers and fall onto the nearby floral and leaf surfaces (Solomon 1979). In the absence of wind, pollen grains fall under the influence of gravity, reaching terminal velocities of 1 to 10 cm/s (Tauber 1965). With a time lag of 15 to 30 minutes, the diurnal pattern in the concentration of airborne pollen tracks the number of flowers that have recently opened in anemophilous plants (Sheldon and Hewson 1960). Warm, dry, and windy weather conditions favor the dispersal of these pollen grains by wind currents. Although pollen grains from deciduous trees have terminal fall velocities in the range of 2 to 6 cm/s, wind speeds are typically several orders of magnitude greater, from 0.5 up to 10 m/s (Tauber 1965). Pollen grains falling through the air can be readily captured by the wind and carried laterally by horizontal currents and vertically by turbulent eddies (Tauber 1965, 1967a,b; Raynor *et al.* 1974a) (Fig. 2.1). Pollen grains that have fallen or impacted against nearby plant or soil surfaces may be swept from those obstacles and resuspended when the velocity of wind near the ground level exceeds 2 to 4 m/s (Solomon 1979). As wind velocities increase from between 4 and 10 m/s, turbulent air flow can transport pollen grains above the wind-baffling influence of the forest and into the atmosphere (Tauber 1965; Solomon 1979). The viability of pollen grains declines markedly within 3 to 8 hr of atmospheric transport (Solomon 1979) as a result of their exposure to ultraviolet radiation (Werfft 1951, cited in Solomon 1979) and rapid water loss from the grains (Harrington and Metzger 1963). Thus, diurnal changes within the local environment of a plant trigger pollen discharge from the flower and directly influence pollen transport as well as the potential success of the pollen germination and transfer of genetic material to a suitable flower stigma.

Two evolutionarily successful strategies are evident in seasonal patterns for the timing of pollen dispersal by wind. Both strategies of pollen release share two elements: (1) plants have successively increased the overall numbers of pollen grains that can be produced at a given time; and (2) the plant's response to an environmental cue results in a peak in concentration of pollen grains released into the air at a time when the stigmas are viable and receptive for the pollen. This latter situation coordinates the time of flowering and triggers a relatively brief episode with cross-fertilization of compatible plant populations (Whitehead 1969).

One seasonal strategy is exhibited by anemophilous, annual and perennial herbs; their mid-summer to early-autumn release of pollen is timed near the end of the growing season (Ogden and Lewis 1960; Whitehead 1969). These wind-pollinated herbs are often characterized as opportunistic, pioneer plants that readily invade recently disturbed sites. They experience rapid growth in spring following the germination of seeds for annual plants or, for perennial plants, following the new growth of shoots emerging from underground root systems. Responding to increasing solar radiation, the production of new leaves reaches a mid-summer peak in aboveground phytomass for photosynthesis and generation of food reserves. The late-summer decline in available solar energy and decreasing day length serve as environmental triggers to these herbaceous species. They reallocate food resources from vegetative growth to maturation of flowers, dispersal of pollen grains and production of viable seed. The seed provides for the continuation of the species following the death of the annual plants (Harper 1977; Fenner 1985).

The second seasonal strategy for wind pollination is utilized by temperate and boreal species of trees and shrubs (Ogden and Lewis 1960; Whitehead 1969). Anemophilous trees and shrubs release their pollen grains in early spring at the beginning of their growing season. Pollen dispersal in spring occurs at a time of high wind velocities, just before the deciduous arboreal species have leafed out, developed a continuous forest canopy, and have effectively reduced wind speeds within the forest (Tauber 1965, 1967a,b). Strong seasonal contrast between cold winters and warm summers (or between dry and wet seasons) and relatively short growing seasons may have selected for the evolutionary development of the deciduous character in temperate woody species (Whitehead 1969, 1983; Regal 1982).

Anemophilous tree species in the northern Temperate Zone can attain reproductive maturity within several years to a few decades following germination. For example, North American species of *Pinus* (pine) can flower and produce seed within 5 to 10 years, *Picea* (spruce) in 10 to 20 years, and *Quercus* (oak) in 15 to 35 years (Fowells 1965). Unless stressed, these trees continue to produce and release pollen for as much as another 150 to 1000 years, depending upon the maximum longevity of individuals (Davis and Botkin 1985). However, flowering intensity and pollen productivity vary from year to year. Andersen (1980a) has demonstrated that, using 10 years of data on modern pollen assemblages collected in traps from within forests, the flowering of deciduous trees such as *Quercus* (oak) and *Fagus* (beech) follows a biennial cycle. Climatic variation distorts this biennial flowering pattern and accentuates variability in the annual productivity of pollen. The yield of pollen is influenced by the weather during the previous summer as well as during the spring flowering season. Warm temperatures and low precipitation in summer favor the initial differentiation of flower buds in oak and beech and determine the maximum pollen yield potentially available for the next pollination season. Soil-moisture levels, recharged by winter precipitation, and early-spring temperatures affect tree growth and indirectly influence the final stages in the development of the pollen grains (Andersen 1980a). Cold or wet spring weather can substantially reduce the quantity of pollen actually dispersed, because episodes of severe frost or intervals of heavy rainfall may damage flowers at a critical stage of development, causing as many as 50% of the anthers to fall unopened from the trees (Solomon 1979). Both climatic variation and biennial cycles in flowering dictate the production of pollen grains, their dispersal, and their overall effectiveness in fertilization of ovaries; this accounts for the typically observed interval, varying from 2 to 5 years, between abundant seed crops for temperate deciduous trees that are wind pollinated (Fowells 1965).

What then is the minimum time interval that is routinely available for examining late-Quaternary changes in modern and fossil pollen assemblages? Seasonal, biennial and decadal variability in both assemblage composition and pollen productivity can be directly sampled in varved sediments. Distinctive laminations of sediments are deposited seasonally, producing a couplet of varves each year in certain kinds of lake systems. Within the northern Temperate Zone, seasonally laminated deposits have been reported from freshwater lakes in the Great Lakes region and New England in the United States and in Fennoscandia, Great Britain, and central Europe (O'Sullivan 1983). Varved sediments accumulate on the bottoms of deep lakes characterized by steep-sided, flat-bottomed basin morphologies. Such lakes are typically meromictic, with weak or incomplete circulation of the water column, little or no disturbance of the sediment surface by water currents, and oxygen-deficient conditions in lower waters of the hypolimnion. The anoxic bottom waters of these lakes inhibit the growth of benthic organisms and consequently limit bioturbation that would otherwise mix the sediments. To the paleoecologist studying meromictic lakes, the limiting factor is not the temporal record available from varved sediments, but rather the ability to recover the material without disturbing it and to sample the varves at the temporal resolution desired. Recent technological advances in coring equipment (summarized in Saarnisto 1979 and in O'Sullivan 1983) now permit the recovery of undistorted, annually laminated sequences spanning as much as 10,000 years (Craig 1972). Varve sequences provide the detailed temporal chronology; time series analysis of their fossil-pollen content offers many possibilites for study of ecological processes at the plant-population level (Watts 1973a; Allison *et al.* 1986).

Physical and biological mixing of lake deposits can blur or reduce the fine-tuned time resolution potentially available from sediments. Many temperate lake systems are dimictic, with complete circulation of water throughout the basin twice each year. In summer and in winter, dimictic lakes develop strong density and temperature gradients, with a pronounced thermocline boundary separating uppermost epilimnetic waters from lower hypolimnetic waters. The density stratification of the lake breaks down in fall with cooling of the surface water; as the uppermost water reaches 4°C (the temperature at which fresh water is densest), circulation of the water column occurs as the denser upper water displaces the lighter bottom water. A second turnover in water circulation occurs in early spring following the melting of the surface layer of winter ice. Davis and Brubaker (1973) compared quantities of pollen grains trapped from the air, from several depths in the water column, and from the bottom sediments of Frains Lake, a temperate dimictic lake in south-central Lower Michigan. The concentration of pollen grains in the lake water is greater during the times of fall and spring overturn than even during the principal times of pollen release by the upland vegetation (Davis 1973; Davis and Brubaker 1973). Large amounts of flocculent surficial sediment and previously deposited pollen are resuspended during the times of increased water circulation during fall and spring. The sediment is reworked more frequently and in larger quantities in the shallow areas of the littoral zone of the lake; as a result, the pollen grains are progressively reworked from the shallower into the deeper portions of the lake central basin (Davis *et al.* 1971; Davis and Brubaker 1973). The quantity of redeposited pollen is approximately four times the new increment of pollen grains added to Frains Lake each year (Davis and Brubaker 1973). Thus, physical processes of enhanced water circulation influence pollen assemblages by mixing the last several years of pollen and mineral-sediment accumulation at a lake such as Frains Lake. This seasonal mixing serves to homogenize recently deposited sediments, erasing the pollen record of seasonal and annual variability and providing an typical pollen assemblage that represents a running average of pollen rain over several years.

The extent of biological activity of benthic communties determincs the extent of bioturbation, or reworking of surficial sediments of lakes. Davis (1967) used detailed pollen analyses of lake cores from Maine, as well as laboratory experiments, in order to quantify the importance of bioturbation in limiting the time resolution available from the pollen record. He collected short cores containing the uppermost 80 cm of sediments from the deepest portions of Long Pond, Little Ossipee Lake, Eagle Lake, and Bracey Pond. The pollen and wood-charcoal content of the cores were studied at sampling intervals as fine as every 4 mm of core depth; benthos were analyzed at 1 cm increments. Sedimentingesting organisms and their burrows were concentrated in the 4 cm immediately below the mud surface; modern biological evidence for burrowing was sparse below 8 cm of the sediment surface. With an average sedimentation rate of 2.0 mm/yr in these lakes, the biologic reworking was found to be most intense in the sediments representing the last 20 years of accumulation; all bioturbation was limited to the uppermost 8 cm, the equivalent of the last 40 years time. In the sediment cores, specific events involving known fires and changes in vegetation composition around the four lakes were recorded by changes in wood-charcoal concentration and in the pollen assemblages. The blurring of these palynological records and the downward displacement of the charcoal peaks confirm a bioturbation effect integrating and smoothing the paleoecological sequences over a 20-year time window. Davis (1967) contends that nutrient inputs from the watershed, in conjunction with basin morphology of the lake and seasonal overturn in water circulation, influence both the quantities of oxygen dissolved within the water column and, consequently, the impact of sediment bioturbation by benthic communities. Times of water turnover result in movement of oxygenated water throughout the lake basin. However, subsequent stratification and development of a thermocline boundary may inhibit transfer of oxygenated upper waters into the lower hypolimnetic zone. Progressive isolation of bottom waters during summer and winter seasons results in lowered levels of dissolved oxygen as it is utilized for decomposition of organic material. Diminished oxygen levels constrain the benthic activity.

Thus, shallow lakes with wind-driven water circulation and high levels of dissolved oxygen and nutrients favor biological reworking of surficial lacustrine sediments (Davis 1967). Bioturbation homogenizes recent sediments of dimictic lakes, continuously integrating pollen rain over a smoothing interval between 4 and 20 years. However, in meromictic lake systems where physical factors predominate, deep lake basins with incomplete circulation of their waters retain oxygen-deficient bottom waters that restrict benthos. Preservation of varved sequences permits investigation of paleoecological phenomena on annual, seasonal, and even diurnal time scales (Simola and Tolonen 1981).

Spatial patterns of pollen dispersal

Contemporary models of pollen dispersal incorporate experimental evidence for both the transport of particulate grains and the source area from which the pollen assemblages are derived. The distance that pollen grains are carried airborne is influenced by the shape, size, and density of the pollen grain; wind velocity and turbulence; and the path of the grains from the source plant through and then over the filtering screen of vegetation.

Tauber (1965, 1967a,b) developed a comprehensive model focusing upon physical mechanisms to characterize aerodynamic transport of pollen grains. Tauber emphasized the horizontal, rather than vertical, movement of pollen within and above the forest. Pollen grains falling because of gravity reach a terminal velocity between 1 and 10 cm/s; this speed is determined by the morphological and buoyancy characteristics of the individual grains (Tauber 1965).

In the absence of wind, their downward movement within the gravitational field is called the C_g or gravity component of pollen movement (vertical dryfall of pollen from local vegetation, as described by Jacobson and Bradshaw 1981) (Fig. 2.2a).

Horizontal wind velocities of 50 to 1000 cm/s greatly exceed the terminal velocities of grains falling vertically. Below the closed canopy of forests, air-borne transfer of pollen occurs as wind currents flow around trunks of trees. This mode of transport is designated as the C_t or trunk-space component (Fig. 2.2a) with pollen carried horizontally through the space between the forest canopy and the ground surface (Tauber 1965). Based solely upon average wind and pollen-grain terminal velocities, Tauber (1965) calculated that hypothetical limits for trunk-space dispersion of grains should range between 200 and 1000 m.

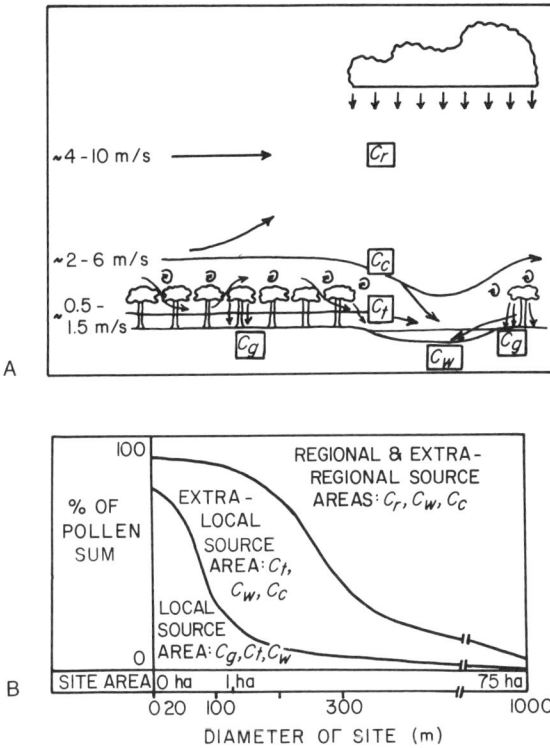

Figure 2.2. a. Tauber model of pollen transport within a forested landscape. Pollen grains are transported by five mechanisms: (1) carried downward by gravity (C_g); (2) carried horizontally through the trunk space below the forest canopy (C_t); (3) carried above and over the canopy (C_c); (4) brought down by rain (C_r); and (5) transported by water within rivers and streams (C_w) (modified from Tauber 1965). b. Relationship between the diameter and area of a site and the pollen source areas and mechanisms of transport for various components of the pollen assemblage (modified from Jacobson and Bradshaw 1981).

As experimentally measured using tracer pollen released within the forest, the airborne concentration of pollen falls markedly over a distance of 100 m (Raynor *et al.* 1974a).The various types of pollen are subject to differential filtration as the grains impact against the obstructing trunks, stems, and leaves of nearby plants. In this situation, wind dispersal selects for the longer transport of small, light grains and against the large, heavy types of grains. Within a stand of temperate deciduous trees, wind velocities are greater for the transport of anemophilous pollen in the early spring before the leaf buds open and leaves grow rapidly. The typical speed of air currents through the trunk space may drop by as much as 50% after the leaves have grown and the forest canopy has closed (Geiger 1966). Trees that pollinate before leafout, such as *Ulmus* (elm), *Acer* (maple), and *Quercus* (oak), possess pollen grains that are carried relatively farther distances by faster air currents. However, for trees that pollinate after leafout, *e.g., Fagus* (beech) and *Tilia* (basswood), pollen tends to be carried only limited distances through the trunk space by the slower winds that are then available for transport (Tauber 1965).

As horizontal air currents flow across the irregular upper surface of tree crowns, turbulent eddies move downward through the forest canopy and into the trunk space. If captured by these turbulent eddies, pollen grains may be raised above the forest canopy by thermal convection. The pollen grains transported by this C_c or above-canopy component (Fig. 2.2a) can be carried over the forest canopy up to several kilometers distance from their starting point (Tauber 1965).

If swept by convection as much 3 km higher into the atmosphere, discrete clouds of pollen grains may drift tens of kilometers (Raynor *et al.* 1974b) to hundreds of kilometers (Barry *et al.* 1981), incorporated by airmasses moving along storm tracks across the continent. Tauber (1965) characterized this long-distance transport of grains from other regions as the C_r or rainout component (Fig. 2.2a). Episodes of rainfall remove pollen from the atmosphere for two reasons: (1) within clouds, the pollen grains represent condensation nuclei for the development of raindrops; and (2) falling raindrops impact against airborne pollen and scavenge them from the sky (McDonald 1962). Pollen grains may undergo several events of aerial transport before they reach a suitable flower stigma for germination, or they may be destroyed by mechanical breakage or by oxidation (Delcourt and Delcourt 1980; Hall 1981; Havinga 1984), or else they finally reach an appropriate depositional evironment where they are incorporated and preserved within sediments.

At an intermediate stage in their transport, grains may impact against plant or soil-litter surfaces or settle directly onto a stream or lake surface. Pollen grains may subsequently be reworked, eroded by intense rainfall, and transported farther overland and downslope by sheetwash. The quantity of pollen supplied by sheetwash to lakes and rivers is largely determined by: (1) the character of the rainfall regime, its seasonality (frequency), intensity, and total annual amount; (2) the erodibility of the soil surface (this is influenced by the quantity of precipitation that percolates below ground relative to the amount of water that flows overland as surface runoff); (3) the terrain topography; (4)

the distribution and density of the vegetation that may filter out the pollen; and (5) the length of time that the grains are exposed to oxidation on the landscape.

The final element of Tauber's (1965) model is the C_w or waterborne component of pollen, that which is carried by sheetwash or transported by flowing water in streams or lakes (Fig. 2.2a). Pollen grains may be eroded from the watershed and funneled through riverine corridors, sometimes deposited in alluvial swamps, backwater sloughs, and, in floodstages, even in oxbow lakes isolated from the main river channel (Delcourt and Delcourt 1980). Pollen accumulation is generally limited in alluvial environments; the moving water tends to flush the pollen through the fluvial system and, should the pollen grains be deposited, they are seasonally subject to erosion by floods (Crowder and Cuddy 1973; Delcourt et al. 1980). Within a stream, pollen grains may be carried in any combination of three ways: (1) buoyant pollen types such as conifer grains with air-filled wings tend to float downstream, caught by water tension on the air–water interface; (2) grains may be immersed by turbulent stream movement, suspending them in the flowing water; and (3) they may adhere to mineral particles and travel along the channel bottom with the bedload (Brush and Brush 1972). One potential problem in interpreting fluvially modified pollen assemblages is that differential sorting of the pollen types occurs as they are carried as floaters, as suspended load, or as bedload (Brush and Brush 1972). In the turbulent stream flow, pollen grains respond as the hydraulic counterparts of medium- and coarse-grained silt particles of quartz. Both kinds of organic and mineral particles are deposited in quiet waters where the water velocity and turbulence are diminished.

As streams contribute pollen to lakes, rates of pollen accumulation in lacustrine basins are often high and variable. An example is that of the St. Croix River, which enters Lake St. Croix along the mutual boundary of eastern Minnesota and western Wisconsin. During the last 1000 years, as the river delta has built into the lake, the pollen accumulation rates have increased to between 100,000 and 169,000 grains of pollen added to each square centimeter of lake bottom each year. In this lake, the river contributes approximately 90% of the pollen assemblage and atmospheric inputs account for the remainder (Eyster-Smith 1977). The riverine input of pollen varies between 15% of the pollen assemblage in Bellham Tarn of the English Lake District (Bonny 1978) to as high as 97% in man-made reservoirs in northeastern England (Peck 1973). In the context of the total pollen assemblage preserved in lacustrine sediments, the relative importance of the waterborne component may increase following the passage of major storms through the watershed (O'Rourke 1976). Accelerated sheetwash erosion and increased streamflow may also follow forest clearance or other anthropogenic disruption of the landscape (Davis 1976a; Davis et al. 1984).

In depositional sites where stream input or outflow does not occur, the pollen assemblage reflects the airborne pollen rain emitted from the surrounding vegetation. Janssen (1966, 1973, 1984) proposes an alternative model to Tauber's that emphasizes the source area from which the pollen grains are derived. Janssen (1967a) interprets different synecological groupings of pollen types as reflecting plant communities within the vegetational mosaic. In order to relate modern pollen rain to vegetational patterns, it is necessary to understand the spatial patterns of dispersal of the key pollen types and to relate them to the proximity of their source plants.

Four groupings of pollen types observed in the pollen assemblage can be considered in terms of their source area: pollen sources that are local, extralocal, regional, and extraregional (Janssen 1966, 1973, 1984). In a closed forest, local pollen is contributed from within a 20-m to 30-m radius (Andersen 1970; Bradshaw 1981a,b; Jacobson and Bradshaw 1981) and is characterized by high percentage values in the overall pollen assemblage. The highest values observed for the local pollen types are to be expected near the source plants. However, their pollen values may fluctuate widely over distances as short as a few meters, depending upon the concentration of anthers of local plants in the sample analyzed (Andersen 1970). The local source area contributes pollen from a relatively large number of both herbaceous and woody insect-pollinated plant species that may represent small populations growing near the depositional site. The complement of local pollen types may provide insights into successional changes within aquatic and bottomland communities (Janssen 1984) as well as gap-phase replacement of canopy trees within closed forest stands (Andersen 1984).

Extralocal pollen deposition is provided by plants within a 20-m to 200-m radius (Jacobson and Bradshaw 1981; Janssen 1984); extralocal pollen percentages are more consistent but generally lower that the peak values for the local pollen types. The prominence of extralocal pollen types is conditioned by the density of shrubs and saplings in thickets and by the topography in the vicinity of the site. Where poorly drained bottomlands and mesic lower slopes occur within several hundred meters of the site, the presence of brushy shrub thickets and poorly stratified swamp forests may influence the pollen spectrum by including species not growing at the coring site or by providing a physical barrier to transport of pollen grains otherwise carried through the trunk space from upland forests (Tauber 1965; Currier and Kapp 1974).

Regional deposition of pollen types reflects the abundance of the major populations of wind-pollinated tree species within the area extending from a 200-m radius (Jacobson and Bradshaw 1981; Janssen 1984) to an outer limit of approximately 200 km (Prentice 1985). The regional pollen component provides a characteristic pollen spectrum that integrates the total pollen contribution from the mosaic of plant communities within major vegetation types.

In Janssen's (1984) model, the extraregional component is comprised of pollen grains that have been transported long distances from beyond the vegetational formation in which the coring site is located. Prentice (1985) suggests that this equates with a dispersal distance typically greater than 200 km from the depositional site where the pollen assemblages have been analyzed.

Jacobson and Bradshaw (1981) developed a synthetic spatial model for pollen dispersal, integrating Tauber's model for physical mechanisms of pollen dispersal with Janssen's model for pollen source area. For a specific site of pollen accumulation, the relative contribution of each of the components of pollen rain is partly determined by the type and size of the site. The blend of pollen-rain components varies with the physical and aerodynamic characteristics of the pollen-collecting basin; the distribution, structure and density of vegetation surrounding the site; streams that flow into or out from the site; and the biennial sequence of climatic conditions that may constrain pollen productivity in the nearby vegetation.

Very small sites (less than 0.5 ha) beneath closed forest canopies are represented by humus-rich soil profiles (Iversen 1969; Andersen 1970, 1984; Stockmarr 1975) and shallow pools occupying woodland hollows (Bradshaw 1981a,b; Davidson 1983; Heide 1984). The pollen assemblage is dominated by local pollen deposition primarily through gravitational (C_g) and trunk-space (C_t) transport (Fig. 2.2b). The relative contribution of the waterborne (C_w) component is related to the extent of pollen erosion and transport with overland flow of water by sheetwash or in channels of ephemeral streams. Because the pollen concentration diminishes rapidly with increasing distance from the source plant, the local pollen deposition masks out the other components.

As the canopy of the forest opens and the size of the depositional basin increases, ponds and small lakes receive pollen rain predominantly from extralocal and regional sources (principally the trunk-space and above-canopy components) (Fig. 2.2b). Tauber (1967a,b) demonstrated that the trunk-space component predominated in small lakes upto 2 ha in size; at Lake Gantekrogso in Zealand, Denmark, the absence of a thicket bordering the lake margin maximized the extralocal influx of pollen from upland deciduous forests. However, at Davis Lake, a small lake 0.8 ha in area in central Lower Michigan, detailed sampling of modern pollen rain confirms that the majority of grains are contributed by the above-canopy (C_c) component. With the horizontal drift of air currents across the open water of the lake basin, turbulent eddies in the lake basin decrease wind velocities and increase the transit time for pollen grains to settle out of the airstream. At Davis Lake, the relative contribution of the trunk-space component of pollen to the lake is minor. The pollen swept through the open trunk space of upland deciduous forests is filtered out by the barrier imposed by a double screen consisting of, first, the dense swamp forest of conifers and hardwoods and, then, a shrub thicket of *Chamaedaphne* and *Vaccinium* (leatherleaf and blueberry) that border the lake (Currier and Kapp 1974).

With progressively larger lake basins, primary constituents of the pollen rain shift to the above-canopy and waterborne components derived from regional and extraregional source areas (Fig. 2.2b). As transport time for pollen increases with the fetch across a given lake, more grains transported from considerable distances are likely to settle into the lake. Very large lakes, such as Lake Ontario, receive pollen assemblages obtained from regional and extraregional source areas. Major storm systems passing over the lake contribute pollen by the rainout (C_r) component. Major rivers drain into the great inland lakes and consequently add their waterborne (C_w) component of pollen to the overall assemblage (McAndrews and Power 1973).

What is the spatial resolution available from late-Quaternary sequences of fossil pollen grains and spores? The experienced palynologist first identifies a specific research question. For example, what is the nature of vegetational dynamics observed in a particular vegetational unit (a unit potentially as fine-grained as a forest stand or as coarse-grained as a formational zone, Fig. 1.1). The research question thus mandates the spatial scale of resolution for the research design of the problem. Based upon an understanding of spatial patterns of pollen dispersal, this in turn requires the selection of an appropriate type and size of depositional site in order to generate paleoecological data relevant to the question asked. Time series of fossil-pollen assemblages preserved in forest-soil profiles or in woodland-hollow pools provide paleoecological data over hundreds to thousands of years, recording micro-scale forest succession within a 20-m radius of the site. Many paleoecologists are intrigued by the developing character of vegetation, responding to climatic change and to the arrival of new plant species invading a deglaciated landscape. As a consequence, they routinely examine the plant-fossil records available from moderate-sized lakes (10 to 100 ha) that receive their regional pollen rain primarily from within a 20 km to 30 km radius. Other palynologists explore evidence of evolutionary changes within the terrestrial flora and vegetation in the hemispheric and global context of the mega-scale—the corresponding paleoecological records are preserved in sediments of great inland lakes and from the oceans (Traverse 1982).

Calibration of the relationship of modern pollen assemblages to extant forests

Calibration methods have been developed in order to translate pollen spectra into quantitative estimates of the population abundances for the major tree species. This "individualistic" or taxon approach attempts to isolate the particular contribution of pollen by populations of each plant taxon to modern (or past) pollen assemblages. Taxon calibrations are applied to pollen assemblages in order to correct biases among tree taxa based upon their differences in pollen productivity (Davis 1963) and in the dispersal potential of their pollen grains (Andersen 1970; Webb et al. 1981; Prentice 1982, 1986; Delcourt et al. 1983a, 1984). The pollen values for a given taxon (A) are expressed either as a percentage typically calculated from the sum of arboreal pollen (AP) or as pollen accumulation rates (PAR) measured in terms of the number of the taxon's pollen grains that accumulate on a square centimeter of collecting surface each year ($gr \cdot cm^{-2} \cdot yr^{-1}$). When the source area for a modern pollen assemblage corresponds directly with the area for which the vegetation has been quantitatively sampled, the pollen percentage for a tree species (p_a) is usually proportional to its population size within the vegetation (v_a). When a number of sites are examined, the empirically determined values (the R value *sensu* Davis 1963) are not necessarily constant for the ratio of p_a/v_a (Comanor 1968). The variability observed for these ratios is attributed to uncertainty concerning the boundary limits of the specific area to be sampled; spatial heterogeneity or patchy distribution of the plant species' populations surrounding the site; the structure, composition and pollen production of the vegetation; and errors introduced for the low percentages of minor pollen types when limited counts are used for the total number of pollen grains tallied (Janssen 1967b; Comanor 1968; Livingstone 1968, 1969).

The ratios of pollen percentage to dominance within forests for many individual tree taxa have been compiled for hundreds of paired sample sites (literature summarized in Delcourt et al., 1984). When these values are graphed on scatter plots (the x-axis with 0 to 100% forest composition such as measured by growing-stock wood volume of trees (%GSV); 0 to 100% arboreal pollen (%AP) on the y-axis), they display a linear trend with a positive correlation between the taxon's pollen and vegetation percentages (Livingstone 1968). Regression techniques, such as ordinary linear regression (Andersen 1970, 1973, 1984), geometric-mean linear regression (Webb et al. 1981; Delcourt et al. 1983a, 1984), and maximum likelihood estimation (Parsons and Prentice 1981; Prentice 1982; Prentice and Parsons 1983), have been used successfully to summarize this linear relationship. Prentice and Webb (1986) demonstrate that the results of linear regression techniques and maximum likelihood estimation give comparable estimates using percentage data.

The best-fit regression line corresponds with the following generalized equation:

$$p_a = (v_a \cdot r) + p_o$$

The slope of the regression line is designated by r; the value calculated for the line slope quantifies the relationship between the populations of taxon A in the vegetation and their relative pollen production. The slope value also reflects the preservation properties of the taxon's pollen grains, the mode of pollination (Delcourt et al. 1983a), and the suite of taxa specifically included in the arboreal pollen sum (Wright and Patten 1963). The symbol p_o is the intercept value where the regression line intersects the y-axis. The p_o is a pollen threshold for dispersibility; thus, p_o is an empirical measure for the extent that pollen grains of taxon A are transported to the depositional site from beyond the vegetation calibration area.

Pollen representation is easily understood for the special case where the pollen source area is identical to the area of sampled vegetation. In this situation, the regression line extends through the point of origin of the graph (x = 0.0, y = p_o = 0.0) and r = p_a/v_a (Fig. 2.3a–d). A tree species is considered to be equally represented if its observed pollen percentages are generally comparable with its values for percent forest composition (Fig. 2.3a). The line slope r varies between 0.9 and 1.1 for such equably represented taxa (Delcourt et al. 1983a, 1984). In the second situation, for an over-represented taxon where r is greater than 1.1, the pollen values are consistently greater than the taxon's dominance within the nearby forest communities (Fig. 2.3b). A taxon is often described as under-represented should its pollen percentages consistently be lower than its population abundance in the vegetation (r < 0.9, Fig. 2.3c). However, these traditional concepts of pollen–vegetation representation (Davis 1963; Faegri

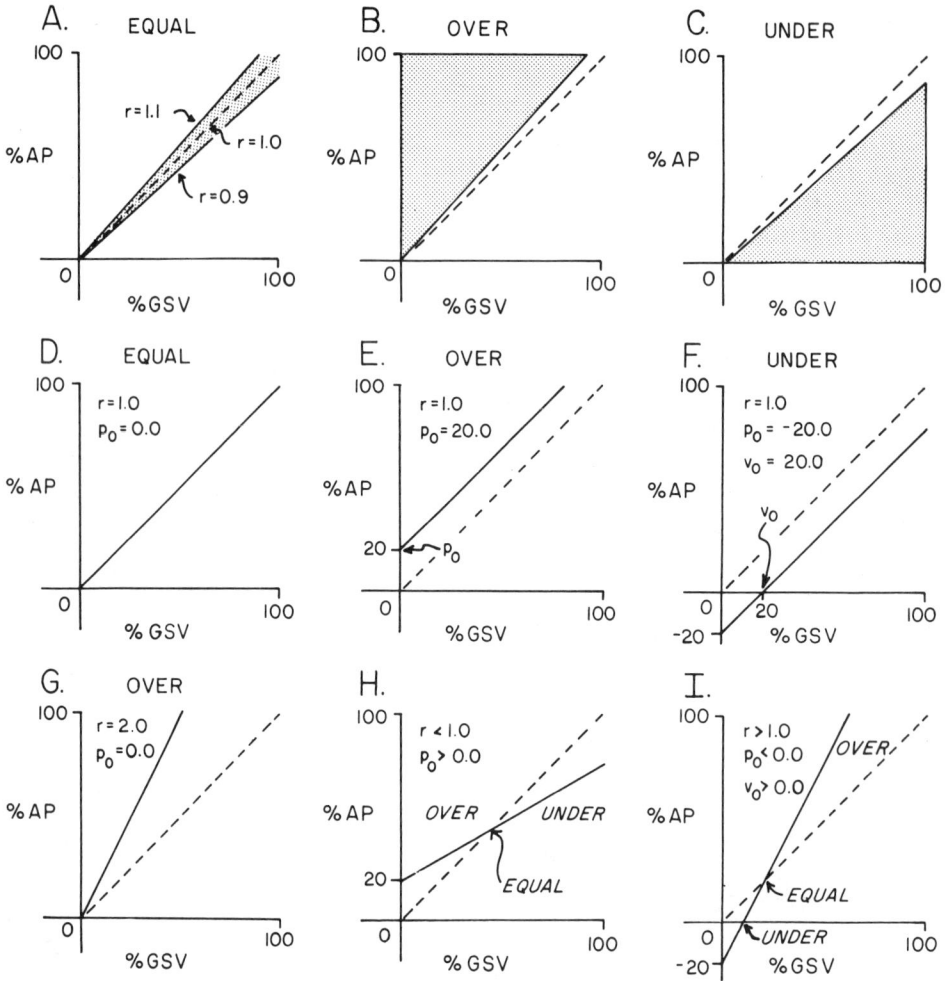

Figure 2.3. Idealized x–y plots illustrating concepts of over-, equal-, and under-representation of vegetational composition (expressed as %GSV or percent of growing-stock wood volume) by arboreal pollen percentages (%AP) (modified from Delcourt *et al.* 1983).

and Iversen 1975) must be refined to include additional situations for which p_o does not equal zero on the y-axis (Andersen 1970; Webb *et al.* 1981; Delcourt *et al.* 1983a, 1984).

    The empirical value for p_o is greater than zero when the pollen source area for a taxon is greater than the vegetational area surveyed. The positive value for p_o constitutes a minimum threshold of pollen abundance that must be exceeded before the observed pollen percentage (p_a) for taxon A can be used to estimate plant-population size based upon its taxon calibraton. The taxon is clearly over-represented by its pollen when the values for p_a exceed the positive threshold value of p_o and when the slope of the regression line either approximates 1.0 (Fig. 2.3e) or is > 1.0. Where statistically distinct from zero, the positive value for the pollen threshold provides a quantitative index for the relative pollen-dispersal characteristics for the pollen grains of each taxon. Should the p_a value be less than the positive p_o, tree-population estimates cannot be made with accuracy, *i.e.*, if $p_a < +p_o$, then we assume $v_a = 0.0$. This uncertainty may be caused by any of three situations: (1) if the contribution of taxon A to the pollen spectrum is solely from long-distance sources beyond the area of sampled vegetation; (2) if the limited percentage for p_a is dispersed from restricted populations of the tree species growing within the surveyed vegetation area; or (3) if the pollen grains have been reworked from older deposits and the taxon's populations no longer grow within the nearby forests.

As calculated by regression analysis, a negative value for p_o cannot be directly monitored using samples of modern pollen rain and forest composition. Rather, a reasonable alternative to the negative pollen threshhold is provided by a corresponding and complementary vegetation threshold, v_o, where:

$$v_o = (p_a - p_o)/r \quad \text{and} \quad p_a \geqslant p_o.$$

As the regression line intersects the x-axis, v_o is the x-intercept (Fig. 2.3f); note that $v_o > 0.0$ for the case that $p_o < 0.0$. The v_o is a critical minimum size required for a species' population within the study area that must be exceeded before a consistent linear relationship can be established between its pollen and vegetation percentages. A taxon is under-represented with respect to its pollen percentages if the $r = 1.0$ but $v_o > 0.0$ (Fig. 2.3f). Such a critical population size exists for an under-represented tree species (*i.e.*, $v_o > 0.0$) if (1) its restricted populations are distributed in patches across a vegetational mosaic; (2) its production of pollen is low relative to the total production of the surrounding vegetation; (3) its pollen grains are poorly dispersed to the depositional site sampled for the pollen spectrum; or (4) its pollen grains are poorly preserved (Delcourt *et al.* 1983a).

Contemporary definitions of fundamental pollen–vegetation relationships require several specific criteria that include explicit designation of the spatial scale of resolution for the pollen source area and sampled vegetation as well as empirical estimates for both the line-slope and axis intercepts as quantified by linear regression analysis (Delcourt *et al.* 1983a, 1984). A tree taxon is **equally or proportionally represented** when its percentage contribution to the modern arboreal-pollen rain is directly comparable to its population abundance within the forest. Three criteria must be satisfied: (1) the area of vegetation sampled is the same as the pollen source area from which the pollen assemblage has been derived; (2) the slope of the regression line has a value in the range of 0.9 to 1.1; and (3) The value for the y-intercept of the regression line is not statistically distinguishable from zero at the probability level of 0.05 (Fig. 2.3a,d). A tree taxon is **over-represented** when its values in the modern arboreal-pollen rain are consistently greater than the taxon's percent composition in the forest surveyed. The regression slope must be typically greater than 1.1, and the pollen threshold must be greater than or equal to zero (Fig. 2.3b,e,g). **Under-represented** tree taxa are characterized by percentages in the modern arboreal-pollen rain that are consistently less than their corresponding population abundance within the area of forest sampled. The regression slope is less than 0.9, and the vegetation threshold is greater than or equal to zero (the pollen threshold is less than or equal to zero) (Fig. 2.3c,f).

Taxon calibrations can be generated for additional plant species with regression lines not conveniently characterized by our standard concepts of over-, equal-, and under-representation. By using quantitative, empirically determined values for r, p_o, and v_o, the pollen–vegetation representation can be characterized for other taxa. For example, if $r < 1$ and $p_o > 0$, the regression line may pass through the region of over-representation for low population levels to the region of under-representation for its high population sizes (Fig. 2.3h). In a second case, where the slope > 1 and $v_o > 0$, the regression line passes from regions of under- to over-representation as a taxon's importance increases within the vegetation (Fig. 2.3i).

The representation of temperate and boreal tree taxa in modern pollen samples and forest vegetation

The nature of correspondence between modern pollen assemblages and vegetation is evaluated usefully on several spatial scales. Pollen assemblages from the surface sediments of individual sites can be compared with the vegetation of the surrounding watershed (Davis and Goodlett 1960; Comanor 1968; Currier and Kapp 1974). Transects may be established across vegetational ecotones (Janssen 1984) or along elevational gradients (Davis 1984; Delcourt and Pittillo 1986). Along such transects, comparison of pollen assemblages from surface samples with corresponding samples of community composition helps to define the dispersal area for individual taxa, to delineate local from regional components of the pollen rain, and to determine the characteristic "signatures" of different community types represented in pollen assemblages. In order to determine the fundamental pollen–vegetation relationship for a species or genus, however, it is necessary to develop larger data sets over a broader scale that describe the variability in both the plant populations and the pollen rain. Such comparisons can be accomplished by mapping the patterns of distribution and dominance across broad landscapes and by making statistical comparisons of representation for individual taxa based upon a geographic array of corresponding pollen and vegetation samples (*e.g.*, Birks *et al.* 1975).

The degree of observed correspondence between pollen percentages and dominance within the vegetation is influenced by the pollen production and dispersal characteristics of the taxon, and it differs with geographic area as well as vegetation type. The relationship of the pollen production and dispersal of a tree taxon to its dominance in the vegetation is in part conditioned by the structure and composition of the surrounding vegetational mosaic. In regions with sparse forest cover, airborne pollen grains may travel long distances. In the Great Plains of central North America, for example, *Pinus* is over-represented in modern pollen samples because substantial quantities of pine pollen grains can be transported hundreds of kilometers by prevailing winds (Birks *et al.* 1975; Bernabo and Webb 1977; Webb and McAndrews 1976; Webb *et al.* 1981). In contrast, in the southeastern United States, a region with approximately 65% forest cover, the contoured isopolls for *Pinus* in the modern pollen rain closely parallel the contours of dominance of *Pinus* trees mapped from continuous forest inventory data (Delcourt *et al.* 1983a).

In order to recognize both the potential uses and the limitations of pollen–vegetation calibrations, it is necessary to compare and contrast calibrations derived from comparable data sets in different geographic regions as well as to evaluate the changes that result from expanding those data sets from a regional to a subcontinental scale of coverage.

Regional pollen–vegetation relationships in eastern North America

The primary tree taxa for which pollen–vegetation relationships have been examined on a regional scale in the midwestern and northeastern United States and adjacent Canada include *Abies, Acer, Betula, Carya, Fagus, Fraxinus, Juglans, Juniperus/Thuja, Larix, Ostrya/Carpinus, Picea, Pinus, Populus, Quercus, Salix, Tilia, Tsuga,* and *Ulmus* (Livingstone 1968; Webb 1974a; Davis and Webb 1975; Webb and McAndrews 1976; Webb *et al.* 1981). These taxa are today the primary dominant and subdominant trees in the vegetation of that region, as represented in the continuous forest inventories of the United States and Canadian Forest Services (Halliday and Brown 1943; literature cited in Delcourt *et al.* 1984). Only a few of these taxa, however, are monospecific in that region of North America, *i.e., Abies balsamea, Fagus grandifolia, Larix laricina, Tilia americana* and *Tsuga canadensis* (Little 1971). Because the pollen grains of most of the other arboreal taxa cannot be routinely identified to the level of species, the calibrations for their pollen representation necessarily represent the combined relationships for populations of two or more species.

Comparison of the mapped patterns of pollen percentages dominance (*e.g.,* basal area or growing-stock wood volume) for each tree taxon generally shows broad correspondence in the overall distributional patterns. The first study in North America to compare quantitative forest composition based upon forest inventory records with pollen assemblages from surface lake sediment samples was Livingstone's (1968) analysis using estimates of commercial timber volume from 22 counties and forest districts in Nova Scotia, New Brunswick, and Maine. For each tree taxon, Livingstone calculated the ratio of its percentage in the pollen rain to its percentage in the vegetation. He then normalized the ''R'' values and applied them to the pollen percentages from long sediment sequences in order to reconstruct the changes in population abundance of each tree taxon through time (Livingstone 1968).

Webb (1974a) used principal components analysis to demonstrate that the north–south gradient in composition from mixed conifer–northern hardwoods forest to deciduous forest across the ecological tension zone in Lower Michigan; the principal components of pollen assemblages from surface sediments of 64 lakes demonstrated comparable gradients across the region. In most cases, the distribution of pollen percentages for individual taxa also tended to parallel their gradients in population dominance reflected in the vegetational samples. The general correspondence in mapped distributions of pollen percentages and relative dominance (in this case, percent basal area) was further demonstrated on a broader spatial scale for *Picea*, *Pinus*, and *Quercus* using basal area data from 69 counties and forest-inventory units compared with pollen data from 400 surface samples taken from lakes distributed from Manitoba and Minnesota eastward to the Atlantic Coastal Plain (Bernabo and Webb 1977).

Using geometric-mean linear regression, Webb *et al.* (1981) and Heide and Bradshaw (1982) examined the quantitative relationships between percent arboreal pollen (AP) and percent basal area for a number of cool-temperate tree taxa in Wisconsin and Upper Michigan. They grouped the tree taxa into several categories according to their representation in the modern pollen rain. *Pinus* and *Betula* both had high y-intercepts and slopes, with large scatter of data points along the y-axis (high variability in %AP at low percentages of basal area), indicating that both taxa are generally over-represented in the pollen rain of the Upper Midwest region. *Abies*, *Acer*, *Fraxinus*, and *Populus*, in contrast, were under-represented by pollen, with high scatter of points along the x-axis (low values of %AP across a wide range of basal area values). The pollen percentages of *Tsuga*, *Juglans*, *Quercus*, *Carya*, *Picea*, and *Ulmus* were represented proportionately with respect to their basal areas in the region (Webb *et al.* 1981). Based upon 167 paired observations in Wisconsin and the Upper Peninsula of Michigan, Webb *et al.* (1981) concluded that the relationship of %AP vs % basal area was linear for the major tree taxa evaluated.

Calibration results compared for a series of nested scales in Wisconsin and Michigan (Bradshaw and Webb 1983) show clearly that not only do the slopes and intercepts for regression lines of quantitative pollen–vegetation calibrations change with differences in the scale on which the samples are collected, but that it is important to pair samples that represent comparable scales (Parsons *et al.* 1980). The sediments of moderate-sized lakes tend to spatially smooth incoming airborne pollen assemblages on a regional scale, averaging the contribution of a broad landscape mosaic of vegetation over an area $\geqslant 100 \text{ km}^2$ around each lake (Davis and Goodlett 1960; Webb *et al.* 1978). Spatial smoothing to a scale approximating the average size of midwestern counties gave optimal calibration results for those tree taxa that produce abundant pollen that is wind dispersed. These taxa include *Pinus* and *Quercus*, whose pollen tended to disperse 30 km or more beyond its source, as well as *Betula*, *Tsuga*, and *Ulmus*, whose pollen disperses as far as 5 to 30 km from its source (Bradshaw and Webb 1985). In contrast, pollen samples collected from moss or soil samples within forest plots record primarily the variation among trees within the local stand (Webb *et al.* 1978). Forest-inventory data summarized over smaller areas, representing forest stands on watersheds immediately surrounding the lake sites sampled for pollen, yielded better correlations for taxa whose pollen is dispersed more locally, including *Fagus* (< 3 km), *Acer* (< 1 km), and *Tilia* (< 100 m) (Bradshaw and Webb 1985).

Delcourt *et al.* (1983a) mapped and calibrated the modern pollen–vegetation relationships of 19 arboreal taxa characteristic of forests within the southeastern United States, based upon 250 modern pollen samples and growing-stock volume data obtained from nearly 900 county-level continuous forest inventories (Delcourt *et al.* 1981). As in the study by Webb *et al.* (1981), geometric-mean linear regression was used to characterize the pollen–vegetation relationship. The regional study of southeastern trees provides a direct comparison with the relationships derived for the Upper Midwest for several broadly ranging tree taxa as well as establishing the pollen–vegetation relationships for a number of tree taxa whose distributions are restricted to the southeastern United States. In the northern and southern calibration sets, certain taxa such as *Quercus, Carya,* and *Ulmus* exhibited similar values for both the slope of the regression line and the y-intercept (p_o) even though there were pronounced regional differences in the number and taxonomic composition of *Quercus* species present. In other cases, for example, *Betula, Fraxinus, Juglans,* and *Pinus,* the northern and southern groups of species showed distinctly different patterns of representation in which the slopes and intercepts differed considerably. These differences were attributed to different pollen productivity of northern and southern species relative to that of the other taxa growing in the respective regions (Delcourt *et al.* 1983a).

The smaller the region to which pollen–vegetation calibrations are restricted, the less likely it will be that a broad enough range of combinations will be studied in order to characterize the total variability of a taxon's pollen productivity and dispersal characteristics. For example, in the case of *Fraxinus* in the Upper Midwest, only relatively low percentages are found in the modern pollen rain, and thus *Fraxinus* appears to be severely under-represented with respect to basal area in the forests (Webb *et al.* 1981). The genus reaches maximum percentages of 20% AP in the bottomlands of the Lower Mississippi Alluvial Valley, where *Fraxinus* trees reach up to 10% of the growing-stock volume in the forests (Delcourt *et al.* 1981, 1983a). In the data set from the southeastern United States, *Fraxinus* therefore appears much more proportionately represented in the pollen rain (Delcourt *et al.* 1983a). This case illustrates that calibrations based upon data sets collected from relatively small geographic subsets of the taxon's distribution may include neither the population centers nor range margins and therefore may not be representative of the range of variability of the taxon. Optimally, the representation for each taxon should be based upon data collected throughout the entire range of the species or species group.

Regional studies elsewhere in the northern Temperate Zone

In Europe, paleoecological reconstructions based upon calibration of modern pollen–vegetation relationships on a regional scale are limited because the modern vegetation is managed to the extent that relatively few examples remain for natural vegetation. Although forest inventory records are available from areas of managed forests (Armentano and Ralston 1980), in many cases the forest stands are monospecific or vary little in composition across large areas.

Studies of pollen representation are possible on a local scale (Tauber 1967a; Andersen 1980a; Bradshaw 1981b) within nature preserves such as the Draved Forest of Denmark. As demonstrated by Prentice (1983) in south and central Sweden, the patterns of pollen percentages for major taxa reflect a combination of variation in forest composition and human impacts on the landscape. Reconstructions of past vegetation are necessarily qualitative for most of Europe and are based upon the assumption that mapped patterns in pollen adequately represent changing patterns in distribution and dominance of the major taxa (Huntley and Birks 1983). Where available, local pollen–vegetation relationships are used to constrain interpretations based upon the maps of changing relative pollen abundances through time (Huntley and Birks 1983).

Only one set of studies within Finland and Norway (Prentice 1978; Parsons et al. 1980) has attempted to compare directly the quantitative variation in forest composition with that of pollen from surface lake sediments using methods comparable to those developed within eastern North America. The forest ecosystem studied in Fennoscandia contained four principal commercially important woody taxa, *Pinus sylvestris, Picea abies, Alnus (A. glutinosa* and *A. incana),* and *Betula (B. pubescens* and *B. pendula)* (Prentice 1978). In Fennoscandia, as southern pine–birch forest thinned northward to pine forest, changes in the forest structure resulted in changes in pollen representation of birch and pine. Pine was over-represented in north-central Finland and Norway because of long-distance transport of pine pollen from the south (Parsons *et al.* 1980).

Subcontinent-scale pollen–vegetation calibrations in eastern North America

Comprehensive calibrations are essential for making reconstructions of past forest composition. Calibrations based upon a restricted modern data set, one that is small in sample number or geographically restricted, are limited in application to the fossil record. For example, the composition of vegetation may have differed in the past from that occurring today within the area for which calibration data sets are compiled (Heide and Bradshaw 1982; Heide 1984). Application of calibration equations based upon geometric-mean linear regression or maximum likelihood estimation to reconstruct past vegetation are invalid if used in situations where the forest assemblage contained taxa not present in the modern calibration data set or if the percentages of constituent taxa in past forests were far beyond the range of values included in the calibration data set (Parsons and Prentice 1981).

For example, in middle latitudes of eastern North America, from Missouri to lower elevations in Tennessee and the Carolinas, the landscape has been continuously forested over the past glacial–interglacial cycle (Delcourt 1979). However, the composition of the vegetation in that region has changed from boreal forest to mixed conifer–northern hardwoods forest to warm-temperate deciduous forest over the past 20,000 years (Delcourt and Delcourt 1981). Neither the pollen/calibrations based upon the data set from the Upper Midwest nor those made for the Southeast region are sufficient alone to base quantitative reconstructions of forest composition from full-glacial times to the present at paleoecological sites such as Anderson Pond, Tennessee. Such interpretations require an extensive calibration data set based upon corresponding modern pollen and vegetation samples distributed throughout boreal and temperate regions and that include most or all of the distributional ranges of constituent taxa.

In order to provide a more comprehensive data set for calibrations of arboreal pollen percentages and percent dominance of principal forest trees, we combined the available modern pollen data and expanded the computerized forest-inventory data set for the boreal and temperate regions of eastern North America (Delcourt et al. 1984). The combined calibration data set included 1742 forest-inventory summaries representing counties and forest-inventory units, as well as 1684 modern-pollen samples, situated between 25°N and 60°N latitude and extending from 50°W to 105°W longitude. This study area covered the distributional ranges for species characteristic of the southeastern evergreen forest, the eastern deciduous forest, and the mixed conifer–northern hardwoods forest regions, as well as the eastern half of the boreal forest (Little 1971). The data base included 24 taxa of trees that were resolvable to comparable taxonomic levels in both the forestry and the pollen data. These taxa represented the dominant, subdominant, and common species and species groups within eastern North America and (listed alphabetically) included *Abies, Acer, Betula, Carya, Celtis,* Cupressaceae plus Taxodiaceae, *Fagus grandifolia, Fraxinus, Ilex, Juglans, Larix laricina, Liquidambar styraciflua, Liriodendron tulipifera, Magnolia, Nyssa, Picea, Pinus, Platanus occidentalis, Populus, Quercus, Salix, Tilia, Tsuga,* and *Ulmus.* Detailed contoured maps of percent arboreal pollen (isopoll maps) and of percent growing-stock volume (isophyte maps) for these taxa are presented in Delcourt et al. (1984). Full documentation of data sources, as well as procedures for compilation, mapping, and calibration, is available in the Atlas published by the American Association of Stratigraphic Palynologists (Delcourt et al. 1984). Based on the subcontinental array of paired pollen and forestry samples, each representing approximately the county-scale of spatial resolution, statistically significant geometric-mean linear regressions were attained for 19 taxa (Tables 2.1 and 2.2).

Taxa for which calibrations were not possible with the available data sets included *Liriodendron tulipifera, Magnolia, Platanus occidentalis,* and *Liquidambar styraciflua. Liriodendron tulipifera* (tuliptree), today widespread and abundant in successional stands throughout the eastern deciduous forest (Delcourt et al. 1984), produces relatively little pollen, which is both insect-dispersed and easily degraded by biological or chemical oxidation (Delcourt et al. 1983a). Species of *Magnolia* occur throughout the southern Appalachian mountains (deciduous species including *M. acuminata, M. fraseri,* and *M. macrophylla*) and are particularly characteristic of mesic to hydric sites in the Gulf and southern Atlantic Coastal Plains (predominantly evergreen species *M. grandiflora* and *M. virginiana*). As in *Liriodendron tulipifera,* all species of *Magnolia* are insect-pollinated and are poorly represented in the regional pollen rain. *Platanus occidentalis* (sycamore) is locally abundant along floodplains of streams

Table 2.1. Tree taxa represented by equivalent groups of commercially important,
native eastern North American species inventoried by the United States and
Canadian Forest Services (US Forest Service 1967; Little 1971) and pollen
morphological types identified with light microscopy by pollen analysts. The list
includes the 19 plant taxa resolvable in both forestry and pollen data that were
selected for mapping and calibration in Delcourt *et al.* (1984).

Plant Taxa	Pollen Type

Abies (fir)
   A. balsamea (L.) Mill. (balsam fir)
   A. fraseri Poir. (Fraser fir)
	Abies

Acer (total maple)
   Hard-maple group:
     A. saccharum complex:
       A. barbatum Michx. (Florida maple)
       A. nigrum L. (black maple)
       A. saccharum Marsh. (sugar maple)
   Soft-maple group:
     A. negundo L. (boxelder)
     A. pensylvanicum L. (striped maple)
     A. rubrum L. (red maple)
     A. saccharinum L. (silver maple)
     A. spicatum Lam. (mountain maple)
	Acer total
	   Hard-maple group:
	     A. saccharum type
	
	   Soft-maple group:
	     A. negundo
	     A. pensylvanicum
	     A. rubrum
	     A. saccharinum
	     A. spicatum

Betula (birch)
   B. alleghaniensis Britt. (yellow birch)
   B. lenta L. (sweet birch)
   B. nigra L. (river birch)
   B. papyrifera Marsh. (paper birch)
   B. populifolia Marsh. (wire birch)
	Betula

Carya (hickory)
   C. aquatica (Michx. f.) Britt. (water hickory)
   C. cordiformis (Wangenh.) K. Koch
   (bitternut hickory)
   C. glabra (Mill.) Sweet (pignut hickory)
   C. illinoensis (Wangenh.) K. Koch (sweet pecan)
   C. laciniosa (Michx. f.) Loud (shellbark hickory)
   C. myristicaeformis (Michx. f.) Nutt.
   (nutmeg hickory)
   C. ovata (Mill.) K. Koch (shagbark hickory)
   C. tomentosa Nutt. (mockernut hickory)
	Carya

Celtis (hackberry)
   C. laevigata Willd. (sugarberry)
   C. occidentalis L. (hackberry)
   plus *Maclura pomifera* (Raf.) Schneid.
   (Osage orange)
	Celtis type (includes
	 Maclura)

Cupressaceae plus Taxodiaceae
   Cupressaceae:
     Chamaecyparis thyoides (L.) B.S.P.
     (Atlantic white cedar)
	Cupressaceae plus Taxo-
	diaceae undifferen-
	tiated (intact pollen of
	Chamaecyparis,

Table 2.1. (*continued*)

Plant Taxa	Pollen Type

Cupressaceae:
  Juniperus (juniper):
    J. silicicola (Small) Bailey
    (southern red cedar)
    J. virginiana L. (eastern red cedar)
    Thuja occidentalis L. (northern white cedar)
Taxodiaceae:
  Taxodium distichum (L.) Rich. (bald cypress)
	Juniperus? Thuja, and
	broken grains of
	Taxodium)
	
	Taxodium distichum
	(identified from
	unbroken papillate
	pollen grains)

Fagus grandifolia Ehrh. (American beech)	*Fagus grandifolia*

Fraxinus (ash)
  Fraxinus americana L. (white ash)
  F. profunda Bush. (pumpkin ash)
  F. pennsylvanica Marsh. (green ash)
  F. nigra Marsh. (black ash)
  F. quadrangulata Michx. (blue ash)
	Fraxinus total
	  F. americana-
	    pennsylvanica type
	
	  F. nigra-quadrangulata
	  type

Juglans (walnut)
  J. cinerea L. (butternut)
  J. nigra L. (black walnut)
	Juglans total
	  J. cinerea
	  J. nigra

Larix laricina (Du Roi) K. Koch (tamarack)	*Larix*

Nyssa (tupelo)
  N. aquatica L. (water tupelo)
  N. ogeche Bartr. (Ogeechee tupelo)
  N. sylvatica Marsh (black gum)
	Nyssa

Picea (spruce)
  P. glauca (Moench.) Voss (white spruce)
  P. mariana (Mill.) B.S.P. (black spruce)
  P. rubens Sarg. (red spruce)
	Picea
	  P. glauca
	  P. mariana
	  P. rubens

Pinus (pine)
  Subgenus Diploxylon:
    P. banksiana Lamb. (jack pine)
    P. clausa (Chapm.) Vasey (sand pine)
    P. echinata Mill. (shortleaf pine)
    P. elliottii Engelm. (slash pine)
    P. glabra Walt. (spruce pine)
    P. palustris Mill. (longleaf pine)
    P. pungens Lamb. (Table-Mountain pine)
    P. resinosa Ait. (red pine)
    P. rigida Mill. (pitch pine)
    P. serotina Michx. (pond pine)
    P. taeda L. (loblolly pine)
    P. virginiana Mill. (Virginia pine)
  Subgenus Haploxylon:
    P. strobus L. (eastern white pine)
	Pinus total
	  Pinus Diploxylon
	
	Pinus Haploxylon

Table 2.1. (*continued*)

Plant Taxa	Pollen Type
Populus (aspen or cottonwood) 　*P. balsamifera* L. (balsam poplar) 　*P. deltoides* Bartr. (eastern cottonwood) 　*P. grandidentata* Michx. (bigtoothed aspen) 　*P. heterophylla* L. (swamp cottonwood) 　*P. tremuloides* Michx. (quaking aspen)	*Populus*
Quercus (oak) 　Subgenus Lepidobalanus (white oak) 　　*Q. alba* L. (white oak) 　　*Q. bicolor* Willd. (swamp white oak) 　　*Q. durandii* Buckl. (Durand oak) 　　*Q. lyrata* Walt. (overcup oak) 　　*Q. macrocarpa* Michx. (bur oak) 　　*Q. michauxii* Nutt. (swamp chestnut oak) 　　*Q. muhlenbergii* Engelm. (chinkapin oak) 　　*Q. prinus* L. (chestnut oak) 　　*Q. stellata* Wangenh. (post oak) 　　*Q. virginiana* Mill. (live oak) 　Subgenus Erythrobalanus (red and black oak) 　　*Q. coccinea* Muenchh. (scarlet oak) 　　*Q. ellipsoidalis* E.J. Hill (northern pin oak) 　　*Q. falcata* Michx. (southern red oak) 　　*Q. ilicifolia* Wangenh. (bear oak) 　　*Q. imbricaria* Michx. (shingle oak) 　　*Q. laevis* Walt. (turkey oak) 　　*Q. laurifolia* Michx. (laurel oak) 　　*Q. marilandica* Muenchh. (blackjack oak) 　　*Q. nigra* L. (water oak) 　　*Q. nuttallii* Palmer (Nuttall oak) 　　*Q. palustris* Muenchh. (pin oak) 　　*Q. phellos* L. (willow oak) 　　*Q. rubra* L. (northern red oak) 　　*Q. shumardii* Buckl. (shumard oak) 　　*Q. velutina* Lam. (black oak)	*Quercus*
Salix (willow) 　*S. amygdaloides* Anderss. (peachleaf willow) 　*S. nigra* Marsh. (black willow)	*Salix*
Tilia (basswood) 　*T. americana* L. (American basswood) 　*T. heterophylla* Vent. (white basswood)	*Tilia*
Tsuga (hemlock) 　*T. canadensis* (L.) Carr. (eastern hemlock) 　*T. caroliniana* Engelm. (Carolina hemlock)	*Tsuga*
Ulmus (elm) 　*U. alata* Michx. (winged elm) 　*U. americana* L. (American elm) 　*U. crassifolia* Nutt. (cedar elm) 　*U. rubra* Muhl. (slippery elm) 　*U. serotina* Sarg. (September elm) 　*U. thomassii* Sarg. (rock elm)	*Ulmus*

Table 2.2. Geometric-mean linear regression results for 19 tree taxa. A * denotes statistical significance at the 0.05 level; ** denotes statistical significance at the 0.01 level. In each group, taxa are listed in order of decreasing values for regression slopes [From Table 1 of Delcourt *et al.* (1984)].

Arboreal Taxa	Slope (r)	Standard error of r	Y-intercept (p_o)	Standard error of p_o	Pearson product-moment correlation coefficient (R)
Over-Represented Taxa:					
Betula	2.33**	0.07	1.15	0.89	0.37**
Salix	1.57**	0.10	1.02*	0.41	0.88**
Pinus	1.23**	0.03	12.46**	0.86	0.52**
Quercus	1.20**	0.03	1.22	0.71	0.79**
Juglans	1.18**	0.07	0.55**	0.15	0.67**
Equally-Represented Taxa:					
Ulmus	0.87**	0.03	−0.29	0.24	0.57**
Under-Represented Taxa:					
Nyssa	0.81**	0.08	−3.39*	1.30	0.21*
Abies	0.44**	0.02	−2.35**	0.41	0.30**
Tilia	0.25**	0.02	−0.36**	0.13	0.31**
Acer	0.24**	0.01	−0.87**	0.16	0.41**
Populus	0.09**	0.004	0.00	0.12	0.15**
Other Taxa:					
Fagus	1.81**	0.09	−2.29**	0.40	0.27**
Tsuga	1.73**	0.09	−2.07**	0.58	0.30**
Cupressaceae plus Taxodiaceae	1.31**	0.05	−1.40**	0.41	0.69**
Picea	1.10**	0.03	−5.28**	0.88	0.80**
Fraxinus	0.92**	0.03	−0.64**	0.16	0.34**
Carya	0.62**	0.03	0.60**	0.20	0.56**
Larix	0.50**	0.03	0.43**	0.06	0.24**
Celtis type	0.39**	0.02	0.43**	0.09	0.96**

throughout southeastern and east-central portions of North America, but its pollen is not widespread in the regional pollen rain. In the case of *Liquidambar styraciflua* (sweetgum), both trees and pollen are represented in the regional pollen rain, but regression analysis showed only poor correspondence between %AP and %GSV. Sweetgum is today successional in the understory of coastal-plain pine forests and is thus recorded in the continuous forest inventories consistently across the southeastern United States. Although its pollen is widespread throughout the region, the peak dominance values of 25% for sweetgum trees are not paralleled by high values for sweetgum pollen, apparently because sweetgum is managed as an understory tree across much of its range and may produce less pollen than if it were allowed to dominate; in addition, the many species of *Pinus* and *Quercus* are much more prolific pollen producers and dilute the contribution of sweetgum to the regional pollen rain. *Ilex* is represented in northeastern North America primarily by species of shrubs that are only locally common within *Sphagnum* peatlands. However, *Ilex opaca* (American holly), reaches tree stature (Fowells 1965) and is recorded today in low abundances in the continuous forest inventories throughout its range in the southeastern United States (Delcourt *et al.* 1981). Species of *Ilex* are insect-pollinated, and their pollen is not dispersed far from its source. Although the regression analysis for *Ilex* had statistically significant results, we elected not to use these results for paleoecological reconstructions because of the sporadic representation of holly in the fossil pollen record.

Contemporary distribution patterns of calibrated tree taxa

For the 19 tree taxa for which subcontinent-scale calibrations are now available (Table 2.2), the patterns of percent growing-stock volume (%GSV) and percent arboreal pollen (%AP) tend to correspond both in overall distributional limits of the constituent taxa and with respect to the gradients from their range margins to their population centers. These gradients are interpretable in terms of environmental gradients and known habitat requirements of the taxa. Because of the broad spatial scale on which they are summarized, the contemporary isophyte maps, as presented in Delcourt *et al.* (1984), offer ecological insights that are not apparent from compositional analysis of individual forest stands. Because they present quantitative data over the ranges of the constituent species, these maps yield more information concerning the distribution/dominance patterns of taxa than do presence–absence maps (Little 1971, 1977).

In the southeastern United States, Cupressaceae plus Taxodiaceae and *Nyssa* illustrate the patterns of dominance for warm-temperate tree taxa characteristic of coastal-plain swamps. The primary Cupressaceae/Taxodiaceae contributor to the pollen rain in the Southeast is *Taxodium distichum* (bald cypress), reaching values of 40 to 60% of the GSV in the Everglades region of south-central Florida and in the Lower Mississippi Alluvial Valley of southern Louisiana (Delcourt *et al.* 1984). *Chamaecyparis thyoides* (Atlantic white cedar) occurs in coastal Virginia, the eastern Carolinas, and southern Alabama but reaches values of no more than 6% of the GSV. *Juniperus* (juniper) is a minor taxon in uplands of the Southeast, with percentages generally < 1%. Three species of *Nyssa* (tupelo) occur in the Southeast; *Nyssa aquatica* (water tupelo) and *Nyssa ogeche* (Ogeechee tupelo) are characteristic of poorly drained swamps and sloughs, codominating with values from 20 to 80% GSV along with *Taxodium distichum* in southern Louisiana. *Nyssa sylvatica* (black gum) occupies upland as well as bottomland habitats and ranges much more broadly northward in eastern North America. Pollen percentages of *Nyssa* tend to be very high in sites where tupelo trees are locally present, and the pollen is not generally evenly distributed throughout the present range of the genus.

Pinus (pine) is a genus with a relatively large number of species that display several patterns of distribution and relative dominance in the East. A group of seven species of pine in the Diploxylon subgenus dominates the forests of the

Gulf and southern Atlantic Coastal Plains, ranging from 40% to nearly 100% GSV. Several species of southern pines (*Pinus serotina, P. taeda, P. palustris*) are tolerant of high soil moisture but predominate on sandy soils of coastal-plain uplands where (in presettlement times) frequent fires or (now) short-term clearcutting and rotation of managed stands promote their regeneration. Southern "coastal-plain" pines reach their northern limits of distribution between 35°N and 40°N latitude. The representation of pine in the pollen rain parallels dominance patterns in %GSV across the southern pine belt of the coastal plains; however, to the north of 35°N, a number of species of hardwoods including *Quercus* (oak) form closed forests into which little pine pollen is dispersed long-distance.

Several hardwood taxa reach peak dominance within forests along riparian corridors such as the Central and Lower Mississippi Alluvial Valley. *Salix* (willow) is a wide-ranging genus in eastern North America, including a number of species of shrubs characteristically found in wetlands from temperate to arctic regions. Only two willow species are commercially important trees, primarily within the Central and Lower Mississippi Alluvial Valley. There, the population centers are strongly reflected in the %AP values. *Celtis* (hackberry) is a warm-temperate genus that is also quantitatively important in the forests of the Mississippi River system, as well as locally in riparian forests of southeastern Texas. *Celtis* is not as well-represented by pollen as is *Salix*, but overall the distribution of the *Celtis* pollen type is generally confined to temperate regions. *Fraxinus* (ash) is a genus (Table 2.1) that is widely distributed in temperate eastern North America as well as occurring as far north as the southern boreal forest region. In hydric sites of the Lower Mississippi Alluvial Valley, as well as gallery forests along alluvial valleys of eastern Texas, *Fraxinus* (including *Fraxinus profunda*, pumpkin ash) is an important contributor to the %GSV and %AP. Other species of ash in the southeastern states occupy mesic environments but together generally constitute less than 10% of the %GSV and %AP.

Quercus (oak), *Carya* (hickory), and *Juglans* (walnut) are warm-temperate to cool-temperate taxa that are dominant within the heartland of the eastern deciduous forest region. In eastern North America, 25 species of *Quercus* (Table 2.1) are considered commercially important. Today, their collective center of dominance (40 to 80% of the total GSV) generally coincides with the geographic center of distribution for the genus, primarily in a belt that extends from the Ozark and Ouachita Plateaus of Missouri and Arkansas eastward to the central and southern Appalachian Mountains. This is directly paralleled by the distribution and abundance of *Quercus* pollen. *Carya* is important primarily within the same latitudinal belt as *Quercus*, reaching up to 30% of the GSV and 10 to 20% of the AP in the Interior Low Plateaus of Kentucky and Tennessee. There, oak and hickory share dominance with blue ash (*F. quadrangulata*) on upland sites. *Juglans* also reaches its greatest importance within the Interior Low Plateaus as well as westward through the Ozark Plateaus of Missouri. *Juglans* pollen occurs at low percentages throughout its distributional range.

Species of *Acer* (maple) are widespread in temperate regions of eastern North America but are particularly important commercially (up to 40% GSV) in the forests of the north-central Great Lakes region and the northern Appalachian Mountains of New England. Pollen percentages of *Acer* tend to be locally variable but, on a broad regional basis, are highest in the tree-population centers in the Great Lakes and southern New England regions. The distribution of *Ulmus* (elm) complements that of *Acer*, reaching high values of up to 20% GSV and 20 to 40% AP in the western portion of the Great Lakes region. In part, the very low values of *Ulmus* in New England today may reflect the recent elm decline due to Dutch elm disease (Davis 1981b). *Tilia* (basswood) is also a genus that is more important in the western Great Lakes region than farther east, reaching over 10% GSV and up to 5% AP in Minnesota and Wisconsin.

Tsuga (hemlock) and *Fagus* (American beech) are characteristic of mesic forests occurring at mid- and high elevations within the Appalachian Mountains as well as across the mixed conifer–northern hardwoods forest extending from Maine to Minnesota. *Tsuga* is typically 5 to 10% both GSV and AP throughout this region. *Fagus grandifolia* generally occurs at < 10% GSV and AP throughout the eastern deciduous forest region east of the Mississippi River, but it is most prominent in both %GSV and %AP within the central and eastern Great Lakes region.

Within the central and northern Appalachian Mountains, a group of Diploxylon pines (*P. rigida, P. pungens, P. virginiana*) occurs within a region otherwise predominantly consisting of mixed hardwood forest (Little 1971). Although these pines are generally minor constituents of the %GSV, they contribute 10 to 20% of the AP in the Appalachian region (Delcourt and Pittillo 1986). One species of Haploxylon pine, *Pinus strobus* (eastern white pine), also occurs in the Appalachian Mountains, but it ranges much farther north and west than the other mountain pines and is also characteristic of mixed conifer–northern hardwoods forests of the Great Lakes region (Little 1971).

Several hardwood taxa are predominant in the area of transition between cool-temperate mixed conifer–northern hardwoods forest and boreal forest within the Great Lakes region. The species of ash that ranges farthest to the north in Canada is *Fraxinus nigra* (black ash), occupying primarily lowland sites (Fowells 1965; Little 1971). The northern limit of distribution of *Fraxinus* pollen coincides with the sharp latitudinal boundary of the northern range margin of black ash. Four species of tree birch (*Betula*, Table 2.1) are particularly important within the forests and the pollen rain of the northern Great Lakes region. *Betula alleghaniensis* (yellow birch) and *B. lenta* (sweet birch) are important in successional forests of mid- and high elevations within the central and southern Appalachian Mountains, and their ranges extend westward through New England to the eastern Great Lakes. *Betula papyrifera* (paper birch) is characteristic of mixed conifer–northern hardwoods forest from northern New England to Minnesota. Paper birch ranges northward into the boreal forest region of Canada, where it and *B. populifolia* (wire birch) constitute up to 30% of the GSV. *Betula* is represented by generally 20 to 40% of the AP in the Appalachian Mountains as well as throughout the northern Great Lakes region and southern Canada.

Populus (aspen or cottonwood) is a genus with several species (*P. balsamifera, P. grandidentata*, and *P. tremuloides*) characteristically growing in successional forests in the western Great Lakes region (10 to 20% GSV) and dominant within central Canada (40 to 80% GSV) as aspen parkland vegetation. The main center of distribution for the genus within eastern North America is within the southern boreal forest, although two southern species (*P. deltoides* and *P. heterophylla*) reach dominance values of 10 to 20% GSV within the Lower Mississippi Alluvial Valley and in the riparian forests of eastern Texas. Occurrences of pollen of *Populus* in surface samples are quite variable and in general under-represent the importance of this genus in the vegetation region of southern and central Manitoba.

Two species of northern Diploxylon pines (*Pinus resinosa* and *P. banksiana*) are characteristic of the mixed conifer–northern hardwoods forest region and adjacent boreal forest region. *Pinus resinosa* (red pine) was an important constituent of Great Lakes forests prior to EuroAmerican settlement and logging that has reduced its seed source and hence occurrence within successional forests to low levels today (Ahlgren and Ahlgren 1983; Flader 1983). *Pinus banksiana* (jack pine) is an important constituent of boreal forests and their arboreal pollen rain, particularly north and west of the Great Lakes region.

Thuja occidentalis (northern white cedar) typically constitutes between 1 and 14% GSV of the mixed conifer–northern hardwoods and southern boreal forests. The principal population centers of *Thuja* ($\geqslant$ 10% GSV) occur in northern Minnesota, Michigan, Ontario, and Maine. Pollen of *Thuja* is included within the group of Cupressaceae and Taxodiaceae and comprises up to 20% AP across the Great Lakes and New England regions.

One deciduous conifer, tamarack (*Larix laricina*), is found occasionally in lowland sites and bogs throughout the southern boreal forest region. Dominance values for tamarack are generally less than 10% as recorded by forest inventories throughout its eastern range in North America. Pollen of tamarack occurs at values generally of 1 to 2% of the AP in surface samples throughout its range, reflecting the local occurrence of small populations in wetlands across the region.

Picea and *Abies* are two genera dominant in the boreal forest region across large areas of northern North America. The southern range limits for these two genera occur within the southern Great Lakes region, with outliers of *Picea rubens* (red spruce) and *Abies fraseri* (Fraser fir) on high peaks in the southern Appalachian Mountains. *Picea rubens* and *Abies balsamea* (balsam fir) reach highest importance values in the maritime provinces of eastern Canada and adjacent Maine, where balsam fir is 40 to 60% of the GSV. Balsam fir diminishes westward to 10 to 20% of the forests of east-central Canada, where it is replaced by spruce, jack pine, and aspen. *Picea mariana* (black spruce) and *P. glauca* (white spruce) grow far to the west across Canada, and together they reach %GSV values up to 60 to 80% in central and western Ontario and Manitoba. Increasing percentage values northward for spruce populations in part reflect the decreasing species richness in the boreal forest. In general, the %AP values for spruce and fir parallel their dominance in the vegetation. *Abies* tends to be under-represented in the regional pollen rain. In the northern boreal forest region, sporadic occurrences of the pollen of *Salix* and *Betula* in surface samples reflect the contribution to the pollen rain of the several species of shrub willow

and dwarf birch growing in taiga and tundra environments near Hudson Bay
and in northern Labrador (Short and Nichols 1977; Elliott-Fisk *et al.* 1982;
Lamb 1984).

Pollen–vegetation relationships

The pollen–vegetation relationships (Table 2.2) are calculated from paired, positive values of %GSV and %AP for the taxa; as such, the calibrations are not
biased by complications that would result from the contributions of shrub species, particularly in the northern Canadian boreal forest. Nineteen tree taxa
(Table 2.2) had both positive, significant slopes for the geometric-mean regression line and positive, significant Pearson product-moment correlation coefficients based on the subcontinent-scale data set.

Based upon geometric-mean linear regression of %GSV and %AP, estimates
and standard errors of the slope and intercept allow the tree taxa to be grouped
into several categories of representation. A number of the taxa conform to
traditional definitions. Over-represented taxa, that is, taxa whose observed
pollen values are consistently greater than the tree percentages measured from
the same geographic area, include *Betula, Juglans, Pinus, Quercus,* and *Salix.*
Only one taxon, *Ulmus,* was equally represented, having %AP values equal to
its %GSV. Five taxa, *Abies, Acer, Nyssa, Populus,* and *Tilia,* were under-
represented with respect to their pollen values. The remaining taxa were
"problematic" in that, although the relationships characterized by linear
regression were statistically significant, the relationships determined for a subcontinental data set are more complex than observed previously using smaller
regional data sets. In one group that included *Fagus, Tsuga,* and Cupressaceae
plus Taxodiaceae, regression slopes were greater than 1.1 but p_o values were
much less than zero. For *Carya, Celtis* type, and *Larix,* slopes were less than
0.9 with p_o values statistically greater than zero. *Fraxinus* and *Picea* had slopes
between 0.9 and 1.1 and p_o values statistically less than 0.0 (Table 2.2).

Conclusions

1. The factors controlling the production, dispersal, deposition, and preservation
   of pollen grains and spores are relatively well-known. Both temporal and
   spatial scale represented by a depositional site must be considered in developing and testing hypotheses concerning long-term vegetational changes
   using time series of data from Quaternary paleoecological records.
2. Calibrations of modern forest composition with surface samples containing
   arboreal pollen assemblages show that both the dispersibility and productivity
   of pollen grains varies among tree taxa as well as between geographic regions
   with different physiognomy and composition of vegetation. For reconstructing forest history on a subcontinent scale, it is essential to use comprehensive
   calibrations based upon coverage across the distributional ranges of the taxa
   examined.

3. For the eastern portions of the United States and Canada, comprehensive calibrations of the modern pollen–vegetation relationship have been developed for 19 major tree taxa that include the primary dominants and subdominants within temperate and boreal forests. These calibrations provide the basis for quantitative reconstruction of past forest composition from late-Quaternary fossil-pollen sequences preserved in medium-sized (from 25 to 100 ha) lakes across the eastern half of North America.

3. Reconstructing Long-Term Forest Changes from Fossil-Pollen Data

Comparison of quantitative techniques for reconstruction of past vegetation

With the recent development of modern-pollen and forestry data sets on the regional and subcontinental scales, it is now possible to examine a variety of quantitative techniques for reconstructing long-term changes in forest composition. Refinement and testing of quantitative techniques is a prerequisite for interpretation of long-term forest patterns and processes. Direct comparison of the results of different quantitative methods aids in the identification of the strengths and limitations, as well as the complementary insights they offer.

Quantitative reconstruction of past vegetation can be approached using a number of different numerical techniques (Birks and Gordon 1985). Pollen percentages for individual taxa can be calibrated against measures of dominance in the vegetation (Andersen 1973; Webb et al. 1981; Delcourt et al. 1984; Prentice 1986). These individual taxon calibrations then can be applied to percentages of each taxon in sequences of fossil-pollen assemblages to quantify the changes in dominance through time (Livingstone 1968, 1969; Andersen 1973; Delcourt and Delcourt 1984). In a second approach to reconstructing past vegetation, entire fossil-pollen assemblages can be compared with modern-pollen assemblages using a variety of dissimilarity measures (Prentice 1982; Overpeck et al. 1985). Quantitative vegetation data from the geographic area of closest modern analogues can then be used to represent the former vegetation that produced the past pollen assemblages (Delcourt and Delcourt 1985a). A third fundamental approach to quantifying past vegetation is achieved through computer models that simulate the composition of vegetation under given environmental conditions (Solomon *et al.* 1980, 1981; Davis and Botkin 1985; Delcourt and Delcourt 1985a; Solomon and Webb 1985).

Taxon calibrations

The taxon calibration approach relies upon three primary assumptions: (1) the fundamental relationship of the modern-pollen rain to forest composition exhibits a systematic linear (or curvilinear) relationship that can be quantified using regression techniques; (2) the fossil pollen spectra include the same taxa within the same range of variation in abundance as the set of modern-pollen data; (3) the regression values for line slope and axis intercepts are constant through time for a given taxon. Calibrations based on modern samples can be applied back in time only as far as the modern flora has existed. On the generic level, these calibrations should not be applied to fossil-pollen spectra beyond 10 million years ago, the time by which 100% of the modern genera of higher vascular plants had evolved (Traverse 1982). The regression values are presumed not to be influenced either by changing vegetation composition or by changes in a taxon's pollen productivity through time. In a study from the Draved forest of Denmark, Andersen (1980b) demonstrated that the relative pollen productivity of common European trees remained constant through at least the last 10,000 years of the Holocene interglacial.

In the development of a calibration data set, several considerations concerning both spatial and temporal scales of resolution are important in the research design. The surface samples of modern-pollen rain should be collected so that they are comparable in pollen source area to the paleoecological sites to which they will be applied. The source area for the modern pollen should also be comparable to the area of vegetation sampled. This optimizes the degree of correspondence between a taxon's importance in the vegetation and its relative abundance within the modern-pollen spectrum. The surface samples of modern-pollen assemblages should correspond in temporal resolution to the time interval over which the vegetation data are valid. Spacing of paired pollen–vegetation samples throughout the geographic region of calibration should contain few overlaps in source area, so as to constitute discretely different pairs of samples for statistical evaluation.

For each calibration of a taxon, the pollen and forest-composition values should cover a relatively broad range of percentages. To ensure this, the vegetational area sampled should cross one or more major ecotones and include at least one population center and one distributional limit for each taxon. Calibrations can be further optimized if the following conditions are met: a spatially heterogeneous vegetational mosaic is surveyed across a diverse landscape; the forest region is dominated primarily by wind-pollinated tree species that produce abundant pollen grains that are widely dispersed and preserve well in lake sediments; the geographic region is covered by standardized, continuous forest inventories (for methods used in CFI surveys by the United States and Canadian

Forest Services, see Delcourt *et al.* 1984) or other vegetation samples with fine-grained spatial control and high taxonomic resolution; and large numbers of paired data for corresponding pollen–vegetation values are obtained prior to statistical evaluation (Davis *et al.* 1973; Webb *et al.* 1978; Prentice 1982, 1986).

To apply the taxon-calibration approach to reconstruct the composition of past vegetation, pollen percentages (p_a) are first calculated for each taxon on the sum of the total arboreal pollen in the fossil-pollen assemblage. Then, based on the geometric-mean linear regressions derived from the modern paired data sets, the parameters of slope (r) and y-intercept (p_o) are used to transform the p_a values into preliminary estimates of past dominance of the taxon in the vegetation (v_a). Because the sum of the v_a values may not equal 100%, they are restandardized to produce corrected values of the past dominance of each taxon (v_c) within the reconstructed forest (Prentice 1982, 1986; Delcourt and Delcourt 1984, 1985a):

$$v_a = (p_a - p_o)/r, \text{ where } p_a > p_o$$

$$\text{If } p_a \leq p_o, \text{ then } v_a = 0.0$$

$$v_c = [v_a/(\text{the sum of } v_a \text{ values for all taxa})] \cdot 100$$

Modern analogue methods

Modern analogue methods rely on several key assumptions. The foremost of these is that the modern data set is all-comprehensive and includes all possible present and past combinations of interacting plant-species populations, as well as their relative abundances and their structural aspects within forests. Second, if a good match is found between a fossil-pollen assemblage and a modern-pollen assemblage included in the data set, it must be assumed that the vegetation producing the closest modern analogue is the same as that producing the fossil-pollen assemblage. A perfect analogue is that in which the fossil-pollen assemblage is identical to one or more modern-pollen samples. A "no-analogue situation" is one in which the fossil sample is completely different in both taxonomic composition and (therefore, necessarily) abundances of pollen types. A difficulty arises as to how "good" or how "poor" an analogue is when fossil-pollen spectra are only partially similar to available modern-pollen spectra.

In practice, the underlying assumptions of the analogue approach may not be met because of limitations in both modern and fossil data sets. Apparently poor analogues can arise because of an insufficient modern-pollen data base, or because the vegetation of the past may have differed both qualitatively and quantitatively from any communities known today. Alternatively, the fossil-pollen assemblage may have been altered from its original composition by selective deterioration (comparable problems in interpreting analogues for fossil marine assemblages are described in Hutson 1977). Even if good modern analogues exist, they may not be detected if modern and fossil samples differ in spatial scale of pollen source area or in terms of their environment of deposition. In the case of apparently good modern analogues, extrapolating similarities in pollen assemblages to similarities in vegetation composition may be further complicated because more than one vegetation type may produce the same pollen assemblage (particularly at the taxonomic level of genus that is usually resolvable by pollen analysis). Pollen spectra are unlikely to include all taxa represented within the vegetational mosaic. Some plant taxa may be "silent", *i.e.*, not represented in the forest reconstruction because they produce limited quantities of pollen, their pollen grains are dispersed by animal agents, or the pollen is rarely preserved within the sedimentary sequences. Inability to identify the silent taxa, or to quantify their population sizes, results in blind spots within vegetational reconstructions (Davis 1963; Birks 1973; Terasmae 1976).

The modern analogue method is a conceptually simple means of relating past plant assemblages to the great diversity of situations available from modern forest communities (however, subjective or intuitive selection of possible analogues may be largely constrained by the experience base of the individual investigator). Decisions concerning the pollen data subtly influence the kind of modern analogues that are sought. For example, when pollen percentages for a tree species are calculated using an arboreal pollen sum, the comparison of past and present pollen spectra focuses the selection toward possible analogues based upon changes in forest composition. If pollen percentages are determined using the total number of grains counted for all upland plant species, then the analogues sought reflect the landscape mosaic of both forest and nonforest communities. Unfortunately, the modern pollen spectra based upon an upland pollen sum contain relatively high percentages of weedy plants such as *Ambrosia* (ragweed) that graphically demonstrate the magnitude of historic human activities in expanding cultivated fields at the expense of forests (Bassett and Terasmae 1962; Bernabo and Webb 1977).

With the compilation of computerized data bases for modern pollen spectra, palynologists have explored a variety of quantitative indices in order to more objectively identify modern analogues for fossil assemblages (Ogden 1969; King 1973; Davis *et al.* 1975; Ritchie and Yarranton 1978; Whitehead 1981; Prentice 1982, 1986; Lamb 1984; Overpeck *et al.* 1985; Liu and Lam 1985; Delcourt and Delcourt 1985a). Since many coefficients of dissimilarity and of similarity can be used in contemporary paleoecological research (Prentice 1980), no unique method exists by which to measure similarity among modern and fossil pollen assemblages. The closest modern analogue for a particular spectrum of fossil pollen may vary depending upon the coefficient used. In conjunction with an international effort to standardize evaluation of analogues for late-Quaternary sequences of fossil pollen, Prentice (1982, 1986) has recommended the application of one specific coefficient from each of the three principal groupings of dissimilarity coefficients: Euclidean distance; Chord distance; and Standardized Euclidean distance. The coefficient of dissimilarity is zero when the composition matches perfectly between fossil- and modern-pollen spectra, and the coefficient value is a larger positive number as the composition of their assemblages becomes increasingly different.

Euclidean distance is an example of the simple class of dissimilarity coefficients where the taxa are not deliberately weighted. Euclidean distance, a measure commonly used in numerical techniques such as Principal Component

Analysis (Prentice 1980), emphasizes the percentage contribution of the dominant taxa in comparisons among fossil and modern spectra. The formula for Euclidean distance is:

$$d(j,k) = \sqrt{\Sigma_i \, (p_{ij} - p_{ik})^2}$$

where $d(j,k)$ = coefficient of dissimilarity between fossil-pollen spectrum j and the modern-pollen spectrum k

p_{ij} = pollen percentage of taxon i in the fossil-pollen spectrum j

p_{ik} = pollen percentage of taxon i in the modern-pollen spectrum k

Σ_i = summation for all of the taxa

For each of the taxa occurring in either the fossil pollen spectrum j or in the modern spectrum k, the respective value of any taxon i in the fossil spectrum is subtracted from its corresponding value in the modern spectrum; the difference is squared for taxon i. The squared differences are then added together for all of the taxa and, finally, the square root is taken for this sum. The result, the coefficient for Euclidean distance, preferentially reflects the larger percentages contributed by the dominant taxa that compose the pollen assemblages.

The second coefficient, **Chord distance**, is an example of the type of dissimilarity coefficient that measures the primary signal, rather than random variation or noise, in the differences between any two pollen spectra. The formula for Chord distance is:

$$d(j,k) = \sqrt{\Sigma_i \, (\sqrt{p_{ij}} - \sqrt{p_{ik}})^2}$$

To calculate Chord distance, the square root is taken for the fossil and modern percentage values for each taxon and their square roots are subtracted. This quantity, *i.e.*, the difference between the square-root values for each taxon, is multiplied by itself and then added together with the squared values for all of the other taxa present in either the fossil- or the modern-pollen spectrum. The square root is taken of the final summation for all the taxa; this value is the coefficient for Chord distance. This coefficient decreases the relative weighting given to the dominants and increases the relative influence of the subdominant and rarer taxa in the pollen assemblages. The highest value for Chord distance, approximately 14.1 or the square root of 200, is achieved if no compositional overlap exists between the fossil- and modern-pollen assemblages that are compared (Prentice 1982, 1986).

The third coefficient of dissimilarity, **Standardized Euclidean distance,** illustrates the class of dissimilarity measures that gives equal weighting to all taxa by standardizing their variability to unit standard deviation. In this case, the relative influence of the taxa is not based upon their abundance. The standard deviation for a given taxon i (SD_i) is determined from the total variability of the taxon observed in the combined fossil and modern data sets used in a particular study. Standardized Euclidean distance is determined by:

$$d(j,k) = \sqrt{\Sigma_i \, [(p_{ij} - p_{ik})/SD_i]^2}$$

Because rare pollen types generally possess a relatively narrow range of variability, this coefficient accentuates the relative influence of the minor taxa in the assemblages (Prentice 1980, 1982).

Overpeck *et al.* (1985) used the modern-pollen data set assembled by Delcourt *et al.* (1984) in the only published study that has attempted to examine the degree of dissimilarity in pollen assemblages from within given vegetation types and across broad latitudinal and longitudinal gradients in vegetation. They compared results for a number of different dissimilarity measures in order to define preliminary "rational limits" or cutoff values for determining goodness of analogue that could be applied to results of comparisons of fossil-pollen assemblages with the modern data set. On a latitudinal transect at about 85°W longitude (Fig. 3.1), they used several measures of dissimilarity to compare a sample from the southeastern pine–oak forest (Fig. 3.1a–c) with other samples from the southeastern forest region, as well as from deciduous forest, mixed conifer–northern hardwoods forest, and boreal forest regions; then this test was repeated using a sample from the mixed conifer–northern hardwoods forest region (Fig. 3.1d–f). Results were broadly similar for all dissimilarity coefficients used in that study. Low dissimilarity values occurred when pollen samples from the same vegetation formation were compared with each other, the lowest values (*i. e.*, the closest modern analogues) obtained within a single forest type. Abrupt changes in dissimilarity values between formations indicated that on the modern landscape, pollen assemblages accurately distinguish between extant vegetation formations (Overpeck *et al.* 1985). Although no formal statistical basis exists for quantitatively comparing results of these analogue measures, the study of Overpeck *et al.* (1985) establishes a basis for making objective judgments concerning the closeness of analogue of fossil-pollen assemblages to modern pollen assemblages.

Forest simulation models

Forest simulation models have been developed for a variety of spatial scales, spanning as small an area as 0.01 ha within forest stands to as large as 10^7 ha for forest mosaics within dynamic landscapes (Botkin 1973; Botkin *et al.* 1972a,b; Shugart and West 1977, 1979; Shugart *et al.* 1981; Shugart 1984). Forest simulation models both integrate and test our current understanding of ecological principles with respect to the processes of forest dynamics. These models quantify relationships involving environmental tolerances, competitive strategies, and life-history characteristics of plant species and incorporate stochastic disturbances within the simulated forest systems. Initially constructed from empirical data based upon modern tree populations, forest models have been extended to simulate forest dynamics on two temporal scales: to examine short-term, micro-scale plant succession with simulation runs ranging from 50 to 500 years (Botkin 1973; Shugart and West 1977) and long-term vegetational change simulated over millennia on the micro- and macro-scale (Solomon *et al.* 1980, 1981; Davis and Botkin 1985).

North American research with forest-stand simulation models has focused upon two key models and their subsequent refinements: JABOWA and FORET.

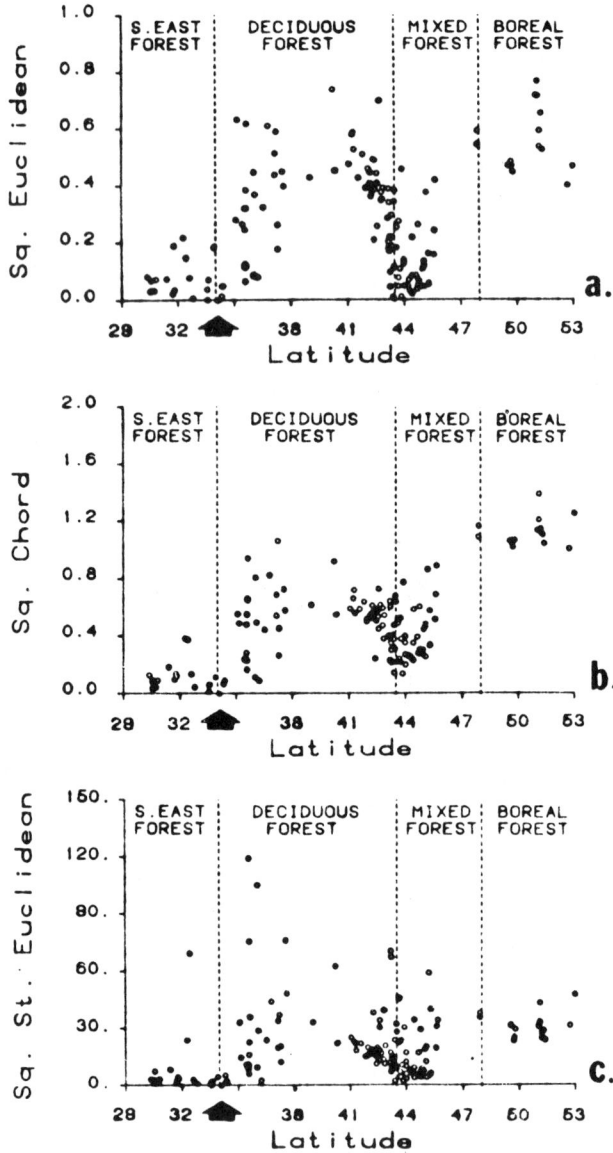

Figure 3.1. Scatter diagrams of latitude versus the values of three dissimilarity measures between a modern-pollen sample (marked by the large arrow) and all of the other modern pollen samples along a north–south transect at about 85°W. Squared Euclidean distance, squared Chord distance, and squared Standardized Euclidean distance were used to compare, first (a–c) a sample from the southeastern pine–oak forest with all of the other samples, and (d–f) a sample from the mixed conifer–hardwoods forest with all others (from Overpeck *et al.* 1985).

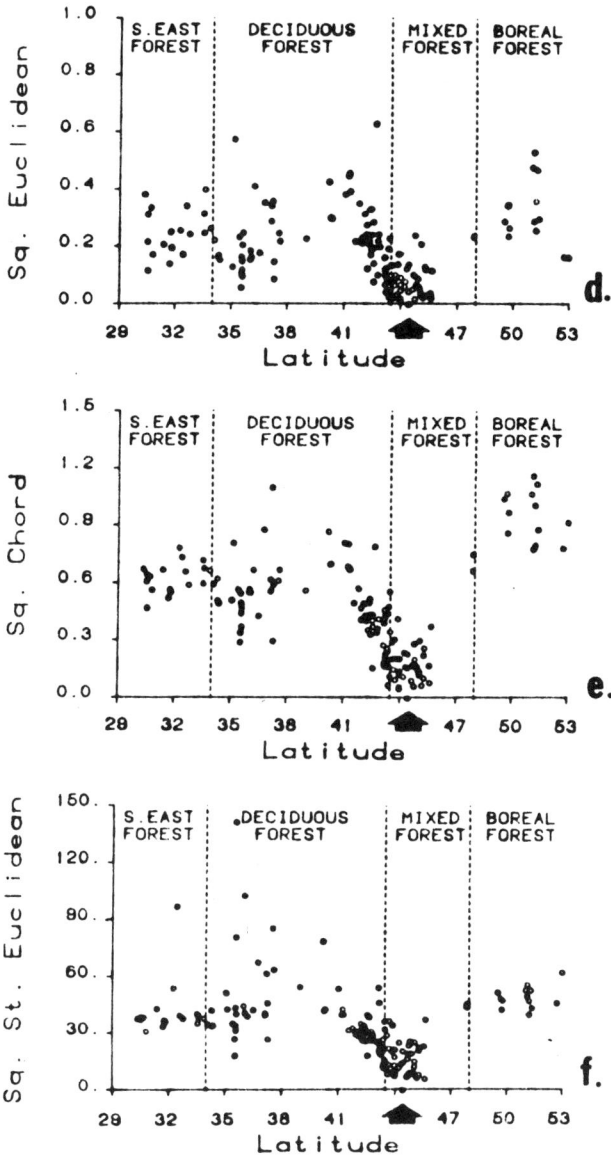

Figure 3.1 *Continued*

JABOWA represents the collaboration in 1970 of three individuals (J. F. *Ja*nak of IBM, D. B. *Bo*tkin then at Yale University, and J. R. *Wa*llis at IBM). The JABOWA model simulated plant succession in mixed conifer–northern hard-woods forests in the northeastern United States (Botkin 1973; Botkin *et al.* 1972a,b). The JABOWA model was constructed from vegetational and envi-ronmental information gained from experimental plots of the Hubbard Brook

Ecosystem Study in the White Mountains, northern New Hampshire. Initially developed using 13 tree species (Botkin *et al.* 1972a, b) and later expanded to include 40 native tree species (Davis and Botkin 1985), the JABOWA model was used to explore gap-phase replacement simulated in forest stands on 0.01-ha plots. The FORET model was modified from JABOWA by H. H. Shugart, Jr., and D. C. West, both at the University of Tennessee and Oak Ridge National Laboratory (Shugart and West 1977). The FORET model, the name an acronym for *For*ests of *E*ast *T*ennessee, simulates the competitive interactions of 65 tree species on 0.08-ha plots in mesic habitats in the middle latitudes of the southeastern United States (Shugart 1984).

These simulation models for forest-stand dynamics share five underlying assumptions:

1. The fundamental ecological processes involved in vegetation change can be expressed in mathematical relationships that accurately characterize the nature of both plant and animal interactions as modulated by their physical environment.
2. The spatial and temporal scale of resolution for the simulation run of a particular forest plot is appropriate for mimicking the nature of dynamics observed in comparable, extant forest ecosystems.
3. All key parameters that control the germination, establishment and growth, reproduction, and death of individual plants within forest populations are known, accurately quantified, and properly implemented within the simulation model.
4. The introduction of stochastic perturbations or the statistical properties that characterize specific disturbance regimes (including the kind, frequency, and intensity of disturbance for a particular disturbance regime) elicit predictable biotic responses in the simulation model that are validated independently within documented sequences of secondary plant succession with comparable disturbance histories.
5. The large number of successive computational iterations required by the model is either feasible and practical within the context of current computer systems.

The last assumption presumes that critical components of the simulation model can be either ignored or simplified in order to permit acceptable approximations.

Both the JABOWA and FORET models deal with competition among individual trees on plots of restricted size (from 0.01 to 0.08 ha) for vital resources including nutrients, space, and light. Both models are run on the size of plot that emphasizes gap-phase replacement, reflecting a balance for a particular tree's effectiveness in competing for light within the forest canopy and its relative success in capturing mineral resources with its underground root system (Shugart and West 1979). The models include computational subroutines for tree growth, death, and regeneration (birth or sprout).

A case study in comparison of taxon calibrations, modern analogue techniques, and simulation models as different approaches for quantifying past vegetation

In order to assess the reproducibility of late-Quaternary vegetational reconstructions, we compared the results of three fundamentally different techniques applied to the same fossil-pollen record. In this study (Delcourt and Delcourt 1985a), we compared the arboreal-pollen record from Anderson Pond, Tennessee, with reconstructed forest composition based upon (1) taxon calibrations as published in Delcourt *et al.* (1984); (2) vegetation composition associated with closest modern-pollen analogues for past arboreal-pollen assemblages, using three different dissimilarity coefficients (Chord distance, Euclidean distance, Standardized Euclidean distance); and (3) simulations of changes in forest-stand composition based upon the FORET model (Solomon *et al.* 1980).

The late-Quaternary pollen record from Anderson Pond

The pollen record from Anderson Pond, Tennessee, continuously spans the time interval from 19,000 yr B.P. to the present (Delcourt 1979). The pollen sequence represents changing vegetation from the last full-glacial interval to the present day in a region that was continuously forested. Full-glacial arboreal pollen assemblages (Fig. 3.2a,b) were dominated by northern Diploxylon *Pinus* (about 80% of the AP), with *Picea* as a subdominant, and only trace percentages of pollen of deciduous forest trees. Beginning by 16,500 yr B.P., *Pinus* began to diminish and the percentages of deciduous trees increased to over 10% of the AP. The first deciduous trees to increase included *Quercus*, *Fraxinus*, and *Ostrya/Carpinus*, followed after about 13,000 yr B.P. by *Acer* and *Fagus*. By 12,500 yr B.P., boreal conifers were diminished to about 30% of the total arboreal pollen, with the remainder representing cool-temperate northern hardwoods. After 9000 yr B.P., warm-temperate taxa such as *Liquidambar, Carya, Castanea,* and *Ulmus* increased, and *Quercus* dominated the AP throughout most of the Holocene interval.

The Anderson Pond sequence reflects changes in regional vegetation from (1) closed, coniferous boreal forest during the full-glacial interval, to (2) cool-temperate mixed conifer–northern hardwoods forest in which the composition continued to change throughout the late-glacial interval from 16,500 to 12,500 yr B.P., and finally to (3) warm-temperate deciduous forest that developed its modern aspect after 9000 yr B.P.

Results of taxon calibrations applied to Anderson Pond

The curves plotted for corrected (v_c) values for calibrated tree taxa in general tended to parallel those of the original AP percentages. Large differences in abundance between arboreal-pollen percentages and forest dominance reconstructed by the taxon approach were obtained only for those taxa with large

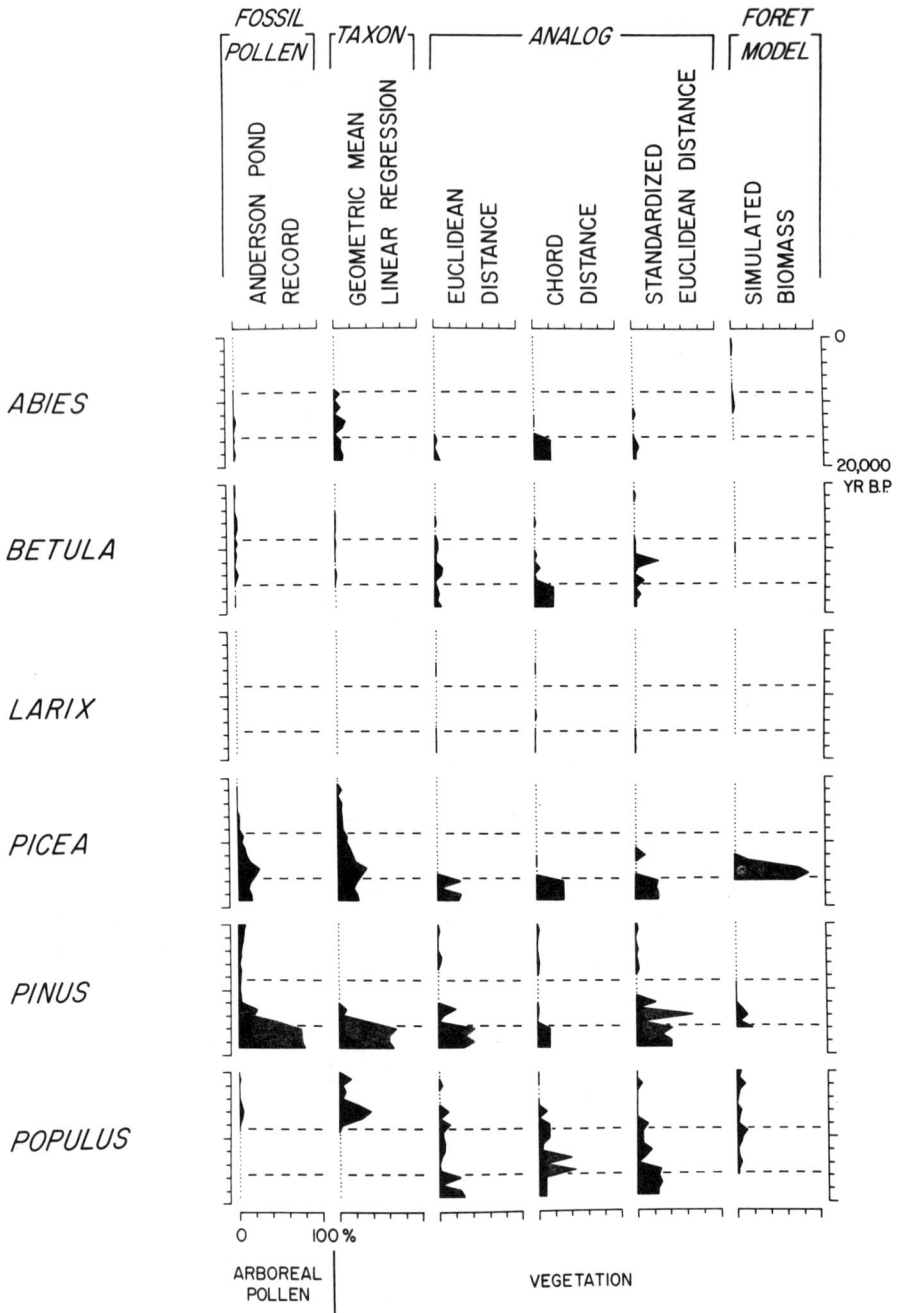

Figure 3.2. a. Curves for percent arboreal pollen and percent dominance in reconstructed vegetation for *Abies, Betula, Larix, Picea, Pinus, Populus* (left panel), *Acer, Fagus, Fraxinus, Tilia, Tsuga,* and *Ulmus* (right panel) for the last 19,000 years at Anderson

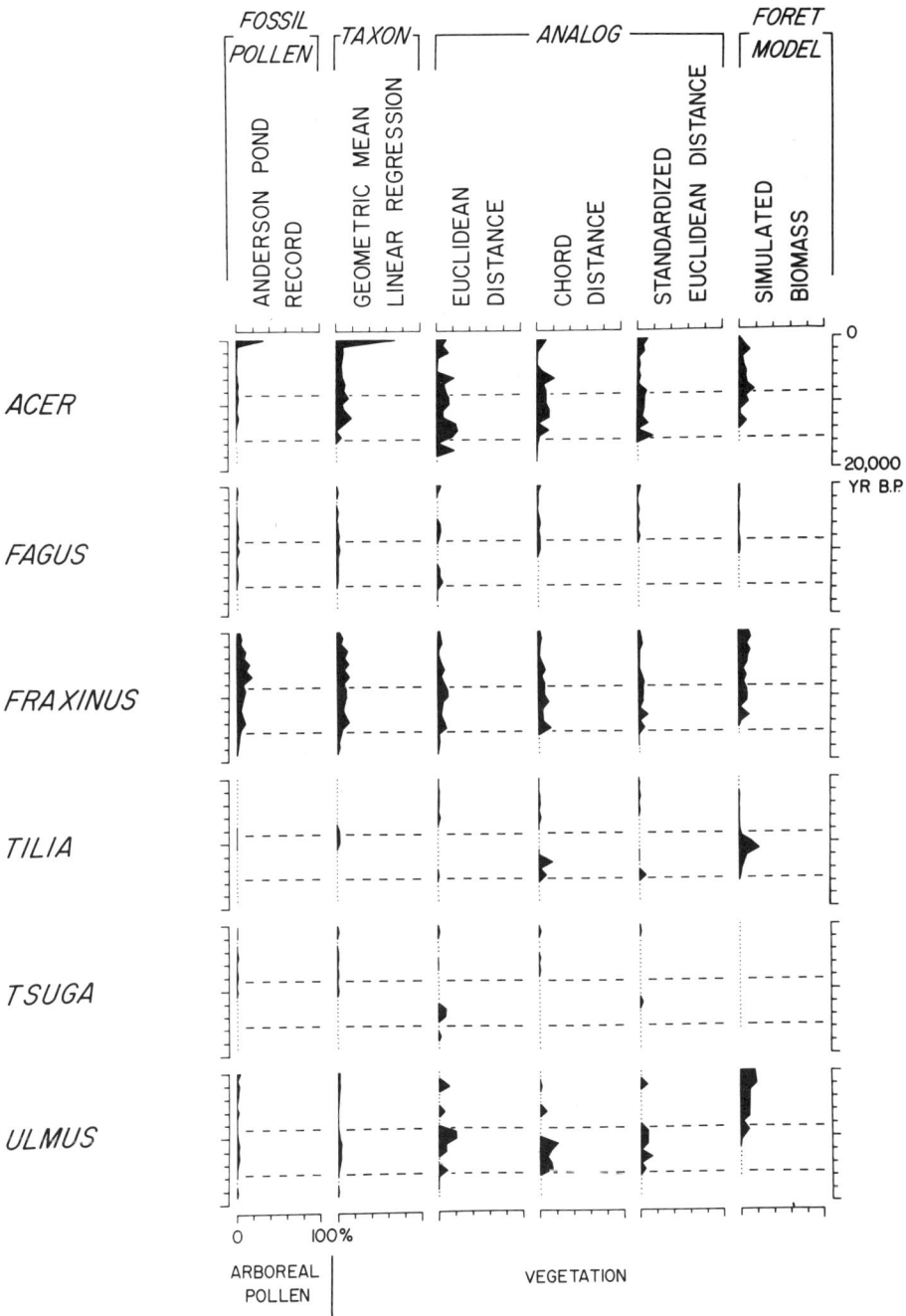

Figure 3.2 *Continued*
Pond, Tennessee (from Delcourt and Delcourt 1985a, © John Wiley & Sons, Ltd, reprinted with permission).

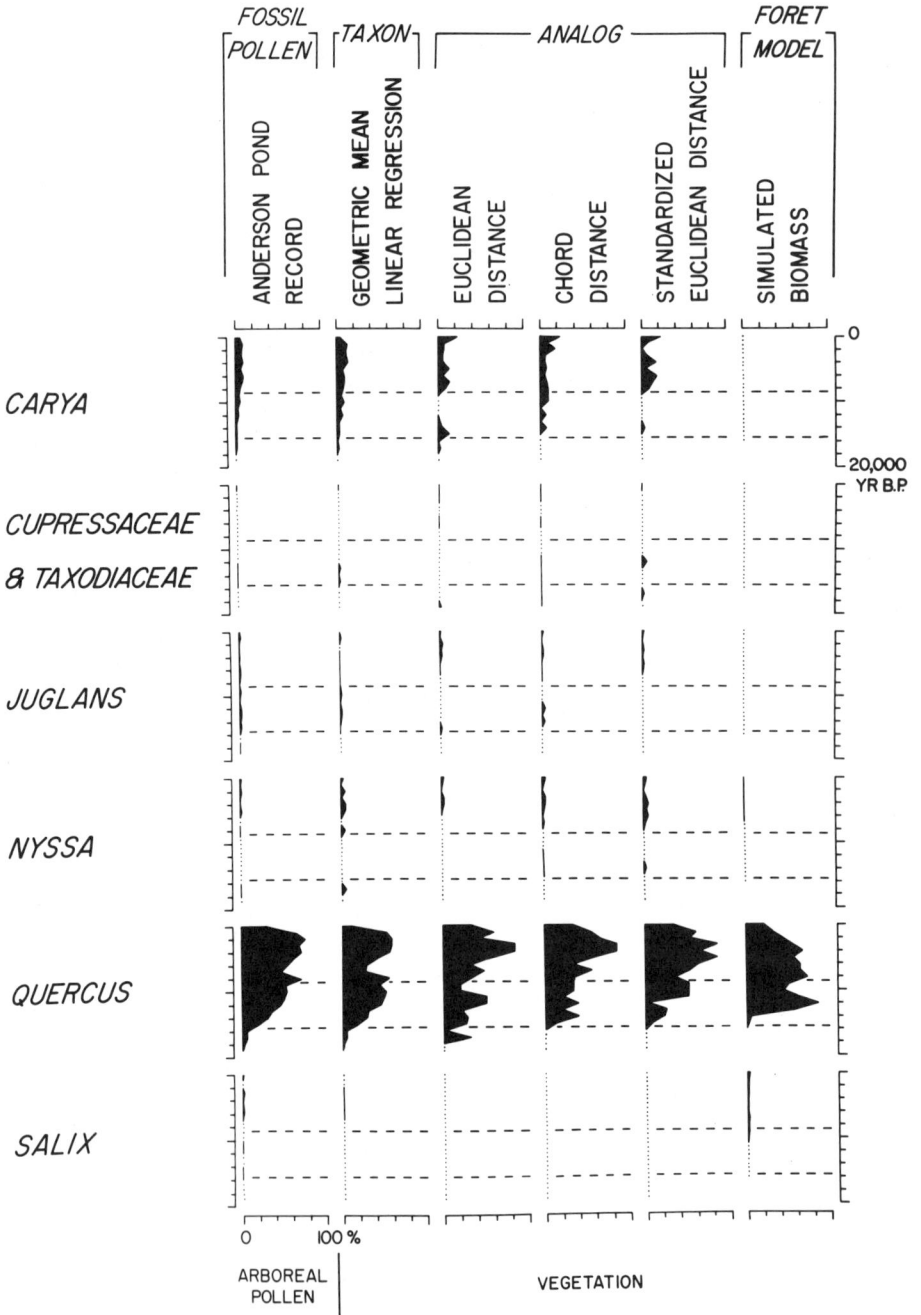

Figure 3.2. b. Curves for percent arboreal pollen and percent dominance in reconstructed vegetation for *Carya*, Cupressaceae and Taxodiaceae, *Juglans, Nyssa, Quercus, Salix* (left panel), *Castanea, Liquidambar, Liriodendron, Ostrya* and *Carpinus, Platanus*, and

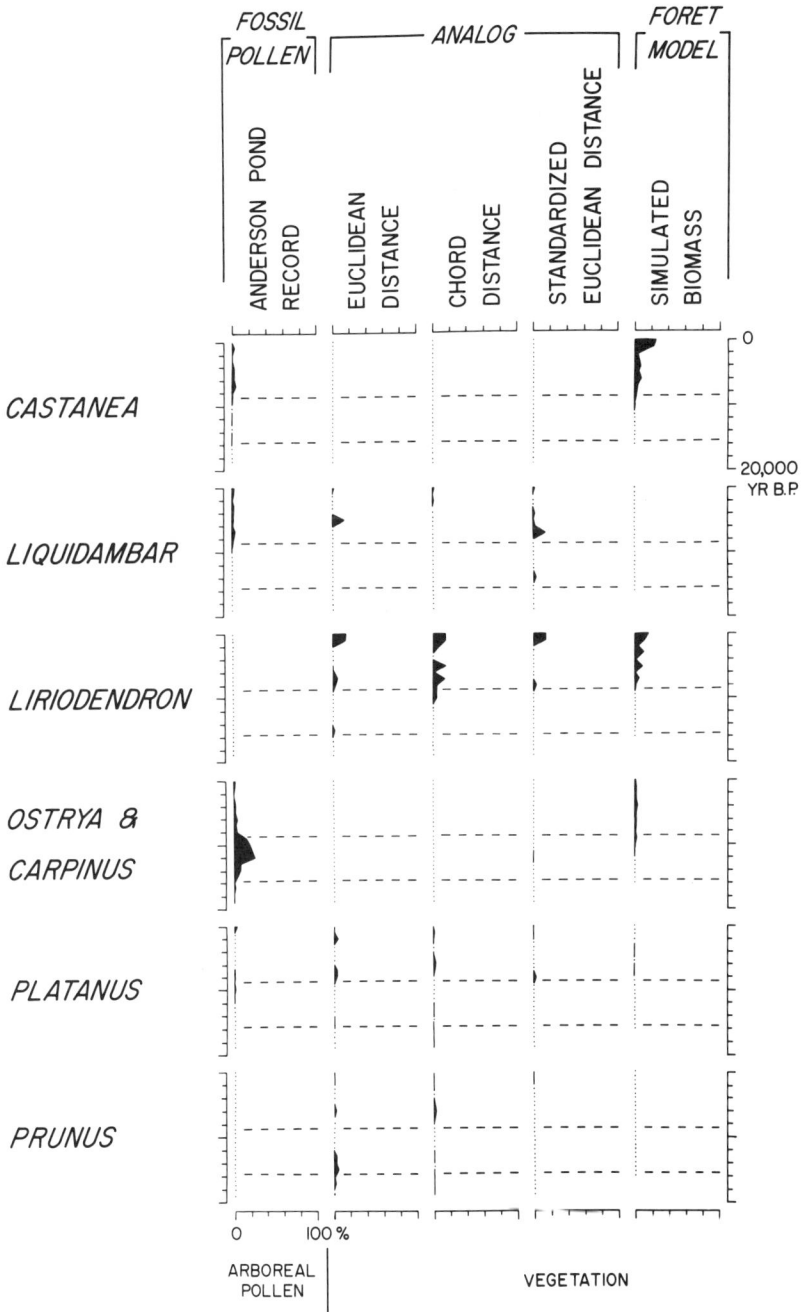

Figure 3.2 *Continued*
Prunus (right panel) for the last 19,000 years at Anderson Pond, Tennessee (from Delcourt and Delcourt 1985a, © John Wiley & Sons, Ltd, reprinted with permission).

y-intercept values (such as *Pinus*) or with regression slopes that were less than 0.9 or greater than 1.1 (*e.g.*, *Populus*). For such taxa, the corrected values take into account the over- or under-exaggerated importance as portrayed by the pollen percentage diagram. Of the arboreal taxa consistently represented in the pollen record, only *Castanea* and *Ostrya/Carpinus* were not included in the taxon-approach results. Thus, the estimates of other taxa during the Holocene are exaggerated to the extent that *Ostrya/Carpinus* was apparently a major forest constituent during the early Holocene and *Castanea* was present in at least minor amounts throughout the mid- and late-Holocene intervals.

The forest reconstructed by the taxon-calibration method represents a spatially averaged composition on the landscape within 30 km radius of the site, representing the sum of forest stands occupying habitats characteristic of both the rolling karst terrain of the eastern Highland Rim and steep slopes of the adjacent Cumberland Plateau.

Full-glacial boreal forests of Middle Tennessee (19,000 to 16,500 yr B.P.) reconstructed by the taxon approach were dominated by about 60 to 70% *Pinus*, with *Picea* as a subdominant at about 20%, and with *Abies* representing as much as 5 to 10% of the overall forest composition (Fig. 3.2a). During the full-glacial interval, very small populations of deciduous trees may have been locally present, including *Fraxinus*, *Ulmus*, *Carya*, *Quercus*, and *Prunus* (Fig. 3.2a,b). Alternatively, results of the taxon approach may have exaggerated their occurrence in the full-glacial vegetation of Middle Tennessee. It is possible that the trace amounts of pollen of these taxa in full-glacial pollen spectra from Anderson Pond may have been transported over relatively long distances from more southerly refugia.

During the late-glacial interval (16,500 to 12,500 yr B.P.), the percentage values of deciduous trees and the number of tree types represented increased in the taxon-based reconstructions. *Quercus* increased rapidly along with *Fraxinus* and *Carya*, and the pollen record shows that *Ostrya/Carpinus* was probably also an important forest constituent during the late-glacial and early-Holocene intervals. *Fagus*, *Betula*, and *Acer* entered the forest after about 16,000 yr B.P. *Acer* may have constituted as much as 15 to 20% of the forest by 12,000 yr B.P., when *Tsuga* became established (Fig. 3.2a). During the late-glacial interval, *Pinus* diminished rapidly, and populations of northern pines were probably locally extinct by 12,000 yr B.P. (Fig. 3.2a). *Picea* increased during the late-glacial interval to about 30% of the reconstructed forest composition at 14,000 yr B.P., then diminished along with *Abies* to small populations that may have persisted locally at the site into the early-Holocene interval.

After 9000 yr B.P., warm-temperate tree taxa became important in the forests surrounding Anderson Pond. These taxa included *Populus* (probably *P. deltoides*), *Nyssa*, and, according to the pollen record, *Liquidambar styraciflua* and *Castanea* (Fig. 3.2a,b). *Quercus* dominated at 30 to 60% of the reconstructed forest throughout the Holocene, with subdominants of *Carya* (15%) and *Fraxinus* (15%).

Results of analogue measures applied to Anderson Pond

The quantitative composition of each of 20 pollen spectra representing approximate 1000-year intervals from Anderson Pond was compared with each of the 1684 pollen surface samples using three dissimilarity measures, Chord distance, Euclidean distance, and Standardized Euclidean distance. In each case, the modern pollen sample with the lowest dissimilarity value was selected, and the corresponding forest-inventory sample was located and used to represent the forest composition for the fossil-pollen spectrum (Fig. 3.2).

Similar estimates for past forest composition were obtained for the three dissimilarity measures for two time intervals, 19,000 to 15,000 yr B.P. (the full-glacial and very early late-glacial intervals), and 9000 to 0 yr B.P. (the mid- to late-Holocene interval). During those two time intervals, the dissimilarity measures tended to choose either the same geographic location or nearby ones to represent the closest modern analogues to the fossil-pollen spectra (Fig. 3.3). All three dissimilarity measures identified best modern analogues for full-glacial pollen assemblages in the region from central to western Ontario, within the boreal forest region today dominated by *Pinus banksiana* and/or *Picea*. Best modern analogues for mid- and late-Holocene pollen assemblages from Anderson Pond were located in the central to southern Midwest region. The numerical values of the dissimilarity coefficients were lowest during the full-glacial and Holocene time intervals (Fig. 3.4), indicating that the best-fit modern-pollen spectra were relatively close in composition, and therefore provide excellent modern analogues to the fossil samples. In addition, two or three of the dissimilarity measures tended to choose the same modern-pollen sample (out of 1684 possibilities) as the closest quantitatively to the fossil samples from those two time intervals (Fig. 3.4).

In contrast, during the late-glacial and early-Holocene intervals (15,000 to 9000 yr B.P.), the techniques differed markedly in terms of the geographic areas in which they chose closest modern-pollen analogues (Fig. 3.3). For example, at 14,000 yr B.P., the best modern analogue found by Euclidean distance was located in Connecticut, that chosen by Chord distance was in southeastern Wisconsin, and Standardized Euclidean distance picked a sample from Arkansas. The forests in these three regions are quite different from each other in floristic composition today, perhaps most obviously in the species of oaks and pines that are characteristic of New England, the Midwest, and the southern Ozarks. However, the level of taxonomic resolution of these groups in the pollen record is not sufficient to distinguish between northern and southern, or eastern and western, species of *Quercus* or of *Pinus*. Interpretations of late-glacial forest composition of Tennessee based on analogue measures therefore include a wide range of possibilities, many of which seem not to be plausible but are rather an artifact of the insensitivity of the technique (Delcourt and Delcourt 1985a). The numerical dissimilarity values between fossil and best modern-pollen samples for the late-glacial interval were high compared with those obtained for full-glacial and Holocene intervals (Fig. 3.4), indicating that the best available

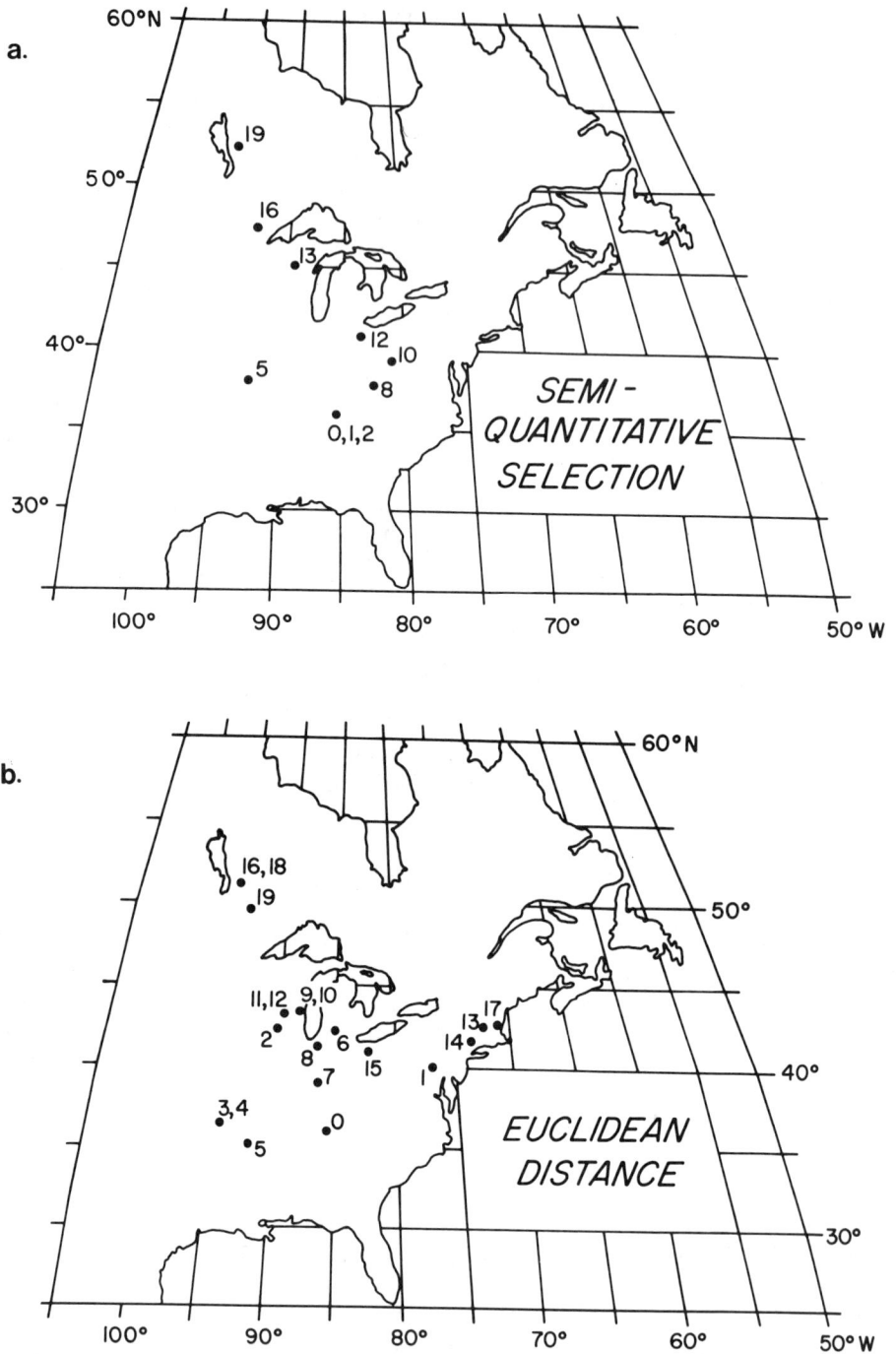

Figure 3.3. Geographic locations for best modern analogues for fossil-pollen spectra at 1000-year intervals from 19,000 yr B.P. to the present at Anderson Pond. Numbers represent age ($\cdot\ 10^3$ yr B.P.). Figure 3.3a is redrafted from Figure 13 in Delcourt (1979) (from Delcourt and Delcourt 1985a, © John Wiley & Sons, Ltd, reprinted with permission).

c.

d.

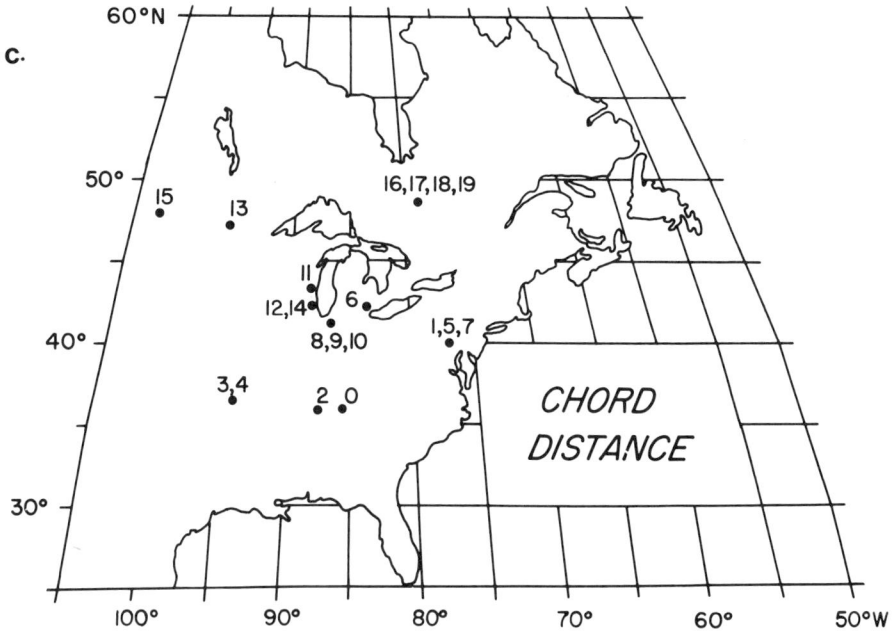

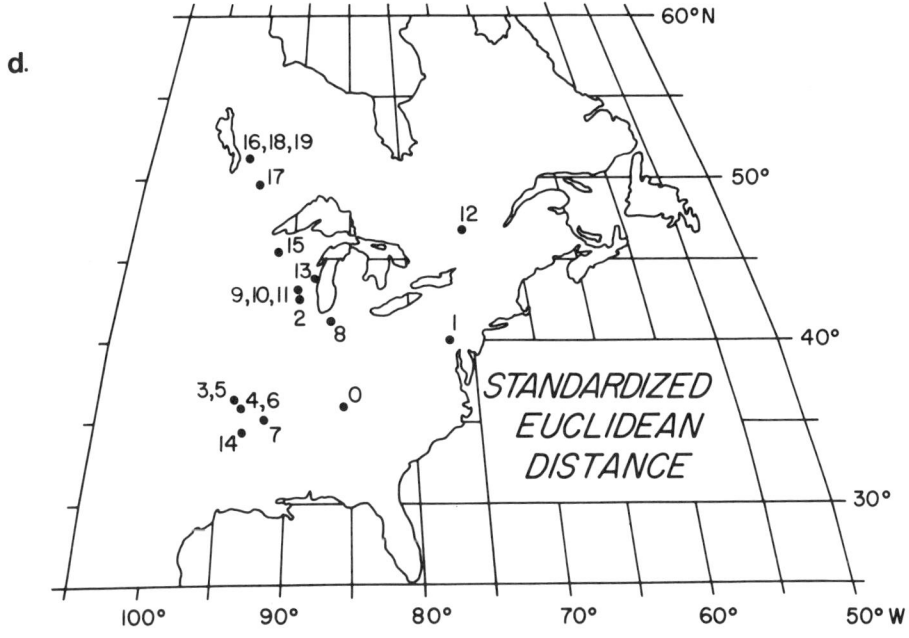

Figure 3.3 *Continued*

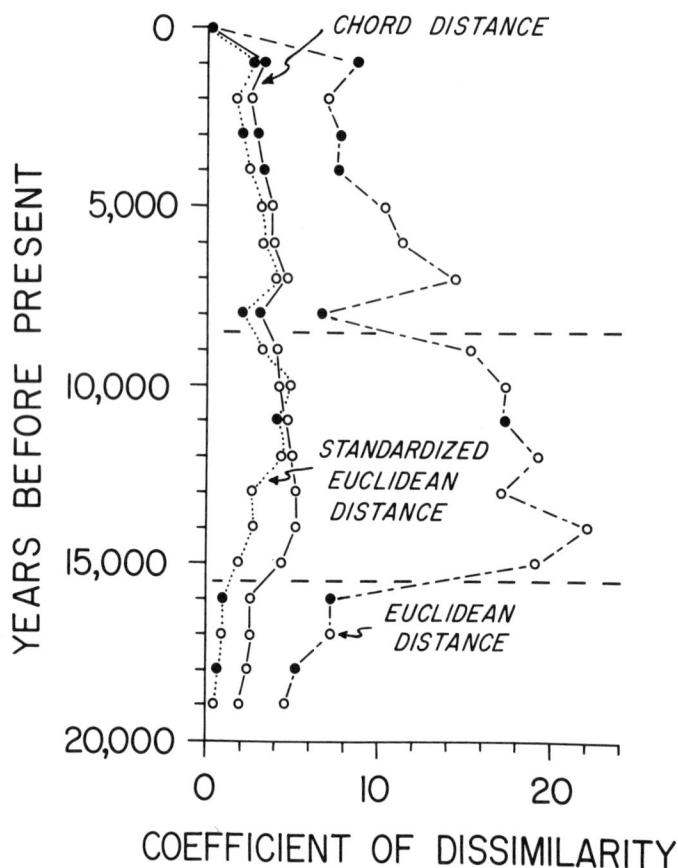

Figure 3.4. Coefficients of dissimilarity calculated for best modern analogues based upon three dissimilarity measures. For the times represented at 1000-year intervals at Anderson Pond, solid dots indicate the selection of the same modern-pollen spectrum as best modern analogue by two or more dissimilarity measures (from Delcourt and Delcourt 1985a, © John Wiley & Sons, Ltd, reprinted with permission).

modern analogues do not resemble closely the fossil-pollen assemblages from the late-glacial and early-Holocene time periods.

The results of this test of the modern analogue approach reveal that for certain times and places for which plant communities were similar to those of today, the analogue approach can be used successfully. This method is not appropriate for vegetation reconstruction specifically for those past intervals of time during which species assemblages occurred that are very different from those of today. Comparing the analogue results at Anderson Pond with that site's position on the late-Quaternary latitudinal gradient of environmental and vegetational change (Delcourt and Delcourt 1983; Fig. 1.4), it is apparent that the utility of analogue measures is greatest during times of relative stability in climate and vegetation. However, during past times of major changes in environment, species migrations, and plant-community composition, vegetational assemblages were least likely to resemble those of today. For the Great Lakes region, good modern analogues are likely to be found only for the last 5000 to 10,000 years of the Holocene (Overpeck *et al.* 1985).

Results of the FORET simulation model at Anderson Pond

In a pioneering study of the use of simulation models to emulate the fossil-pollen record, the forest-stand model, FORET, simulated late-Quaternary changes in biomass of trees growing on a 1/12-ha lower-slope plot in the vicinity of Anderson Pond (Solomon *et al.* 1980, 1981). The model contained life-history parameters for 65 tree species characteristic of boreal and temperate forests of eastern North America; all were assumed to be continuously available for occupation, with no migrational lags built into the model. Compositional change was driven by changes in climate, with changes in annual growing-degree days estimated at 500-year intervals from a curve of mean annual temperature derived from postulated geographic analogues for the fossil-pollen assemblages (Delcourt 1979; Delcourt and Delcourt 1985a; Fig. 3.3a).

In broad outline, the FORET model results resemble those of the taxon and analogue methods (Fig. 3.2), in that the model successfully simulated the changeover from boreal coniferous to temperate deciduous forest at about 13,000 yr B.P. This transition reflects the fundamental thresholds of temperature tolerance for the respective taxa (Davis and Botkin 1985). On close inspection, however, the details of changes in forest composition simulated by the model are substantially different from those estimated by the other quantitative measures (Fig. 3.2). For example, the dominance of *Pinus* during the full-glacial and late-glacial intervals was greatly under-estimated by the model, and that of *Picea* was greatly over-estimated. These results probably follow from the assumption within the model that long-term forest dynamics were taking place principally on a local, lower mesic to hydric slope, whereas the pollen record evaluated by the taxon and analogue methods represented regional forests averaged over a landscape that included xeric uplands as well as mesic and hydric sites. Differences in reconstructions of past forest composition between the taxon-calibration method and the FORET model simulations for mesic taxa such as *Fagus* and *Tilia* can also be attributed to differences in spatial scale reflected in the different methods.

In a second study using the FORET model, forest composition at 16,000 yr B.P. was simulated on different site types at Anderson Pond using pollen analogues to estimate climatic conditions (Shugart 1984; Solomon and Webb 1985). *Pinus* dominated on steep south-facing slopes, was a subdominant on level to rolling terrain with loam soils, and was a minor constituent of forest stands on steep north-facing slopes. In contrast, *Picea* dominated rolling loamy uplands but was nearly excluded on both steep north-facing and south-facing slopes. Hardwoods including *Acer, Betula, Populus, Tilia,* and *Ulmus* occupied steep slopes (Shugart 1984; Solomon and Webb 1985). The results of this simulation study (Shugart 1984; Solomon and Webb 1985) are consistent with taxon-calibration reconstructions of overall forest composition in the early late-glacial interval, as deciduous trees invaded into a landscape patchwork on favorable microsites (Delcourt and Delcourt 1985a; Fig. 3.2a,b).

Advantages and disadvantages of quantitative methods for forest reconstruction

The choice of appropriate quantitative method to use for reconstructing long-term forest dynamics is conditioned by the spatial and temporal scale of resolution desired, as well as by restrictions imposed by the inherent limitations of each available method. Each of the approaches has advantages and disadvantages. Many of their potentially useful insights as well as their biases become apparent only by comparing them with each other in test cases such as at Anderson Pond.

Curves of changing relative dominance for tree populations derived from applying taxon calibrations to fossil-pollen spectra tend to track changes in abundance of each taxon. They thus emphasize continuity in progressive changes of dominance through time. In contrast, the curves of forest composition that result from analogue measures display greater apparent variability in the abundance of taxa from one time to the next, particularly during times of poor analogue when the dissimilarity measure may select analogues for temporally adjacent fossil spectra from modern samples that are widely separated geographically. The population curves for the taxa based on this approach appear in many cases to increase and decrease erratically through time in a manner that is not intuitively expected from basic ecological knowledge of population dynamics.

Analogue measures help to identify times and places when past vegetation was compositionally dissimilar to that of today. During times of good analogue, the analogue measures can identify modern geographic areas with vegetation that may have been similar to that of the past, and either the taxon calibrations or analogue measures can be used to reconstruct vegetation. However, during times of poor analogue, dissimilarity coefficients should not be used to reconstruct past vegetation, and use of either taxon calibrations or some alternative, such as forest simulation models, is preferred. As during times of good analogues, use of taxon calibrations during times of poor analogues must be constrained so as not to extrapolate beyond the variability represented in the modern calibration data set.

Even during times of good analogues, it cannot be assumed that apparently analogous pollen assemblages or vegetational assemblages mean that the climate of the past was the same as today. The simulation-model approach suffers from the necessity of relying on the plant-fossil record to supply it with a proxy record of climatic change. If, in turn, the climate record is based upon the analogue approach, it may be invalid for large portions of the time interval over which simulations are run. Quantitative climatic records that are independent of the plant-fossil evidence are rarely available in the same geographic area. In addition, finer-grained disturbance regimes, such as fire frequency or recurrence intervals of wind throw, are superimposed upon the regional climatic trends, but they are important on the local spatial scale relevant to simulation results and may be difficult to derive independently from the fossil-pollen record.

Overall, the taxon calibration approach has several advantages to recommend its use over the other methods evaluated at Anderson Pond for broad-scale vegetation reconstructions (Delcourt and Delcourt 1985a). First, given sufficient areal coverage, such calibrations can provide a first approximation for factoring out fundamental differences in pollen production and dispersal for taxa over their range of variability. Thus, calibrations remove the key sources of distortion in the pollen record. Since the population curves for reconstructed dominance tend to parallel those for pollen percentages, their mapped values can be used to illustrate trends in the relative size of tree populations across a region. With due care, reconstructions of Quaternary forest composition based upon taxon calibrations are appropriate even during times of relatively poor analogues.

Generating late-Quaternary plant-ecological data sets

Pollen percentage diagrams are analogous to paleo-releves, in that they represent a sequence of floristic lists for the area studied and provide a semiquantitative estimate of the relative abundances of taxa at selected time intervals. By adjusting for sources of distortion in the pollen record, corrected paleovegetational (v_c) data yield quantitative measures of a calibrated taxon's relative dominance. Forest composition (v_c) diagrams based upon taxon calibrations are therefore more analogous to a sequence of resurveys using either plotless (*e.g.*, point-quarter) or plot sampling within a known area around the fossil-pollen site. Neither pollen percentage diagrams nor forest composition (v_c) diagrams yield absolute measures of tree population sizes independent of the percentage constraint. Diagrams of pollen accumulation rates (PAR) represent absolute measures of dominance, basal area, volume, or biomass of individual tree populations. PAR diagrams can be used to evaluate trends in population growth as well as intraspecific and interspecific competition within forest communities.

Use of pollen accumulation rate (PAR) data for paleoecological analysis is still in a preliminary stage, with relatively few fossil-pollen sites yet studied for PAR estimates. Most contemporary studies of Quaternary paleoecology are based upon interpretation of pollen percentage data, and this is a constraint that will be prevalent in Europe for some time to come (Huntley and Birks 1983). Forest composition (v_c) estimates can now be made for late-Quaternary pollen sequences from across eastern North America, based upon the taxon calibrations compiled in Delcourt *et al.* (1984). The forest composition data can then be mapped and analyzed by a variety of ecological techniques in order to provide new insights into pattern and process in long-term vegetational dynamics.

Several complementary approaches are available for using pollen percentages, reconstructed forest composition data, and PAR estimates to interpret vegetational history. Percentage composition of pollen assemblages that characterize modern vegetation types and formations can be used in a general way by analogy to map changes in the distribution of broad vegetation units, or biomes, through time (Delcourt and Delcourt 1981; Chapter 4). Shifting temporal and geographic patterns of major, physiognomically distinctive types of vegetation, such as boreal coniferous forest, cool-temperate deciduous forests, warm-temperate coniferous and deciduous forests, and nonforest types such as prairie or tundra provide insights into the nature of broad changes in climate on a subcontinent scale (Delcourt and Delcourt 1984).

For examination of the directions and rates of spread of individual taxa from full-glacial refuge areas, quantitative data of past forest composition can be mapped (Chapter 5). Contoured maps of dominance for each taxon at specified intervals of time depict both changes in positions of range margins and centers of dominance. Rates of spread of taxa can be measured from isochrone maps, depicting the changing locations of leading or retreating range margins through time (Chapter 6). The three-dimensional shape of the "wave front" of migration may give insight into the relationship between life-history strategy and the pattern of advance of a taxon's populations into new areas following deglaciation. PAR data permit quantitative evaluation of the growth rates of invading populations (Bennett 1983) as well as the extent to which competition shapes the structure of the plant community at any given time (Chapter 7). Direct gradient analysis of change in dominance across the ranges of taxa through time helps to assess the nature of ecotones between adjacent broad vegetation types. Gradients in the abundance of taxa across a region can be analyzed (Chapter 8) using plant-ecological techniques such as Detrended Correspondence Analysis (Gauch 1982).

Conclusions

1. Three fundamentally different approaches are currently available for the quantitative reconstruction of past forest communities. Application of these approaches to one test case emphasizes the extent to which the reconstructions converge and accents both strengths and weaknesses of the techniques.
2. The forest simulation-model approach is most appropriate for characterization of micro-scale forest dynamics within individual stands and is compatible with paleoecological records from woodland hollows and forest soils. This approach suffers from the necessity to input time series of quantitative climate and disturbance histories, which are difficult to obtain independently from the paleoecological records that are being simulated.
3. During times of climatic and vegetational stability, the results reconstructed from taxon-calibration and modern analogue approaches are similar. During times and in geographic areas of major environmental change, the taxon-calibration approach provides ecologically plausible results; however, in such situations, the results from the analogue approach are highly variable.
4. The taxon-calibration approach to vegetational reconstruction is superior to either modern analogue techniques or forest simulation models for the generation of late-Quaternary plant-ecological data sets on a broad geographic scale.

4.   Vegetation Map Patterns at the Biome Level

Constructing paleogeographic maps

In order to examine the response of vegetation to changing environmental conditions, it is critical to develop paleogeographic base maps that accurately depict the positions of prominent physical barriers to and corridors for plant invasions. For our base maps, we used a sinusoidal equal-area projection of eastern North America covering the geographic region extending from 25°N to 60°N, and from 50°W to 100°W. This map projection, generated by the Harvard-based CAM (Computer-Assisted Mapping) program, depicts the position of the modern shorelines for the Atlantic Ocean, Gulf of Mexico, and Hudson Bay, as well as the Great Lakes and Lake Winnepeg. We used a grid of lines at 5° latitude by 5° longitude, centered on the 85°W meridian (Delcourt *et al.* 1984).

Physical geography

In mapping the late-Quaternary positions of glacial-ice margins, we used the regional syntheses of glacial geology as published in Clayton and Moran (1982), Davis and Jacobson (1985), Denton and Hughes (1981), Elson (1967), Fastook and Hughes (1982), Fulton (1984), Hughes *et al.* (1985), Mickelson *et al.* (1983), Prest (1969, 1970), Saarnisto (1974, 1975), and Teller (1985). The continuous border of the ice limit is typically interpreted from the positions of terminal and recessional moraines produced by minor stillstands of glacial ice during its retreat (Prest 1969). The chronologies for establishing the position of the ice terminus are indirectly tied to radiocarbon dates on organic material either incorporated within glacially derived sediments (Clayton and Moran 1982) or contained within the oldest organic sediments preserved in kettle-lake basins on the surface of the moraine (Davis and Jacobson 1985). In glaciated terrain, radiocarbon dates on basal organic lake sediments represent the minimum age of deglaciation because stagnant ice left behind in ice-cored moraines may require as much as several thousand years to melt and become kettle lakes (Wright 1972). We mapped the position of the Laurentide Ice Sheet margin on base maps at 1000-year intervals from 20,000 yr B.P. to 6000 yr B.P.

Changes in areal extent of proglacial lakes (those which bordered the ice front), postglacial lakes, and marine embayments were mapped on the paleogeographic maps following the reconstructions of Clayton and Moran (1982), Davis and Jacobson (1985), Denton and Hughes (1981), Elson (1967), Mickelson *et al.* (1983), Prest (1970), Richard (1977, 1985), Saarnisto (1974, 1975), Teller (1985), Teller and Clayton (1983), and Webb *et al.* (1983b). Detailed environmental characterization and radiocarbon dates on submerged peat deposits along the Atlantic Coast provided evidence for late-Quaternary fluctuations in sea level. Field *et al.* (1979) integrated this evidence and prepared curves that trace changing positions of sea level from full-glacial times to the present. During peak glacial times, mean sea-level position along the Atlantic Coast was approximately 130 m below the present position, rising to about -30 m by 12,000 yr B.P., and reaching modern position in the past 4000 years. We mapped the position of marine shoreline during the last 20,000 years using bathymetric maps for the continental margin of eastern North America (Belding and Holland 1970). Late-glacial and early-Holocene fluctuations of the marine coastline in southern New England resulted from the glacial-age submergence of the terrain because of the weight of the nearby glacial ice (Hughes *et al.* 1985); this shoreline was uplifted during the postglacial interval with subsequent isostatic rebound (Clark *et al.* 1978; Bloom 1983). In order to portray this regional anomaly, we follow the detailed mapping of the New England marine shoreline as summarized for the interval from 14,000 to 9000 yr B.P. by Davis and Jacobson (1985).

Site selection and chronologies

We identified all late-Quaternary fossil-pollen sites available as either published or unpublished records east of 100°W and between 25°N and 60°N in eastern North America. Within this area, we evaluated the sites occurring in each 1° latitude by 1° longitude block in order to select those sites that provided the best quality of data on a geographic array with relatively uniform spatial resolution. Where possible, the paleoecological sites we selected were medium-sized lakes, for which the regional pollen component was dominant. Where a variety of sites from different-sized lakes was available, we excluded both very small and very large lake basins. For a given grid block, if lake sites were not available, we used data collected from bogs, and, if necessary, from alluvial environments. We selected sites with the most complete pollen data in terms of taxonomic detail, total number of pollen grains and spores included in each stratigraphic sample, and temporal resolution as indicated by the number of pollen samples counted. In addition, from the available records we chose those with the best pollen preservation and most continuous sediment sequences, without evidence of significant temporal hiatuses or major lithologic changes in environment of deposition.

In order to map paleovegetation on an absolute time scale, we used data from only those sites for which radiocarbon dates were available. At least two dates were required as a minimum in order to estimate ages for each pollen sample obtained from sediment cores from a site. In each 1° by 1° block, we preferentially chose the site with the greatest number of radiocarbon dates spanning the longest interval of time. Where the original investigator identified specific dates that were inconsistent with others from the site, we excluded the anomalous dates from the site chronologies. We used linear interpolation between adjacent pairs of dates in order to establish the relationship between age and depth of sediment for each site. If, however, the original investigators used curvilinear regression in order to establish a sedimentation-rate curve, we followed their chronology (*e.g.*, Whitehead 1981). Where the basal sediments were not dated directly, we extrapolated from the sedimentation rate determined from the oldest pair of available radiocarbon dates in order to calculate the age of the oldest sediments. In deglaciated regions, the extrapolated age of the basal lake sediments was set to be consistent with the regional time of deglaciation as determined by glacial geology. Where δ ^{13}C analyses were available, they were used to correct the original radiocarbon dates. As a rule, we did not identify regional time lines based on pollen-stratigraphic boundaries in order to constrain the site chronologies; such artificial constraints make it impossible to distinguish changes in vegetation that are synchronous from those that are differential due to the individualistic nature of species responses to environmental changes. The only exception to this procedure is the use of the "*Ambrosia* rise", a well-documented increase in ragweed pollen that occurred because of deforestation and conversion to agricultural land throughout eastern North America after EuroAmerican settlement (Bassett and Terasmae 1962). Even in this case, we set the time of the *Ambrosia* rise according to the specific time of historic settlement near each site. If this date was not known, we used the pattern of settlement mapped by Marschner to determine this age (see Fig. 9 in Marschner 1959). In certain cases, the *Ambrosia* rise was dated directly with radiocarbon dates that were anomalously old [contamination from radioactively "dead" carbon contained in the fossilized tissues of aquatic plants apparently was used in obtaining the radiocarbon date (Birks and Birks 1980)]. In such cases, we calculated the difference in time and used it as a correction factor for all older dates at the site.

All of the data, including pollen tallies, radiocarbon dates, and site information for each site used in this study, was compiled in a computerized data base for subsequent analysis (as presented in Chapters 5, 6, and 8). The Quaternary paleoecological data base at the University of Tennessee was supplemented by data exchange with Thompson Webb III's Quaternary palynological data base at Brown University, Providence, Rhode Island.

The documentation of the 162 fossil-pollen sites used in this study is included in the Appendix to this volume. This listing includes the name of each site, its location by latitude and longitude, time range represented, name of pollen analyst, and citations for published literature.

Paleovegetation maps for eastern North America: 20,000 yr B.P. to the present

Paleovegetation maps provide a context for evaluating vegetational response to a dynamically changing environment (Delcourt and Delcourt 1981). The average composition of pollen assemblages within plant formations and large vegetation types can serve as a qualitative guideline for mapping changes in the boundaries of major physiognomic units through time.

In recent studies (Richard 1977; Delcourt and Delcourt 1979, 1981; Grichuk 1984; Khotinskiy 1984; Webb *et al.* 1983b; Tsukada 1985; Davis and Jacobson 1985), several constraints have been recognized in preparing generalized maps depicting changes in vegetation patterns through the late Quaternary. For proper interpretation of changes in vegetation boundaries, the maps should depict changes in physical geography, including locations of glacial margins, proglacial and postglacial lakes, and marine shoreline positions. The map scale used must reflect the density of available fossil-pollen sites, as well as the geographic extent and landscape heterogeneity (*e.g.*, topographic relief) of the area mapped (Tsukada 1985; Davis and Jacobson 1985). The times chosen for mapping will influence the patterns resolved. For example, maps drawn at 18,000 yr B.P. and 6000 yr B.P. will emphasize the extremes in biogeographic patterns between maximum full-glacial and maximum interglacial conditions (Delcourt and Delcourt 1981). In contrast, a series of maps drawn at close time intervals for the late-glacial period will tend to emphasize rapid changes in locations of vegetation boundaries (Davis and Jacobson 1985).

In order to depict the changes in locations of major vegetation types through time across regions in which the density of fossil-pollen sites is low, certain assumptions are appropriate for the mapping procedure (Delcourt and Delcourt 1981): (1) once the fossil-pollen spectra have been interpreted in terms of vegetation type at the location of each available site, the vegetation types can be generalized from the array of site locations, using boundaries of discrete physiographic regions or elevational contours; and (2) if no paleobotanical information is available for a region, it may be necessary to extrapolate from adjacent regions. For example, full-glacial and late-glacial alpine tundra is generalized within the central Appalachian Mountains. Where few sites were available within a physiographic region, we used information from sites in adjacent regions to infer changes in vegetation. An example is the extrapolation of the full-glacial and late-glacial distributions of boreal conifers within the Central and Lower Mississippi Alluvial Valley based on sites along the margins of that distinctive physiographic province. The fewer the available fossil sites for a given time interval, the more speculative become the mapped patterns, but even maps based upon reasoned speculation (Deevey 1949; Martin 1958; Whitehead 1973) can serve as important hypotheses to be tested given new paleoecological data from critical areas.

In mapping paleovegetation, an implicit assumption is that broadly similar assemblages can be found on today's landscape for the major physiognomic or taxonomic groupings of plants to be mapped through time. Recent studies have demonstrated the lack of good modern analogues for specific compositions of forest communities at certain times in the late Quaternary (Overpeck *et al.* 1985; Delcourt and Delcourt 1985a). Even certain biomes, such as the Arctic steppe–tundra, may have changed character or ceased to exist since the last glacial maximum (Hopkins *et al.* 1982; for opposing arguments that support the persistence of tundra communities in the Canadian Arctic throughout the late Quaternary, see Cwynar and Ritchie 1980 and Ritchie 1984). In such cases, this presents an additional complication to be considered when constructing generalized paleovegetation maps.

Preparation of broadly defined, largely qualitative paleovegetation maps is a productive exercise that is complementary to that of examining distributions and dominance patterns for individual taxa. Specific plant taxa may increase in their abundances and spread from refugial areas in directions and rates that are individualistic according to their environmental tolerances and competitive abilities (Davis 1976b, 1981a, 1983). The process of change in geographic distribution for individual species has been suggested to occur as a "diffusion" outward from refugial populations (Pielou 1979; Bennett 1985), or as "osmosis" where physical barriers and biological competition pose differential resistance to the spread of the species (Rapoport 1982). To the extent that more than one species exhibits comparable limits in climatic tolerance, species groups remain together within biome-level assemblages even during times of environmental change.

The suite of procedural decisions concerning map scale, choice of mapped time planes, and degree of continuity of major physiognomic units through time can be evaluated on a time series of paleovegetational maps, based on 162 fossil pollen sites distributed across eastern North America (Fig. 4.1). Figures 4.2a, 4.3a,b, and 4.4a,b are maps of paleovegetation summarized on the level of formation and large vegetation types for selected times in the past 18,000 years. These maps have been updated and generalized from those published previously (Delcourt and Delcourt 1979, 1981, 1984).

Characterization of modern vegetation by pollen assemblages

The modern pollen rain was characterized for major vegetation types of eastern North America by Davis and Webb (1975). We calculated the mean percent (based on the sum of the arboreal pollen), one standard deviation from the mean, and the range in percentage for each tree type, as well as total nonarboreal pollen percentages, from samples within (1) tundra, (2) boreal forest, (3) mixed conifer–northern hardwoods forest, (4) deciduous forest, (5) southeastern evergreen forest, and (6) prairie (after Delcourt and Delcourt 1981).

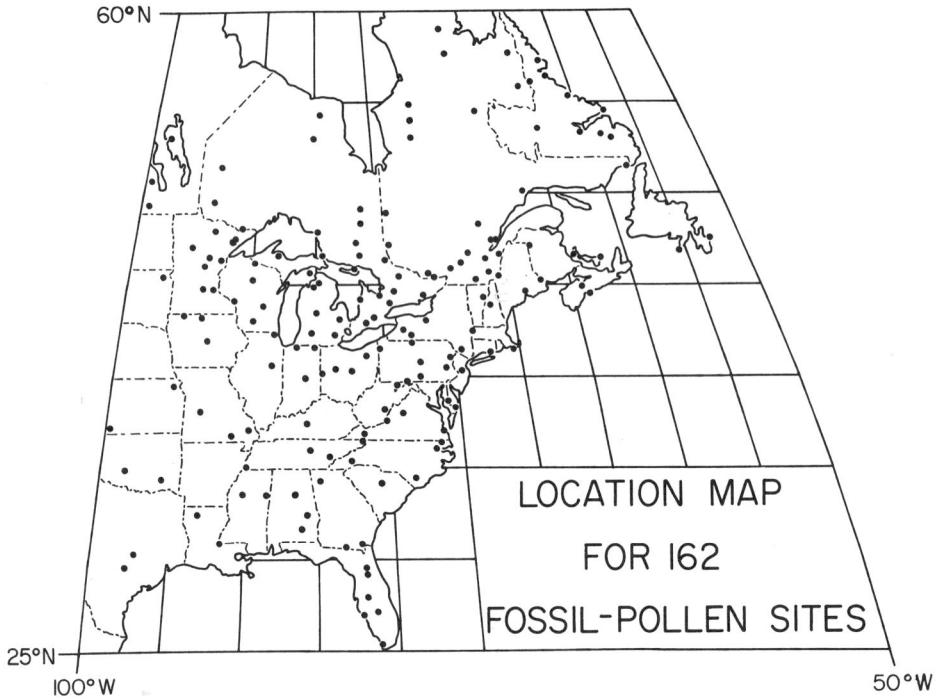

Figure 4.1. Location map for 162 fossil-pollen sites used in this study (documentation concerning site characteristics and corresponding literature citations are in the Appendix).

Modern tundra environments in northern Canada are characterized by low total PAR (less than 2000 gr·cm^{-2}·yr^{-1}) (Ritchie and Lichti-Federovich 1967; Ritchie 1984) and by pollen assemblages with total percent of Gramineae, Cyperaceae, Ericales, and *Salix* (particularly the dwarf willow, *Salix herbacea*) ≥ 25% (Elliott-Fisk *et al.* 1982; Webb *et al.* 1983b; Lamb 1984). Pollen of *Picea*, shrub birch (*Betula*), and green alder (*Alnus crispa*) may locally dominate the pollen percentage spectra because of the low pollen productivity of herbaceous tundra plants. Indicator taxa found within modern- and fossil-pollen samples within tundra environments include pinks (Caryophyllaceae), dryas (*Dryas*), crowberry (*Empetrum nigrum*), *Koenigia*, mountain sorrel (*Oxyria digyna*), Jacob's ladder (*Polemonium*), bistort (*Polygonum bistorta*), pearlwort (*Sagina nodosa*), and purple mountain saxifrage (*Saxifraga oppositifolia* type) (Ogden 1966; Birks 1976, 1981a).

Farther south in central and eastern Canada, within the boreal forest region, pollen percentages of herbs are much lower, and total PAR is generally greater than 2000 gr·cm^{-2}·yr^{-1} (Ritchie and Lichti-Federovich 1967; Ritchie 1984). PAR values in modern, open boreal woodland range between 2000 and 5000 gr·cm^{-2}·yr^{-1}. PAR exceeds 5000 gr·cm^{-2}·yr^{-1} consistently only in areas of more dense, closed boreal forest. Dominant and characteristic arboreal taxa in

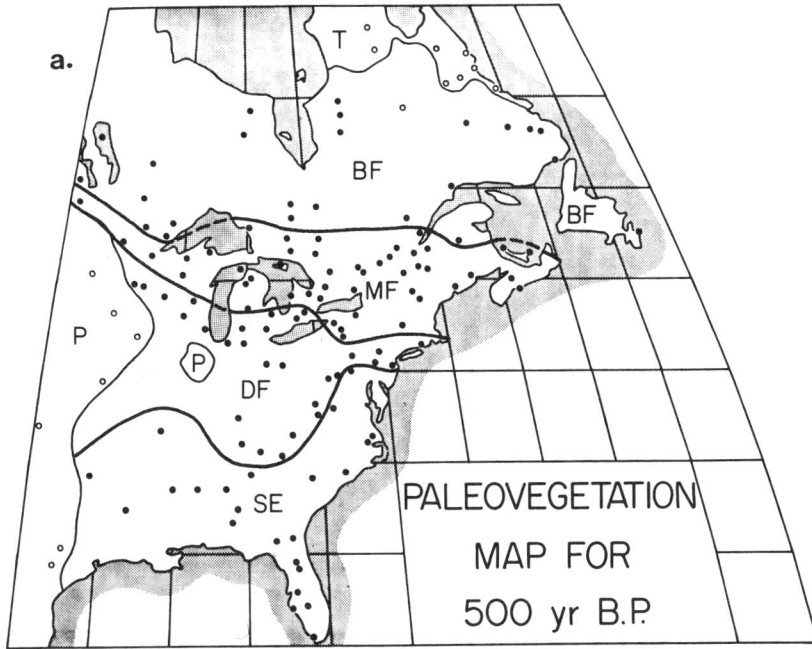

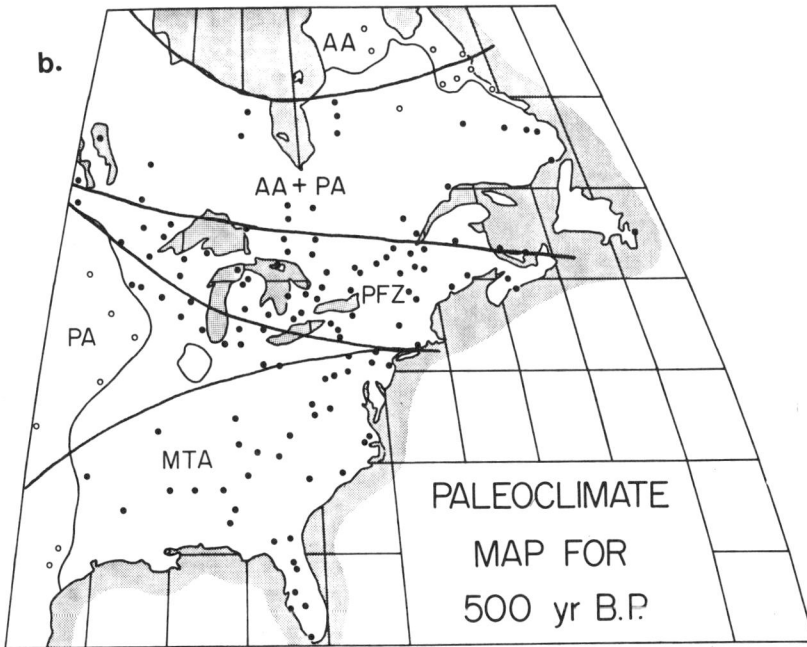

Figure 4.2. a. Paleovegetation map for the late-Holocene interval, 500 yr B.P. b. Paleoclimate map for the late-Holocene interval, 500 yr B.P. Abbreviations for vegetation and airmasses: T, tundra; P, prairie; BF, boreal forest; MF, mixed conifer–northern hardwoods forest; DF, deciduous forest; SE, southeastern evergreen forest; SS, sand dune scrub; AA, Arctic Airmass; PA, Pacific Airmass; MTA, Maritime Tropical Airmass; PFZ, Polar Frontal Zone (generalized from Delcourt and Delcourt 1981, 1984).

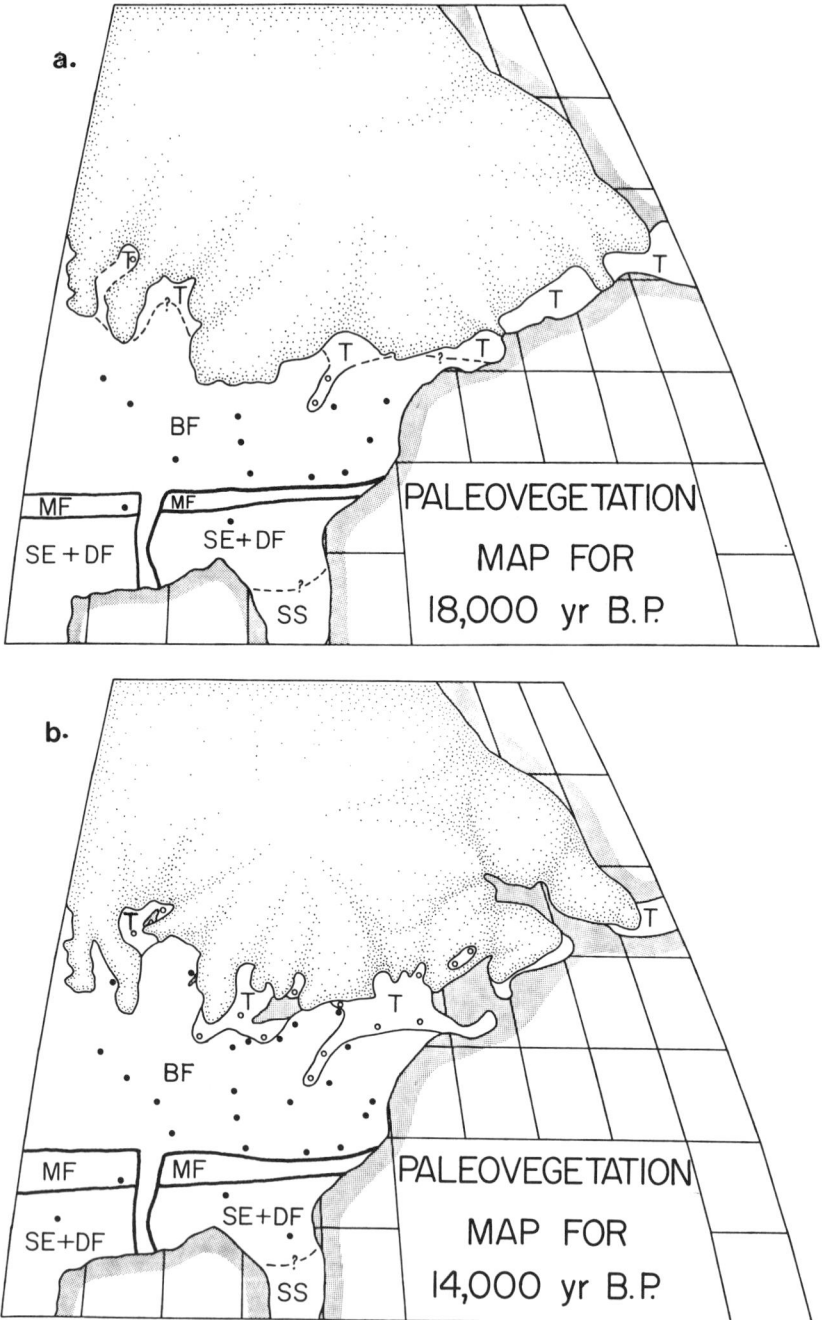

Figure 4.3. Paleovegetation maps for the full-glacial and late-glacial intervals: a, 18,000 yr B.P.; b, 14,000 yr B.P. Abbreviations as defined in Fig. 4.2 (generalized from Delcourt and Delcourt 1981).

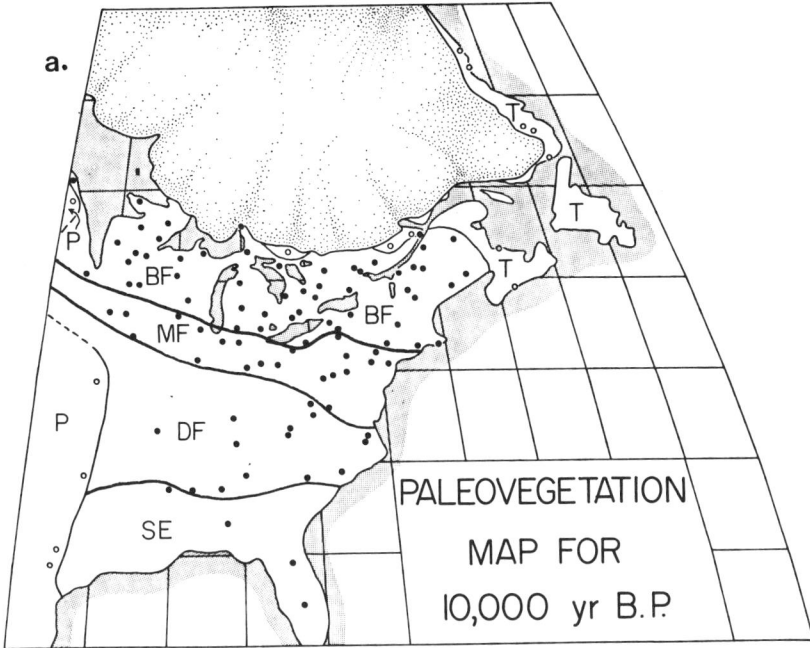

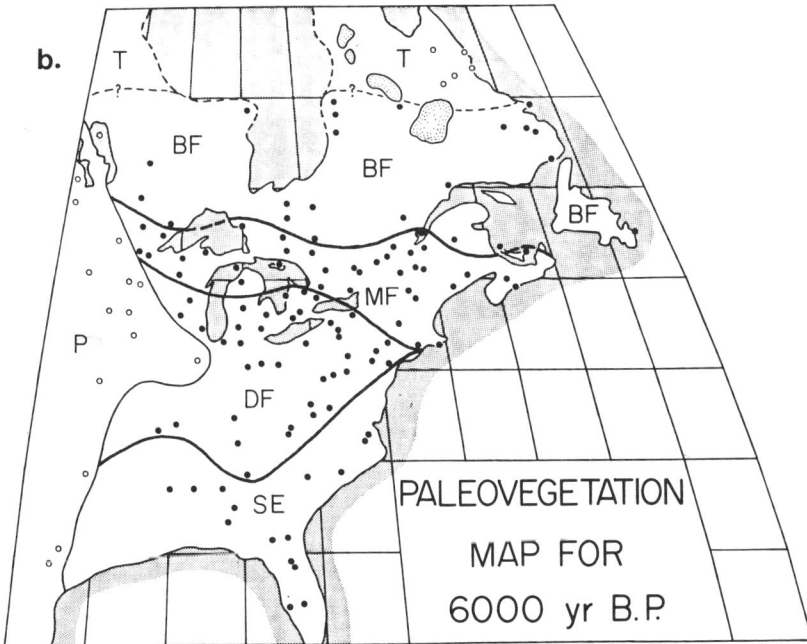

Figure 4.4. Paleovegetation maps for the early- and mid-Holocene intervals: a, 10,000 yr B.P.; b, 6000 yr B.P. Abbreviations as defined in Fig. 4.2 (generalized from Delcourt and Delcourt 1981).

the pollen rain include *Picea* (mean of 40%, range from 6 to 74%), northern Diploxylon *Pinus* (mean of 25%, range from 3 to 77%), *Betula* (represented primarily by tree birch and, to a much lesser extent, by shrub birch; mean of 22%, range from 0 to 58%), *Abies* (mean of 7%, range from 0 to 46%), and *Larix* (mean of 0.2%, range from 0 to 2%) (Delcourt and Delcourt 1981). In coniferous forests dominated by jack pine (*Pinus banksiana*) or red pine (*Pinus resinosa*), pollen values for northern Diploxylon pines tend to be greater than 40% of the AP sum, whereas boreal forests in which spruce is dominant typically contain Diploxylon pine percentages between 25% and 40%, along with high values for spruce.

A distinctive pollen assemblage characterizes the mixed conifer–northern hardwoods forest region today. Arboreal constituents include hemlock, Haploxylon pine (represented in eastern North America solely by eastern white pine, *Pinus strobus*), Diploxylon pine, spruce, fir, oak, birch, elm, ash, hornbeam, maple, and beech.

The eastern deciduous forest is represented by pollen assemblages dominated by oak and includes numerous additional taxa of deciduous trees. In this deciduous forest region, hickory pollen is typically greater than 2.5% of the AP, and pine pollen is less than 15%. Deciduous tree taxa include maple, beech, basswood, elm, and walnut (*Juglans cinerea* and *J. nigra*) in the pollen rain.

Within the southeastern evergreen forest region, Diploxylon pine pollen reaches peak values, but additional characteristic tree taxa distinguish this warm-temperate forest type from northern pine-dominated forests. Indicator taxa include tupelo, bald cypress, and sweetgum. Subtropical hardwoods, such as occur in southern peninsular Florida, are characterized by pollen of mangroves (primarily *Avicennia*, *Conocarpus*, *Laguncularia*, and *Rhizophora*), buttonwood, holly, and willow (Riegel 1965).

Several nonforest vegetation types in addition to tundra can be recognized by distinctive pollen assemblages and low total PAR (less than 2000 $gr \cdot cm^{-2} \cdot yr^{-1}$). Presettlement (pre-agricultural) pollen spectra from the prairie region of the Great Plains are characterized by combined percentages of $\geqslant 20\%$ for prairie forbs in the families Amaranthaceae, Chenopodiaceae, and Compositae (Webb *et al.* 1983a). Indicator taxa of prairie vegetation include leadplant (*Amorpha*), gaura (*Gaura*), and prairie clover [*Petalostemon (Dalea)*]. Sand dune scrub, today confined to coastal areas and xeric portions of the southern peninsula of Florida, has a modern-pollen rain composed of at least 40% herbs, including heliophytes such as rosemary (*Ceratiola*), myrtle (*Myrica*), jointweed (*Polygonella*), and spikemoss (*Selaginella*) (Watts 1975).

Selecting time planes for mapping

We selected five time planes for mapping of broad patterns in paleovegetation (Figs. 4.2a, 4.3a,b, 4.4a,b). Reconstructions for these times illustrate marked contrasts in vegetation patterns for the time period since the last glacial maximum. The map for 18,000 yr B.P. (Fig. 4.3a) represents vegetation at the peak of continental glaciation. The late-glacial map for 14,000 yr B.P. (Fig. 4.3b) illustrates the character of vegetation response with the onset of limited climatic warming and the beginning of deglaciation. The early-Holocene map at 10,000 yr B.P. (Fig. 4.4a) reflects the vegetation after the transition from glacial to interglacial climatic regimes. The 6000 yr B.P. map (Fig. 4.4b) shows vegetation during the mid-Holocene interval of maximum interglacial warmth and aridity in the Midwestern region of central North America. The map of presettlement vegetation (500 yr B.P.; Fig. 4.2a) illustrates late-Holocene vegetation prior to significant disturbance by EuroAmericans.

Interpretation of paleovegetation maps

18,000 yr B.P.

The paleovegetation map for 18,000 yr B.P. illustrates the predominantly latitudinal orientation of five major plant formations and large vegetation types (Fig. 4.3a). A discontinuous zone of tundra is mapped as isolated patches bordering the southern and eastern border of the Laurentide Ice Sheet. Evidence for full-glacial tundra at Wolf Creek, Minnesota (Birks 1976), confirms the persistence of treeless plant communities growing near the glacial margin, particularly in the reentrants bounded on three sides by the terminus of major glacial lobes and ice surges. The southeastern margin of the Laurentide Ice Sheet probably flowed directly into the North Atlantic Ocean only in major ice streams. In the absence of specific paleoecological data, the intervening exposed shelf bordering the Atlantic is speculatively mapped as occupied by tundra communities. Tundra extended south of the ice sheet along the crest of the Appalachian Highlands to at least 37°N, for example, in the Alleghenies as far south as Buckle's Bog, Maryland (Maxwell and Davis 1972) and Cranberry Glades, West Virginia (Watts 1979).

Boreal forest extended across the region from generally 41°N to 34°N. In some areas, particularly west of the Appalachians, coniferous trees may have grown adjacent to the margin of the Laurentide Ice Sheet (Wright 1981); wood fragments found within glacial tills have been radiocarbon-dated from full-glacial times (Dreimanis 1977). The closest site to the glacial margin with evidence for full-glacial boreal forest is Jackson Pond, located 190 km south of the glacial margin in central Kentucky (Wilkins 1985). Based on pollen accumulation rates and the relative percentage of nonarboreal pollen (NAP), boreal forests were most open in the northern third of the full-glacial boreal forest region, within the continental interior west of 95°W, and along the central Atlantic Coastal Plain. Elsewhere, boreal forests were relatively closed in forest canopy. Along the southern margin of the boreal forest, one major southward extension of boreal conifers is speculatively mapped within the Lower Mississippi Alluvial Valley.

The full-glacial transition from boreal to temperate forests occurred along a narrow ecotone from 34°N to about 33°N. One site, situated on Rayburn's Salt Dome in north-central Louisiana (Kolb and Fredlund 1981), documents the full-glacial location of mixed conifer–northern hardwoods forest west of the Mississippi River. East of the Mississippi River, the transition zone was bounded to the north by several sites containing boreal forest assemblages, and to the south by temperate forest.

The temperate forest region was apparently restricted to the Gulf and southern Atlantic Coastal Plains during the full-glacial interval. Warm-temperate southeastern evergreen forest communities would have been favored on fire-prone, xeric, sandy uplands across the region. Temperate deciduous forest communities would have found refuges only in fire-protected, mesic habitats with rich soils, such as river bluffs (Delcourt and Delcourt 1975; Delcourt 1980) and dissected topography around deep sinkholes (Watts and Stuiver 1980).

In the central and southern portion of the Florida Peninsula, open scrub vegetation is mapped within full-glacial sand dune fields. Evidence for sand-dune scrub vegetation, however, remains tentative and indirect. Paleoecological sites from that region commonly have gaps in their records from the time interval from about 30,000 yr B.P. to between 13,000 and 10,000 yr B.P. (Watts 1980a, 1983). In deposits predating 30,000 yr B.P., as well as in those dating from the late-glacial and early-Holocene intervals, the fossil pollen assemblages depict open, herbaceous and shrubby vegetation indicative of shifting sand dunes.

To date, no definitive paleoecological evidence is available to document the full-glacial locations of coastal warm-temperate or subtropical swamps in eastern North America.

14,000 yr B.P.

During the late-glacial interval, glacial ice retreated substantially across the Great Lakes region from Minnesota to southeastern Ontario and southern Quebec, as well as across New England (Fig. 4.3b). Newly deglaciated terrain was occupied by tundra, particularly in the northern Appalachians, New England, and the central Great Lakes region. Ice-cored recessional moraines, in which differential ice melting produced topographic reversals and unstable landscapes, provided favorable habitats for occupation by tundra communities persisting for millennia after initial ice wastage (Wright 1980).

Boreal forest expanded northward during the late-glacial interval, particularly into central Wisconsin, northwestern Iowa, and into eastern Ohio and southwestern Pennsylvania. The tundra–forest border was stationary within the central Appalachians and to their east through the central Atlantic states. The southern margin of the boreal forest shifted northward to about 34.5°N.

The region of mixed conifer–northern hardwoods forest expanded northward during the early late-glacial interval, reflecting the decline of boreal coniferous forest and its replacement by mixed forest.

The mosaic of southeastern evergreen forest and temperate deciduous forest persisted through the late-glacial interval across the Gulf and southern Atlantic Coastal Plains, documented by paleoecological sites extending from southeastern Texas to north-central Florida. The boundary between closed forest and more open sand-dune scrub remains tentatively mapped across central Florida.

10,000 yr B.P.

Major changes in the ice-sheet boundary as well as in the position of major vegetation types occurred in the interval between 14,000 and 10,000 yr B.P. By the early Holocene, represented in Fig. 4.4a by the map for 10,000 yr B.P., the Laurentide Ice Sheet had retreated northward from Iowa and the Dakotas to central Manitoba and Ontario as well as from New England and Newfoundland across the Gulf of St. Lawrence to southeastern Quebec and Labrador. Two large proglacial lakes bordered the ice margin. Glacial Lake Agassiz occupied southern Manitoba and adjacent portions of southwestern Ontario, eastern North Dakota and northwestern Minnesota. Lake Superior had developed across the border between western and central Ontario and the Upper Peninsula of Michigan. Other great lakes formed in the topographic depressions carved by retreating ice lobes. These included the postglacial predecessors of the modern lakes Michigan, Huron, Ontario, and Erie. By 10,000 yr B.P., marine waters extended inland as far as northern New York state along the St. Lawrence River Valley, producing the Champlain Sea.

Tundra occurred in three general areas at 10,000 yr B.P. The largest contiguous areas were located along the coast of Labrador, the island of Newfoundland, Nova Scotia, and coastal areas of New Brunswick. Smaller, discontinuous patches of tundra occurred along the retreating edge of the ice sheet in southwestern Quebec and in south-central Ontario.

In the early Holocene, the northern limits of boreal forest were bounded primarily by proglacial lakes to the northwest, by glacial ice to the north, and by tundra to the northeast. The southern margin of the boreal forest had retreated north to about 40°N along the Atlantic Coast and beyond 45°N in the continental interior. Between 14,000 and 10,000 yr B.P., the width of the latitudinal band occupied by boreal taxa diminished radically and the contiguity of forest across the deglaciated landscape was broken by major lakes.

The entire zone of boreal forest shifted northward by at least 7° latitude in only 4000 years. The region dominated by boreal forest at 14,000 yr B.P. was occupied at 10,000 yr B.P. by a complex mosaic of mixed conifer–northern hardwoods forest, temperate deciduous forest, and prairie.

The mixed conifer–northern hardwoods forest region both broadened and shifted northward as a contiguous arc, situated at 10,000 yr B.P. between 37°N and 41°N along the central Atlantic Seaboard and extending northwestward to between 42°N and 44°N in the western Great Lakes region.

By the early Holocene, deciduous forest had segregated out as a coherent vegetation type mappable on a subcontinent scale. The deciduous forest region occupied the region between about 75°W and 95°W, and from 34°N to the southern limit of the mixed conifer–northern hardwoods forest region.

Prairie appeared as a mappable formational unit first in southeastern Texas at about 12,000 yr B.P. (Bryant 1977). By 10,000 yr B.P., prairie vegetation extended eastward in the study area from eastern Texas and Oklahoma to northeastern Kansas, with an outlier in southern Manitoba.

In the early Holocene, the southeastern evergreen forest region was replaced by prairie to the west of 97°W, but it had expanded into southern Florida as sand-dune scrub vegetation diminished in areal extent. The northern boundary of the southeastern evergreen forest spread northward to about 34°N.

6000 yr B.P.

Between 6000 and 5000 yr B.P., the last fragments of Laurentide Ice disappeared from the high plateaus of northern Quebec and western Labrador. The latitudinal position of Arctic tundra extended as far south as about 55°N in the upland regions surrounding Hudson Bay. The boreal forest extended northward into deglaciated terrain, with its southern margin at 6000 yr B.P. located to the north of its former (10,000 yr B.P.) northern margin (Fig. 4.4b). By 6000 yr B.P., boreal forest extended from the maritime provinces of Canada, including the island of Newfoundland, westward through Ontario to central Manitoba.

The region of mixed conifer–northern hardwoods forest occupied the northern Great Lakes region, extending from southeastern Manitoba to northern New England and Nova Scotia by 6000 yr B.P. Between 10,000 and 6000 yr B.P., the breadth of this vegetation zone expanded from 2° latitude in width to about 6°.

The deciduous forest region expanded northward to southern Minnesota, central Wisconsin and Michigan, and to the southern New England region. The boundaries of this plant formation retreated on the east, south, and west, however, because of changes in distribution of both the prairie and the southeastern evergreen forest regions. To the west, closed deciduous forest became more open savanna near the prairie–forest border.

During the mid-Holocene interval, prairie extended from 29°N to about 50°N across the continental interior and eastward as a ''Prairie Peninsula'' to central Illinois (King 1981).

The southeastern evergreen forest region expanded northward during the mid-Holocene interval into the Ozarks of Missouri west of the Mississippi Alluvial Valley and northeastward on the Atlantic Coastal Plain as far north as New Jersey. Although not resolvable at this map scale of resolution, the mid-Holocene rise in sea level resulted in the development of extensive coastal swamps and marshes along the Atlantic and Gulf Coastal Plains.

500 yr B.P.

The southern boundary of the tundra extended to its northernmost limit at about 59°N, east of Hudson Bay, by about 4000 yr B.P. It then shifted progressively southward, especially along the coast of Hudson Bay and the northwestern Atlantic Ocean. By 500 yr B.P., maritime tundra extended as far south as 54°N along the coast of Labrador (Fig. 4.2a).

The northern limit of boreal forest expanded from 6000 yr B.P. to 4000 yr B.P. and then readjusted southward in eastern Canada. In the late Holocene, the southern margin of the boreal forest shifted southward in east-central and western Ontario, but remained relatively stable in the continental interior and the maritime provinces. The mixed conifer–northern hardwoods forest was, in turn, displaced approximately 1° southward through the central and western Great Lakes region between 6000 and 500 yr B.P.

The deciduous forest expanded westward in the late Holocene, displacing prairie from all but a major outlier mapped across central Illinois. The southern and eastern margins of the deciduous forest remained relatively stationary from 6000 to 500 yr B.P.

Prairie was displaced by forest along its northern and eastern margins. The magnitude of displacement was about 70 to 100 km along the prairie–forest border in Minnesota and Wisconsin, as much as 400 km along its eastern axis through the Prairie Peninsula of Iowa and Illinois, and about 100 km on its southeastern border through Missouri, Arkansas, and Oklahoma.

The southeastern evergreen forest region remained relatively constant in its position, expanding only on its northwestern boundary during the last 3500 years as prairie retreated.

Patterns of paleoclimate inferred from maps of paleovegetation

In a modification of the approach of Bryson (1966), Bryson and Hare (1974), and Bryson and Wendland (1967), we compared mean positions of airmasses with vegetation boundaries on the modern landscape. We then used geographically coherent patterns in the vegetation reconstructions to infer broad changes in paleoclimate over the past 20,000 years (Delcourt and Delcourt 1984).

The relationship of modern climatic regions to vegetation

On the subcontinental scale, major vegetational regions (Fig. 4.2a) correspond extremely well with the geographic boundaries of climatic regions (Fig. 4.2b). Climatic regions today are defined by the annual sequence of dominant airmasses and the bounding positions of major climatic frontal zones (Bryson 1966; Bryson and Hare 1974). Three fundamental airmasses shape contemporary climatic patterns in eastern North America: (1) the Arctic Airmass of frigid, dry, polar air generated in the northern polar region; (2) the Pacific Airmass, representing prevailing westerlies that flow across the Pacific Ocean but become dry when they lose moisture as they rise over the western North American mountain ranges and across the Great Plains; and (3) the Maritime Tropical Airmass, which transports warm, moist air from the Gulf of Mexico and the equatorial Atlantic Ocean.

Today, the Arctic tundra region in eastern North America is characterized by year-round dominance (10 to 12 months) of the Arctic Airmass (Bryson and Hare 1974). Tundra vegetation predominates in the region where the mean temperature of the warmest summer month is less than 10°C, the limiting threshold for growth of arborescent woody plants (Billings and Mooney 1968; Chapin and Shaver 1985).

The northern boundary of the boreal forest of central and eastern Canada (the tundra–boreal forest transition) coincides with the summer position of the

Arctic Frontal Zone (Bryson and Hare 1974). The southern boundary of the boreal forest region corresponds to the mean winter position of the Arctic Frontal Zone. Thus, the climatic region today associated with the boreal forest is bounded by the mean position of major climatic frontal zones. This boreal climatic region is dominated by the Arctic Airmass from early winter through spring (between 3.5 and 10 months each year) and generally by the Pacific Airmass for the remainder of each year.

The mixed conifer–northern hardwoods forest region is associated with the Polar Frontal Zone that separates subpolar climatic regions to the north from more temperate climatic regions to the south. This climatic transition zone is characterized by several months of Arctic Air influence during the late winter and spring, followed by incursions of the Maritime Tropical Airmass bringing moisture in summer, and the Pacific Airmass dominating seasonally from autumn through early winter (Bryson and Hare 1974). Incursions of the Arctic Airmass, resulting in episodes of extreme winter cold, exceed the supercooling threshold of $-41°C$ and kill all but the most cold hardy of deciduous forest species (Burke *et al.* 1975).

The deciduous forest region receives moisture throughout the spring and summer growing season from the Maritime Tropical Airmass (Bryson and Hare 1974). In the autumn and winter, this region is typically under the relatively weak influence of the Pacific Airmass as well as the southern anticyclone. The southeastern evergreen forest region is under the influence of the Maritime Tropical Airmass nearly throughout the year and receives substantial moisture from both tropical depressions and hurricanes.

The prairie is in the area influenced for much of the year by the relatively dry Pacific Airmass, with prevailing westerlies prevailing for 4 to 6 months from autumn through winter and with more limited incursions of the moisture-bearing Maritime Tropical Airmass occurring from spring through summer.

Biotic responses to changing atmospheric circulation patterns during the late Quaternary

Bryson and Wendland (1967) applied the modern relationships of vegetation to climatic regions in order to prepare subcontinental climatic reconstructions over the last 13,000 years for central and eastern North America. Environmental interpretations of fossil faunal and floral assemblages were used to identify shifting boundaries separating major biotic regions. This approach involved three key assumptions:

1. Climate represents the ultimate ecological control on the macro-scale, and cuts across differences in geologic substrate, soils, and hydrology, factors that on a finer spatial scale would influence species distributions on a landscape mosaic.
2. Glacial and nonglacial climatic regimes have shifted atmospheric states abruptly, and, therefore, climatic changes observed at a given site tend to occur rapidly as a result of the passage of major frontal zones or because of a major change in seasonal distribution of predominant airmasses.

3. The correspondence between modern climatic regimes and vegetation can be used to infer changes in climate based upon fossil-plant assemblages.

The third assumption is based on the geologic principle of uniformitarianism and implies that if past vegetation were comparable to that of the present, then the climatic conditions associated with that vegetation were comparable. However, past vegetation types that differed markedly from those known today would have reflected different climatic conditions with airmasses occurring in different frequencies and annual sequence than today. The nature of biological response to climatic change would be observed first along sensitive ecotones between major biotic regions and later in the geographic centers of the biotic regions (Bryson and Wendland 1967).

These assumptions have been tested subsequently in several ways by studies of later workers. Quantitative paleoclimatic reconstructions can be generated using "transfer functions" or "response surfaces" between pollen assemblages and specific climatic parameters of temperature and precipitation (Howe and Webb 1983; Bartlein et al. 1984, 1986), as well as analyses of airmass frequencies (Webb and Bryson 1972). Considerable debate continues to question the degree to which climate is the "ultimate ecological control" on the biota, and the degree to which vegetation is in equilibrium with climate (Davis 1978; Webb 1980; Birks 1981b; Delcourt and Delcourt 1983). This controversy persists, in part, because of the general lack of quantitative paleoclimatic records of changes in temperature and precipitation that are *independent* of the pollen records used to infer past vegetation; this situation precludes making definitive tests of the inherent assumptions concerning cause and effect in biotic responses to late-Quaternary climatic change. Differences in interpretation are scale dependent, often arising from differences in resolution level of evidence concerning the relative influence of climate and other environmental factors, dispersal of propagules, and competitive interactions on biotic change.

At the biome level, the terrestrial record of biotic response to late-Quaternary climatic change can be compared directly with the marine record based on foraminiferal assemblages. In such a comparison, fundamental correspondence in the position of the Polar Frontal Zone has been demonstrated for eastern North America and the western North Atlantic Ocean (Delcourt and Delcourt 1984) as well as for the eastern North Atlantic Ocean and western Europe (Duplessy et al. 1981; Van Campo 1984).

We extended the paleoclimatic mapping back to 18,000 yr B.P. (Figs. 4.5 and 4.6, revised and generalized from Delcourt and Delcourt 1984), based upon inferences from paleovegetation maps (Figs. 4.2a, 4.3, and 4.4) and the independent marine record documented by Balsam (1981) and CLIMAP (1981). Following McIntyre et al. (1976), the position of the marine Polar Frontal Zone over the Atlantic Ocean was identified between the 14°C and 16°C winter isotherms, along a steepened gradient in sea-surface temperature separating subpolar and subtropical foraminiferal assemblages.

During full-glacial times, 18,000 yr B.P. (Fig. 4.5a), the climatic regime was

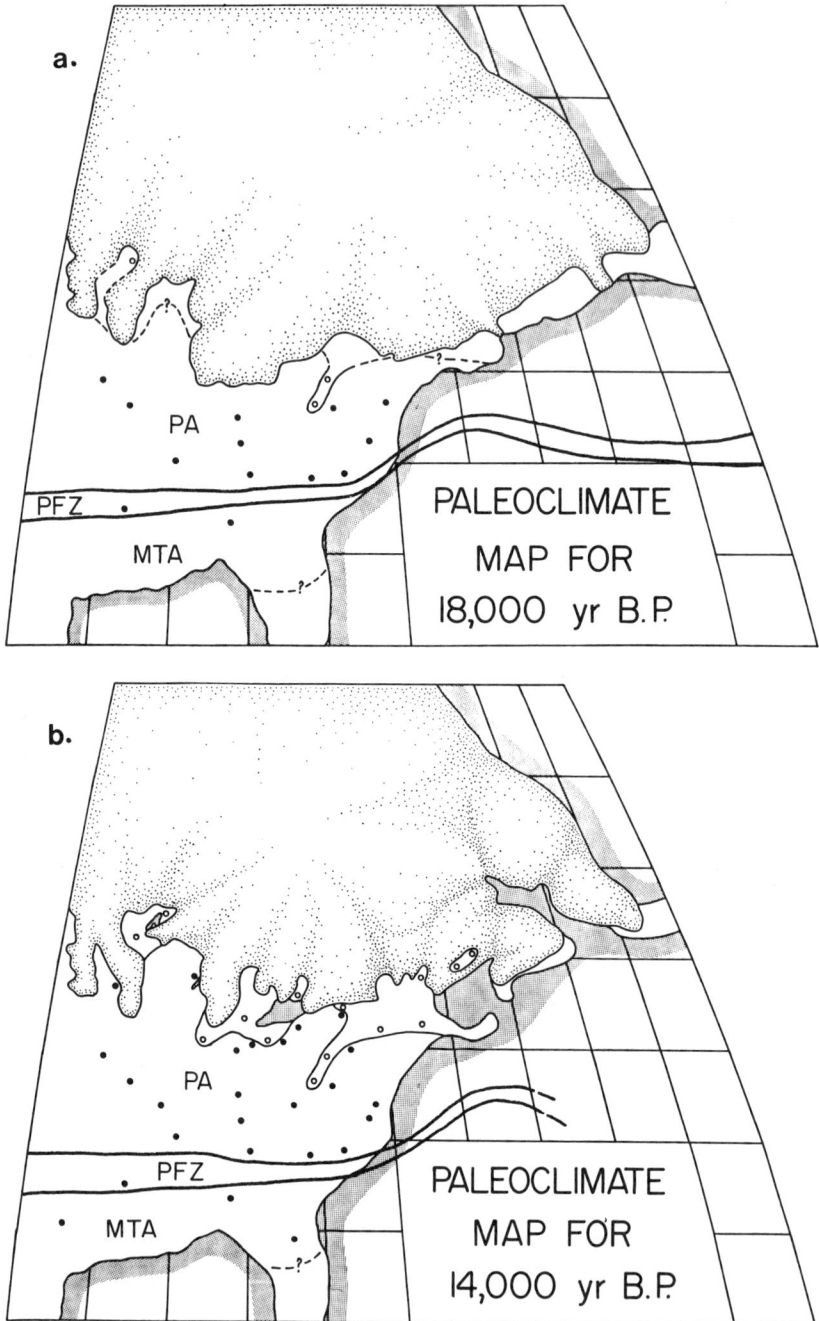

Figure 4.5. Paleoclimate maps for the full-glacial and late-glacial intervals: a, 18,000 yr B.P.; b, 14,000 yr B.P. Abbreviations as defined in Fig. 4.2 (generalized from Delcourt and Delcourt 1984).

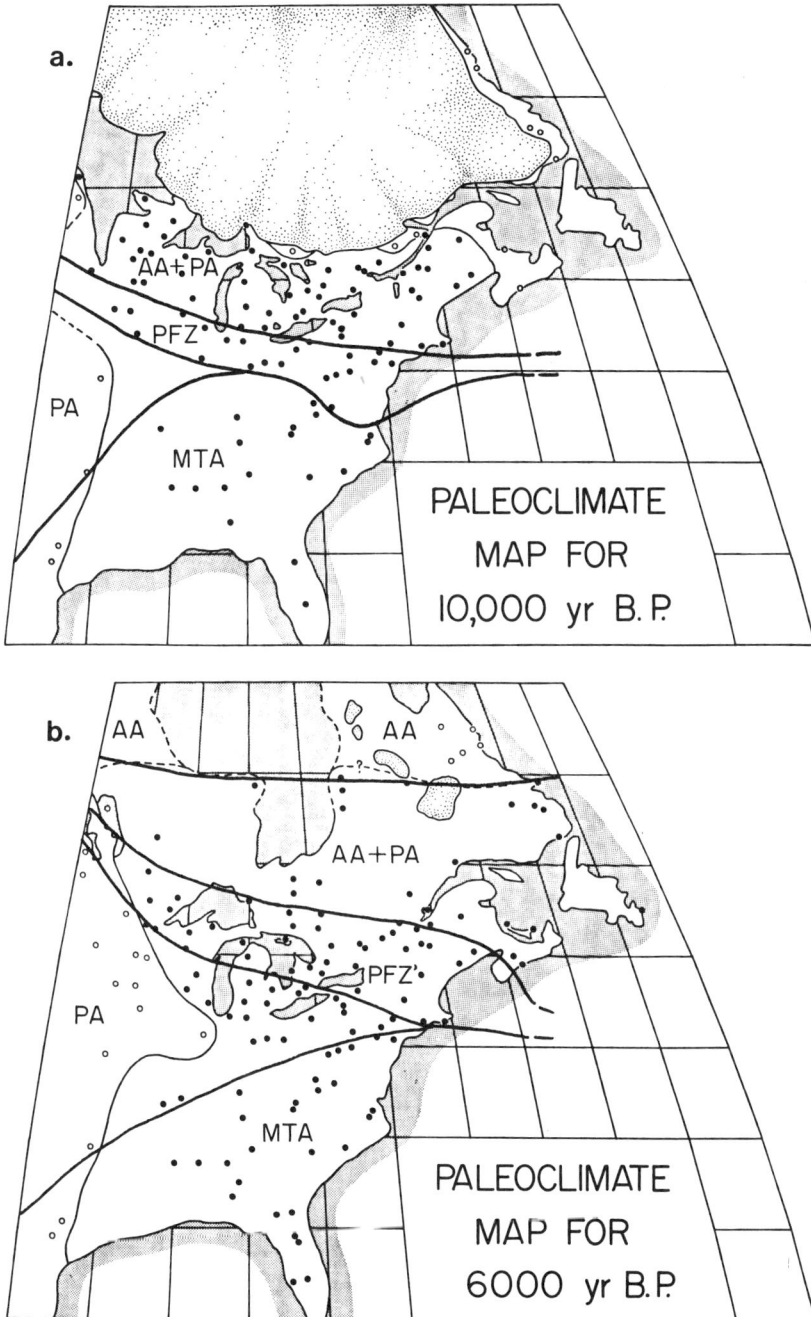

Figure 4.6. Paleoclimate maps for the early- and mid-Holocene intervals: a, 10,000 yr B.P.; b, 6000 yr B.P. Abbreviations as defined in Fig. 4.2 (generalized from Delcourt and Delcourt 1984).

characterized by west-to-east or zonal flow of prevailing westerlies across the middle latitudes (Kutzbach and Wright 1985). This dominant influence of the Pacific Airmass extended across the boreal forest region from 34°N north to the southern ice limit. We interpret that the ecotone between boreal and temperate forests (Fig. 4.3a) coincided with the position of the Polar Frontal Zone. The mean position of this major storm track was anchored between 33°N and 34°N over the midcontinent and extended northeastward from Cape Hatteras to a position between 35° and 37° N in the western North Atlantic Ocean (Fig. 4.5a). During peak glacial times, the Arctic Airmass remained in the north polar region, north of the Laurentide Ice Sheet; the high ice barrier served to block the southward penetration of the Arctic Airmass. To that extent, the climate of the full-glacial boreal forest region was very different from the seasonal dominance of Arctic and Pacific Airmasses characterizing the boreal forest region today. More equable climates in mid-latitudinal regions, with reduced seasonal contrast in temperatures and absence of extreme winter cold, may account for the restricted occurrence of full-glacial tundra along the southern margin of the continental ice sheet. In contrast, more extensive tundra (possibly grading northeastward to polar desert) may have occurred along the exposed continental shelf along the southeastern border of the Laurentide Ice Sheet because of the maritime effects of winter pack ice and relatively cold sea-surface temperatures offshore along the Labrador Current (CLIMAP 1981; Imbrie et al. 1983). Throughout the full-glacial interval, the Maritime Tropical Airmass prevailed across southeastern North America, the Gulf of Mexico (Brunner 1982), and the equatorial Atlantic Ocean (CLIMAP 1981; Imbrie et al. 1983). However, with reduced sea-surface temperatures in the equatorial Atlantic and cooling of surface waters of the Gulf of Mexico by only 1 to 2°C, both a decrease in evaporation from the ocean surface and a reduction in the frequency of tropical depressions and hurricanes lessened the effective precipitation received on the Gulf and southern Atlantic Coastal Plains relative to that of today (Wendland 1977). This warm, but relatively dry, full-glacial climate favored dominance of xeric oak–hickory forests rather than a mosaic of southern pine forest and mesic deciduous forest.

Between 17,000 and 16,000 yr B.P., increasing solar radiation (Berger 1978) resulted in a minor weakening of zonal atmospheric flow and a strengthening in the Bermuda High Pressure system positioned over the central Atlantic Ocean (Delcourt and Delcourt 1984). The late-glacial intensification of the Bermuda High resulted in more northward extension of the Maritime Tropical Airmass during the summer, reaching the southern flank of the ice sheet (Fig. 4.5b). The increased summer warmth and rain triggered substantial melting and initial retreat of glacial ice. However, the Arctic Airmass was still effectively blocked from reaching the continental interior by the ice dome. The climatic transition zone between Pacific and Maritime Tropical Airmasses broadened along the Polar Frontal Zone, particularly west of the Appalachian Mountains.

By 10,000 yr B.P., during the early Holocene, northward retreat and thinning of glacial ice resulted in the development of an ice-free corridor between the residual Laurentide Ice Sheet in northern Canada and the Cordilleran Ice Sheet in northwestern Canada (Fig. 4.6a). The Arctic Airmass was able to penetrate in the wintertime through the ice-free corridor into the deglaciated region between 40°N and 50°N latitude. The persistence of tundra (Fig. 4.4a) along the coasts of Labrador, Newfoundland, and Nova Scotia may have been due to the proximity of the cold, marine Labrador Current rather than year-round presence of the Arctic Airmass. The boreal forest region was characterized by seasonal dominance of Arctic and Pacific Airmasses. The area over which the Pacific Airmass dominated exclusively was greatly reduced in extent and confined to the developing prairie region in the continental interior. The Maritime Tropical Airmass dominated over the Gulf Coastal Plain year-round and extended in summer northward as far as 40°N. The Polar Frontal Zone shifted northward in the early Holocene to between 39°N and 41°N. The breadth of the Polar Frontal Zone in the central Atlantic states matched both the breadth and position of its oceanic counterpart offshore in the Atlantic. Overall, the meridional atmospheric circulation patterns of the early Holocene were strikingly different from the strongly zonal regimes of the full-glacial and late-glacial intervals.

By 6000 yr B.P., during the mid-Holocene interval, only small remnants of Laurentide ice remained north of 53°N latitude, and they no longer either influenced the position of predominant airmasses or anchored major frontal boundaries (Fig. 4.6b). Arctic tundra occupied the area north of 55°N latitude (Fig. 4.4b), the region dominated year-round by the Arctic Airmass. As in the late Holocene, the boreal forest region was bounded by mean winter and summer positions of the Arctic Airmass. The transitional Polar Frontal Zone expanded across 6° of latitude encompassing the area from the western Great Lakes to New England and extending offshore from Cape Cod to southernmost Nova Scotia. The wedge-shaped zone dominated by the Pacific Airmass expanded northward and eastward across eastern North America, reaching its maximum extent between 7000 and 5000 yr B.P. (Wright 1968). With this intensification of a zonal atmospheric flow regime, the prevailing westerlies effectively restricted the influence of the Maritime Tropical Airmass to the southeastern United States. By the mid-Holocene interval, glacial meltwaters had returned to the oceans, sea level had risen to its modern position, and the sea surface had warmed to its present temperature. These oceanographic conditions favored increased evaporation rates and encouraged the development of tropical storms and hurricanes (Wendland 1977). On the Gulf and Atlantic Coastal Plains, increased available precipitation during the growing season favored the mid-Holocene expansion of southern pine forest across the southeastern evergreen forest region.

During the last 4000 years of the late Holocene, the zonal pattern has weakened, and the meridional character of modern North American climate has intensified (Fig. 4.2b). The Polar Frontal Zone has been constricted as much as 2° latitude southward and has become more sharply defined along its northern border, especially in the Great Lakes region. The Maritime Tropical Airmass has reexpanded northward and northwestward into the region formerly dominated through most of the year by the Pacific Airmass.

Conclusions

1. To the extent that climatic tolerance limits or environmental barriers affect both rate and route of species migrations during times of vegetational disequilibrium, and to the extent that many species are limited by the same fundamental factors, species may tend to remain together through time and space, mappable as discrete assemblages at the biome level.
2. Late-Quaternary changes in the positions of ecotones separating major vegetational types provide a useful basis for broad-scale paleoclimatic reconstruction. Although both its position and breadth have changed during the late Quaternary, the Polar Frontal Zone has been a fundamental climatic boundary between Boreal and Temperate Zones of terrestrial vegetation across eastern North America, as well as separating subpolar from subtropical marine biotas in the northern Atlantic Ocean.
3. Two climatic regimes have substantially influenced the development of late-Quaternary environments both of the eastern United States and of central and eastern Canada. The glacial climatic regime of the late Pleistocene was zonal in character south of the Laurentide Ice Sheet. The interglacial climatic regime has been meridional in atmospheric flow pattern during the Holocene.

5. Tree Population Dynamics During the Past 20,000 Years

Paleovegetation maps compiled at the biome level (Chapter 4) provide a general overview of major trends in change of vegetation patterns through time. In order to evaluate the changes in composition and structure in vegetation during times of environmental changes such as the transition from Pleistocene to Holocene climatic conditions, it is instructive to examine the individualistic nature of the response of specific tree taxa. The nature of the pollen record does not allow mapping of distributional changes in dominance of all species of native eastern North American trees (*e.g.*, Little 1971, 1977). Taxon calibrations available to date on a subcontinental scale, however, do allow us to map changes in relative dominance of many of the dominant, subdominant, and characteristic tree taxa (see discussions in Chapters 2 and 3).

These calibrated taxa represent a mixture of gymnosperm and angiosperm families, genera, and individual tree species. Interpretations of changes in their individual mapped patterns therefore must be generalized to the extent that the ecological tolerances of species within larger taxonomic units differ or overlap. For example, over 40 species of oaks (*Quercus*) are taxonomically distinct within the broad range of the genus in eastern North America (Fernald 1970; Harrar and Harrar 1962; Steyermark 1963). The positions of the species populations on elevational and soil-moisture gradients overlap in a gradational manner in areas of sympatry (Martin 1978). In addition, many of the species are notorious for introgressive hybridization, resulting in many intermediate forms that are difficult to distinguish in the field (Fernald 1970; Steyermark 1963). As a result, pollen morphology also intergrades, making it virtually impossible to key out fossil pollen grains of North American *Quercus* to the species level (Lieux 1980; Solomon 1983a,b). In the fossil record, supporting macrofossil evidence concerning past distributions of particular oak species is scarce (*e.g.*, Baker *et al.* 1980; Delcourt and Delcourt 1977b). Therefore, in this treatment, we consider the distributional history of *Quercus* at the genus level.

Population trajectories through time

Mapping the "trajectories" of tree populations through time involves not only tracing the movements of the leading edges of their migration fronts, but also evaluating demographic changes in dominance structure of their entire populations. Contoured map patterns portrayed on paleogeographic base maps give a visual impression of distributional changes in area and dominance for each taxon through time, and they also provide the basis for subsequent quantitative ecological analysis.

Documentation of changing number and density of paleoecological sites through time

One limiting factor to interpretation of late-Quaternary vegetational history is the number and density of paleoecological sites available at each given time plane of interest. The fewer the sites from which to draw quantitative vegetation reconstructions, the less detailed can be the conclusions drawn. In examining a time series of migration maps for tree taxa, the amount of detail interpreted must be balanced from one time plane to the next according to changes in quantity and density of available sites.

In eastern North America, the available data base for late-Quaternary vegetational reconstructions includes relatively few sites spanning full-glacial and late-glacial time intervals (Fig. 5.1a) in comparison with those that have been studied for their Holocene records. However, many of the Holocene sites are situated within prairie or tundra regions rather than in forested landscapes. Expressed in terms of the amount of total forest area (Fig. 5.1b) relative to the number of available palynological sites through time, forest-site density (Fig. 5.1c) varied from $\frac{1}{250,000}$ km^2 during full-glacial and early late-glacial times to $\frac{1}{50,000}$ km^2 in the middle to late Holocene. This concentration of late-glacial and Holocene sites primarily in the Great Lakes and New England regions reflects in part the longstanding concentration of palynological research in these deglaciated terrains (see regional syntheses in Bryant and Holloway 1985a). The Holocene site density reflects the great abundance of kettle-lake, bog, and swamp environments, sites suitable for pollen preservation, in the extensively glaciated area of central and northeastern North America. The few full-glacial sites that have been studied near the glacial margin were located in glacial-margin reentrants that were ice-free (Birks 1976). South of the direct influence of the Laurentide Ice Sheet, various kinds of depositional environments preserve

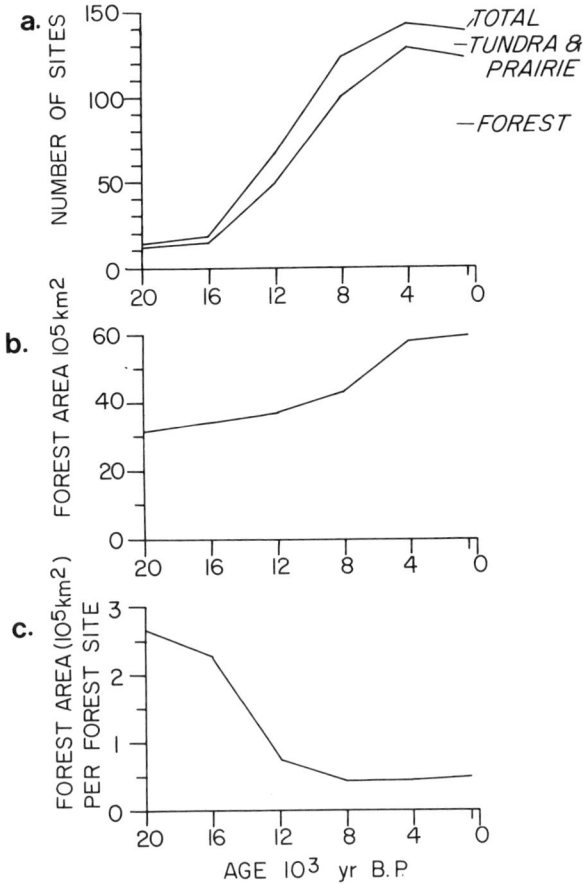

Figure 5.1. a. Number of sites in prairie, tundra, and forest and total number of sites included in the late-Quaternary plant-ecological data set for eastern North America for the interval from 20,000 yr B.P. to the present. b. Changes in total forest area in eastern North America through the past 20,000 years. c. Changes in area/site density for paleoecological sites located in the forested region of eastern North America through the past 20,000 years.

pollen and plant macrofossils dating from as much as the past 20,000 years (Delcourt and Delcourt 1985b). These environments include sinkholes in karst regions such as the Interior Low Plateaus of Kentucky, Tennessee, and Alabama, the Ridge and Valley Physiographic Province ranging from Pennsylvania southwestward to Alabama, the Ozark Plateaus of Missouri and Arkansas, as well as throughout the Florida Peninsula. On the central and southern portions of the Atlantic Coastal Plain, wind-oriented lakes known as "Carolina Bays" date from full-glacial times to the present (Frey 1953, 1955; Watts 1980a,b; Whitehead 1981). To date, fewer of these southern sites have been studied than those farther to the north, however, reflecting the relative youth of the field of Quaternary paleoecology in the Southeast (Delcourt and Delcourt 1985b).

Minimum, maximum, and mean values of dominance through time

As a basis for developing guidelines for choice of contour intervals, we summarized statistics concerning the population trends in reconstructed dominance for each of the 19 tree taxa to be mapped. For each taxon, Statistical Analysis System "PROC MEAN" routines (Statistical Analysis System 1979) were used to calculate the mean (bracketed by one standard deviation about the mean), minimum, and maximum estimates of dominance at 1000-year intervals for the past 20,000 years, based upon all positive values for the array of sites located in forested landscapes at each time mapped (*e.g.*, Fig. 5.2a). Additional population statistics were calculated for selected 500-year intervals during the Holocene. These data gave a late-Quaternary trajectory of change in total population size for each tree taxon.

Determination of contour intervals of dominance scaled to the population characteristic of each taxon

The maximum-dominance curve of the population trajectory was used to determine the highest interval used for contouring the time series of paleo-dominance maps for each taxon. Taxa with dominance values consistently greater than 60% of the forest composition, for example, were contoured at intervals of 20%. Those which reached values greater than 60% only once or a few times during the time series were evaluated by examining individual maps to determine whether only a few outlier values were responsible for the apparent maximal peaks in dominance. Where the mean values were consistently much lower than the sporadic peak values, a lower contour interval such as 10% was used in order to characterize more adequately the regionally coherent trends in paleo-dominance. For these situations, the outlier (peak) points were indicated on the contour maps by their numerical value at their geographic locations. This procedure minimized the generation of artifacts such as tight contour intervals centered on individual datum points, allowing generalization of the dominance patterns more in accord with the broad scale of map resolution justified by the overall density of sites.

The regular intervals chosen for contouring were scaled to the population characteristics of each taxon and then used on all maps of the time series for a particular taxon. In general, dominant taxa were contoured at 20% intervals (*e.g.*, contours of 0, 20, 40, 60, 80, and 100%), subdominants at 10% intervals, and minor taxa at 5%, 2%, or 1% intervals. This ensured comparability in the evenness of contouring across the distributions of all taxa.

Temporal changes in areal extent and dominance structure of tree populations

As the forested sites are not strictly uniform in their distributional spacing for the paleoecological data base of the past 20,000 years, the statistical summary of the population trajectory consisting of values of reconstructed dominance for any specific time is weighted toward those geographic areas with relatively dense clusters of sites. In order to reduce this possible bias in the statistics based upon the total number of dominance values for the forested sites, additional areal measurements were obtained from the time series of contoured maps. These areal data integrate information in terms of both the value of population dominance at each site as well as the areal extent occupied by populations between successive contour intervals (*e.g.*, between 10% and 20% of forest composition).

An electronic digitizer (Model 1224, Numonics Corp., Lansdale, Pa.) was used for 19 major taxa to digitize map areas in order to determine total distributional areas as well as the areas occupied between successive contour intervals. Areal measurements for each taxon were calculated based upon the paleo-dominance maps for six specific times at approximately 4000-year intervals: (1) 20,000 yr B.P.; (2) 16,000 yr B.P.; (3) 12,000 yr B.P.; (4) 8000 yr B.P.; (5) 4000 yr B.P.; and (6) 500 yr B.P. Replicate measurements using the digital analyzer were made on a base map for this equal-area projection using 5° latitude by 5° longitude grid blocks from 25°N to 60°N and from 50°W to 100°W; these areal values were compared with the known (cartographic) areas for each 5° latitude–longitude block (Robinson and Sale 1969). Digital area readings replicated known cartographic values with a precision of ± 2%. Prepared for the six times, histograms of absolute area and relative dominance graphically portray late-Quaternary changes in the dominance structure for each taxon. Cumulative plots of taxon dominance and its areal extent reveal the absolute changes in mapped distribution and population–dominance structure for the past 20,000 years during the last glacial–interglacial cycle.

Contoured paleo-dominance maps for major tree taxa in eastern North America

Population trajectories, contoured maps of paleo-dominance, and area–dominance histograms (which summarize the changes in area occupied within each contour interval) illustrate temporal changes in locations of population centers, range margins, and dominance structure through the last 20,000 years. These analyses offer complementary insights concerning the patterns and processes of changes in whole populations of the 19 calibrated tree taxa across eastern North America. For convenience, in the following discussion the taxa are arranged in alphabetical order by scientific name. The stippled area on the maps represents the Laurentide Ice Sheet; closed circles represent locations of paleoecological sites in forested landscapes; open circles represent paleoecological sites located in prairie (usually to the west on the maps) or tundra (usually to the north on the maps).

Fir (*Abies*)

The late-Quaternary trajectory for fir populations (Fig. 5.2a) exhibits a full-glacial consistency in dominance values, rising through the late-glacial interval to a crest at the Pleistocene–Holocene transition, and reaching a new plateau

Abies

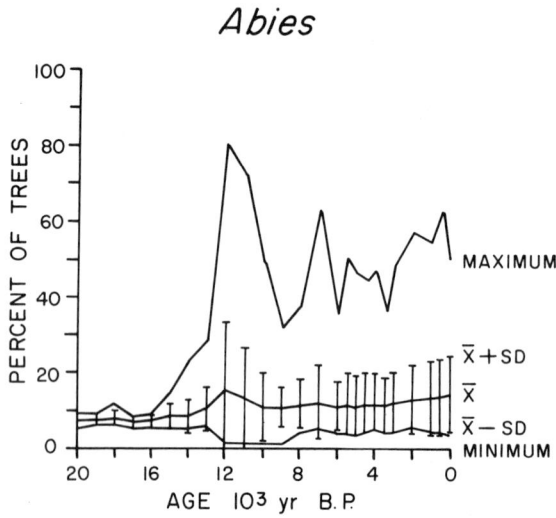

Figure 5.2. Fir (*Abies*). a. Population trajectory, including maximum, minimum, and mean values of paleo-dominance as well as ± 1 SD of the mean, for the past 20,000 years.

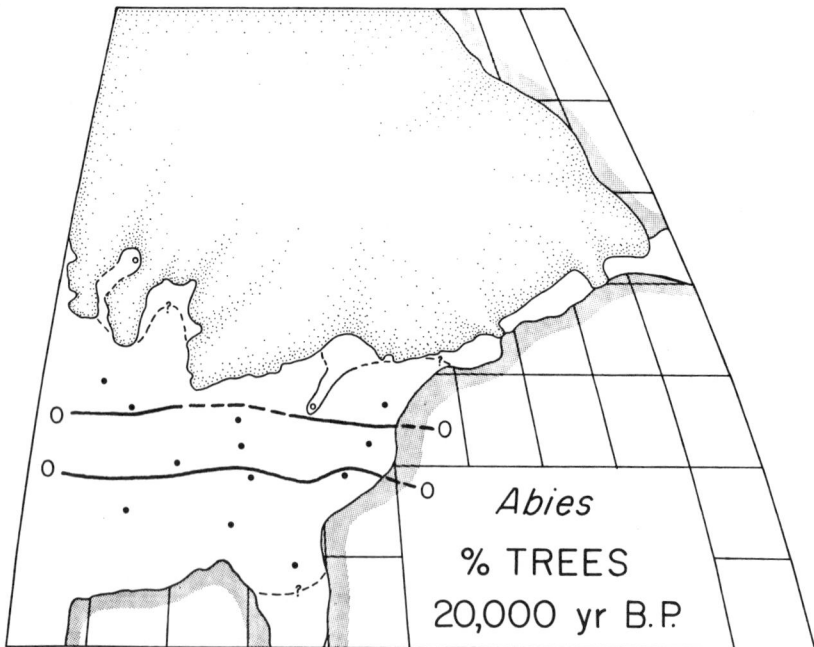

Figure 5.2. Fir (*Abies*). b. Paleo-dominance map for 20,000 yr B.P.

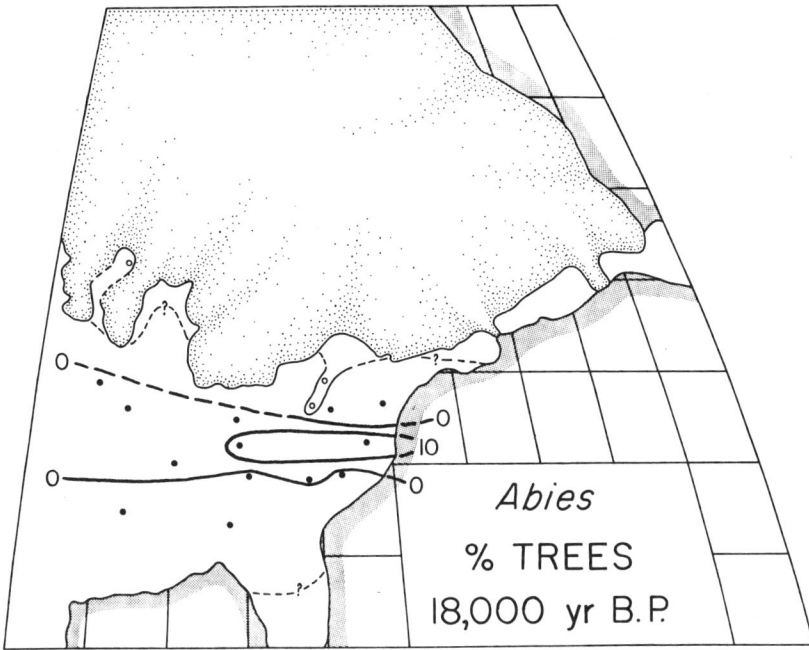

Figure 5.2. Fir (*Abies*). c. Paleo-dominance map for 18,000 yr B.P.

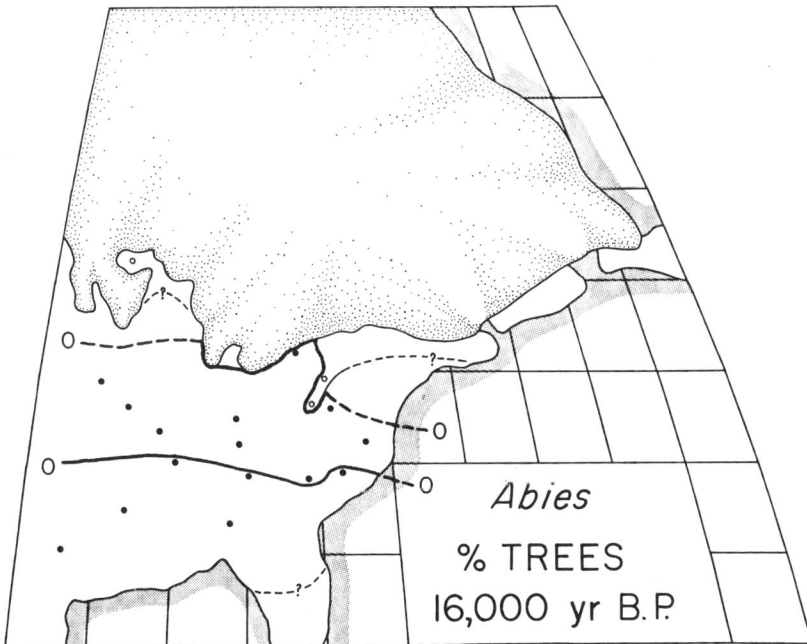

Figure 5.2. Fir (*Abies*). d. Paleo-dominance map for 16,000 yr B.P.

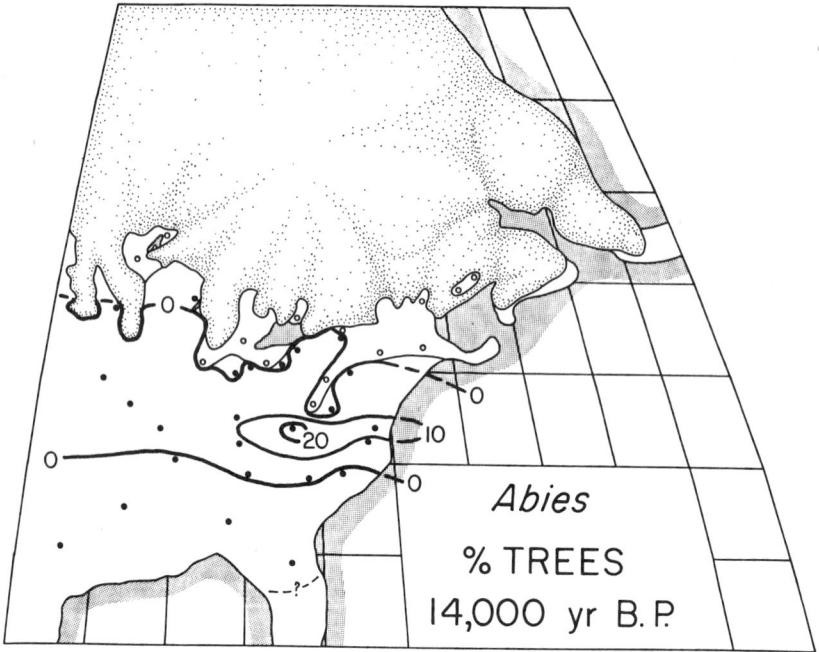

Figure 5.2. Fir (*Abies*). e. Paleo-dominance map for 14,000 yr B.P.

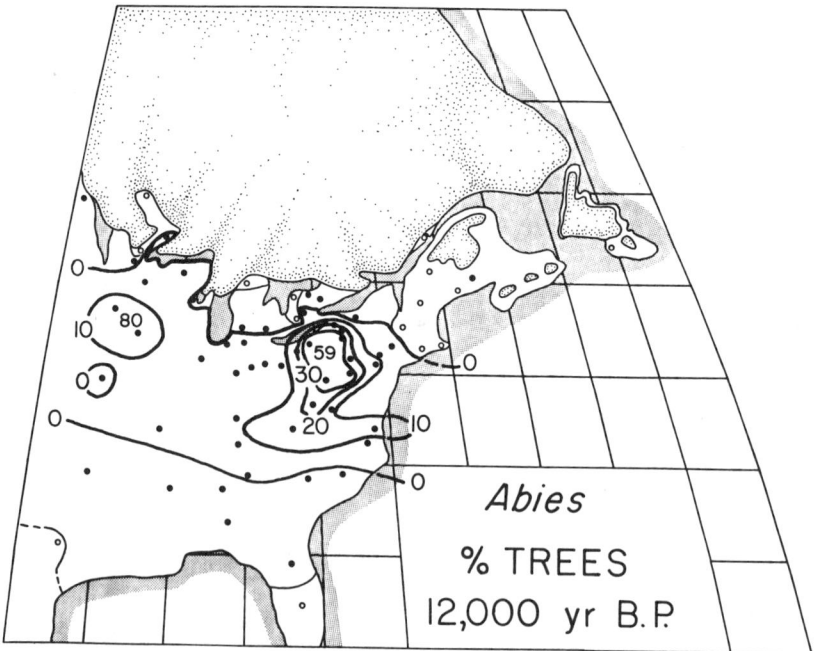

Figure 5.2. Fir (*Abies*). f. Paleo-dominance map for 12,000 yr B.P.

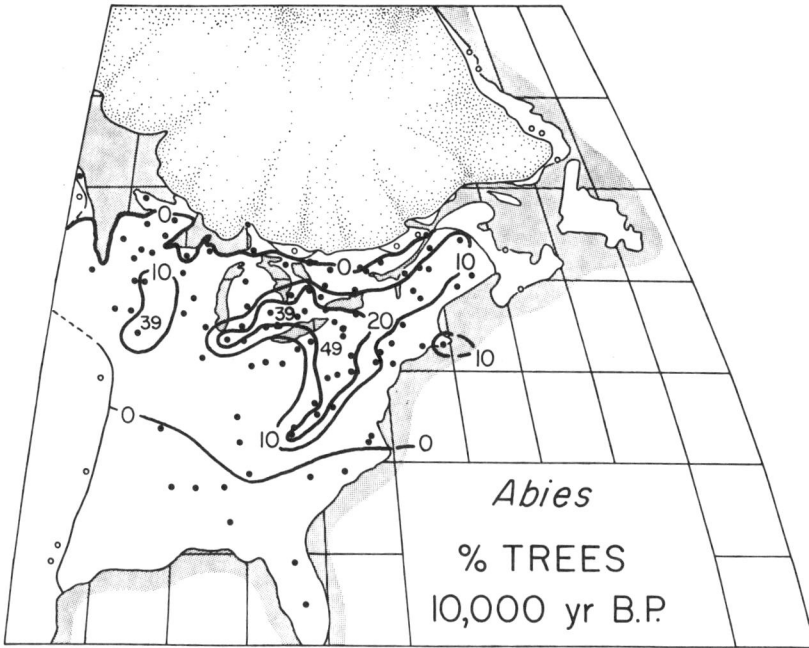

Figure 5.2. Fir (*Abies*). g. Paleo-dominance map for 10,000 yr B.P.

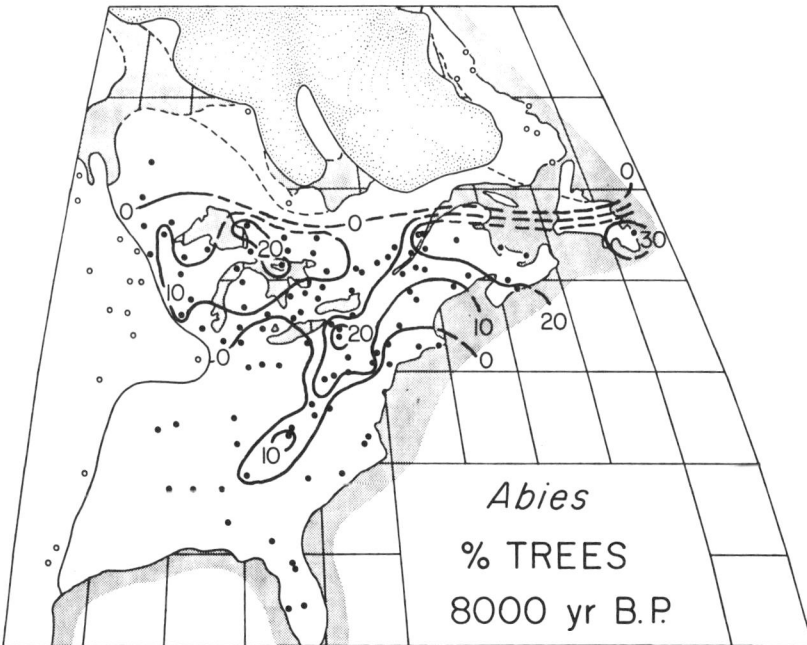

Figure 5.2. Fir (*Abies*). h. Paleo-dominance map for 8000 yr B.P.

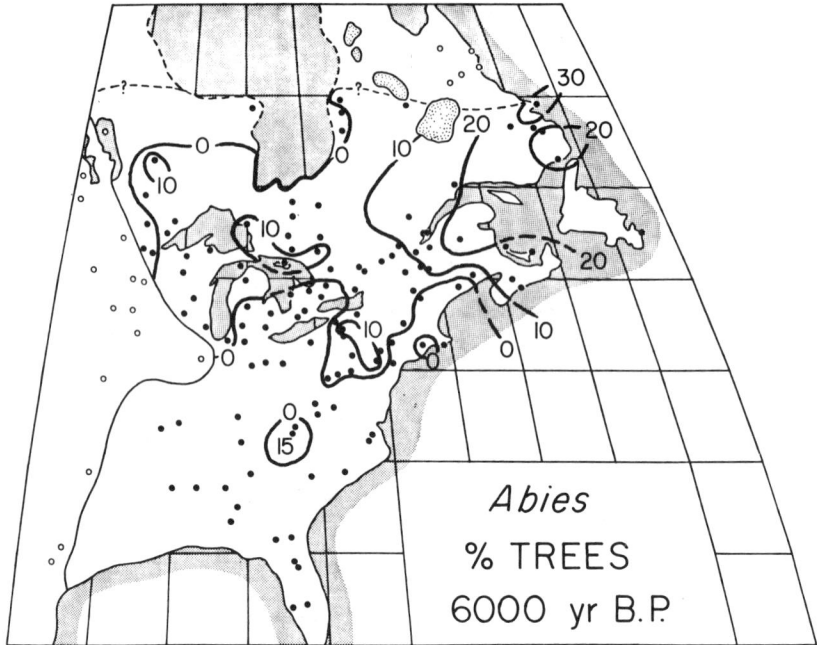

Figure 5.2. Fir (*Abies*). i. Paleo-dominance map for 6000 yr B.P.

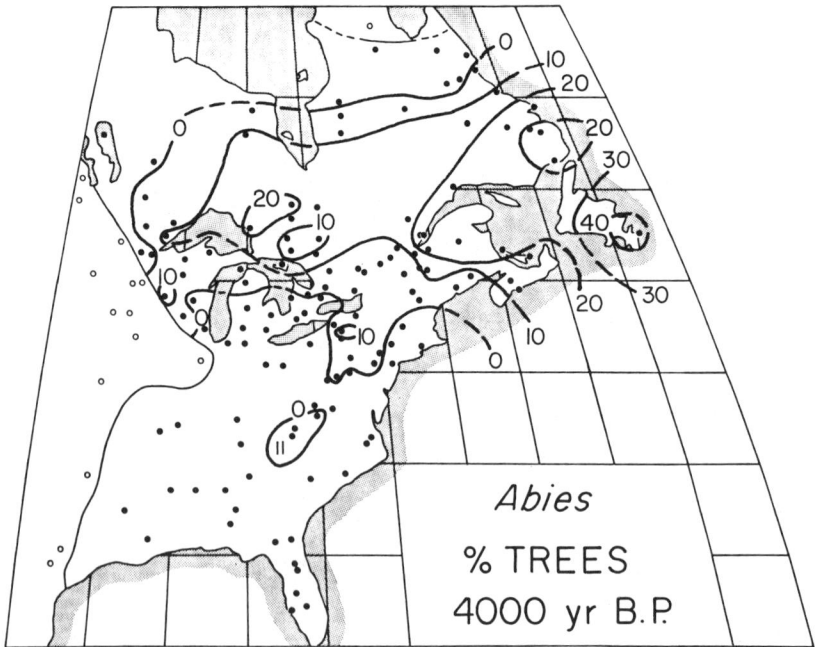

Figure 5.2. Fir (*Abies*). j. Paleo-dominance map for 4000 yr B.P.

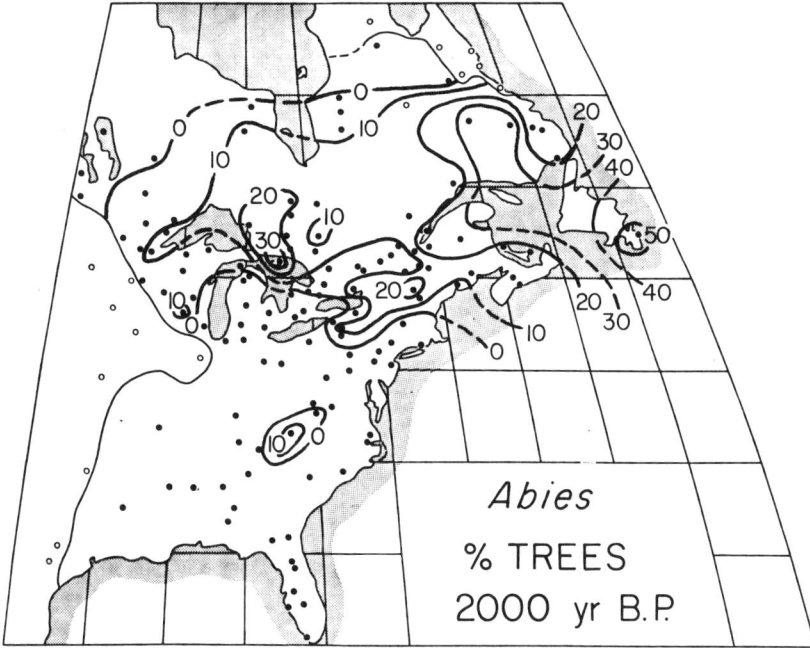

Figure 5.2. Fir (*Abies*). k. Paleo-dominance map for 2000 yr B.P.

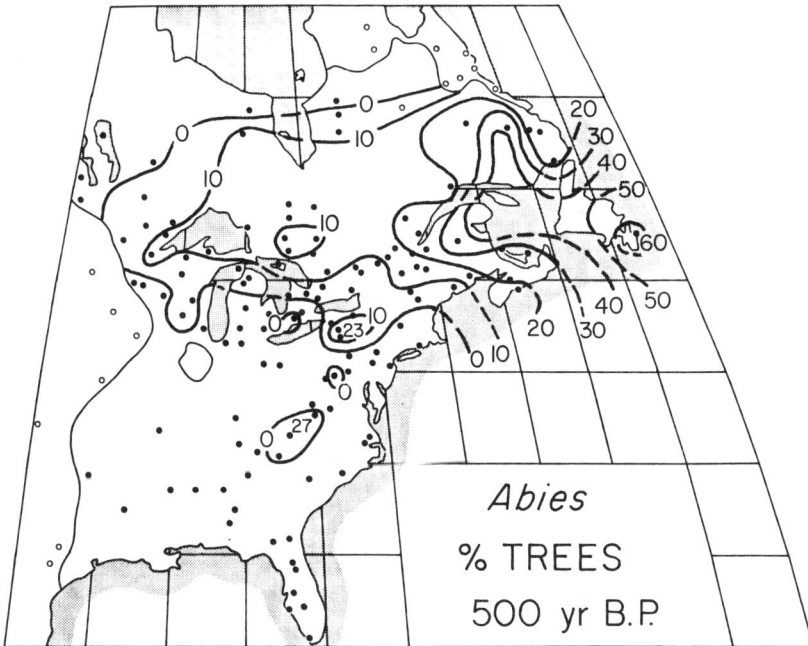

Figure 5.2. Fir (*Abies*). l. Paleo-dominance map for 500 yr B.P.

spanning the present interglacial. Between 20,000 yr B.P. and 17,000 yr B.P., the full-glacial populations of fir possessed a narrow range of both means (7% to 8% dominance, with ± 1 SD of 1% to 2%) and maximum values (8% to 11%). During the late-glacial and early-Holocene intervals, the population mean doubled from 8% at 16,000 yr B.P. to 16% at 12,000 yr B.P.; the maximum values reconstructed for fir dominance increased ninefold (from 9% up to 80% over this 4000-year interval). From 12,000 yr B.P. to 9000 yr B.P., the peak and mean population values were reduced to 32% and 11%, respectively. Following the ecological overshoot of fir during the early Holocene, the mid- and late-Holocene populations readjusted to recurring means of about 12% from 8000 yr B.P. to 3000 yr B.P. and then rose to about 14% during the last several thousand years. For the past 9000 years, the highest reconstructed values fluctuated between 32% and 63%, dropping quasi-periodically with troughs at 9000 yr B.P., 6000 yr B.P., and 3500 yr B.P. The inherent variability of mid- and late-Holocene populations of fir typically is reflected in the standard deviation from approximately 6% to 10% about the mean.

The fir paleo-dominance map for 20,000 yr B.P. (Fig. 5.2b) reveals a broad swath of its full-glacial populations situated between 34°N and 38°N from the coastal portions of Virginia and North Carolina, west to the continental interior. The highest value is 9% dominance reconstructed from Anderson Pond in Middle Tennessee (Delcourt 1979). The fir populations are bounded by the continuous line of the 0% contour; where this contour line is broken into short dashes, this border has been extrapolated across regions characterized by sparse site density.

By 18,000 yr B.P., the contour pattern for relative dominance of fir (Fig. 5.2c) exhibited a prominent latitudinal band of more than 10% centered on 36°N; this elongate population center extended along the southern flank of the boreal forest region from 75°W to 87°W (Fig. 4.3a). In the region west of the Lower Mississippi Alluvial Valley, the southern and northern distributional limits of fir occurred at about 35°N and 46°N, respectively.

With the initial onset of climatic amelioration and glacial retreat during the early late-glacial interval, fir populations advanced along their northern distributional boundary. By 16,000 yr B.P. (Fig. 5.2d), fir migrated northward across the central Atlantic Seaboard of Virginia and the lower–elevation slopes of the central Appalachian Mountains in Maryland and Pennsylvania. The northern limit of fir (0% contour) traced west from the northern reaches of Ohio, Indiana, and Illinois, through Missouri and Nebraska, and probably into southern Iowa and Kansas. The southern limit of fir trees fluctuated about the latitude of 35°N. Although widely distributed, dominance values for fir remained less than 10% of the boreal forest composition.

By 14,000 yr B.P., in the middle of the late-glacial interval (Fig. 5.2e), the latitudinally elongate population center with values of greater than 10% dominance reestablished along 37°N across Virginia. The species-rich boreal forest in the Saltville Valley, located within the Ridge and Valley Province of southwestern Virginia (Delcourt and Delcourt 1986), contained 23% fir. Macrofossils of balsam fir (*Abies balsamea*) dating from 13,500 yr B.P. were recovered from the Allegheny Plateau, at Big Run Bog (980 m elevation) in northern West Virginia (Larabee 1986). East of the Appalachian Highlands, fir advanced to about 35°N in Pennsylvania. West of the Appalachians, fir invaded deglaciated terrain and spread north into the northern boreal forest as far as the forest–tundra ecotone in Ohio and Indiana and across southern Wisconsin to the glacial margin of the Laurentide Ice Sheet in Iowa (Fig. 4.3b). The southern limit of fir maintained its position at 35°N at 14,000 yr B.P., although its populations extended locally southward into northern Georgia and northern South Carolina.

By 12,000 yr B.P. (Fig. 5.2f), the major population center of fir extended west along 37°N from southeastern Virginia to southwestern Virginia and then northeastward along the Appalachian corridor through West Virginia and western Maryland, reaching peak population values of 59% in western Pennsylvania and western New York. A second population center for fir of generally > 20% dominance occurred across Iowa (Brush 1967; Van Zant 1979). The northern fir boundary followed the forest–tundra ecotone in southern New England, extended across the Great Lakes region at about 43°N, and extended into northern Wisconsin and central Minnesota to the glacial margin and shorelines of proglacial lakes. Along its southern border, fir persisted at about 35°N from 75°W to about 87°W. However, west of 87°W, the southern limit shifted north to approximately 37°N within the continental interior.

The paleo-dominance map for fir at 10,000 yr B.P. (Fig. 5.2g) exhibits the northeastward extension of the fir population center along the Appalachians and into the eastern Great Lakes region. Regions with > 20% fir extended from southwestern Virginia northeast to central New York and west into Lower Michigan. The northern limit of fir was generally between 46°N and 48°N, extending to the forest–tundra ecotone to the east and anchored by physical barriers of glacial ice and meltwater lakes on the north and northwest. Along its southern limit, fir became locally extinct through the central Atlantic states and the region west of the Mississippi Embayment.

By 8000 yr B.P. (Fig. 5.2h), the distributional pattern for fir was characterized by a southwest–northeast trend extending along the Appalachians from northern Georgia to Nova Scotia and Newfoundland. This montane distribution of fir intersected a broad latitudinal zone between 42°N and 49°N, in which fir extended across the Great Lakes region west to Minnesota. The southern distributional limit of fir coincided with the prairie–forest boundary to the southwest, arched across the southern Great Lakes region, and paralleled the lower slopes of the Appalachian Mountains. The largest region with > 20% fir extended across the maritime provinces of New Brunswick, Nova Scotia, Prince Edward Island, and Newfoundland. From 12,000 yr B.P. to 8000 yr B.P., the primary population center for fir shifted progressively northeastward along the montane axis of the Appalachians.

During the mid-Holocene interval of peak interglacial warmth (Fig. 5.2i), the principal population center (with > 20% dominance of fir) occupied the maritime provinces of eastern Canada, including peak values of 35% in coastal Labrador and between 20% and 30% in the southern interior of Labrador, easternmost Quebec, New Brunswick, and Newfoundland. By 6000 yr B.P., fir extended across eastern and central Canada and northern portions of the eastern United States. The northern boundary of fir extended to the tundra–forest boundary at about 55°N in the region east of Hudson Bay. However, in central Ontario, west of Hudson Bay, the fir limit stopped at about 52°N, within the boreal forest region. Fir was mapped as a continuous band along the length of the Appalachian Mountains at 8000 yr B.P. (Fig. 5.2h). With the extinction of fir from the lower elevations of the central Appalachians, by 7000 years ago the distributional range of fir was broken into two populations, after which the endemic Fraser fir (*Abies fraseri*) was isolated on high peaks in the southern Appalachian Mountains of western North Carolina, eastern Tennessee, and southwestern Virginia. By 6000 yr B.P. (Fig. 5.2i), the southern continuous distribution of the northern balsam fir extended west to near the prairie–forest ecotone and through the Great Lakes region from southern Wisconsin through central Michigan, southern Ontario, and into the northern Appalachians from Maine southwest to Pennsylvania.

By 4000 yr B.P. (Fig. 5.2j), the population center of fir located in southeastern Canada increased in relative dominance, with peak values of up to 47% in Newfoundland. To the west, fir also increased in dominance, with a second population center > 20% in east-central Ontario and a zone of 10% to 20% extending throughout central and western Labrador, central and southern Quebec, and the region from central Ontario to Minnesota. By 4000 yr B.P., fir had migrated northward along the Labrador coast to 57°N and northwest to approximately 55°N in northern Ontario west of Hudson Bay. The southern distributional limits of fir remained relatively stationary. Fir increased to > 10% dominance in the latitudinal zone from about 45°N to about 54°N and from 53°W to about 88°W (Fig. 5.2j).

During the late Holocene, balsam fir became a subdominant to dominant tree throughout the eastern boreal forest region. In the eastern portion of its range, typical dominance values ranged from 20% to 58% by 2000 yr B.P. (Fig. 5.2k). In central Ontario east of Lake Superior, fir ranged from 20% to 31% of the forest composition at 2000 yr B.P. By 500 yr B.P. (Fig. 5.2l), fir attained relative dominance values up to 63% in Newfoundland. Populations of fir also increased in relative dominance in the Appalachian Mountains in New York State (balsam fir) and in western North Carolina and eastern Tennessee (Fraser fir). By 500 yr B.P., a pronounced longitudinal gradient had developed in the dominance of fir within the eastern boreal forest region, with greatest abundance in the eastern maritime provinces, and decreasing values westward to Manitoba (Fig. 5.2l).

At 20,000 yr B.P., fir occupied $8.3 \cdot 10^5$ km^2, with dominance values less than 10% throughout its full-glacial range (Fig. 5.3a,b). By 16,000 yr B.P., fir increased by 70% in area occupied (Fig. 5.3a), to a total of $14.1 \cdot 10^5$ km^2, although it remained a minor constituent of the boreal forest at < 10% dominance throughout its range. By 12,000 yr B.P., the area of distribution of fir increased to $20.6 \cdot 10^5$ km^2, and it increased in the number of dominance classes (Fig. 5.3a). The lowest dominance class (0 to 10%) continued to expand in area to

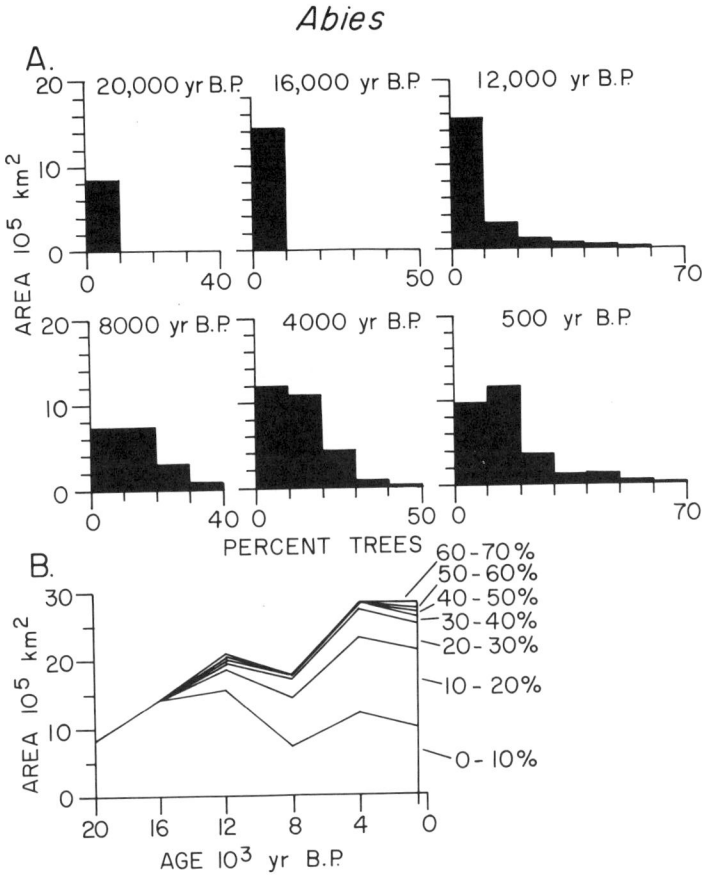

Figure 5.3. Changes in area and dominance of fir (*Abies*). a. Histograms of area–dominance structure at 20,000, 16,000, 12,000, 8000, 4000, and 500 yr B.P. b. Cumulative plot from 20,000 yr B.P. to the present.

$15.5 \cdot 10^5$ km^2; however, higher dominance classes expanded to 25% of the mapped distributional area.

The area–dominance histogram for 8000 yr B.P. (Fig. 5.3a) reflected a progressive shift toward more equable distribution across all dominance classes. The area of < 10% fir dominance occupied $7.2 \cdot 10^5$ km^2, 41% of the total distributional area of fir. The dominance class of 10% to 20% fir occupied an additional 41% of the total area. An area of $2.7 \cdot 10^5$ km^2 was occupied by the 20 to 30% dominance class (16% of the total area). The remaining $0.4 \cdot 10^5$ km^2 was occupied by 30 to 40% fir and represented 2% of the total area.

By 4000 yr B.P. (Fig. 5.3a), the number of dominance classes for fir increased to 5, including the 40 to 50% interval. Each of the two lowest dominance classes increased in area occupied by 60%, to a total of $28.0 \cdot 10^5$ km^2. By 500 yr B.P.

(Fig. 5.3a), the area occupied by the 0 to 10% class and 10 to 20% class remained nearly constant, but the total number of dominance classes increased to 7.

The cumulative area–dominance histogram for fir (Fig. 5.3b) illustrates that from 20,000 yr B.P. to 500 yr B.P., the overall distribution of fir expanded approximately 342%, from an initial value of $8.3 \cdot 10^5$ km^2 to a maximum of $28.4 \cdot 10^5$ km^2. The area of fir increased from full-glacial to early-Holocene times, then decreased during the middle Holocene, and reexpanded during the late-Holocene interval. The reduction in distributional area of fir from 12,000 yr B.P. to 8000 yr B.P. reflected the local extinction of fir populations along the southern margin of its range, and the corresponding decrease in area within the 0 to 10% contour interval. The increase in importance of fir within the Holocene forests of eastern Canada was reflected in increases in both the number of dominance classes > 10% and the area they occupied from 12,000 yr B.P. to 500 yr B.P. (Fig. 5.3b).

The modern population centers of fir in the maritime provinces of eastern Canada coincide with the geographic region that receives > 76 cm of precipitation annually (Halliday and Brown 1946; Fowells 1965). The full-glacial distribution of fir occupied the latitudinal belt between 34°N and 37°N, primarily along the southern edge of the boreal forest region. This east–west axis coincided approximately with the primary climatic boundary of the Polar Frontal Zone (Fig. 4.5a), which would have concentrated storm tracks and precipitation in the region just south of the southern Appalachian Mountains. During the late-glacial and early-Holocene intervals (Figs. 4.5b, 4.6a), the Polar Frontal Zone shifted northward, and its storms would have been intercepted by the Appalachian Mountains because of an orographic effect. By 12,000 yr B.P. (Fig. 5.2f), the primary population center for fir began to build in dominance along the southwest–northeast trend of the Appalachian Mountains. Within the last 8000 years, fir spread into the maritime provinces of eastern Canada and began to increase in dominance there. Subsequent increases in dominance of fir during the mid- and late Holocene occurred in response to storms tracking along the Polar Frontal Zone through the Great Lakes region and New England (Figs. 4.6b and 4.2b) as well as storms tracking northeastward along the Atlantic Seaboard.

Maple (*Acer*)

The population trajectory plot for maple (*Acer*) depicts mean values between 5% and 6% with $\pm$ 1 SD < 3% between 20,000 yr B.P. and 17,000 yr B.P. (Fig. 5.4a). During the full-glacial interval, the peak values reconstructed ranged between 5% and 10% relative dominance. Through the late-glacial and early-Holocene intervals, mean values for maple were consistently about 7%, with maximum values oscillating between 11% and 26% from 16,000 yr B.P. to 10,000 yr B.P. Between 10,000 yr B.P. and 8000 yr B.P., maple percentages increased to an interglacial plateau, with means converging on about 15% and maximum values generally between 45% and 53% relative dominance. During the mid- and late-Holocene intervals, the variability expressed in terms of $\pm$ 1 SD

Acer

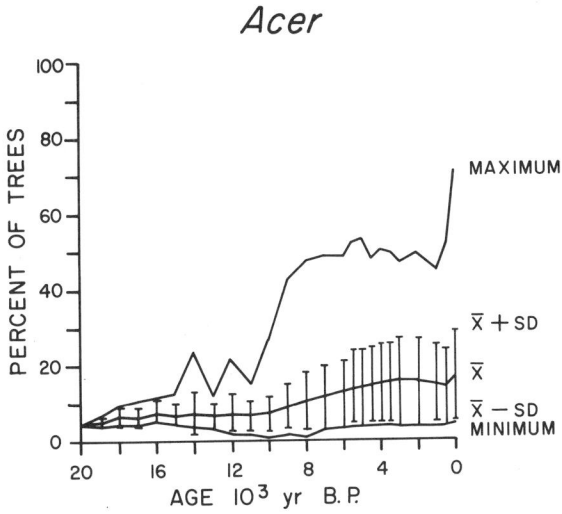

Figure 5.4. Maple (*Acer*). a. Population trajectory, including maximum, minimum, and mean values of paleo-dominance as well as ± 1 SD of the mean, for the past 20,000 years.

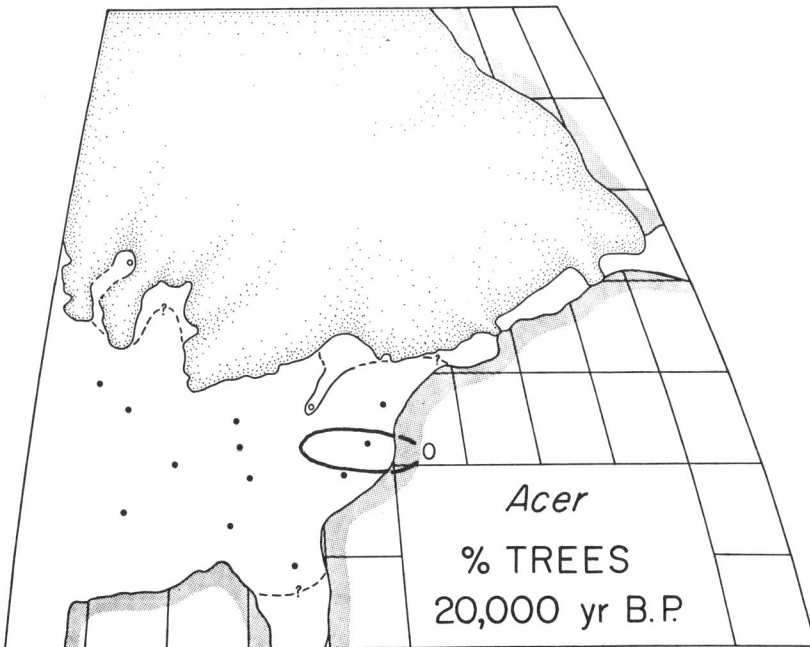

Figure 5.4. Maple (*Acer*). b. Paleo-dominance map for 20,000 yr B.P.

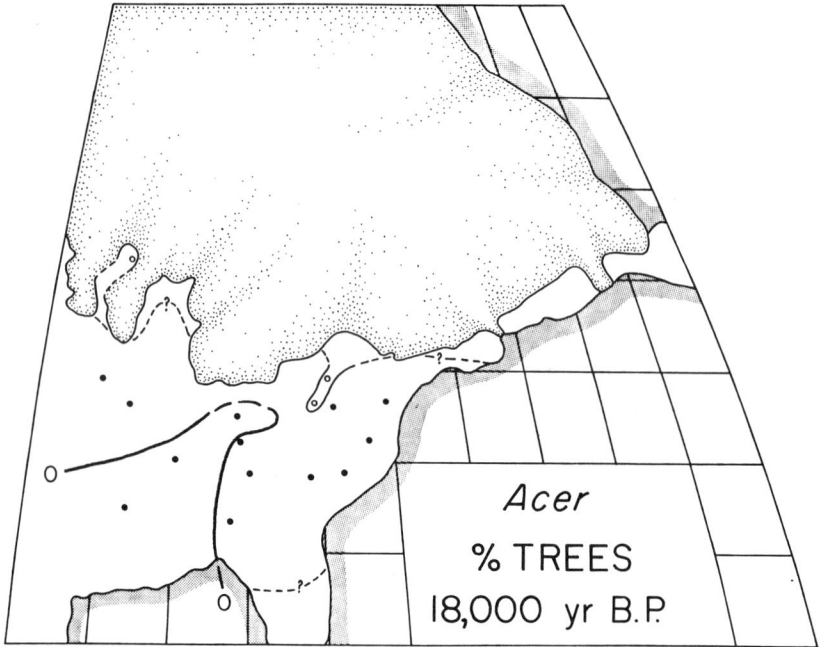

Figure 5.4. Maple (*Acer*). c. Paleo-dominance map for 18,000 yr B.P.

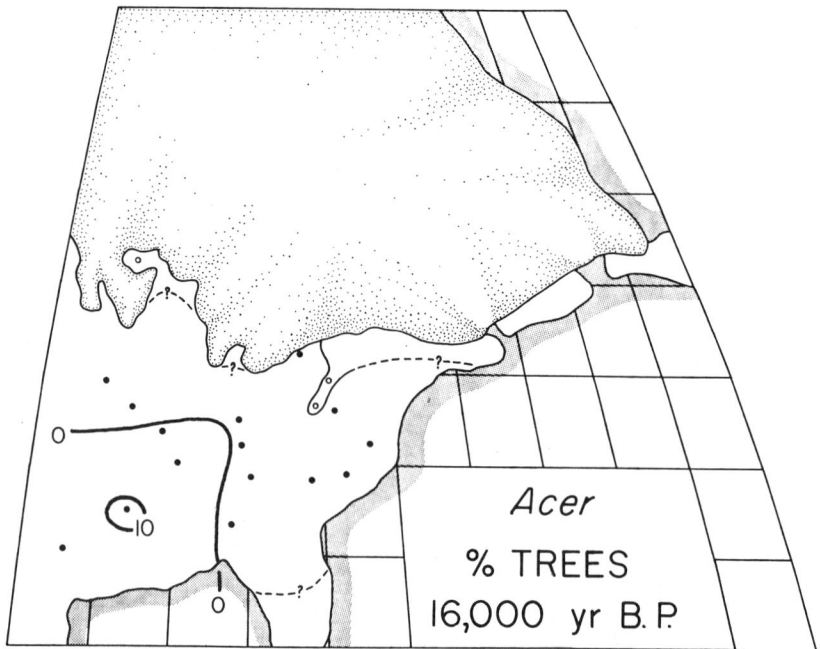

Figure 5.4. Maple (*Acer*). d. Paleo-dominance map for 16,000 yr B.P.

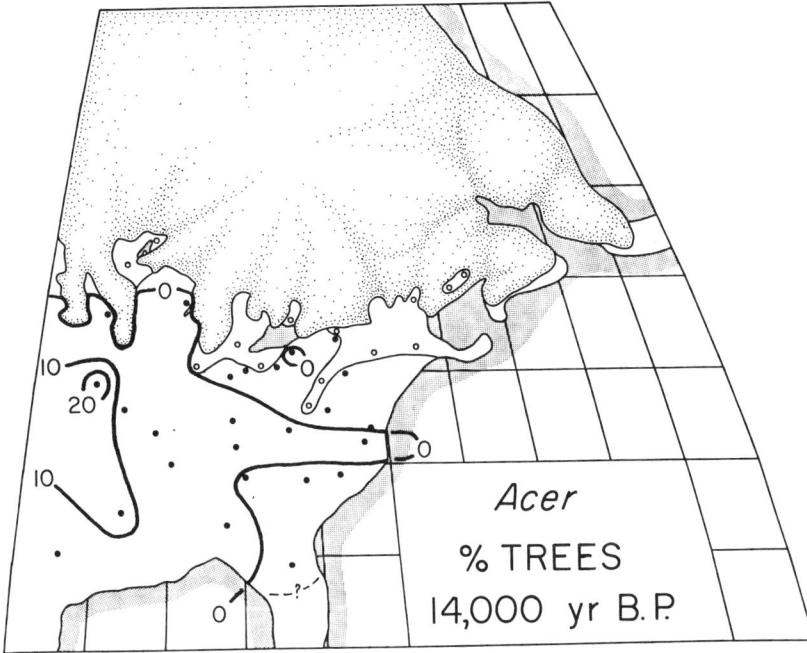

Figure 5.4. Maple (*Acer*). e. Paleo-dominance map for 14,000 yr B.P.

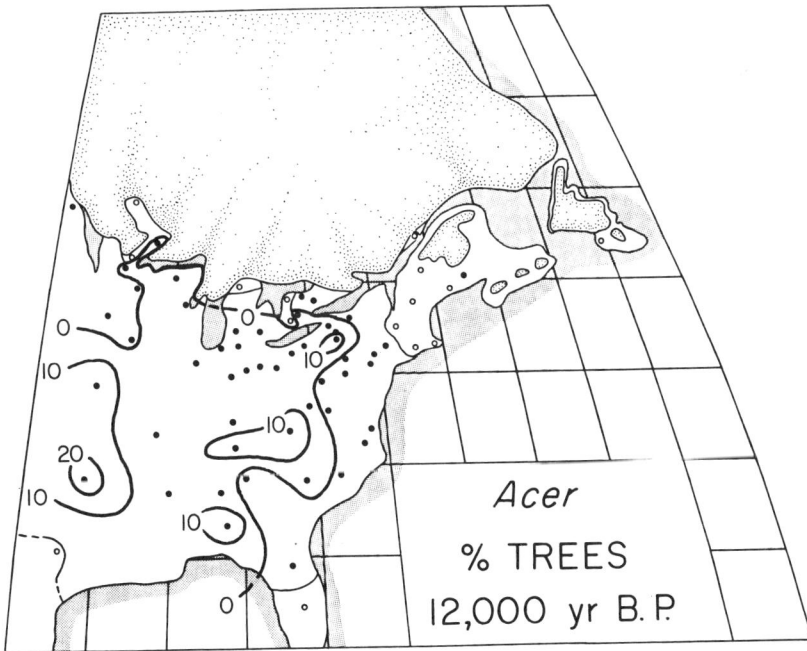

Figure 5.4. Maple (*Acer*). f. Paleo-dominance map for 12,000 yr B.P.

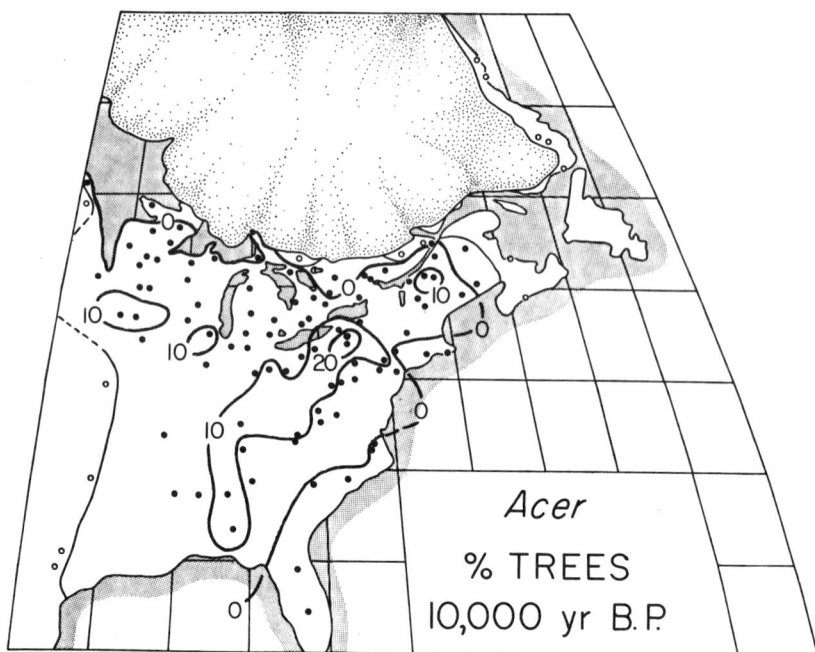

Figure 5.4. Maple (*Acer*). g. Paleo-dominance map for 10,000 yr B.P.

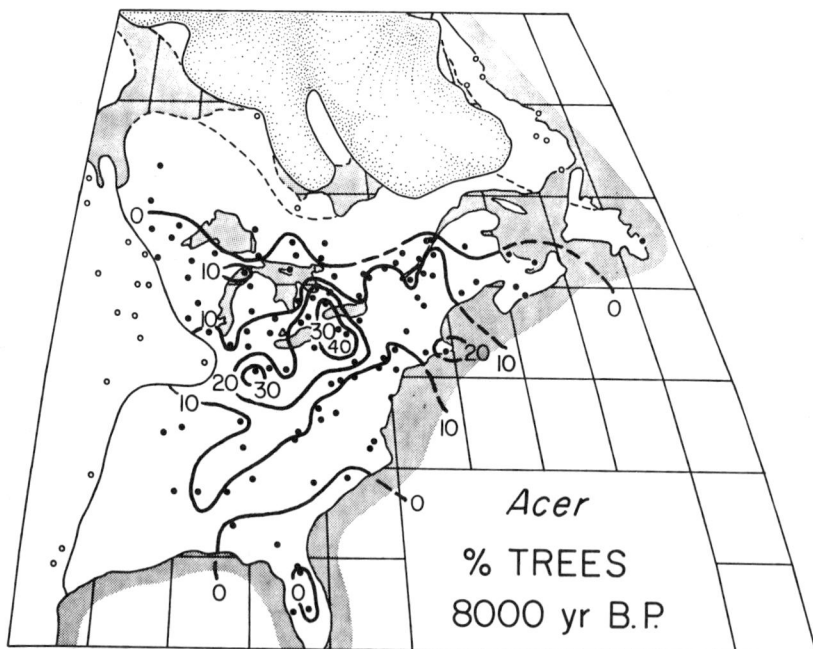

Figure 5.4. Maple (*Acer*). h. Paleo-dominance map for 8000 yr B.P.

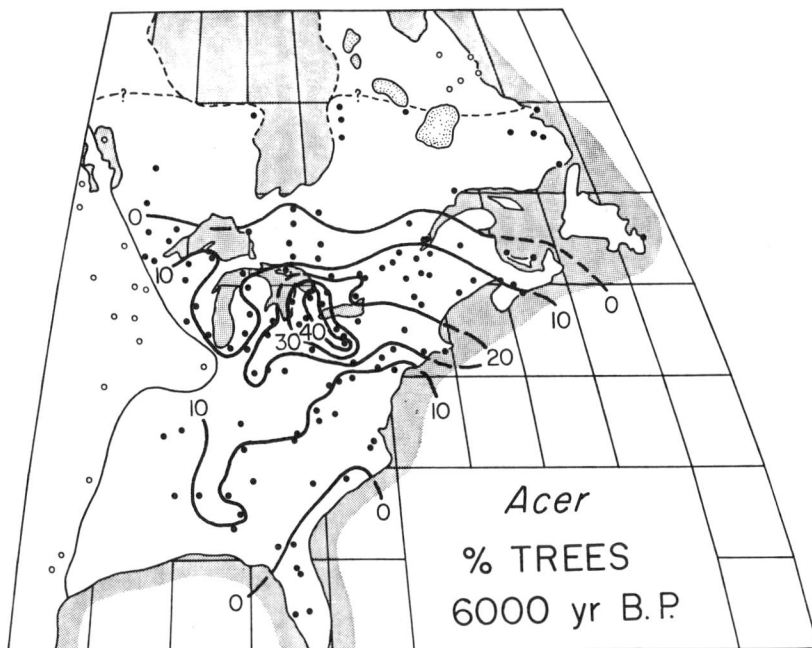

Figure 5.4. Maple (*Acer*). i. Paleo-dominance map for 6000 yr B.P.

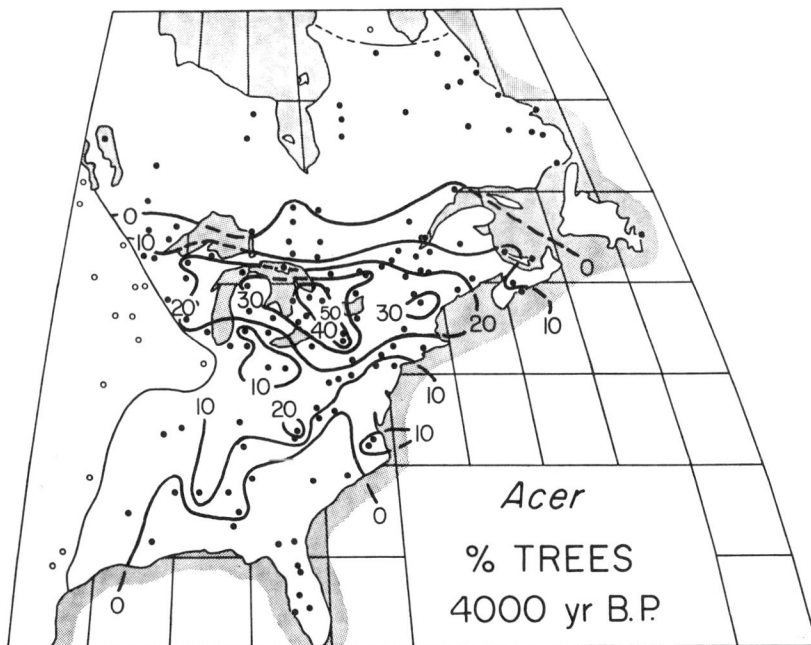

Figure 5.4. Maple (*Acer*). j. Paleo-dominance map for 4000 yr B.P.

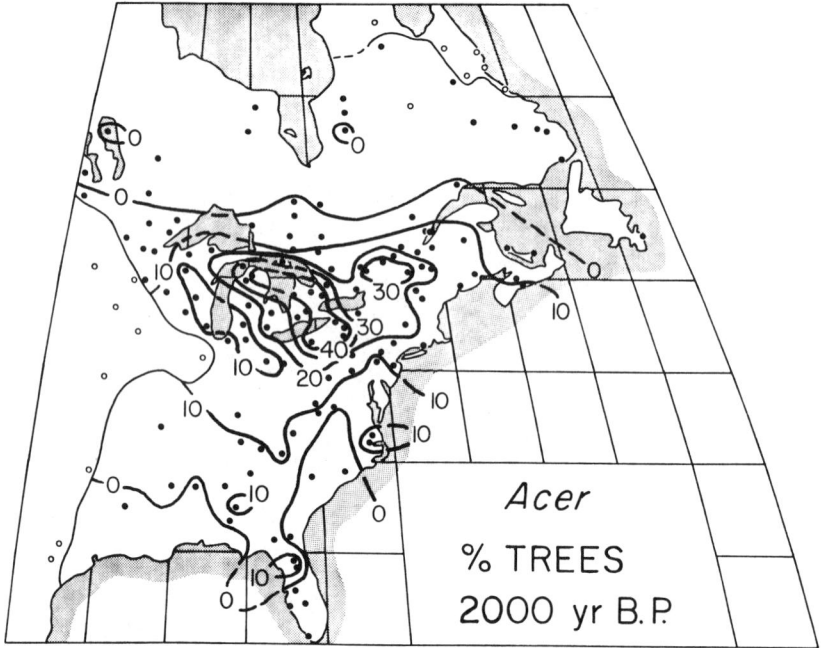

Figure 5.4. Maple (*Acer*). k. Paleo-dominance map for 2000 yr B.P.

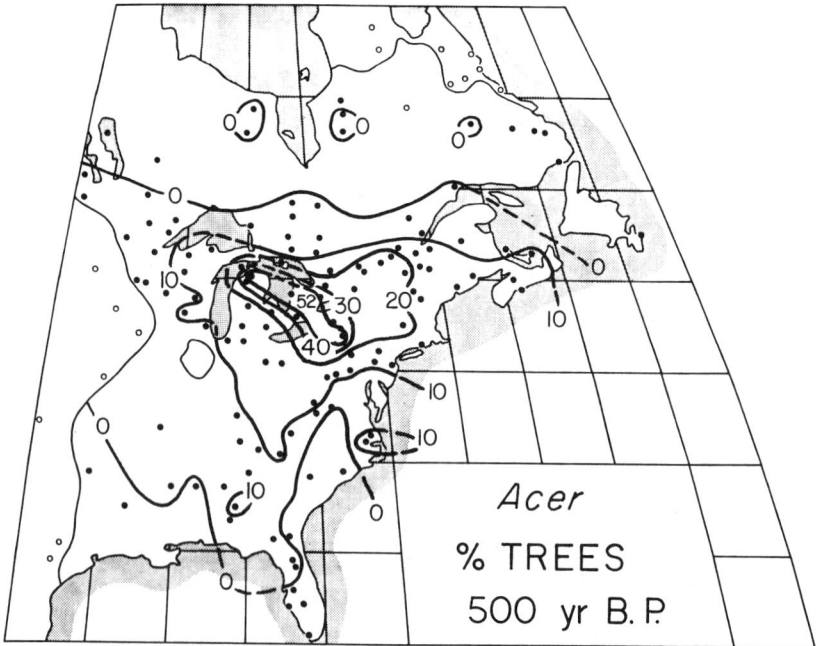

Figure 5.4. Maple (*Acer*). l. Paleo-dominance map for 500 yr B.P.

increased to approximately 10%. The increase to a peak value of 71% in post-settlement times occurred at only one site and is a local situation not representative of regional trends.

During the full-glacial interval, the paleo-dominance maps for maple at 20,000 yr B.P. (Fig. 5.4b) and 18,000 yr B.P. (Fig. 5.4c) illustrate that although maple species were probably widespread throughout the region between 25°N and 38°N, they occurred in low numbers that were close to the lower limit of resolution for paleoecological reconstruction. The map of maple for 20,000 yr B.P. (Fig. 5.4b) identifies one population of about 5% dominance, located near Rockyhock Bay in northeastern North Carolina (Whitehead 1981). The map for 18,000 yr B.P. (Fig. 5.4c) includes additional localities for maple, with dominance values up to 9%, distributed from central Kentucky south to the Gulf of Mexico and across the western Gulf Coastal Plain.

The 16,000 yr B.P. map for maple (Fig. 5.4d) delineates populations of maple that occurred principally in the southwestern quadrant of the mapped region, south of 37°N and west of 87°W. One population center of 11% dominance was located in north-central Louisiana (Kolb and Fredlund 1981). Maple populations began to spread northward and eastward by 15,000 yr B.P., reaching the glacial margin by 14,000 yr B.P. (Fig. 5.4e). Maple also extended eastward across the Interior Low Plateaus, southern Appalachians, and into the central Atlantic Coastal Plain. The population center expanded northwestward from Louisiana, reaching a maximum of 23% reconstructed in northeastern Kansas.

By 12,000 yr B.P. (Fig. 5.4f), maple had extended northeastward through the central and northern Appalachians. The northern perimeter of its distribution extended from the shore of the Champlain Sea in western New York across southern Ontario and central Michigan, northern Wisconsin, and central Minnesota. Minor population centers of > 10% dominance were located (1) in western New York and adjacent Pennsylvania; (2) from southwestern Virginia to Middle Tennessee; and (3) in central Alabama. The main center of population for *Acer* (from 10% to 21%) was located from Kansas to Oklahoma.

By 10,000 yr B.P. (Fig. 5.4g), maple populations had invaded north to within several hundred kilometers of the northern forest limit, limited to the northwest by proglacial lakes such as Glacial Lake Agassiz and to the north by the margin of the retreating Laurentide Ice Sheet. Maple extended to the northeast nearly to the tundra border in coastal Quebec. Southeast of the Champlain Sea, maple advanced into forests of north-central and eastern Maine. Between 12,000 yr B.P. and 10,000 yr B.P., the western limit for maple retracted as prairie expanded northward from central Texas into eastern Nebraska. The western population center diminished in dominance to between 10% and 16% and shifted in location to northern Iowa and southern Minnesota. In contrast, the former small and isolated eastern centers expanded and formed one continuous center of 10% to 26%, located from Alabama north to Tennessee and Kentucky, then northeast across Ohio and West Virginia to southwestern New York. Maples also dispersed eastward onto the eastern Gulf Coastal Plain, as well as the central and northern Atlantic Seaboard.

The 8000 yr B.P. paleo-dominance map for maple (Fig. 5.4h) exhibits sub-

Acer

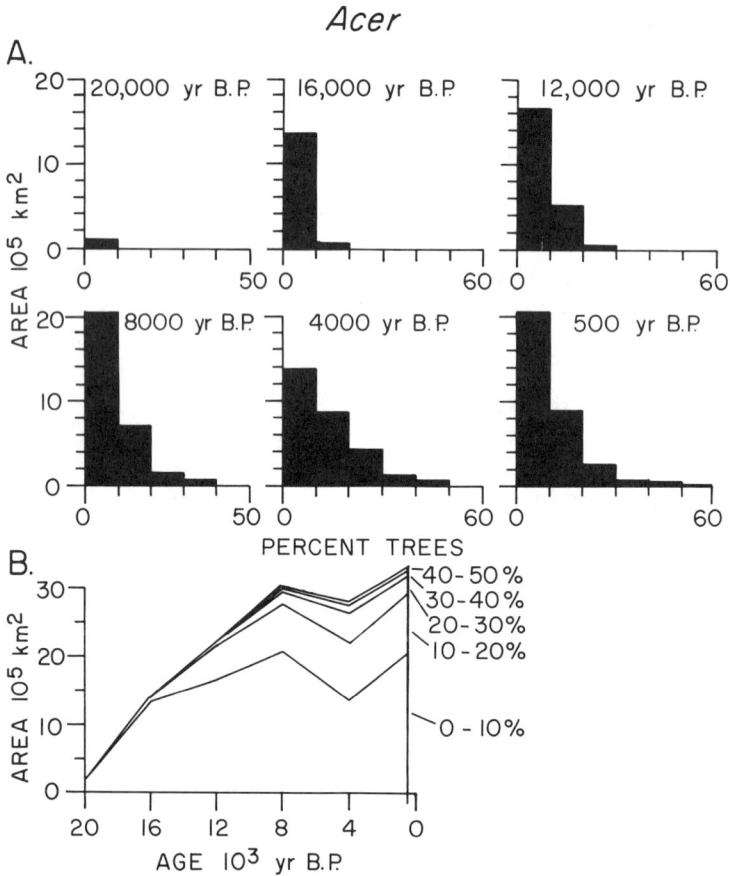

Figure 5.5. Changes in area and dominance of maple (*Acer*). a. Histograms of area–dominance structure at 20,000, 16,000, 12,000, 8000, 4000, and 500 yr B.P. b. Cumulative plot from 20,000 yr B.P. to the present.

stantial increases in population abundances within the central and northeastern portion of its mid-Holocene distribution. The boundaries of the prominent eastern center shifted to the north across Indiana, southern Michigan, and southern Ontario and northeastward across New York from Connecticut into Maine. Peak maple populations occurred in western Ohio (up to 32% dominance) and southwestern New York (up to 48%). The northern border of maple distribution advanced to approximately 47°N from Nova Scotia to eastern Ontario and, west of Lake Superior, to approximately 49°N in northern Minnesota and western Ontario. The western limit of maple coincided with the prairie–forest boundary. A restricted outlier of maple colonized central Florida.

By 6000 yr B.P. (Fig. 5.4i), maple had spread north to about 49°N. The primary population center of > 20% maple, however, was distributed in the latitudinal band from about 41°N to 45°N from southern New England into the central Great Lakes region. More limited populations of 10% to 20% maple grew from northern New England across southern Quebec and Ontario to Wisconsin, south to Mississippi and Alabama, and across Kentucky and Ohio to Pennsylvania.

The maple paleo-dominance map for 4000 yr B.P. (Fig. 5.4j) displays the expansion of the principal center of dominance northeastward across northern New England and into the western Great Lakes region. By 4000 yr B.P., maple populations became more fragmented within the southern Atlantic and central Gulf Coastal Plains. During the late Holocene, maps of maple for 2000 yr B.P. (Fig. 5.4k) and 500 yr B.P. (Fig. 5.4l) show a retraction of its range along the southwestern margin but expansion along its western and northwestern limits. The major population center for maple, with 20% to 50% dominance, remained in the central and eastern Great Lakes region through the middle and late Holocene.

The full-glacial dominance–structure histogram for maple (Fig 5.5a) documents a limited areal extent (in the range between $1.3 \cdot 10^5$ km^2 measured from evidence available at 20,000 yr B.P. and about $15 \cdot 10^5$ km^2 at 18,000 yr B.P.) with dominance values < 10%. However, through the late-glacial and Holocene intervals, the area occupied and number of dominance classes increased progressively to maximum values at 500 yr B.P. (Fig. 5.5b). At its interglacial maximum 500 yr B.P., maple occupied $32.9 \cdot 10^5$ km^2. The lowest dominance class (< 10%) covered 61% of the total distributional area. The 10% to 20% dominance class represented 27% of the area, and the primary population center greater than 20% dominance comprised the remaining 12% (Fig. 5.5a).

Birch (*Betula*)

The birch trajectory (Fig. 5.6a) displays a quasi-sinusoidal curve for population means. Initially, full-glacial values for mean dominance of birch were relatively low (12% based on one site at 20,000 yr B.P. and 3% at both 19,000 yr B.P. and 18,000 yr B.P.). Mean values declined through a relative trough in the early late-glacial interval (the lowest mean of 0.4% occurred at 16,000 yr B.P.) but rose to a crest during the interglacial of 13% (at both 4500 yr B.P. and 3500 yr B.P.) before falling to about 10% by 500 yr B.P. The maximum birch values were relatively low during the full-glacial interval, ranging between 5% and 12%. During the late-glacial interval, maximum values rose from 1% at 16,000 yr B.P. to 10% at 13,000 yr B.P. The highest dominance values reconstructed for birch continued to increase during the early Holocene, reaching 19% at 12,000 yr B.P. and peaking at 61% at 11,000 yr B.P. Birch maxima diminished to 27% at 8000 yr B.P., increased and fluctuated to reach a high of 69% at 4500 yr B.P., and then converged toward about 40% during the late Holocene. The times of relatively accentuated population variability (*i.e.*, ± 1 SD ≥ 8%) occurred during the Holocene at about 11,000 yr B.P. as well as during the last 7000 years.

The *Betula* paleo-dominance maps for 20,000 yr B.P. (Fig. 5.6b) and 18,000 yr B.P. (Fig. 5.6c) highlight one full-glacial refugium for tree birches (with up

Betula

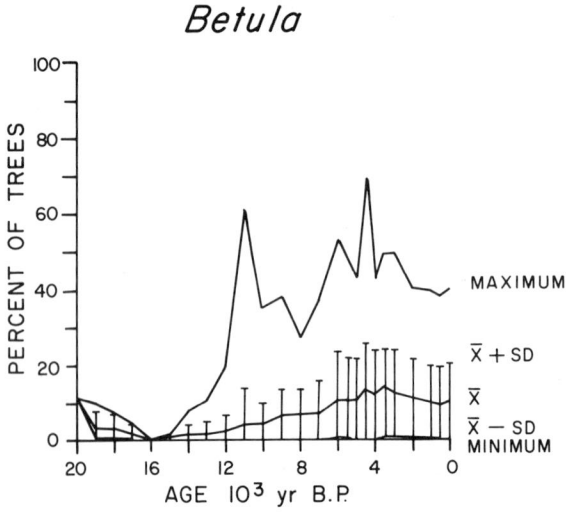

Figure 5.6. Birch (*Betula*). a. Population trajectory, including maximum, minimum, and mean values of paleo-dominance as well as ± 1 SD of the mean, for the past 20,000 years.

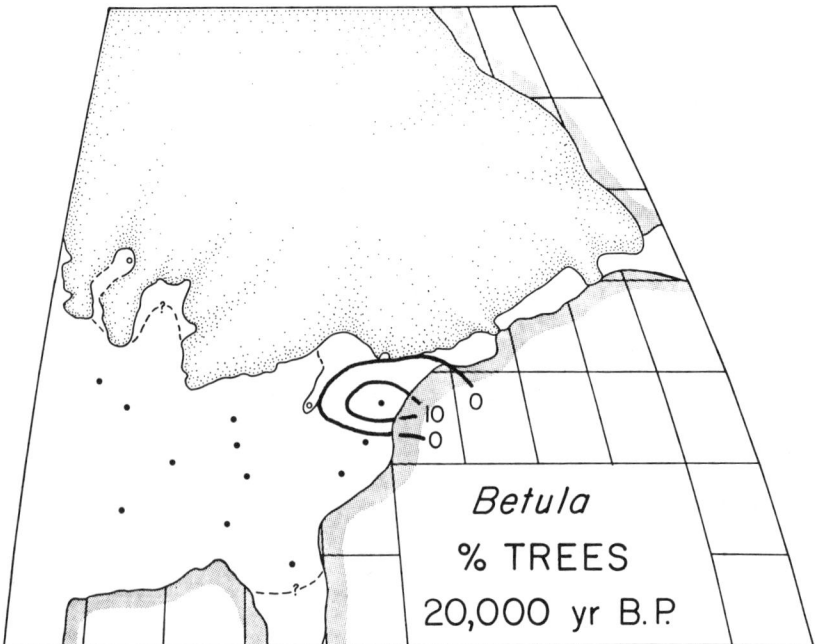

Figure 5.6. Birch (*Betula*). b. Paleo-dominance map for 20,000 yr B.P.

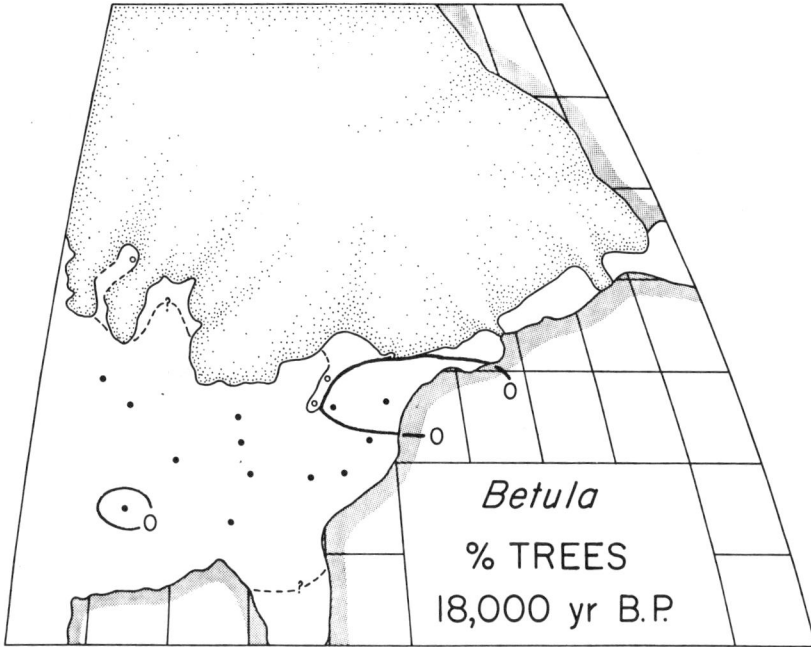

Figure 5.6. Birch (*Betula*). c. Paleo-dominance map for 18,000 yr B.P.

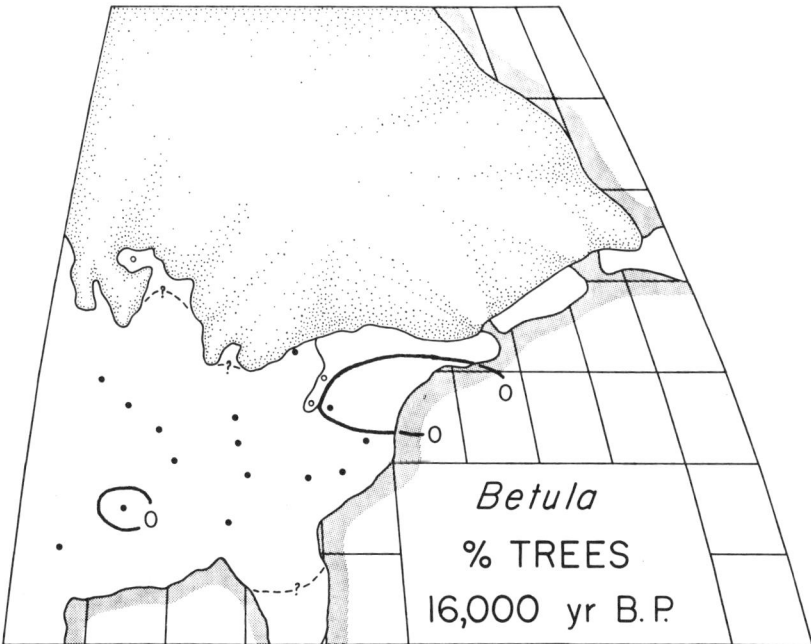

Figure 5.6. Birch (*Betula*). d. Paleo-dominance map for 16,000 yr B.P.

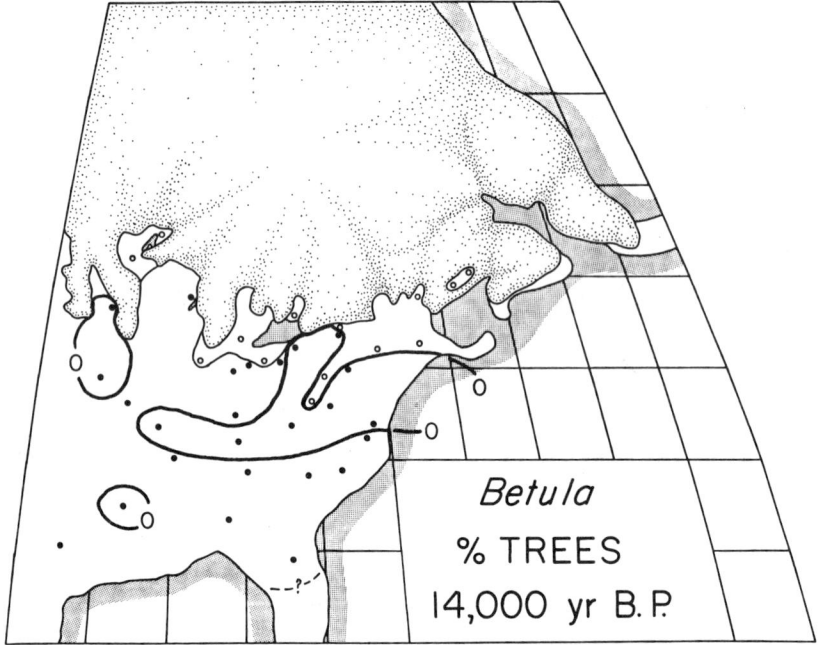

Figure 5.6. Birch (*Betula*). e. Paleo-dominance map for 14,000 yr B.P.

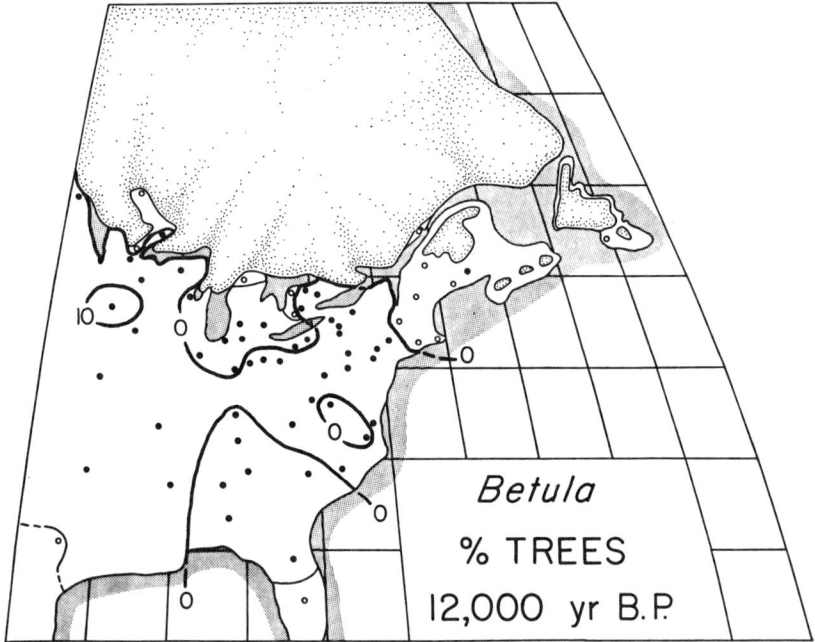

Figure 5.6. Birch (*Betula*). f. Paleo-dominance map for 12,000 yr B.P.

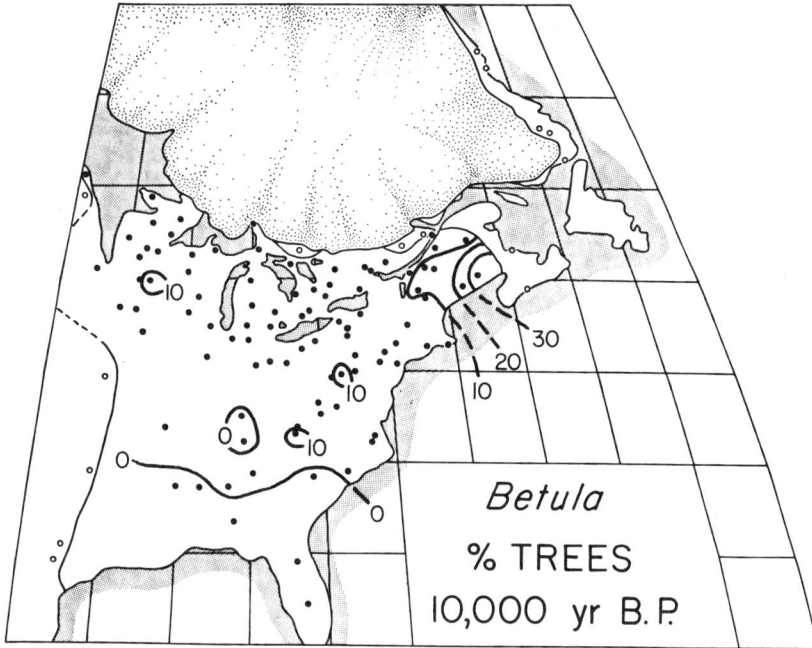

Figure 5.6. Birch (*Betula*). g. Paleo-dominance map for 10,000 yr B.P.

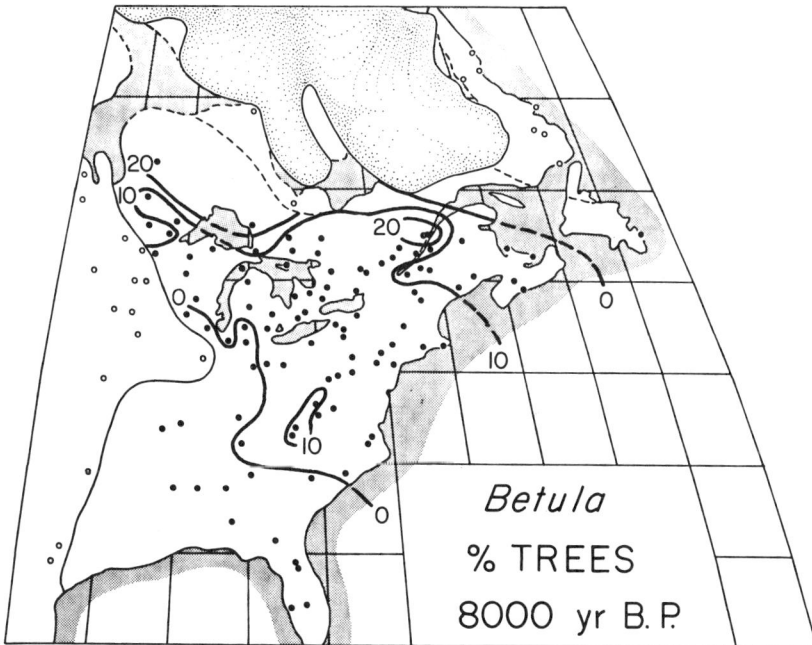

Figure 5.6. Birch (*Betula*). h. Paleo-dominance map for 8000 yr B.P.

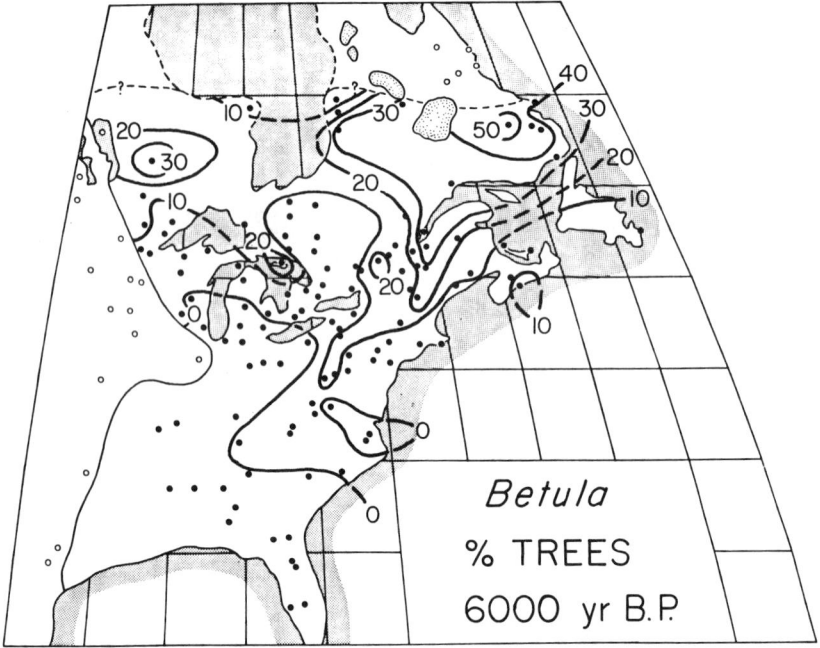

Figure 5.6. Birch (*Betula*). i. Paleo-dominance map for 6000 yr B.P.

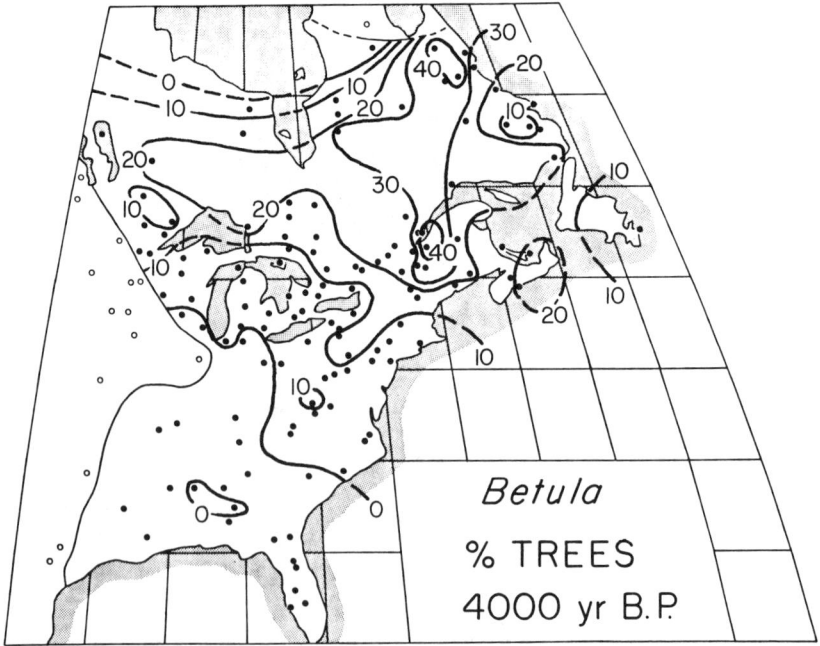

Figure 5.6. Birch (*Betula*). j. Paleo-dominance map for 4000 yr B.P.

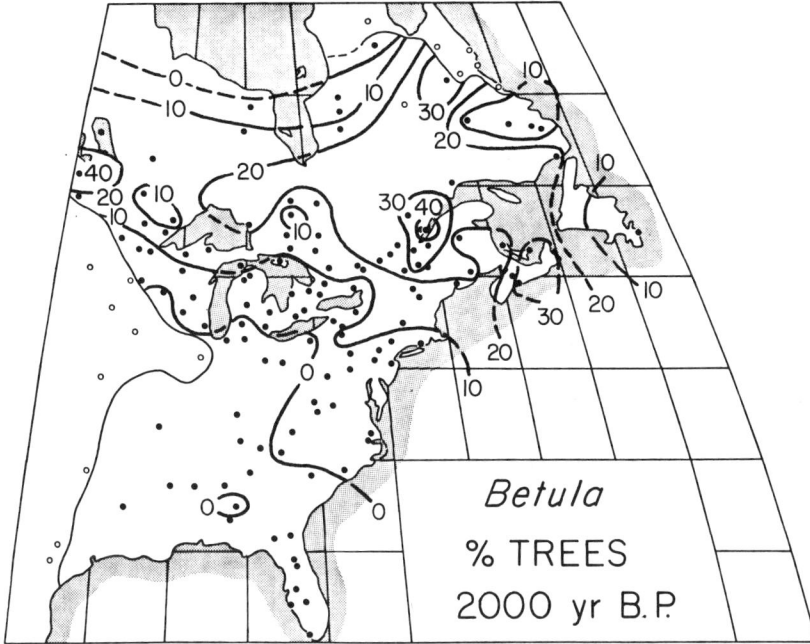

Figure 5.6. Birch (*Betula*). k. Paleo-dominance map for 2000 yr B.P.

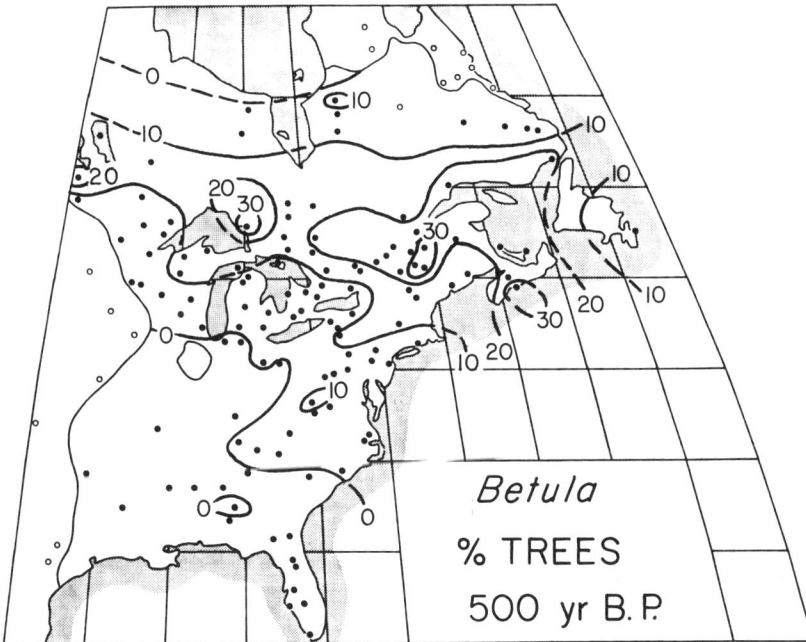

Figure 5.6. Birch (*Betula*). l. Paleo-dominance map for 500 yr B.P.

to 12% dominance) centered along the Delmarva Peninsula of eastern Maryland, Delaware, and Virginia. This portion of the central Atlantic Seaboard was occupied by relatively open boreal forest within a periglacial landscape characterized by considerable sand-dune activity (Sirkin *et al.* 1970, 1977). In addition, the 18,000 yr B.P. map delineates a second minor tree-birch population (bounded by the 0% isophyte contour) situated in north-central Louisiana (Kolb and Fredlund, 1981) along the ecotone between boreal forest to the north and temperate forests to the south (Fig. 4.3a). This dual, full-glacial pattern of restricted birch populations occurring in both southwestern and northeastern populations persisted until 16,000 yr B.P., the early part of the late-glacial interval (Fig. 5.6d).

By 14,000 yr B.P. (Fig. 5.6e), birch dispersed widely to establish three discrete populations. The northeastern population expanded westward from the central Atlantic Coastal Plain and occupied lower and middle elevation slopes of the central and southern Appalachian Highlands. This population, representing < 8% of the forest composition, occupied habitats from Virginia north to Pennsylvania and west to southeastern Missouri and eastern Ohio, respectively. The southwestern population persisted in the western Gulf Coastal Plain of north-central Louisiana. The third, northwestern birch population grew as a minor constituent of the boreal forest from eastern Kansas to western Iowa (Fig. 4.3b).

The scattered late-glacial birch populations continued to expand in area and coalesced, forming one continuous range by 12,000 yr B.P. (Fig. 5.6f). At this early-Holocene time, birch colonized south along the Atlantic Coastal Plain through North Carolina and north to the forest–tundra border in New York. The northern limit for tree birch coincided with the northern forest boundary in New York and southern Ontario. However, by 12,000 yr B.P., forest occurred as far north as 44°N throughout the central Great Lakes region. In contrast, the northern limit of birch swept in an arc south from southern Ontario, across northern Ohio, Indiana, and Illinois, returning northward to the forest border along the glacial margin in northern Wisconsin, Minnesota, and Manitoba. Except in Iowa, birch populations generally remained less than 10% of the overall forest composition. In the western study area, birch trees grew from the shores of the Gulf of Mexico north to the edge of the Laurentide Ice Sheet in central Canada. However, birch had not effectively colonized the southeastern quadrant of unglaciated North America south of about 34°N and east of 88°W.

From 12,000 yr B.P. to 10,000 yr B.P. (Fig. 5.6g), birch trees established populations north to the periphery of the boreal forest. By 10,000 yr B.P., the leading edge of tree-birch populations advanced to about 47°N in the northern Appalachians, to 46°N in the central Great Lakes region, and, west of Lake Superior, invaded to approximately 49°N where it was halted by the migrational barrier imposed by Glacial Lake Agassiz. The primary area of abundant tree birch (10% to 35% dominance) was mapped across the New England states and into the adjacent Canadian Province of New Brunswick. Three additional population centers for birch (with values between 10% and 20%) grew in the southern and central Appalachian Mountains, as well as in the western region of the Great Lakes. The western and southwestern borders of birch distribution retracted as prairie displaced forest in the continental interior. On its southern margin, birch generally retreated north to about 34°N 10,000 years ago.

The birch paleo-dominance map for 8000 yr B.P. (Fig. 5.6h) exhibits an asymmetrical latitudinal pattern with highest dominance levels reconstructed along its northern range limit and its dominance typically dropping to much lower values (< 10%) throughout the southern two-thirds of its distribution. By 8000 yr B.P., key population centers for birch (20% to 27%) occupied two regions: (1) the forested uplands adjacent to the St. Lawrence River Valley in southeastern Quebec; and (2) forests of western and central Ontario, north of Lake Superior. Bordering these two centers on their south, a diagonal band of 10% to 20% birch extended from Nova Scotia, New Brunswick, and Maine, mapped across southern Quebec and southern Ontario and into northern Minnesota. The montane birch center was elongated southwest–northeast along the axis of the southern Appalachian Highlands. On its southern periphery of continuous occurrence, birch persisted at 34°N east of 85°W; however, west of this meridian, the margin of birch was displaced 9° latitude, to about 43°N across the central and western Great Lakes region. Between 10,000 yr B.P. and 8000 yr B.P., the western margin of birch shifted as much as 7° latitude northward and about 6° longitude eastward, the result of the early- and mid-Holocene extension of grassland through the Midwestern Prairie Peninsula.

The postglacial melting of the Laurentide Ice Sheet was nearly complete by 6000 yr B.P. (Fig. 5.6i), with only remnants of stagnant ice masses stranded on high bedrock plateaus in northeastern Canada. The asymmetry in birch's latitudinal pattern of paleo-dominance intensified as birch populations (probably both tree- and shrub-birch species) dispersed into tundra environments of deglaciated central and eastern Canada. A prominent east–west crest of high values for birch dominance occupied the latitudinal zone between about 50°N and 55°N; Hudson Bay separated this northern "ridge" in dominance into two primary population centers for birch. The northwestern birch center (20% to 31%) constituted the arboreal vanguard that colonized western, then northern Ontario during the mid-Holocene interval. The primary northeastern center (20% to 53% dominance) expanded in area from its position in the northernmost Appalachians and St. Lawrence River Valley at 8000 yr B.P. across central Quebec and southern Labrador by 6000 yr B.P. In northeastern Canada, the highest reconstructed values of birch (35% to 53%) bordered the tundra–forest ecotone as birch populations dispersed airborne seeds, then established successful colonies along the retreating margins of the melting blocks of glacial ice in the Ungava Plateau.

With mid-Holocene isostatic uplift of the Hudson Bay region (Andrews 1982), Hudson Bay shrank markedly, abandoning former shoreline positions between 6000 yr B.P. and 4000 yr B.P. Among other tree species, birches invaded the landscape abandoned by the sea. By 4000 yr B.P. (Fig. 5.6j), the 20% contour traced the latitudinal band of birch dominance westward from northern Ontario across west-central Ontario, then curved southeastward through southern Quebec and New England to the shores of the North Atlantic Ocean at about 45°N. The continuous extension of the 20% contour from 55°W to 95°W enveloped both primary northwestern and northeastern population centers by 4000 yr B.P. As shown on the birch maps for 8000 yr B.P. and 6000 yr B.P., the western birch limit was coincident with the prairie–forest ecotone; western birch populations contributed about 13% of the forest composition from southern Manitoba to central Minnesota and comprised less than 5% of the forests of western and southern Wisconsin and southern Michigan. The southern continuous border of birch delineated the birch population within montane areas of the Appalachians and in the central Atlantic Seaboard south to about 34°N.

The birch map for 4000 yr B.P. illustrates outlier, minor populations (< 3%) known from sites located within alluvial valleys of Mississippi (Whitehead and Sheehan 1985) and Alabama (Markewich and Christopher 1982). Based upon this palynological evidence for birch populations and species determinations of late-Holocene seeds of river birch (*Betula nigra*) from southeastern Louisiana (Delcourt and Delcourt 1977b), limited birch populations, including at least river birch, occupied late-Holocene alluvial bottomlands of major river systems traversing the Gulf and southern Atlantic Coastal Plains.

The birch paleo-dominance maps for 2000 yr B.P. (Fig. 5.6k) and 500 yr B.P. (Fig. 5.6l) illustrate that, during the last several thousand years, there has been a progressive reduction in dominance of birch along the northeastern tundra–forest ecotone. As the tundra reexpanded southward, particularly along coastal Labrador (Short and Nichols 1977; Lamb 1980, 1984), the 20% contour shifted from its northern position of 57°N at 2000 yr B.P. to about 52°N at 500 yr B.P. As birch populations diminished and became locally extinct along their extreme northeastern margin, they advanced westward, colonizing former prairie landscapes along their western and southwestern margins. For example, birch invaded into southernmost Manitoba, then increased in population size up to 40% at 2000 yr B.P., then declined to 23% at 500 yr B.P. (Ritchie 1964, 1969). By 500 yr B.P., the latitudinal band of primary birch dominance had constricted to generally between 44°N and 52°N. Three presettlement centers represented > 20% dominance of birch. These included the late-Holocene development of the Manitoba population, the birch population of 20% to 33% northeast of Lake Superior in central Ontario, and the largest center in terms of both area and dominance (20% to 39%) in southern and southeastern Quebec, northern Maine, New Brunswick, Prince Edward Island, and Nova Scotia.

The histograms of dominance structure for birch (Fig. 5.7a) display few (1 to 2) dominance classes during the late Pleistocene and early Holocene, and consistently more classes (from 3 to 5) throughout the mid- and late-Holocene intervals. The lowest dominance class (0% to 10%) comprised 70%, 100%, and 98% of the total area mapped for birch at 20,000 yr B.P., 16,000 yr B.P., and 12,000 yr B.P., respectively. Although consistently < 20% from full-glacial through early-Holocene times, birch populations enlarged their area from $2.2 \cdot 10^5$ km^2 at 20,000 yr B.P. to $3.1 \cdot 10^5$ km^2 at 16,000 yr B.P., and then to $28.1 \cdot 10^5$ km^2 at 12,000 yr B.P. This represented an increase of nearly 1300% area occupied over that 8000-year period (Fig. 5.7b). However, during the Holocene, the total area of birch distribution was nearly the same from 12,000 yr

Betula

A.

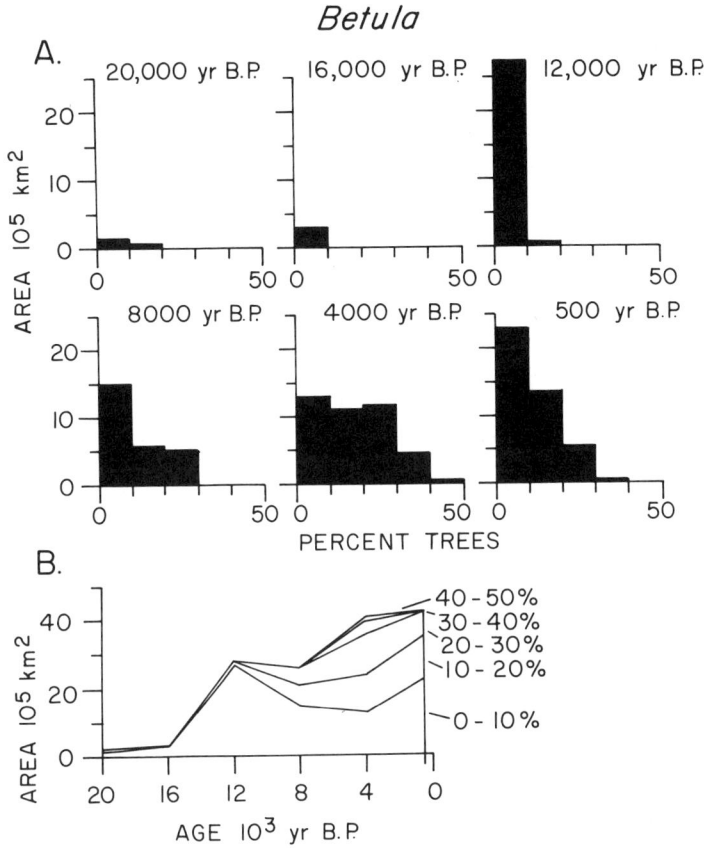

Figure 5.7. Changes in area and dominance of birch (*Betula*). a. Histograms of area–dominance structure at 20,000, 16,000, 12,000, 8000, 4000, and 500 yr B.P. b. Cumulative plot from 20,000 yr B.P. to the present.

B.P. to 8000 yr B.P. (approximately $26.1 \cdot 10^5$ km²), and then increased to about $42.5 \cdot 10^5$ km² at 500 yr B.P. (overall, an increase in area of about 150% from 12,000 yr B.P. to 500 yr B.P.; Fig. 5.7b). The area–dominance histograms for 12,000 yr B.P., 8000 yr B.P., and 4000 yr B.P. (Fig. 5.7a) document the progressive decline in area for the lowest birch–dominance class. Coincident areal increases for higher classes document the increased abundance of birch populations as they became subdominant and dominant components of Holocene boreal forest and mixed conifer–northern hardwoods forest. From 4000 yr B.P. to 500 yr B.P., the population collapse of the northeastern birch center reduced the areal extent of late-Holocene dominance classes > 20%. The late-Holocene population readjustments to lowered birch levels (< 10%) adjacent to the tundra–forest ecotone and the southwestern colonization of new lands along the prairie–

forest ecotone resulted in increased area for the lowest (< 10%) dominance class on the 500 yr B.P. histogram.

Hickory (*Carya*)

The trajectory plot for hickory populations (Fig. 5.8a) exhibits a long-term decline in mean values from about 9% dominance in the full-glacial interval, shifting to 7% for the late-glacial and early-Holocene intervals, then remaining between 5% and 6% for the rest of the Holocene. Maximum reconstructed values for hickory remained between approximately 22% and 26% from 20,000 yr B.P. until 15,000 yr B.P. However, during the last 14,000 years, maximum values for hickory fluctuated widely between a low of 19% at 8000 yr B.P. and the interglacial extreme of 45% at 1000 yr B.P. The population variability in dominance for hickory decreased after 10,000 yr B.P.; from 20,000 yr B.P. to 10,000 yr B.P., ± 1 SD ranged between 8% and 11%. The variability about mean dominance ranged between 5% and 7% during the past 9000 years.

The time series of paleo-dominance maps for hickory (Fig. 5.8b–l) reveal additional, complementary insights concerning hickory's dynamics that are not evident in the statistical summary graphed for the population trajectory (Fig. 5.8a). Full-glacial maps for hickory (20,000 yr B.P., Fig. 5.8b; 18,000 yr B.P., Fig. 5.8c) display its populations distributed across the Gulf and southern Atlantic Coastal Plains generally south of 35°N. To the east, 20,000 years ago, hickory increased southward in dominance from marginal northern populations (*ca.* 1%) near Singletary Bay (Frey 1951, 1953; Whitehead 1973) in North Carolina up to 15% hickory in the vicinity of Sheelar Lake in north-central Florida (Watts and Stuiver 1980). However, a steepened southward gradient occurred in the western part of the full-glacial distribution of hickory, increasing from 1% dominance at Nonconnah Creek in southwestern Tennessee (Delcourt *et al.* 1980) to the primary population center located in the western Gulf Coastal Plain [with 26% hickory reconstructed for forests in north-central Louisiana (Kolb and Fredlund 1981)].

Throughout the late-glacial interval, the northern limit of hickory trees coincided with 35°N. The 16,000 yr B.P. map (Fig. 5.8d) portrayed another steepened gradient of isophyte contours highlighting the western limits of hickory abundance in east-central Texas; the western pattern reflected the important addition of the paleoecological sequence from Boriack Bog, a site with a palynological record spanning the last 16,000 years (Bryant 1977). By 14,000 yr B.P. (Fig. 5.8e), limited hickory populations dispersed into the southern Appalachians of southwestern Virginia. As hickory values in Louisiana decreased from 23% at 16,000 yr B.P. to 9% at 14,000 yr B.P., hickory trees expanded in dominance in a new population center in northern Florida (with values of 20% to 34%). The 10% contour line for hickory swept from about 33°N in the Carolinas to about 32°N in Mississippi and Louisiana.

The paleo-dominance map at 12,000 yr B.P. (Fig. 5.8f) illustrates a major shift in hickory's northern boundary during the late-glacial to early-Holocene transition. By 12,000 years ago, hickory populations had dispersed along the

Carya

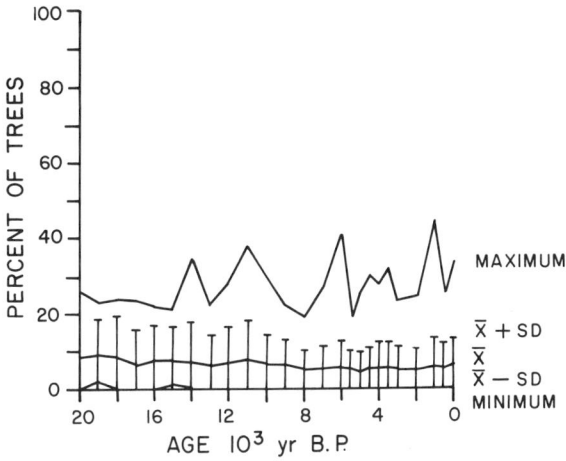

Figure 5.8. Hickory (*Carya*). a. Population trajectory, including maximum, minimum, and mean values of paleo-dominance as well as ± 1 SD of the mean, for the past 20,000 years.

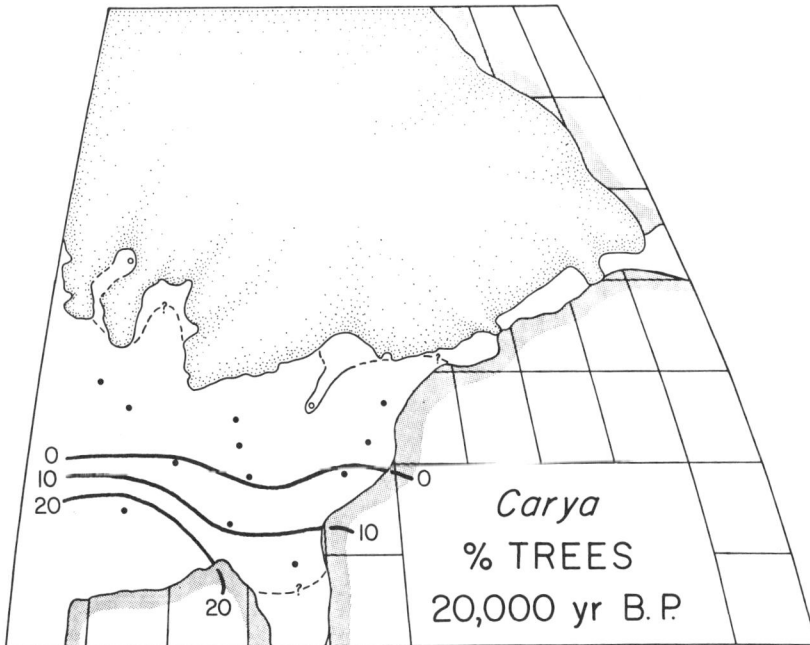

Figure 5.8. Hickory (*Carya*). b. Paleo-dominance map for 20,000 yr B.P.

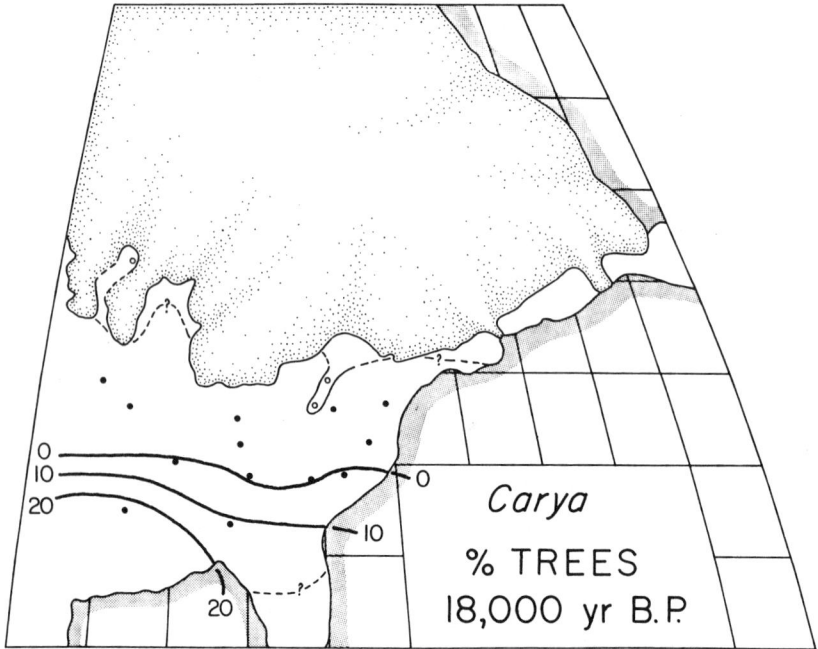

Figure 5.8. Hickory (*Carya*). c. Paleo-dominance map for 18,000 yr B.P.

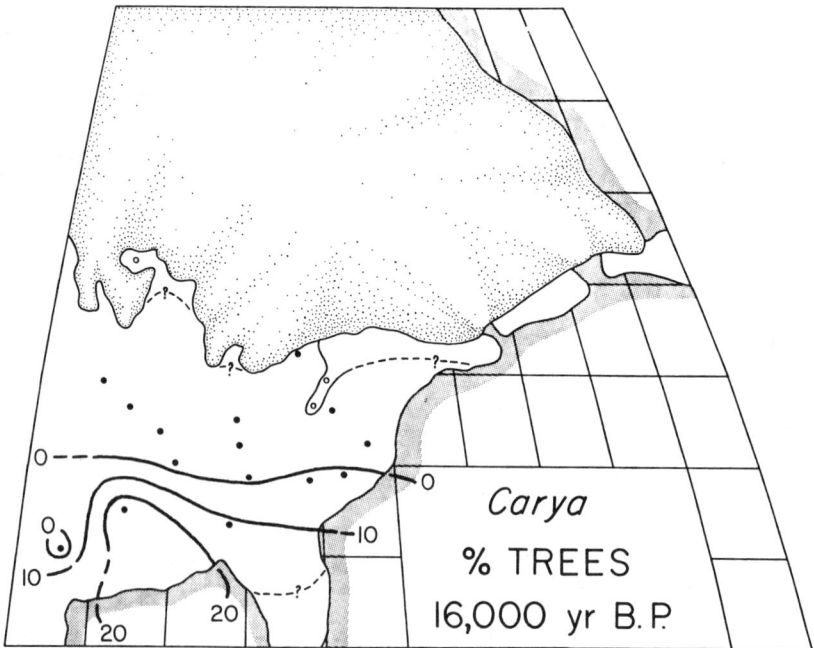

Figure 5.8. Hickory (*Carya*). d. Paleo-dominance map for 16,000 yr B.P.

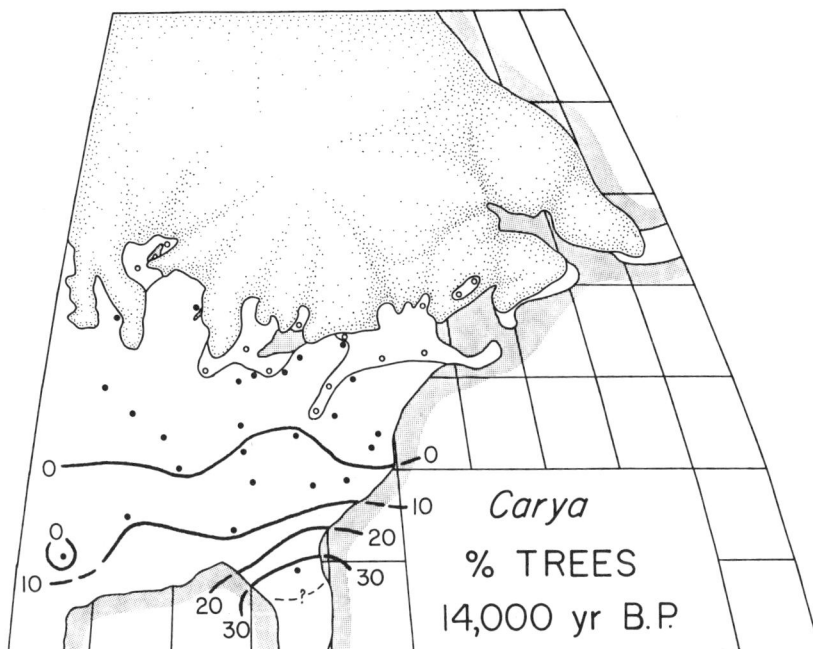

Figure 5.8. Hickory (*Carya*). e. Paleo-dominance map for 14,000 yr B.P.

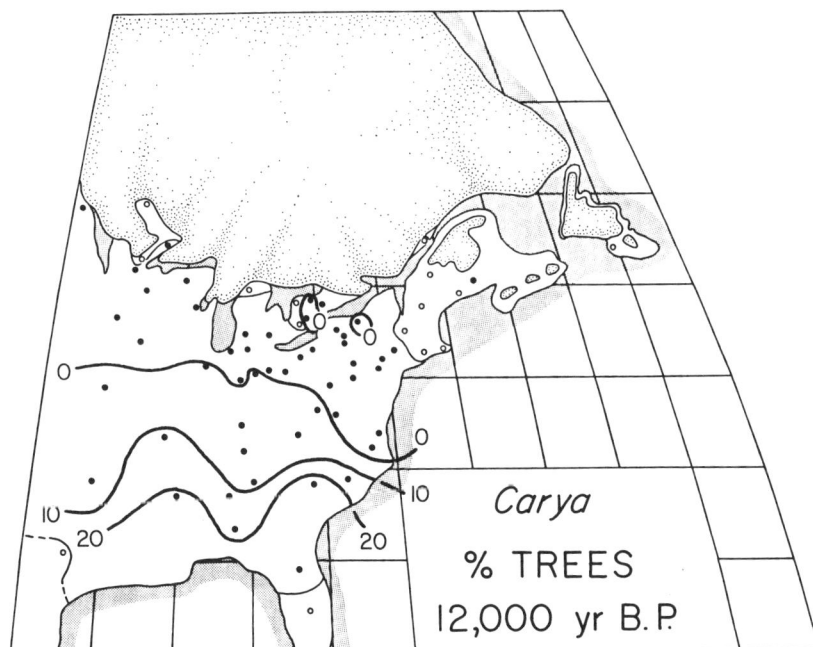

Figure 5.8. Hickory (*Carya*). f. Paleo-dominance map for 12,000 yr B.P.

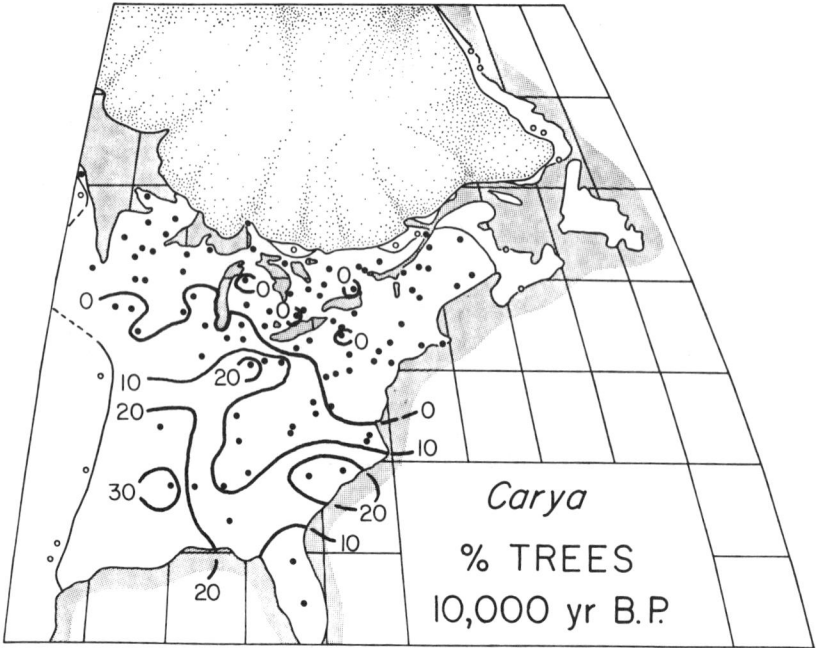

Figure 5.8. Hickory (*Carya*). g. Paleo-dominance map for 10,000 yr B.P.

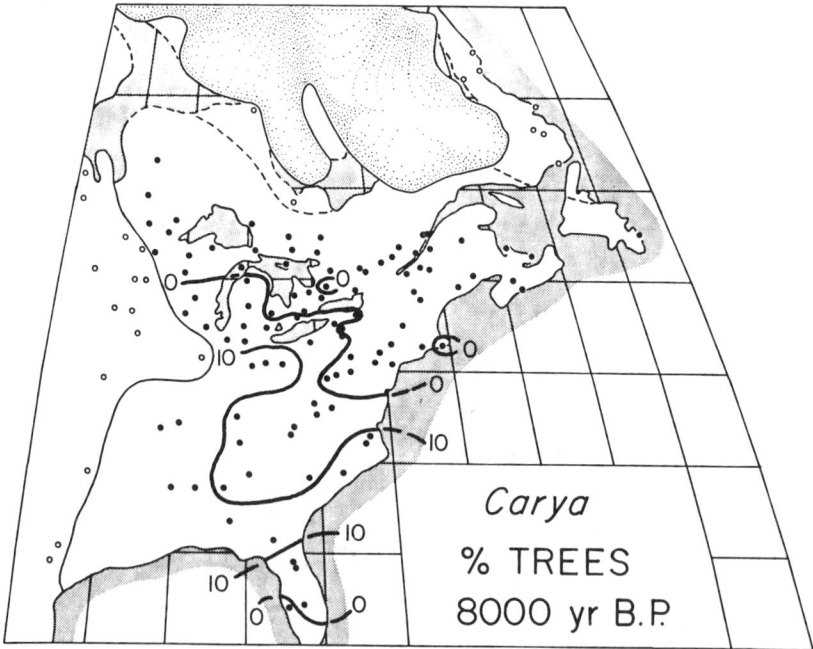

Figure 5.8. Hickory (*Carya*). h. Paleo-dominance map for 8000 yr B.P.

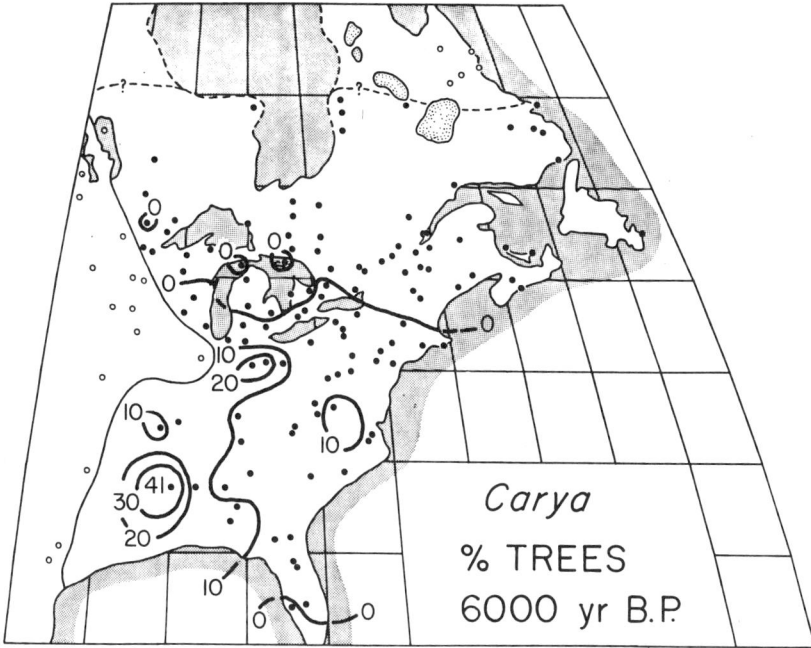

Figure 5.8. Hickory (*Carya*). i. Paleo-dominance map for 6000 yr B.P.

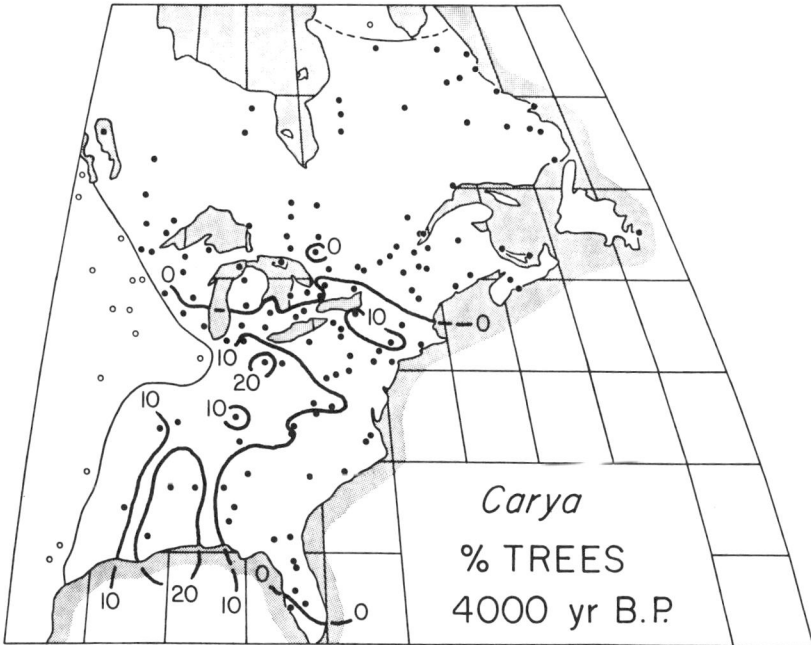

Figure 5.8. Hickory (*Carya*). j. Paleo-dominance map for 4000 yr B.P.

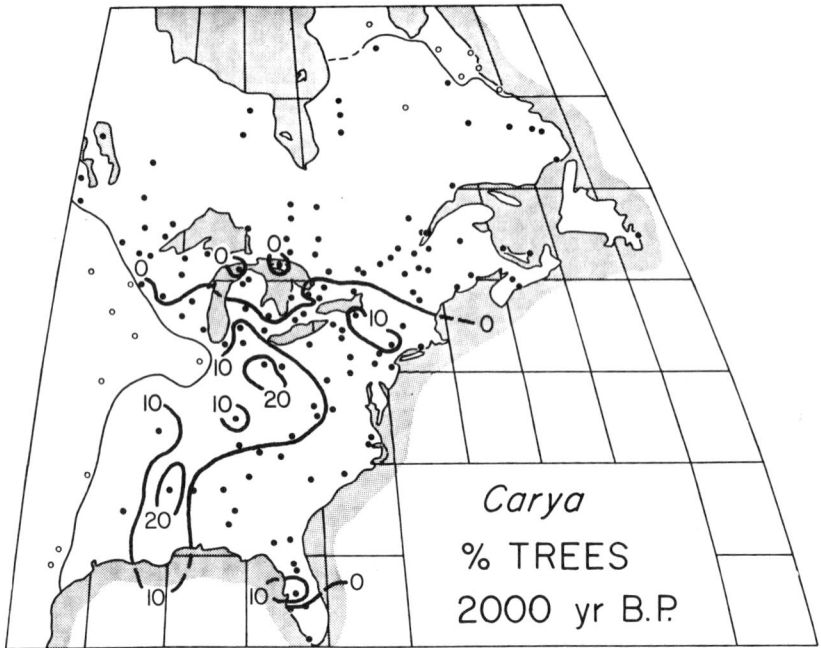

Figure 5.8. Hickory (*Carya*). k. Paleo-dominance map for 2000 yr B.P.

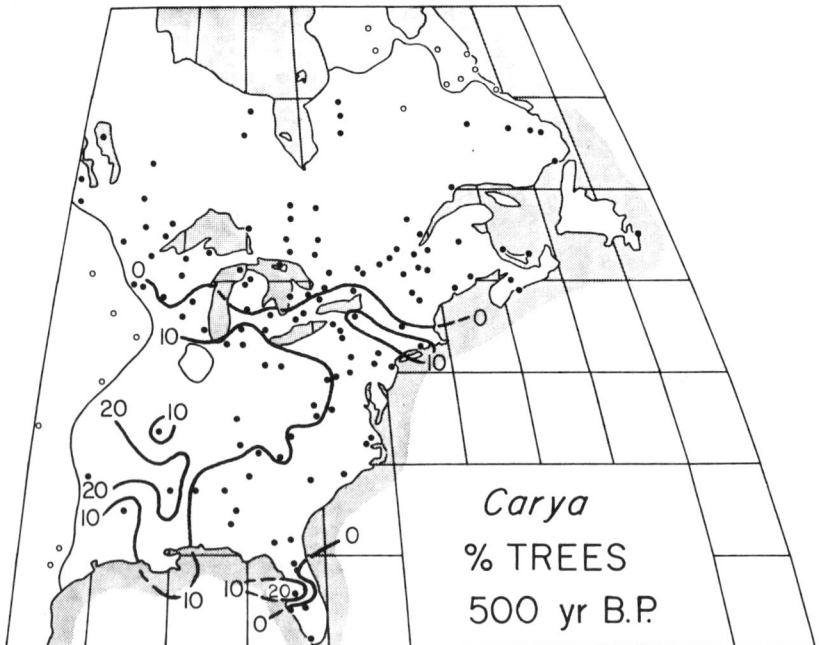

Figure 5.8. Hickory (*Carya*). l. Paleo-dominance map for 500 yr B.P.

Appalachian trend into Maryland, and west along 40°N into Illinois, Iowa, and Kansas. East of the Appalachians, however, hickory populations maintained their northeastern margin in North Carolina. As hickory migrated north, its contoured pattern of dominance retained its latitudinal asymmetry with lowest values (< 10%) generally from 35°N to 40°N and its primary population center (20% to 28%) between 28°N and 34°N. Isolated outliers of hickory (about 1% dominance) are mapped in central New York and southern Ontario.

At 10,000 yr B.P., the dominance map for hickory (Fig. 5.8g) illustrated a fragmentation in the formerly latitudinal zonation of hickory abundance. Populations with low (< 10%) dominance occupied three key distributional areas: (1) the northern irregular perimeter of advancing hickory populations from 40°N to 45°N in Iowa, Illinois, Wisconsin, and southern Michigan; (2) the distributional heartland in Tennessee and Kentucky, extending across the Appalachian Highlands to hickory's northeastern range limit in Maryland and coastal Virginia; and (3) a southeastern expansion into peninsular Florida, which was colonized by hickory by 10,000 yr B.P. Three principal centers of > 20% hickory occupied (1) the temperate forest region west of 89°W from Missouri to the Gulf of Mexico; (2) the Carolinas; and (3) western Ohio. The western hickory border coincided with the prairie–forest ecotone. Northern outliers of hickory were scattered across the eastern Great Lakes and northern Appalachian regions.

The mid-Holocene maps for hickory at 8000 yr B.P. (Fig. 5.8h) and 6000 yr B.P. (Fig. 5.8i) illustrate progressive movement of hickory's continuous northern margin as advance hickory outliers merged with its main range, from the western Great Lakes to southern New England. During this time interval, hickory populations in Minnesota, Iowa, and central Illinois were extirpated, as they were situated in the path of the expanding Prairie Peninsula. The former early-Holocene population centers experienced differential decreases in abundance at 8000 yr B.P., but partially increased in dominance by 6000 yr B.P. in western Ohio and throughout Mississippi and Louisiana.

The paleo-dominance maps for 4000 yr B.P. (Fig. 5.8j), 2000 yr B.P. (Fig. 5.8k), and 500 yr B.P. (Fig. 5.8l) delineate the late-Holocene reexpansion in hickory population centers in the central Gulf Coastal Plain, particularly along the Lower Mississippi Alluvial Valley of Mississippi, Arkansas, and Louisiana. A late-Holocene suite of hickory species, generally delineated by the 10% contour, established across the region mapped as deciduous forest (Fig. 4.2a), extending as a wedge from the prairie–forest boundary east through the central Appalachian Mountains. By 500 yr B.P., hickory was reestablished as a prominent constituent of forests in the highland region including the Ozark and Ouachita Plateaus of Arkansas, Missouri, and eastern Oklahoma.

As the hickory population trajectory portrays a 33% to 44% decline in mean values during the last 20,000 years (Fig. 5.8a), the area–dominance histograms denote a fundamental change in dominance structure through this interval (Fig. 5.9a). The full- and late-glacial histograms illustrate patterns that are either skewed in area toward the larger dominance classes, as shown at 20,000 yr B.P., or, as at 16,000 yr B.P., are equally distributed with comparable areas measured across all dominance classes. The total area for hickory remained

Carya

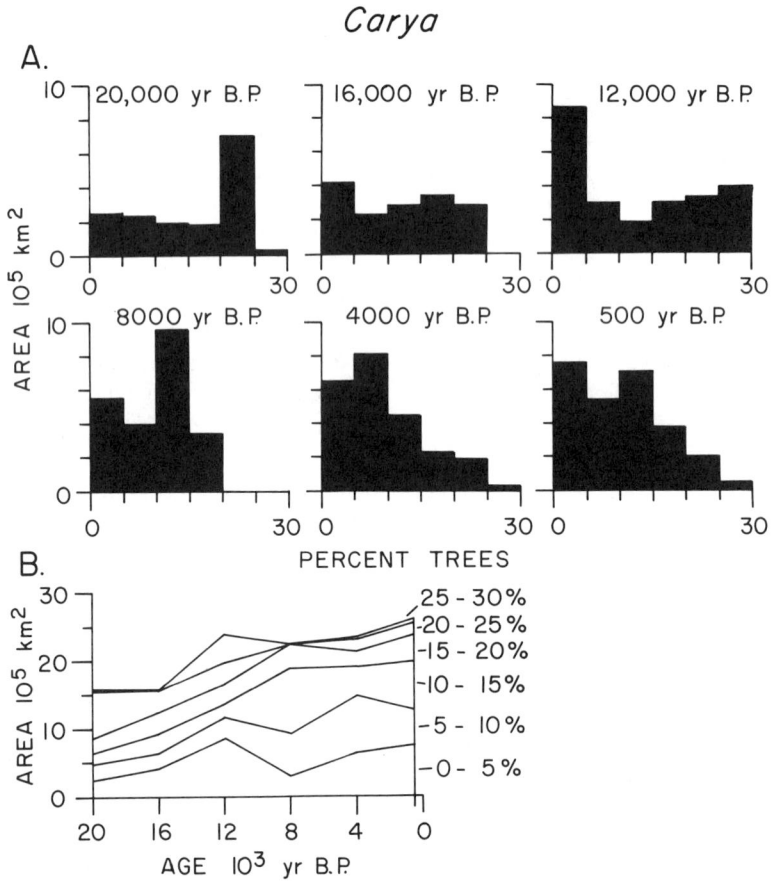

Figure 5.9. Changes in area and dominance of hickory (*Carya*). a. Histograms of area–dominance structure at 20,000, 16,000, 12,000, 8000, 4000, and 500 yr B.P. b. Cumulative plot from 20,000 yr B.P. to the present.

relatively consistent, occupying $15.9 \cdot 10^5$ km^2 at 20,000 yr B.P. and $15.7 \cdot 10^5$ km^2 at 16,000 yr B.P. (Fig. 5.9b). In the histogram for 12,000 yr B.P., hickory populations show a response to postglacial environmental changes, with increases in area of both their highest dominance classes (*e.g.*, the 20% to 25% class and the 25% to 30% class) and in their lowest dominance class (0% to 5%). This dual response represented both hickory's successful competition in the forest canopy within its primary population centers, and its invasion of 52% more area (to a total of $24.0 \cdot 10^5$ km^2) during the interval from 16,000 yr B.P. to 12,000 yr B.P. (Fig. 5.9b). The mid-Holocene histogram for 8000 yr B.P. reflects the loss of the two highest dominance classes and the subsequent increase of hickory populations within the two intermediate classes (from 10% to 15%, and from 15% to 20%) and the two lowest dominance classes (Fig.

5.9a). This change in population dominance structure was accompanied by only a 6% loss in overall area (22.6 · 10^5 km^2) at 8000 yr B.P. During the late Holocene, the area–dominance histograms for 4000 yr B.P. and 500 yr B.P. display areal losses preferentially within the intermediate classes and moderate expansion in both number and area occupied by the highest classes. The major late-Holocene changes in hickory dominance structure are indicated by prominent areal increases for both the two lowest classes. These are accompanied by the maximum interglacial expansion of hickory to its presettlement distributional range of 26.3 · 10^5 km^2 500 years ago.

Hackberry (*Celtis*)

Reconstructions for hackberry include the fossil records of *Celtis* species (*C. laevigata*, *C. occidentalis*, and *C. tenuifolia*) as well as those of Osage orange (*Maclura pomifera*), as their pollen grains are morphologically indistinguishable from each other (Kapp 1969). The modern populations of hackberry are more abundant and more widely distributed than Osage orange (Delcourt *et al.* 1984). Although the extent to which *Maclura* contributed to the fossil pollen record is unknown, we presume that the reconstructed mapped patterns probably primarily reflect the history of *Celtis*. The plot of population trajectory for hackberry (Fig. 5.10a) contained no reconstructed values of *Celtis* for the full-glacial interval between 20,000 yr B.P. and 17,000 yr B.P. However, the late-glacial interval contained sporadically fluctuating values, reaching 14% at 16,000 yr B.P., with no data values represented at 15,000 yr B.P., and maximum values increasing from 3% at 14,000 yr B.P. to its highest reconstructed value of 24% by 13,000 yr B.P. The trajectory for *Celtis* maintained a consistently high mean between 3% and 7% and peak values between 9% and 16% for the early- and mid-Holocene intervals spanning from 12,000 yr B.P. to 6000 yr B.P. Values for hackberry dropped from 6000 yr B.P. to 4000 yr B.P., maintaining relatively minor values for the remainder of the late-Holocene interval. During the last 4000 years, the mean dominance values for hackberry were consistently low, generally < 2%, with maximum values typically < 4%.

The earliest time for which hackberry was recorded from the array of fossil-pollen sites was at 16,000 yr B.P. at Boriack Bog, east-central Texas (Bryant 1977), with a reconstructed paleo-dominance value of 14% (Fig. 5.10b). One plausible full-glacial refuge for *Celtis* may have occurred in central and eastern Texas and possibly extended into the western Gulf Coastal Plain. During the late-glacial interval, *Celtis* was reported in north-central Florida at 14,000 yr B.P. (Fig. 5.10c). By 13,000 yr B.P., hackberry occurred in sites distributed across the Gulf Coastal Plain, reaching dominance values of 24% in the Yazoo Basin within the Lower Mississippi Alluvial Valley (Holloway and Valastro 1983). *Celtis* invasion of the Mississippi River Valley apparently coincided with both the changeover from a braided to a meandering fluvial regime (Saucier 1974) and the local extirpation from the valley of boreal conifers during late-glacial climatic amelioration. Late-glacial values of hackberry reexpanded up to 5% in the east-central Texas region.

Celtis

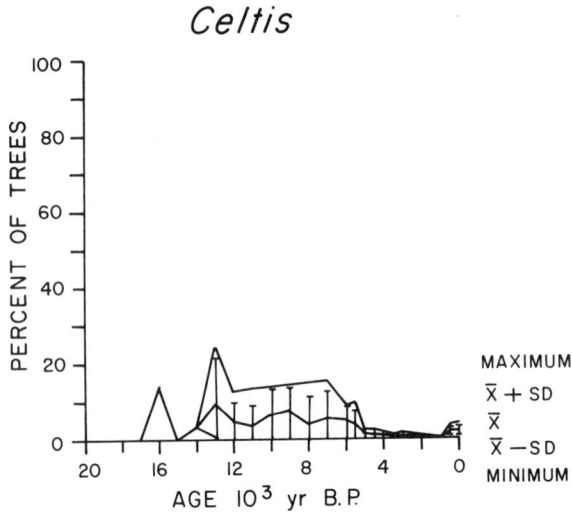

Figure 5.10. Hackberry (*Celtis*). a. Population trajectory, including maximum, minimum, and mean values of paleo-dominance as well as ± 1 SD of the mean, for the past 20,000 years.

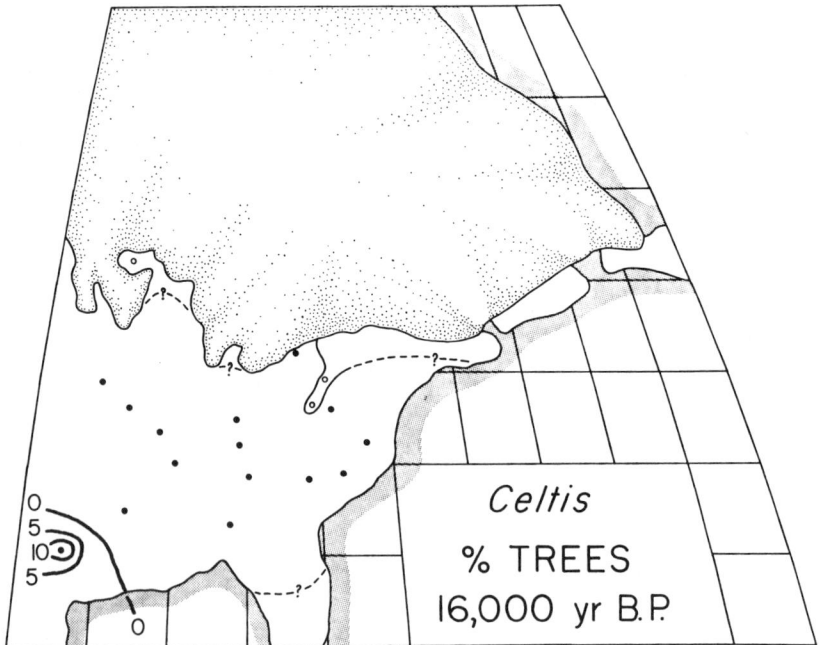

Figure 5.10. Hackberry (*Celtis*). b. Paleo-dominance map for 16,000 yr B.P.

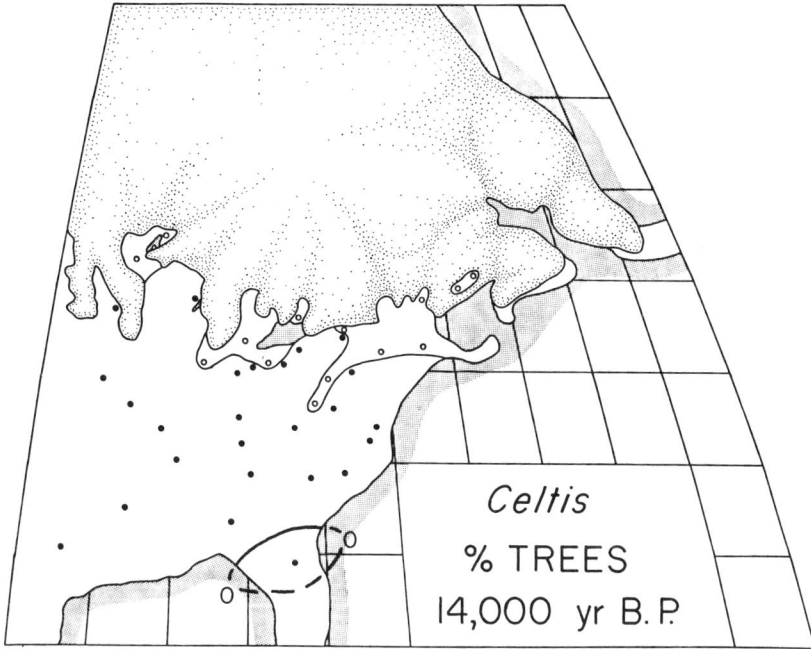

Figure 5.10. Hackberry (*Celtis*). c. Paleo-dominance map for 14,000 yr B.P.

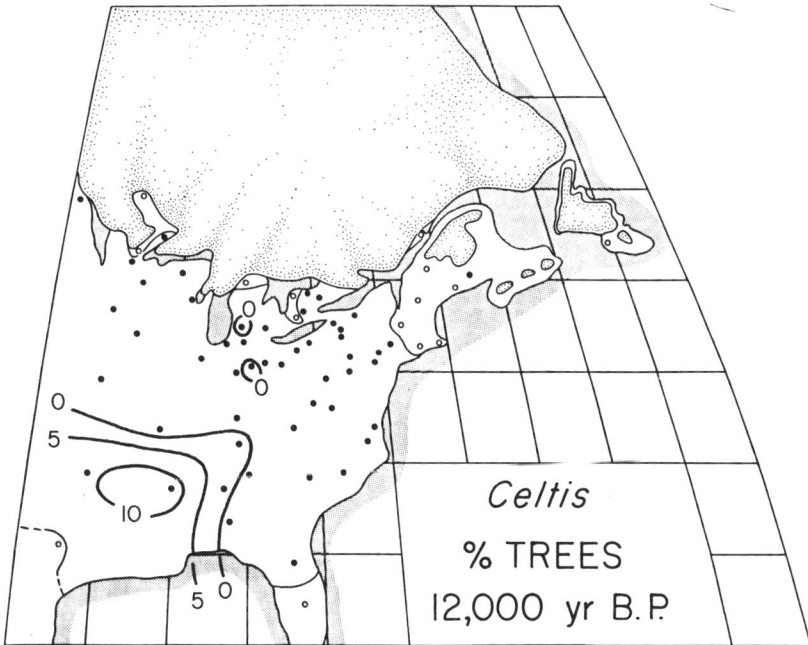

Figure 5.10. Hackberry (*Celtis*). d. Paleo-dominance map for 12,000 yr B.P.

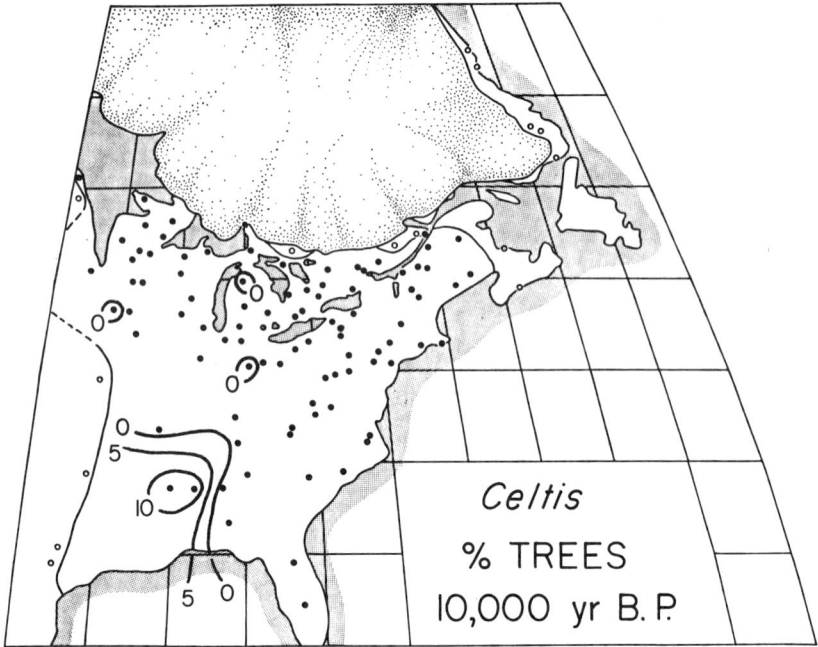

Figure 5.10. Hackberry (*Celtis*). e. Paleo-dominance map for 10,000 yr B.P.

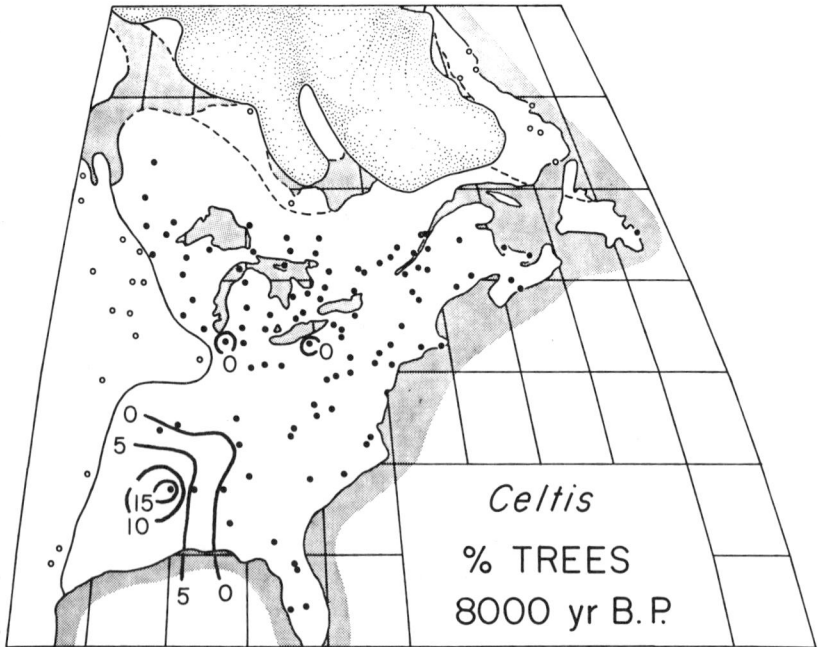

Figure 5.10. Hackberry (*Celtis*). f. Paleo-dominance map for 8000 yr B.P.

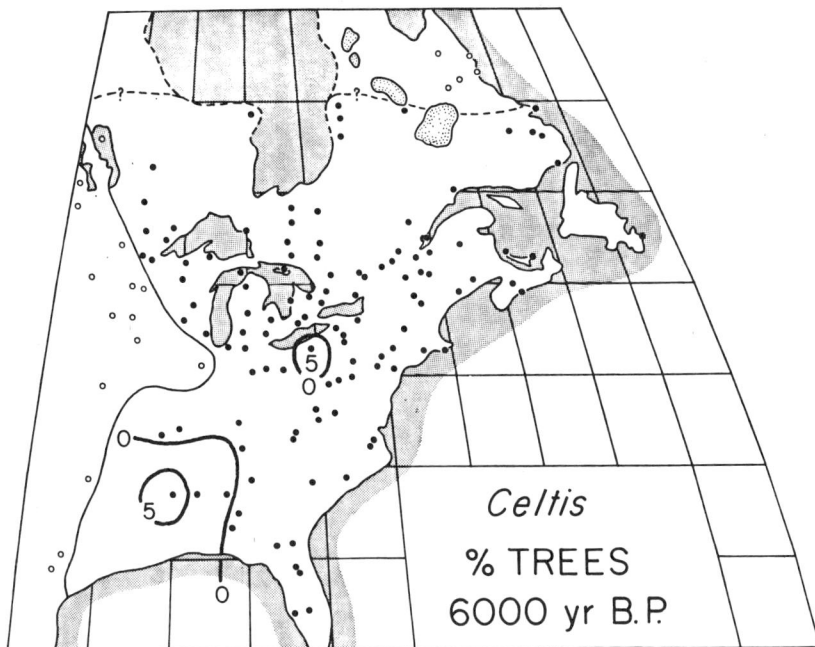

Figure 5.10. Hackberry (*Celtis*). g. Paleo-dominance map for 6000 yr B.P.

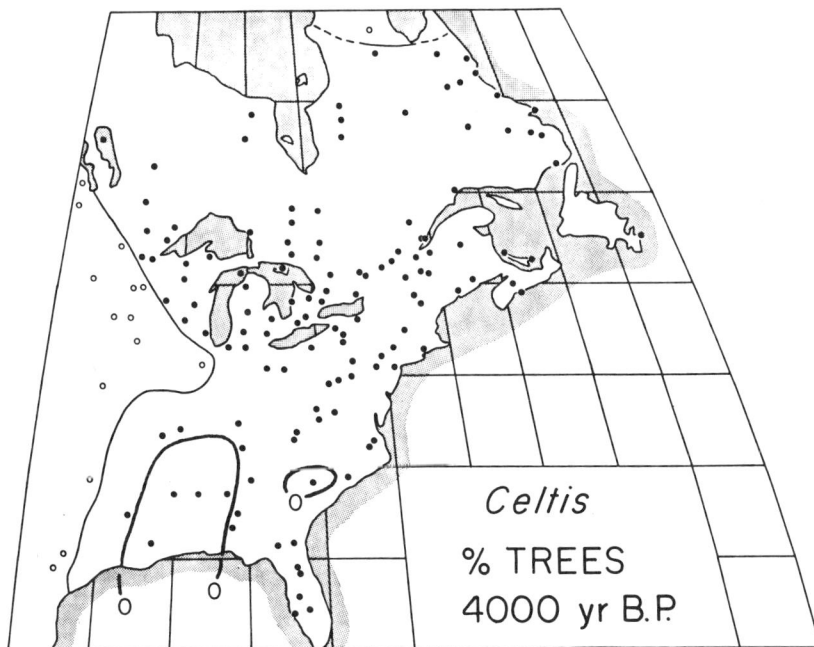

Figure 5.10. Hackberry (*Celtis*). h. Paleo-dominance map for 4000 yr B.P.

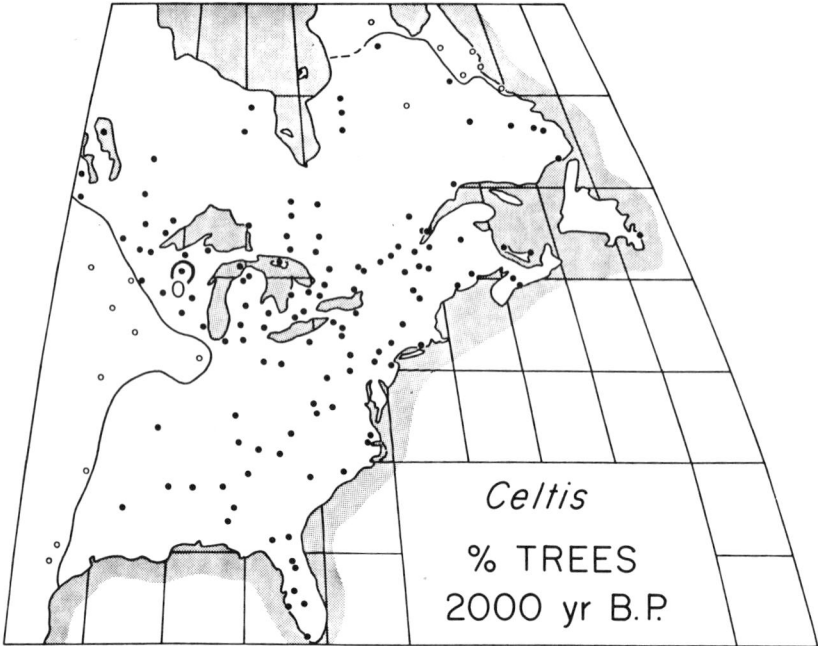

Figure 5.10. Hackberry (*Celtis*). i. Paleo-dominance map for 2000 yr B.P.

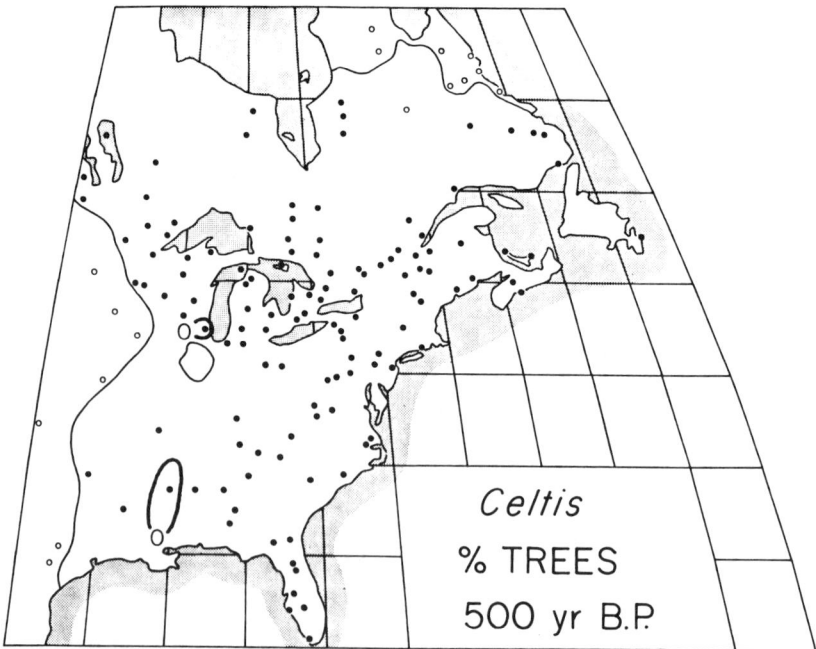

Figure 5.10. Hackberry (*Celtis*). j. Paleo-dominance map for 500 yr B.P.

During the early Holocene, maps for 12,000 yr B.P. (Fig. 5.10d) and 10,000 yr B.P. (Fig. 5.10e) reveal the areal expansion of hackberry across the southwestern sector of the study area. For example, at 12,000 yr B.P., *Celtis* occupied the forested region generally west of 85°W and south of 37°N. Peak values for hackberry occurred in the poorly drained Mississippi Alluvial Valley and across the uplands to the western Ouachita Mountains of eastern Oklahoma (Albert and Wyckoff 1981; Bryant and Holloway 1985b). During the early Holocene, isolated populations of hackberry occurred as far north as 45°N through the central and western Great Lakes region. As the prairie expanded across the continental interior, the western limit of continuous hackberry populations became delineated by the prairie–forest ecotone.

As shown on the maps for 8000 yr B.P. (Fig. 5.10f) and 6000 yr B.P. (Fig. 5.10g), peak population values for hackberry declined from 15% down to 9% within the Mississippi Valley of west-central Mississippi. By 6000 yr B.P., the

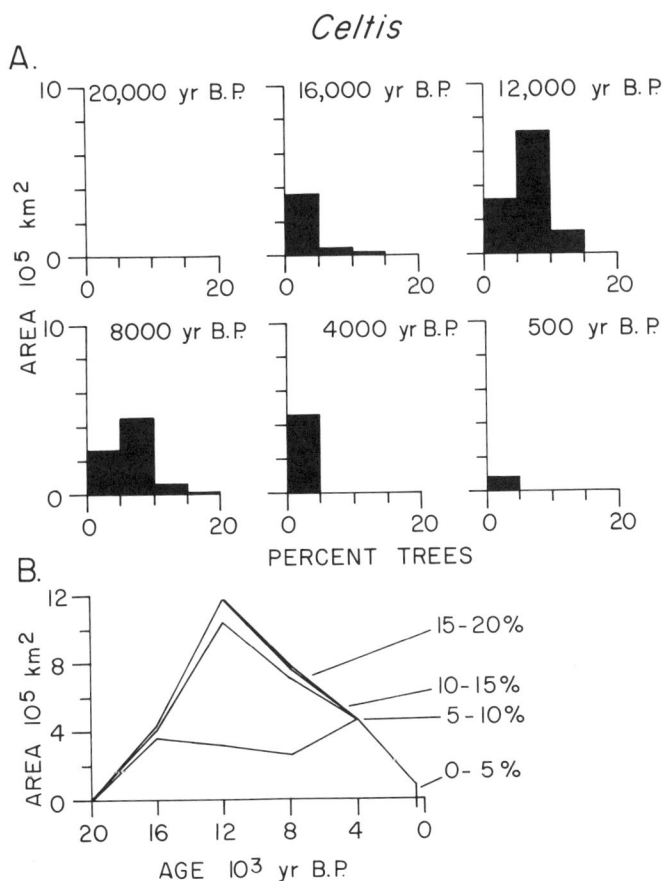

Figure 5.11. Changes in area and dominance of hackberry (*Celtis*). a. Histograms of area–dominance structure at 20,000, 16,000, 12,000, 8000, 4000, and 500 yr B.P. b. Cumulative plot from 20,000 yr B.P. to the present.

region of continuous hackberry distribution was primarily confined to the region between 86°W and 96°W and generally south of 37°N to the shores of the Gulf of Mexico.

During the late Holocene, values of paleo-dominance for hackberry generally dropped below 2% of the forest composition, as shown in the map for 4000 yr B.P. (Fig. 5.10h). The areal distribution of hackberry became progressively more fragmented, occupying the central portion of the Gulf Coastal Plain and portions of the southern Atlantic Coastal Plain. By 2000 yr B.P. (Fig. 5.10i) and 500 yr B.P. (Fig. 5.10j), respectively, hackberry populations occupied geographically restricted areas along the prairie–forest ecotone and within the Lower Mississippi Alluvial Valley.

The area–dominance histograms for hackberry (Fig. 5.11a) display 3 or 4 dominance classes from 16,000 yr B.P. to 8000 yr B.P. and only the lowest dominance class of 0% to 5% for 4000 yr B.P. and 500 yr B.P. The histogram for 16,000 yr B.P. has a total area of $4.2 \cdot 10^5$ km^2, of which 85% is contained within the lowest dominance class. Comparison of the 16,000 yr B.P. and 12,000 yr B.P. histograms defines a major increase in hackberry populations at intermediate dominance classes. During that 4000-year interval, the area between 5% and 10% dominance increased over 1400%, from $0.5 \cdot 10^5$ km^2 to $7.2 \cdot 10^5$ km^2.

The cumulative plot of dominance structure for *Celtis* (Fig. 5.11b) reflects a nearly symmetrical expansion in both area and diversification in dominance classes to a peak at 12,000 yr B.P. and then progressive decline to the present. The histograms display a systematic decrease in both area represented and number of dominance classes throughout the last 12,000 years (Fig. 5.11a). Hackberry populations declined in the area they occupied, from $7.8 \cdot 10^5$ km^2 at 8000 yr B.P. to $4.6 \cdot 10^5$ km^2 at 4000 yr B.P., and then diminished to $0.8 \cdot 10^5$ km^2 by 500 yr B.P.

Cedars and cypress (Cupressaceae and Taxodiaceae)

The population trajectory exhibited by the group of cedars and cypress reflects the pooled characteristics for the tree species of four genera and two families. Within the cedar family (Cupressaceae), three genera are represented: (1) northern white cedar (*Thuja occidentalis*); (2) red cedar (*Juniperus virginiana* and *J. silicicola*); and (3) Atlantic white cedar (*Chamaecyparis thyoides*) (Delcourt *et al.* 1984). Within the cypress family (Taxodiaceae), bald cypress (*Taxodium distichum*) is the arboreal representative within the study area. *Taxodium* can be identified to species by pollen when the grains are preserved with intact papilla; however, broken pollen grains are indistinguishable from each other for the four genera (Kapp 1969). In the mapped reconstructions, the more northern group probably includes primarily *Thuja* and shrub species of *Juniperus*, whereas the more southern and coastal group probably reflects the contributions of the tree species *Juniperus virginiana*, *Taxodium distichum*, and *Chamaecyparis thyoides* (Delcourt *et al.* 1984).

For the last 20,000 years, the population statistics summarized for the cedar and cypress group exhibit a consistently low mean dominance of between 3% and 6% (Fig. 5.12a). The full-glacial values of forest composition include max-

Cupressaceae & Taxodiaceae

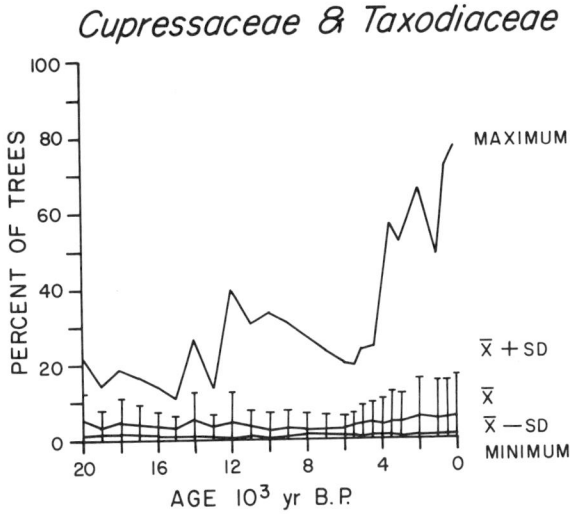

Figure 5.12. Cedars and cypress (*Cupressaceae* and *Taxodiaceae*). a. Population trajectory, including maximum, minimum, and mean values of paleo-dominance as well as ± 1 SD of the mean, for the past 20,000 years.

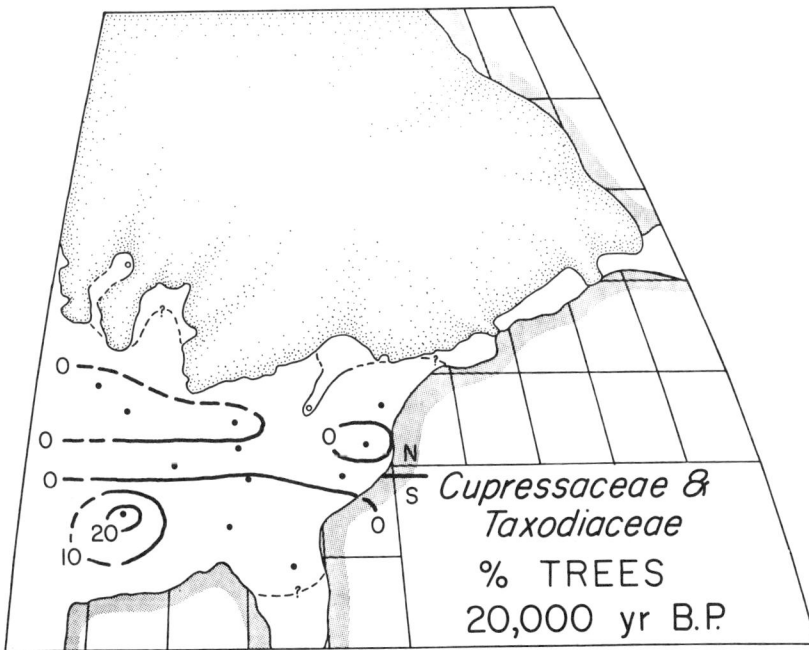

Figure 5.12. Cedars and cypress (*Cupressaceae* and *Taxodiaceae*). b. Paleo-dominance map for 20,000 yr B.P.

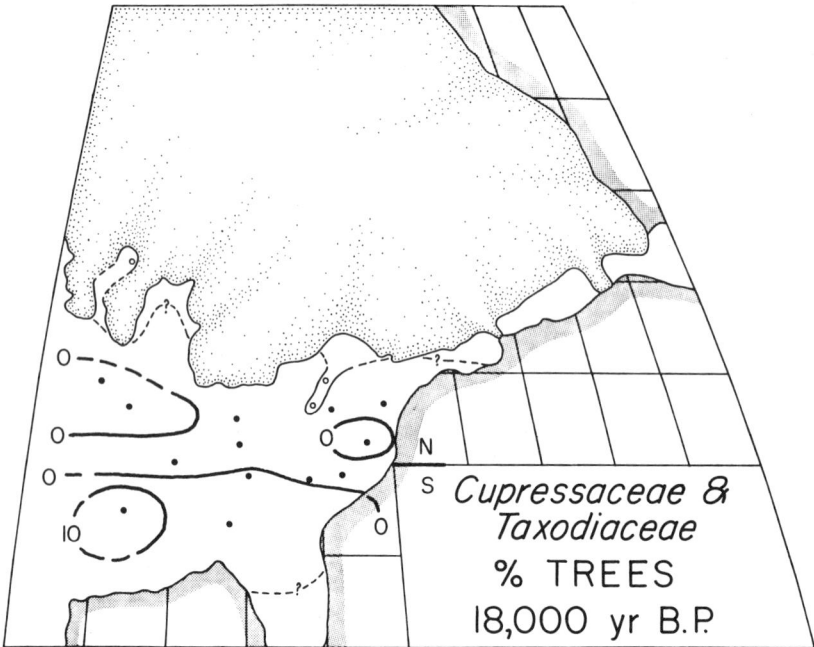

Figure 5.12. Cedars and cypress (*Cupressaceae* and *Taxodiaceae*). c. Paleo-dominance map for 18,000 yr B.P.

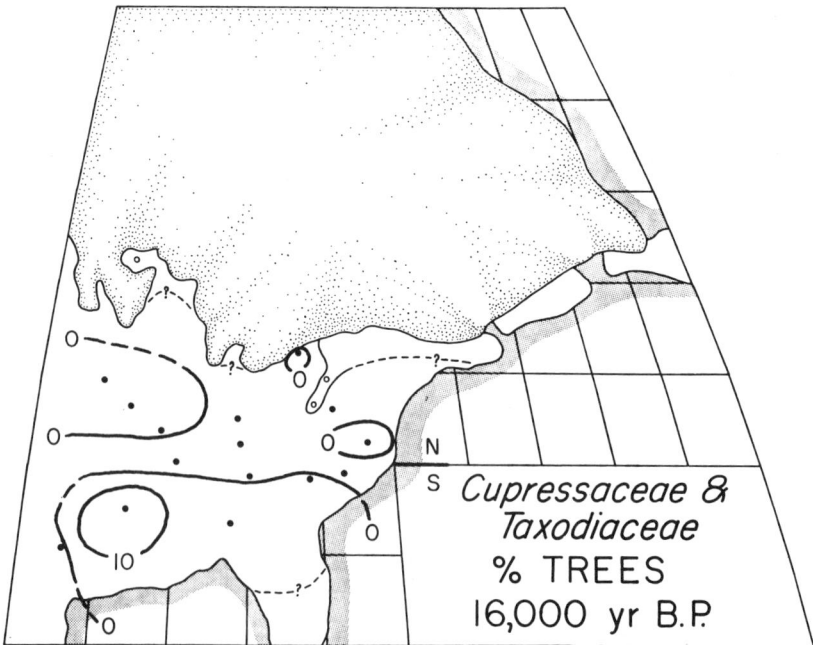

Figure 5.12. Cedars and cypress (*Cupressaceae* and *Taxodiaceae*). d. Paleo-dominance map for 16,000 yr B.P.

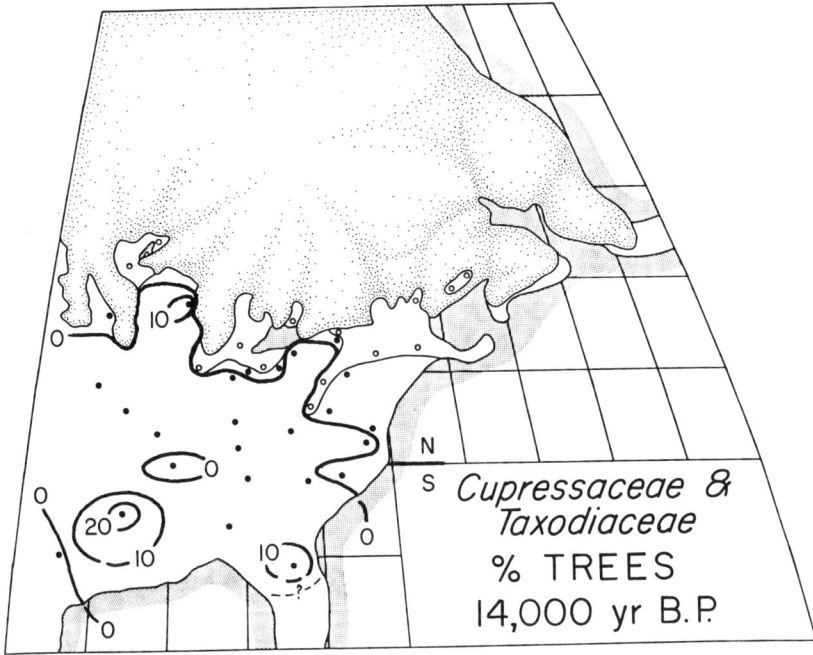

Figure 5.12. Cedars and cypress (*Cupressaceae* and *Taxodiaceae*). e. Paleo-dominance map for 14,000 yr B.P.

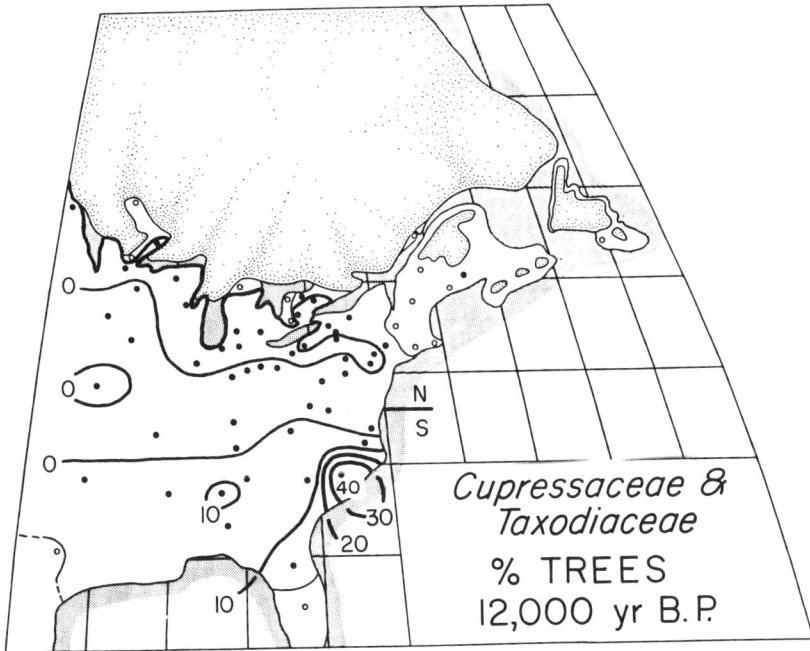

Figure 5.12. Cedars and cypress (*Cupressaceae* and *Taxodiaceae*). f. Paleo-dominance map for 12,000 yr B.P.

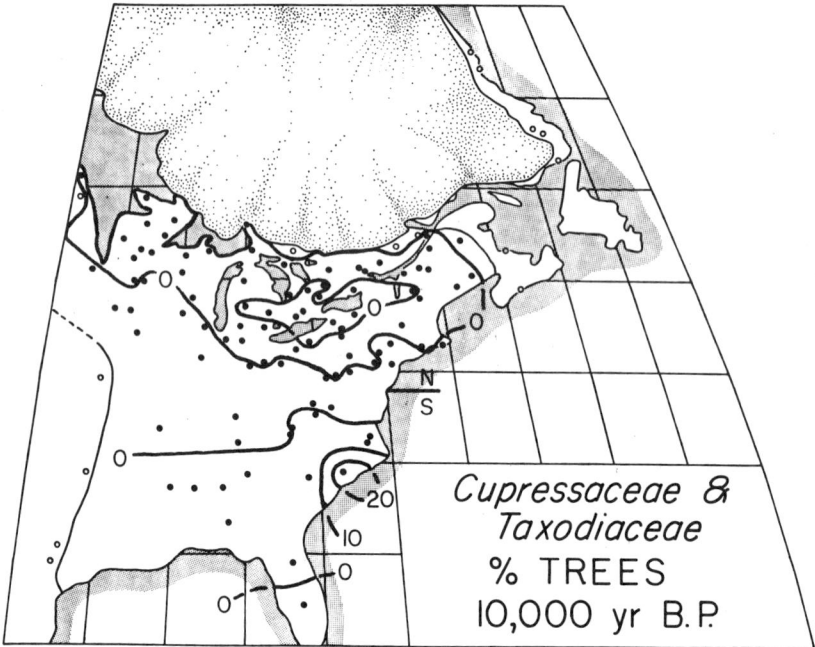

Figure 5.12. Cedars and cypress (*Cupressaceae* and *Taxodiaceae*). g. Paleo-dominance map for 10,000 yr B.P.

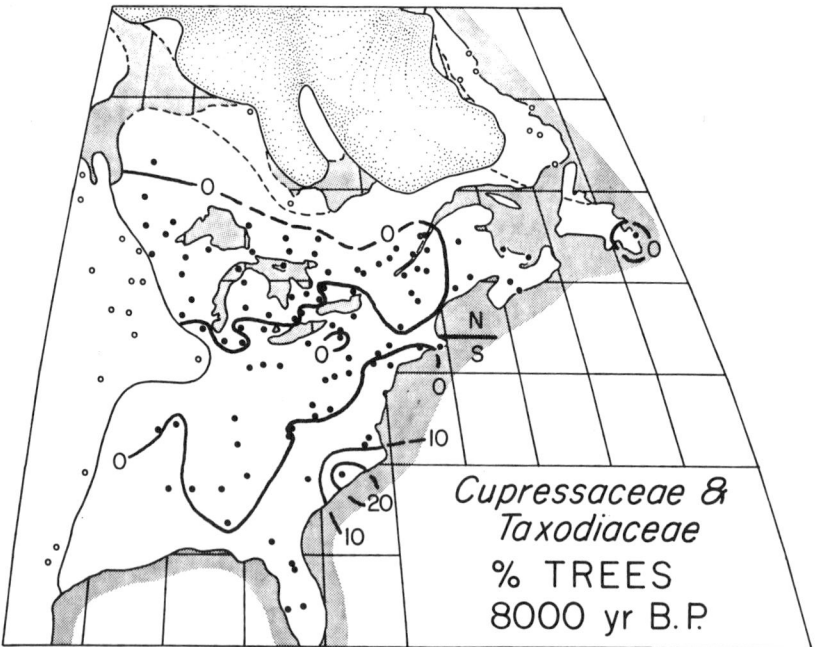

Figure 5.12. Cedars and cypress (*Cupressaceae* and *Taxodiaceae*). h. Paleo-dominance map for 8000 yr B.P.

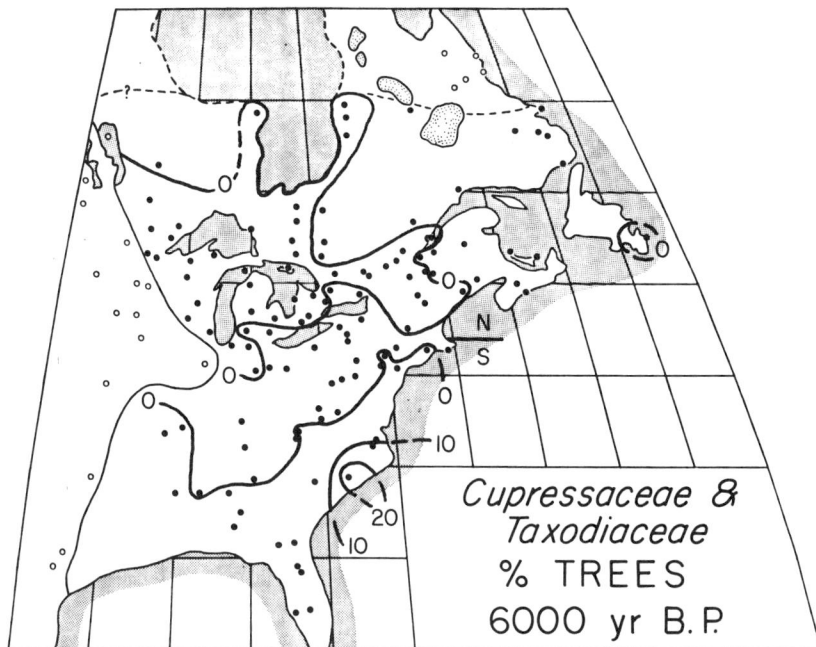

Figure 5.12. Cedars and cypress (*Cupressaceae* and *Taxodiaceae*). i. Paleo-dominance map for 6000 yr B.P.

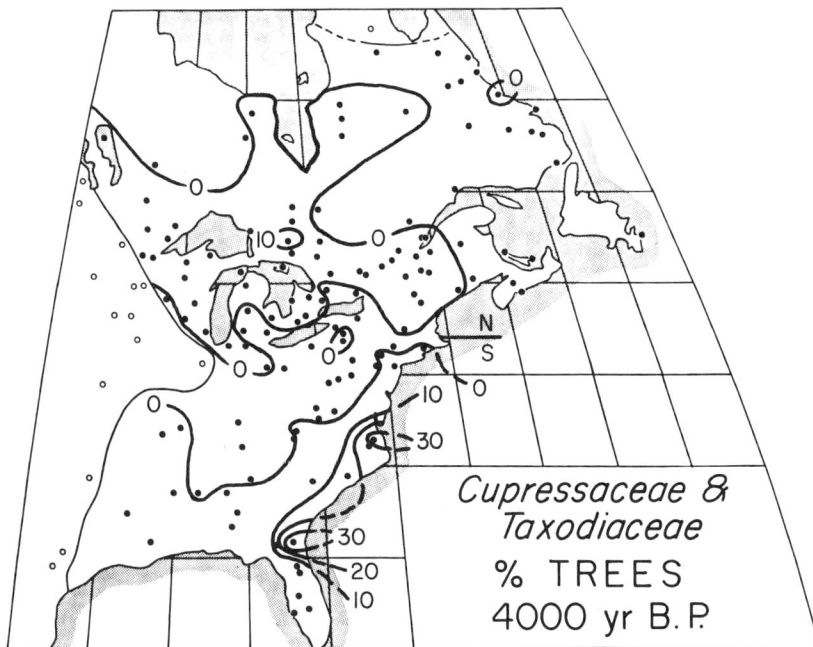

Figure 5.12. Cedars and cypress (*Cupressaceae* and *Taxodiaceae*). j. Paleo-dominance map for 4000 yr B.P.

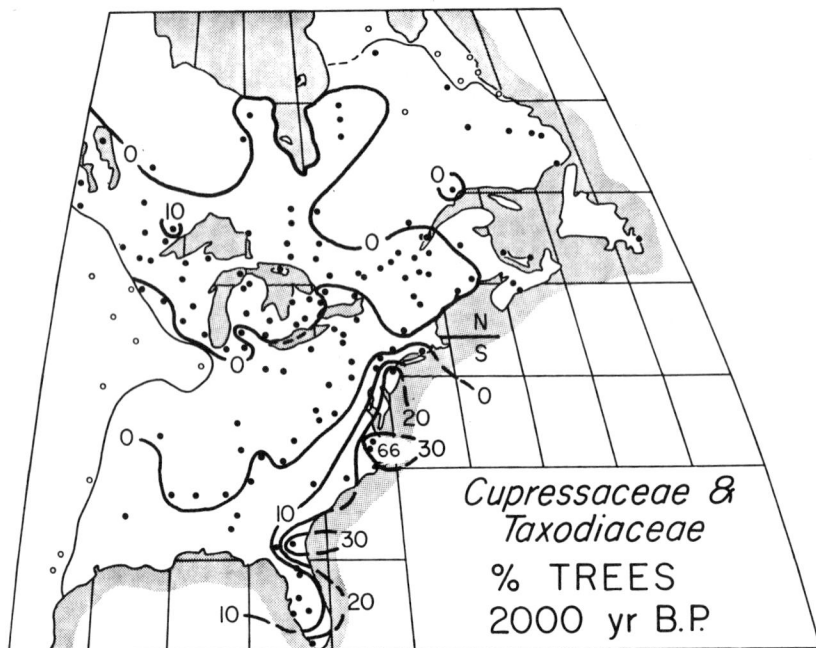

Figure 5.12. Cedars and cypress (*Cupressaceae* and *Taxodiaceae*). k. Paleo-dominance map for 2000 yr B.P.

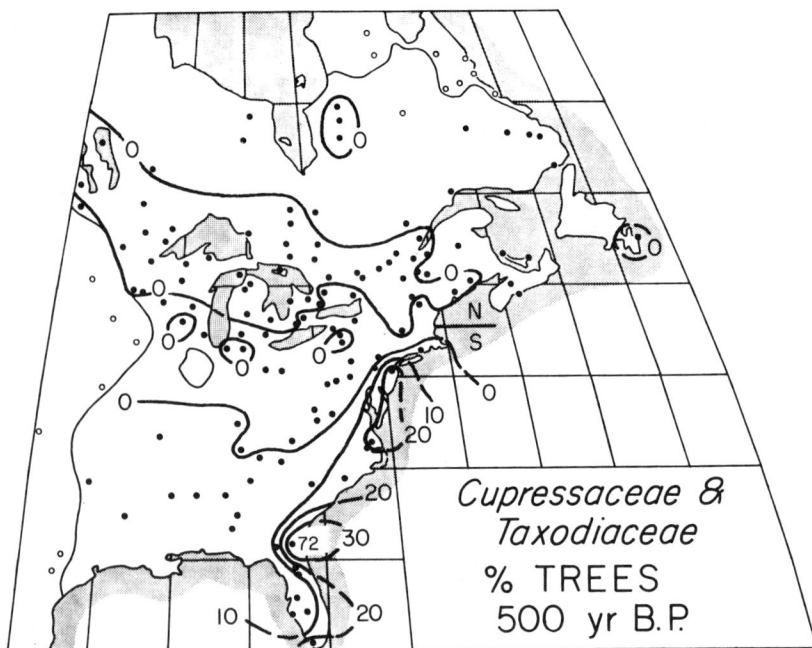

Figure 5.12. Cedars and cypress (*Cupressaceae* and *Taxodiaceae*). l. Paleo-dominance map for 500 yr B.P.

imum values between 17% and 22%, with ± 1 SD typically between 4% and 7% (Fig. 5.12a). Both the maximum values and the population variability expressed as ± 1 SD generally increased to the Pleistocene–Holocene transition at 12,000 yr B.P. and then progressively decreased through the early- and mid-Holocene intervals. For example, the peak values reconstructed for 16,000 yr B.P., 12,000 yr B.P., and 8000 yr B.P., respectively, were 14%, 40%, and 27% (Fig. 5.12a). The lowest Holocene values reconstructed as population maxima occurred between 7000 yr B.P. and 5000 yr B.P., generally at levels of about 20%. Between 5000 yr B.P. and 4000 yr B.P., the means increased along with maximum values and the variability about the mean. During the last 4000 years, maximum values fluctuated between 40% and 77%, and ± 1 SD increased gradually from 6% to 10%.

The paleo-dominance map for cedars and cypress at 20,000 yr B.P. (Fig. 5.12b) displays three populations situated along two latitudinal bands. The southern group, located south of approximately 34.5°N, occupied the southern Atlantic Coastal Plain, the northern portion of the Florida peninsula, and the Gulf Coastal Plain. This southern group displayed a primary population center with a peak value of 22% paleo-dominance in north-central Louisiana (Fig. 5.12b). The occurrence of *Taxodium* wood in sediments dating about 23,000 yr B.P. from Green Pond in northwestern Georgia (Watts 1973b) implies that bald cypress may have been distributed widely through the Gulf Coastal Plain in early full-glacial times. At 20,000 yr B.P., the more northern populations were located at about 36°N along the central Atlantic Seaboard and between about 36°N and 40°N west of the Appalachian Mountains. These populations were generally < 10% of the forest composition in that latitudinal band (Fig. 5.12b).

During peak full-glacial times, as illustrated in the map for 18,000 yr B.P. (Fig. 5.12c), the northern group of cedars was separated from the southern group of cedars and cypress at about 35°N. The northern group retracted westward from the Interior Low Plateau of Middle Tennessee and represented < 2% of total forest composition. The primary center of dominance for cedars and cypress persisted in the western Gulf Coastal Plain, with a value of 19% dominance in north-central Louisiana.

The late-glacial maps for 16,000 yr B.P. (Fig. 5.12d) and 14,000 yr B.P. (Fig. 5.12e) exhibit the delineation of a southwestern boundary in eastern Texas and the progressive northward dispersal into the central and western Great Lakes regions. By 14,000 yr B.P., the previous mutually exclusive boundaries between northern and southern groups at 35°N blurred as the populations of both groups expanded and formed a generally continuous distribution from the Gulf of Mexico north to approximately 45°N in central Wisconsin. Within the uplands surrounding the northern Gulf of Mexico, primary population centers 14,000 years ago (Fig. 5.12e) occurred in both Louisiana (26%) and north-central Florida (15%). The northern boundary of the cedars and cypress group extended from northeastern North Carolina northwest across Virginia and then north through Maryland and Ohio (Fig. 5.12e). This border extended west across Indiana, Illinois, and Wisconsin, and was projected west through Iowa and Nebraska.

The early-Holocene maps for 12,000 yr B.P. (Fig. 5.12f) and 10,000 yr B.P. (Fig. 5.12g) display a pronounced spatial separation of between 3° and 10° latitude separating the distributional limits of the northern and southern groups of cedars and cypress. Between 12,000 yr B.P. and 10,000 yr B.P., the southern group occupied the Atlantic Coastal Plain south of approximately 37°N and extended along the Gulf Coastal Plain generally south of about 35°N. By 10,000 yr B.P., the western boundary of the cedar and cypress group coincided with the prairie–forest border. During this 2000-year interval, the plant-fossil sequence from Cahaba Pond in north-central Alabama (Delcourt *et al.* 1983b) included macrofossils of *Chamaecyparis thyoides* as well as pollen grains of *Taxodium*. The northern cedar group experienced progressive fragmentation and local isolation along its southern margin, with limited outliers persisting through the early Holocene in sites such as Muscotah Marsh in northeastern Kansas (Gruger 1973). With glacial retreat accompanying postglacial warming, the northern cedar group advanced northward, occupying the latitudinal belt between 40°N and 50°N; a mixture of *Juniperus* and *Thuja* probably advanced into the forested regions of New England, the Great Lakes region, and into the central portion of Canada along the margins of Glacial Lake Agassiz. Macrofossils of *Thuja occidentalis* dating from about 10,000 yr B.P. have been recovered from what is now an island in Georgian Bay in southern Ontario (Warner *et al.* 1984). It remains plausible that populations of both *Thuja occidentalis* and shrub *Juniperus* advanced preferentially into distinctive habitats such as along the sandy beaches and dune fields of proglacial lakes. These northern representatives of the Cupressaceae family may have invaded the periglacially modified strips of bare ground and tundra adjacent the retreating ice sheet (Richard 1977, 1985).

The paleo-dominance map for 8000 yr B.P. (Fig. 5.12h) illustrates limited extinctions of the southern cedar and cypress populations within northern Alabama and Georgia, but expansion of their range limits southeastward into Florida, northeastward along the Atlantic Coastal Plain into New Jersey, and northward along the Lower Mississippi Alluvial Valley into southeastern Missouri. Through the early Holocene, one population center ($> 20\%$) persisted in the Carolinas. The northern cedar group retracted on its western boundary in response to the eastward advance of prairie, expanded north into southern Quebec and southern Ontario, and retreated on its southern boundary to positions generally north of 42°N. To the northeast, populations of the northern cedar group extended along the St. Lawrence River Valley and into central New England; a limited population outlier established successfully in Newfoundland (Fig. 5.12h). Although the northern cedar group was widely distributed at 8000 yr B.P., its dominance values were generally $< 10\%$.

By 6000 yr B.P., the southern group of cedars and cypress reexpanded across central Alabama and migrated northwest across the Ozark Plateaus of Missouri and Arkansas to the prairie–forest ecotone (Fig. 5.12i). The distributional limits remained relatively stationary for the mid-Holocene populations of northern cedars except along their northern border. By 6000 yr B.P., peninsular extensions of cedar populations encircled the southern margin of Hudson Bay, extending north to approximately 55°N in northeastern Ontario and northwestern Quebec. The northern and southern cedars and cypress groups were separated by a zone with no overlap, situated between 41°N and 42°N east of the Appalachian Mountains.

The paleo-dominance map for cedars and cypress at 4000 yr B.P. (Fig. 5.12j) reflects the expansion of major population centers for *Taxodium* and probably *Chamaecyparis* in coastal swamps along the Atlantic Seaboard. The increased dominance of the southern cedars and cypress group reflects the Holocene rise in sea level to near-modern position and the inundation of coastal habitats suitable for development of southern swamps, such as the Great Dismal Swamp of Virginia/North Carolina (Whitehead 1972) and the Okefenokee Swamp of southeastern Georgia (Cohen *et al.* 1984). Paleo-dominance values of the southern cedars and cypress group reached up to 33% in the Dismal Swamp and 39% in the Okefenokee Swamp (Fig. 5.12j). For the northern group of cedars, a limited population center of 10% occurred in east-central Ontario; typically, however, cedar values remained less than 3% elsewhere across its broad distributional range. Along the southern perimeter of the northern group, the range margin remained stationary. However, minor range extensions occurred northeastward into central Quebec, with outliers established in coastal Labrador. Within central Manitoba, cedars advanced northwestward in the region of Lake Winnepeg (Fig. 5.12j).

The paleo-dominance maps for 2000 yr B.P. (Fig. 5.12k) and 500 yr B.P. (Fig. 5.12l) reflect the increased abundance of *Taxodium* and *Chamaecyparis* (up to 72%) in southern coastal swamps. Within the last several thousand years, *Juniperus* expanded across the Interior Low Plateaus and along the Ridge and Valley Province (Fig. 5.12l). *Juniperus* is represented abundantly in the plant-macrofossil and wood-charcoal ethnobotanical record from archaeological sites in East Tennessee (Chapman and Shea 1981; Delcourt *et al.* 1986a), indicating that during the late Holocene, Native Americans may have been partly responsible for extending the range limits of red cedar northward on disturbed sites along river valleys. On the other hand, intensive Indian use of *Taxodium* as a sacred wood may have exhausted bald cypress populations and resulted in a southward retraction of its northernmost range limit in the Central Mississippi Valley during the past two millennia (Johannessen 1984).

During the last 2000 years, the northern extensions of cedar populations growing near Hudson Bay became progressively fragmented [note the northernmost, relict populations of *Thuja* and *Juniperus* along Hudson Bay on the presettlement distribution maps published by Little (1971)]. By 500 yr B.P. (Fig. 5.12l), the primary region for northern white cedar and shrub junipers extended in an arc-shaped swath from central New England west across the Great Lakes region and into central Manitoba. The southern boundary of this northern group of species coincided with the ecotone between mixed conifer–northern hardwoods forest and deciduous forest (Fig. 4.2a). The southwestern border advanced in southern Manitoba and Minnesota as the prairie–forest border retreated in the late Holocene.

Because the northern cedar group was geographically separated from the southern group of cedars and cypress throughout most of the last 20,000 years, the dominance structure of the two groups of populations can be examined

Southern Cupressaceae / Taxodiaceae

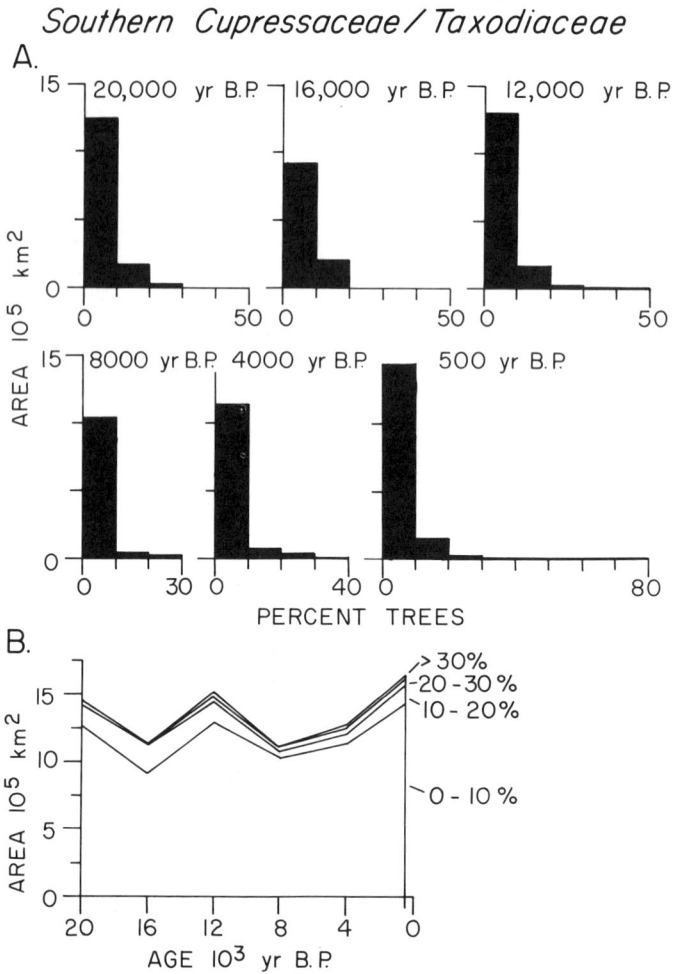

Figure 5.13. Changes in area and dominance of the southern group of cedars and cypress (*Juniperus virginiana*, *Chamaecyparis thyoides*, and *Taxodium distichum*): a. Histograms of area–dominance structure at 20,000, 16,000, 12,000, 8000, 4000, and 500 yr B.P.; b. Cumulative plot from 20,000 yr B.P. to the present.

separately (Fig. 5.13a–d). For the southern cedar and cypress group (Fig. 5.13a,b), the individual area–dominance histograms display a consistent population structure throughout the last 20,000 years, in that the majority of area occupied was characterized by 0% to 10% dominance, with much smaller amounts of area dominated by the 10% to 20% class, and little or no area in higher dominance classes. Only in the last 2000 years has the number of dominance classes increased to 8 as cypress and Atlantic white cedar became dominants locally within southern swamps. From 20,000 yr B.P. to 500 yr B.P., the area occupied by the southern cedar and cypress group has fluctuated be-

Northern Cupressaceae

C.

D.

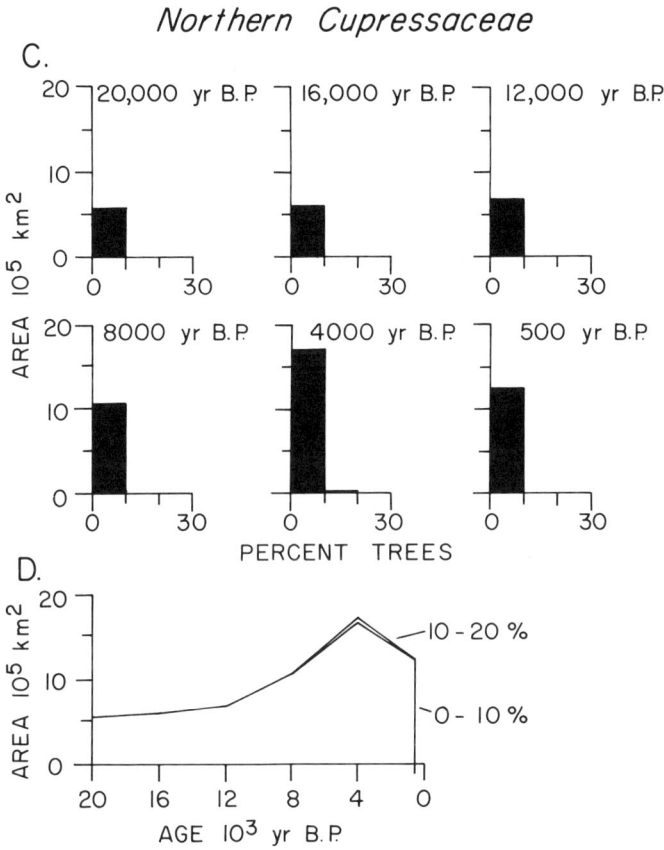

Figure 5.13. Changes in area and dominance of the northern group of cedars (*Juniperus* and *Thuja occidentalis*): c. Histograms of area–dominance structure at 20,000, 16,000, 12,000, 8000, 4000, and 500 yr B.P.; d. Cumulative plot from 20,000 yr B.P. to the present.

tween $11.0 \cdot 10^5$ km^2 and $16.4 \cdot 10^5$ km^2. From peak glacial times to presettlement times, the area occupied by the southern group expanded only 14%, from $14.4 \cdot 10^5$ km^2 at 20,000 yr B.P. to $16.4 \cdot 10^5$ km^2 at 500 yr B.P.

In marked contrast, the northern cedar group (Fig. 5.13c,d) consistently occupied only the lowest dominance class ($< 10\%$) and expanded from a full-glacial minimum in area ($5.6 \cdot 10^5$ km^2) to its interglacial maximum at 4000 yr B.P. ($17.3 \cdot 10^5$ km^2), an increase of over 300%. After 4000 yr B.P., the area occupied by northern cedars diminished to $12.5 \cdot 10^5$ km^2 at 500 yr B.P., a decrease of 28% from the area occupied at 4000 yr B.P.

Beech (*Fagus grandifolia*)

The population trajectory for beech (Fig. 5.14a) exhibits consistently low maximum values from 20,000 yr B.P. to 14,000 yr B.P., progressively increasing

Fagus grandifolia

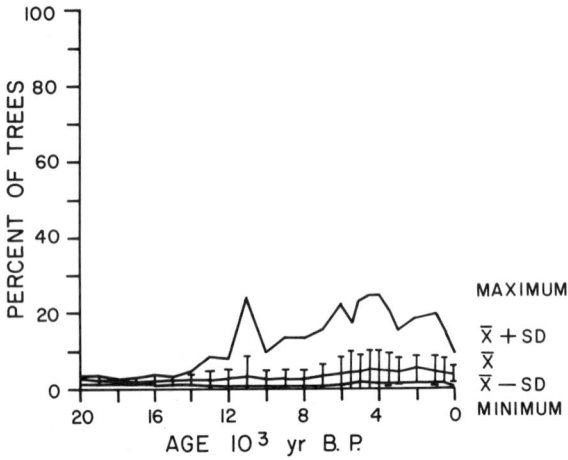

Figure 5.14. Beech (*Fagus grandifolia*). a. Population trajectory, including maximum, minimum, and mean values of paleo-dominance as well as ± 1 SD of the mean, for the past 20,000 years.

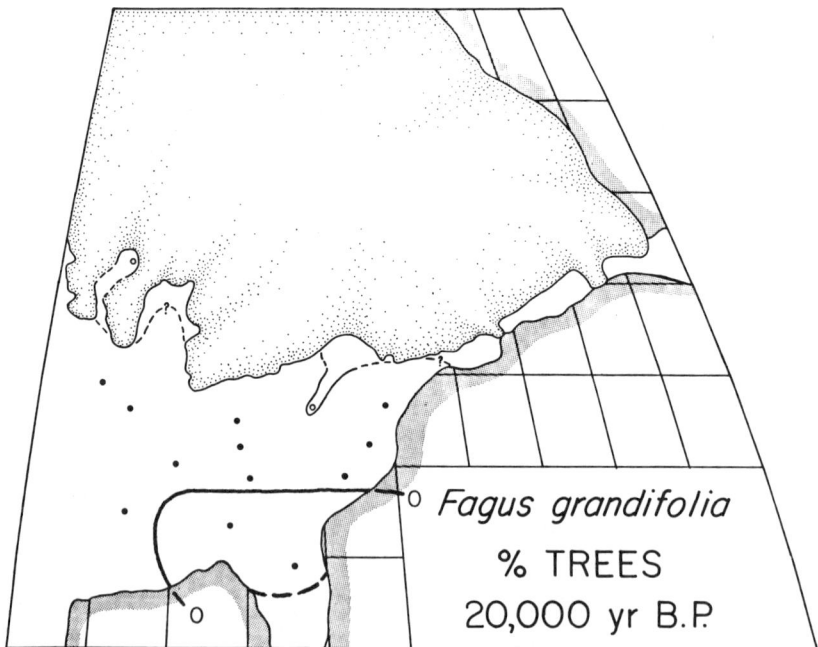

Figure 5.14. Beech (*Fagus grandifolia*). b. Paleo-dominance map for 20,000 yr B.P.

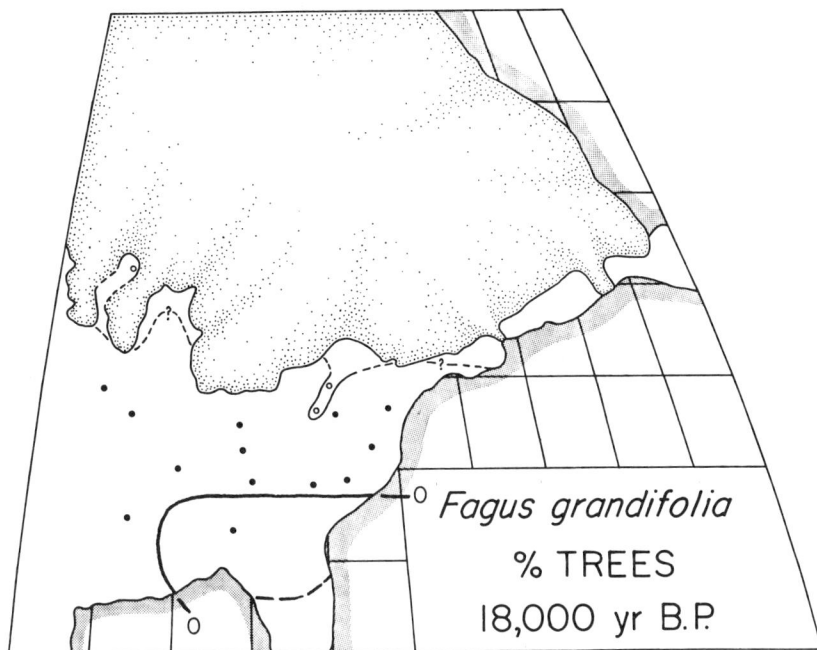

Figure 5.14. Beech (*Fagus grandifolia*). c. Paleo-dominance map for 18,000 yr B.P.

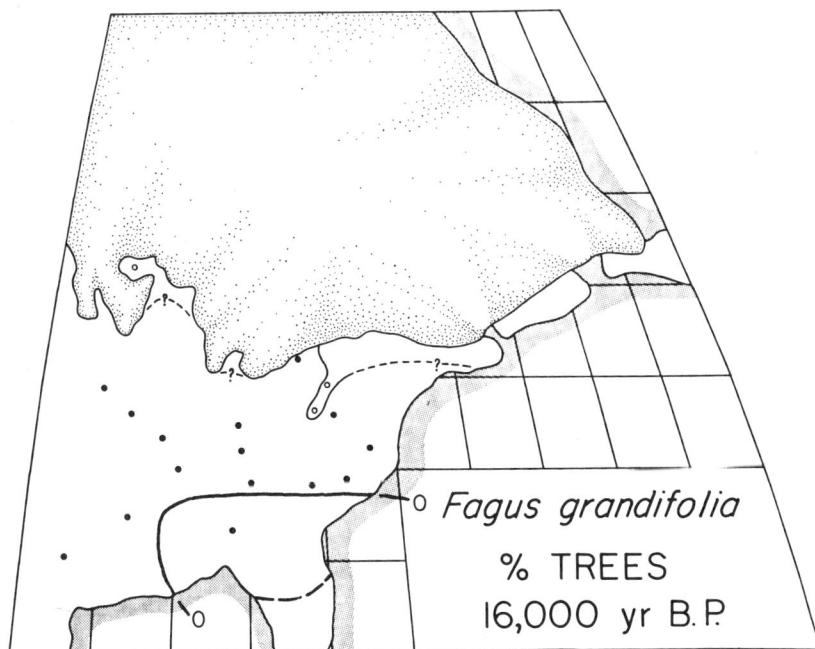

Figure 5.14. Beech (*Fagus grandifolia*). d. Paleo-dominance map for 16,000 yr B.P.

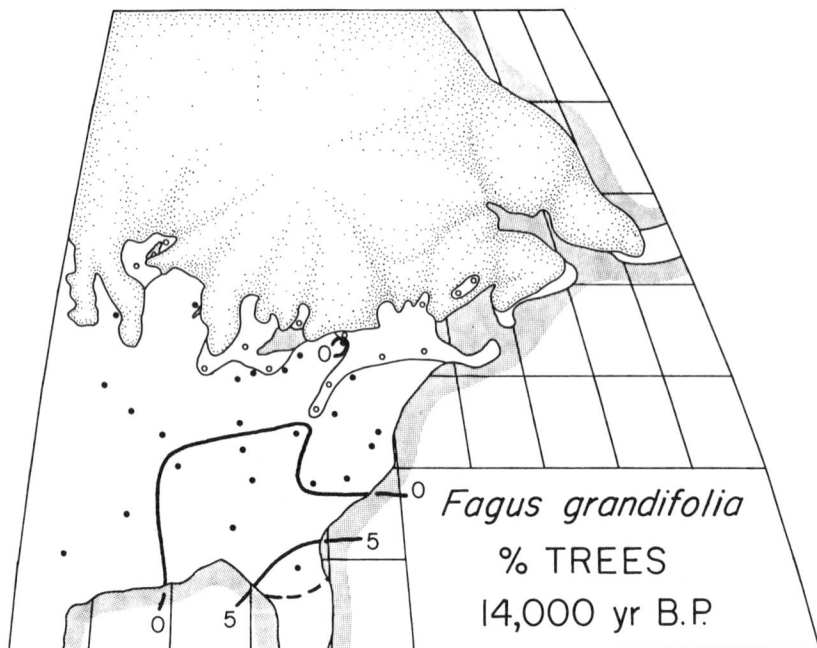

Figure 5.14. Beech (*Fagus grandifolia*). e. Paleo-dominance map for 14,000 yr B.P.

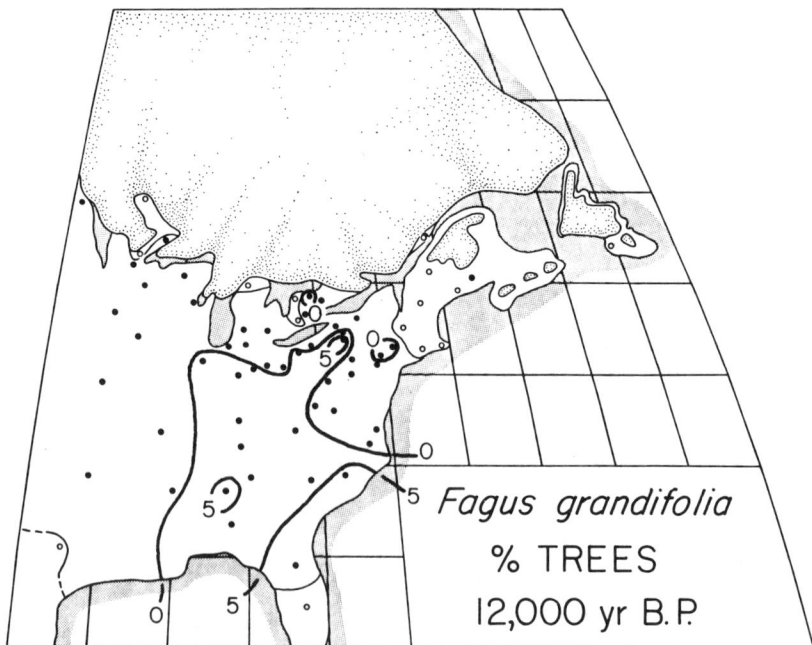

Figure 5.14. Beech (*Fagus grandifolia*). f. Paleo-dominance map for 12,000 yr B.P.

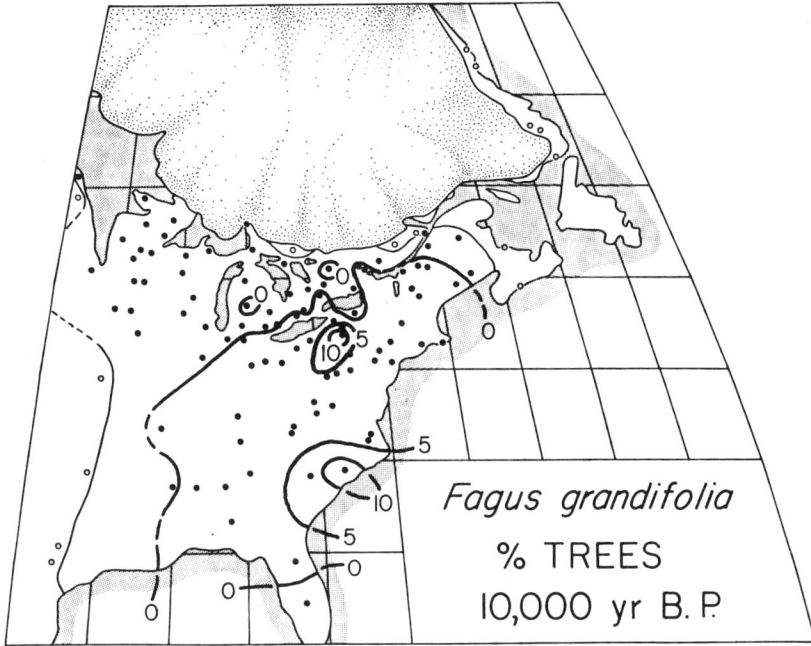

Figure 5.14. Beech (*Fagus grandifolia*). g. Paleo-dominance map for 10,000 yr B.P.

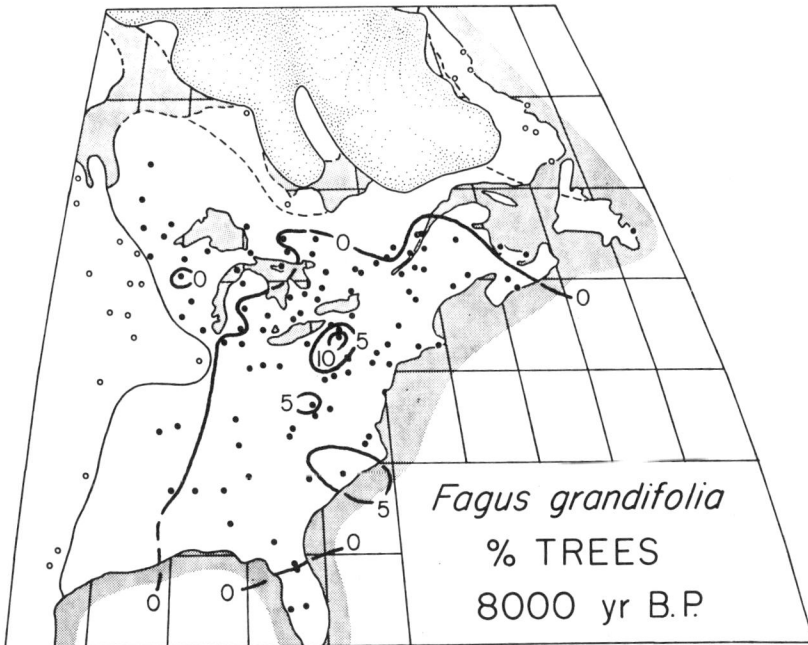

Figure 5.14. Beech (*Fagus grandifolia*). h. Paleo-dominance map for 8000 yr B.P.

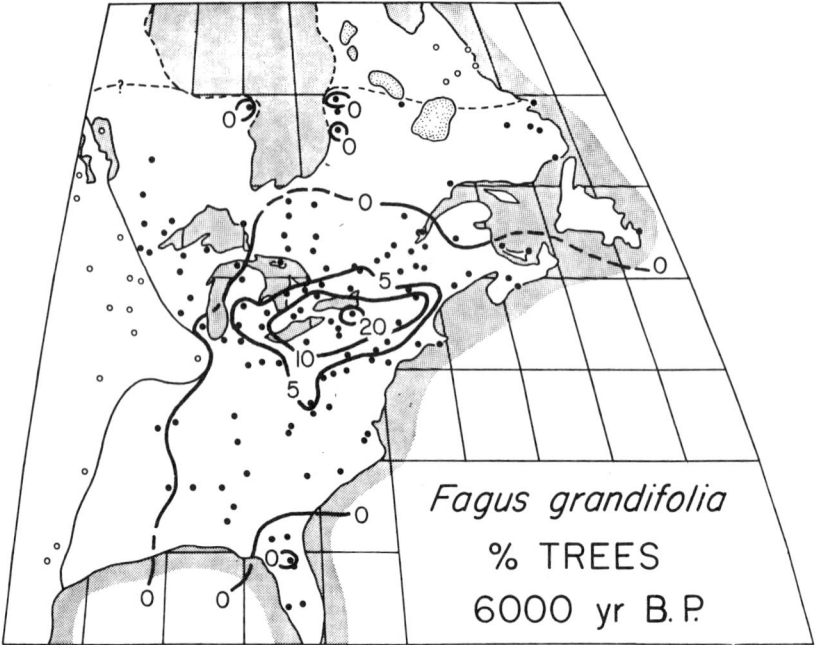

Figure 5.14. Beech (*Fagus grandifolia*). i. Paleo-dominance map for 6000 yr B.P.

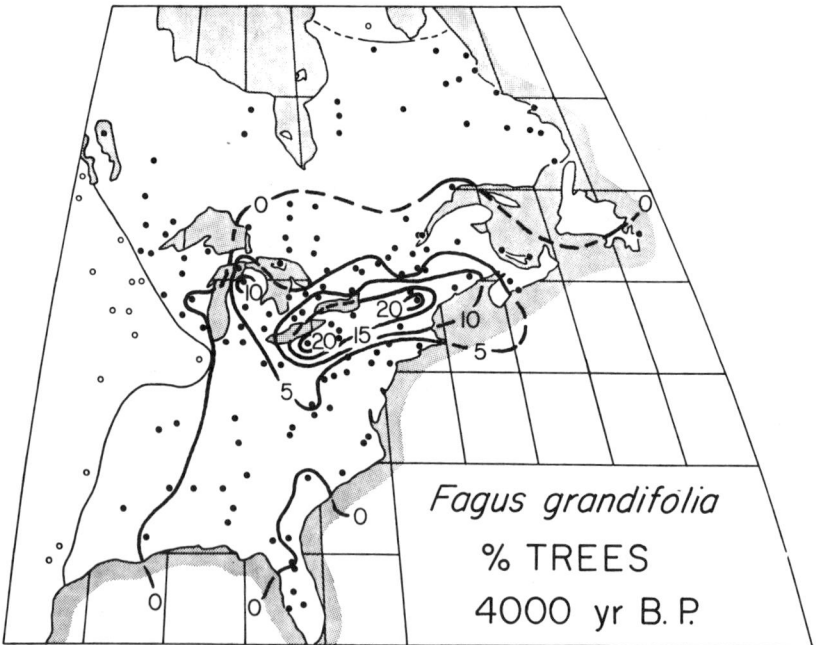

Figure 5.14. Beech (*Fagus grandifolia*). j. Paleo-dominance map for 4000 yr B.P.

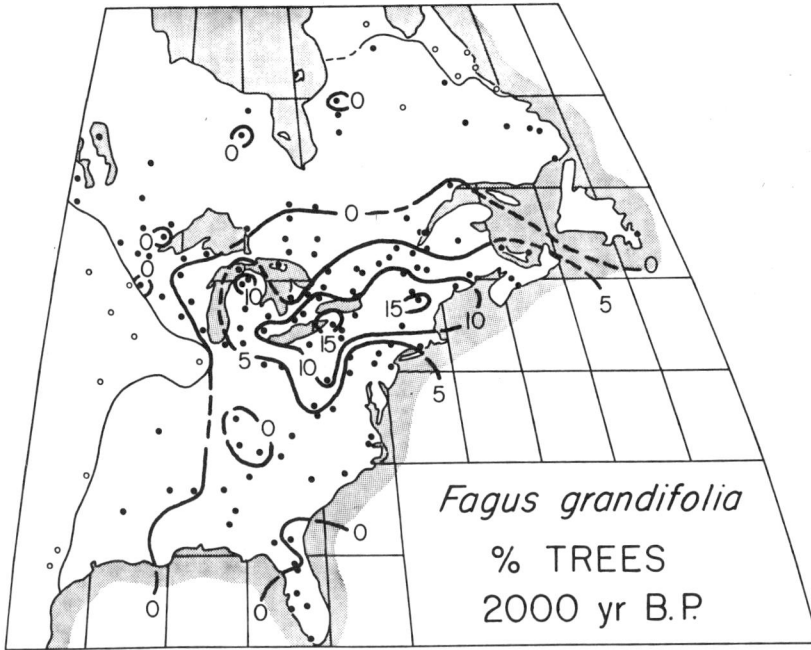

Figure 5.14. Beech (*Fagus grandifolia*). k. Paleo-dominance map for 2000 yr B.P.

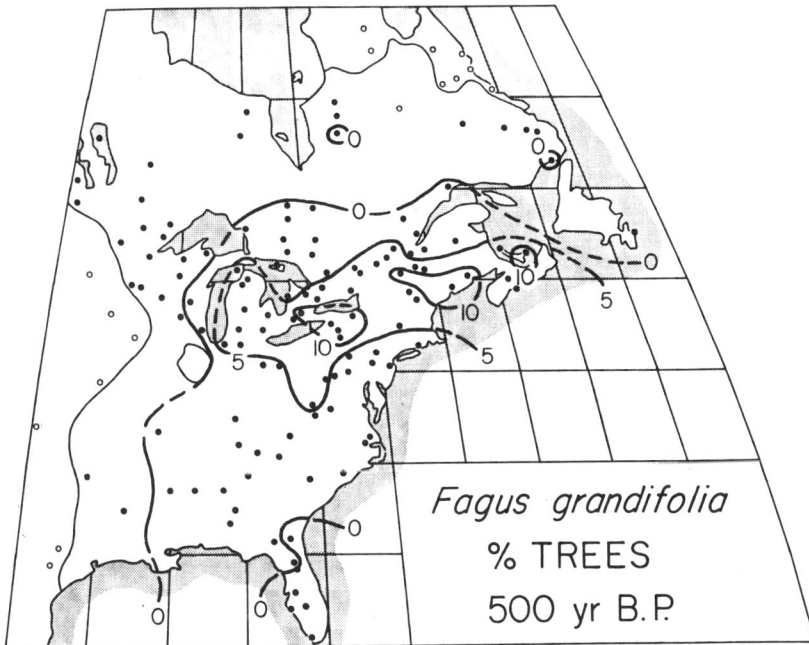

Figure 5.14. Beech (*Fagus grandifolia*). l. Paleo-dominance map for 500 yr B.P.

dominance values to a peak in the early Holocene at 11,000 yr B.P., and then generally declining from 10,000 yr B.P. to 8000 yr B.P. Beech populations then increased to an interglacial maximum between 6000 yr B.P. and 4000 yr B.P., decreasing once again in the last 4000 years. Mean dominance values for beech during full-glacial and late-glacial times were about 2% of the total forest composition. The mean values rose to approximately 3% between 12,000 yr B.P. and 7000 yr B.P. and stabilized at 5% from 5000 yr B.P. to 1000 yr B.P. During the last 1000 years, mean dominance values for beech dropped to 3%. For the maximum values reconstructed for beech, full-glacial values ranged from 2% to 3%. Maximum values rose through the late-glacial and early-Holocene intervals to as much as 24% by 11,000 yr B.P. Maximum dominance values then dropped to 10% by 10,000 yr B.P. and ranged between 10% and 15% between 10,000 and 7000 years ago. Peak values reconstructed for beech rose to 25% by 4000 yr B.P. and then declined to 14% by 500 yr B.P.

The time series of beech paleo-dominance maps for 20,000 yr B.P. (Fig. 5.14b), 18,000 yr B.P. (Fig. 5.14c), and 16,000 yr B.P. (Fig. 5.14d) illustrate limited populations of beech south of 34°N and east of 90°W, primarily located in the eastern and central Gulf Coastal Plain and the northern half of the Florida peninsula. For example, the map for 20,000 yr B.P. identifies areas of up to 3% beech dominance at scattered localities including Sheelar Lake, north-central Florida (Watts and Stuiver 1980) and Goshen Springs, south-central Alabama (Delcourt 1980). Macrofossils and trace amounts of pollen in full-glacial sediments of Nonconnah Creek, Tennessee (Delcourt *et al.* 1980) indicate that beech was present locally in outlier populations northwest of its mapped range between 20,000 yr B.P. and 16,000 yr B.P. (Figs. 5.14b–d).

By 14,000 yr B.P. (Fig. 5.14e), beech populations in Florida increased to 5% dominance. The northern perimeter of beech's distributional range extended northward through Middle Tennessee, across the Interior Low Plateaus, and northeastward along the Ridge and Valley Province to the central Appalachians of southwestern Virginia. The western border of *Fagus* distribution appeared to correspond with the location of loess-covered blufflands of West Tennessee, Mississippi, and southeastern Louisiana; the braided-stream environments of the Mississippi Alluvial Valley may have been a barrier to effective colonization of beech west of the river. A beech outlier of 2% dominance was located in the northern Appalachians, as indicated by pollen evidence from Rose Lake, central Pennsylvania (Cotter and Crowl 1981). This outlier represented the successful establishment of a colony of beech 5° latitude north of its main range at 14,000 yr B.P.

The beech paleo-dominance map for 12,000 yr B.P. (Fig. 5.14f) shows a prominent coastal zone of about 7% dominance extending from South Carolina south through northern Florida. In addition, more geographically restricted population centers were situated within the southern Ridge and Valley Province of central Alabama (6% at Cahaba Pond) and in central Pennsylvania along its northern distributional limit. The continuous 0% contour marking the northern range limit of beech extended from eastern North Carolina, across the Appalachians through Maryland and Pennsylvania, and westward through the central portions of Ohio, Indiana, and Illinois. Two northern disjunct colonies of beech established in eastern Pennsylvania–southern New York and in southern Ontario. The western continuous limit of beech extended to the eastern edge of the Mississippi Alluvial Valley at about 90°W.

Between 12,000 yr B.P. and 10,000 yr B.P., the principal population centers for beech shifted from the Southeast to the Northeast. For example, by 11,000 yr B.P., local population centers developed between approximately 33°N and 35°N, reaching reconstructed population levels of 11% in the uplands near Singletary Lake, North Carolina (Frey 1951; Whitehead 1973) and up to 24% at Cahaba Pond in the southern Ridge and Valley Province of north-central Alabama (Delcourt *et al.* 1983b). By 10,000 yr B.P. (Fig. 5.14g), one primary center for beech occurred in the vicinity of Rose Lake, Pennsylvania (10%), with another remaining in southeastern North Carolina (10%). By 10,000 years ago, beech was geographically widespread, although it rarely constituted > 3% of the total forest composition. Beech invaded coastal-plain habitats from Virginia northeast into central Maine and also colonized along peninsular extensions around Lake Erie and Lake Ontario. The continuous northern border of beech distribution (the 0% contour line in Fig. 5.14g), extended west across the northern New England states, through southern Quebec, and southwestward across southern Ontario, the Lower Peninsula of Michigan, and northern Indiana. Isolated outliers of beech established within several hundred kilometers of its continuous northern border. Beech also colonized the southeastern Ozark Plateau in the vicinity of Cupola Pond (Smith 1984). This may reflect an early-Holocene westward extension of beech beyond its general limits along the eastern side of the Lower Mississippi Alluvial Valley.

By 8000 yr B.P. (Fig. 5.14h), beech populations > 5% dominance occurred within the Carolina Coastal Plain as well as in the central and northern Appalachian Highlands. Beech continued to advance northeastward into Nova Scotia and New Brunswick, along the St. Lawrence River Valley, and north through the central Great Lakes region. The northern limit of beech 8000 yr B.P. extended across southern Quebec and into the region of east-central Ontario bordering the northeastern shore of Lake Huron. Beech populations occupied southern Ontario and the southeastern one-third of the Lower Peninsula of Michigan. With lowered lake levels of Great Lakes such as Lake Michigan at this time, water barriers to dispersal of beech may have been less effective. The isolated population of beech mapped in central Wisconsin (Fig. 5.14h) indicates continued long-distance dispersal and establishment of beech beyond its continuous leading edge of migration (Bennett 1985; Davis *et al.* 1986). During the mid-Holocene interval, beech appears to have retracted eastward in distribution, abandoning the Ozark Plateaus and becoming restricted generally to the east of the Mississippi River.

The paleo-dominance map for beech at 6000 yr B.P. (Fig. 5.14i) exhibits progressive fragmentation and local extinction of beech along its southeastern margin, shifting from northern Florida into southeastern Georgia. By 6000 yr B.P., a major beech population center with values up to 22% developed across the southern New England states, the northern and central Appalachians, and

Fagus grandifolia

A.

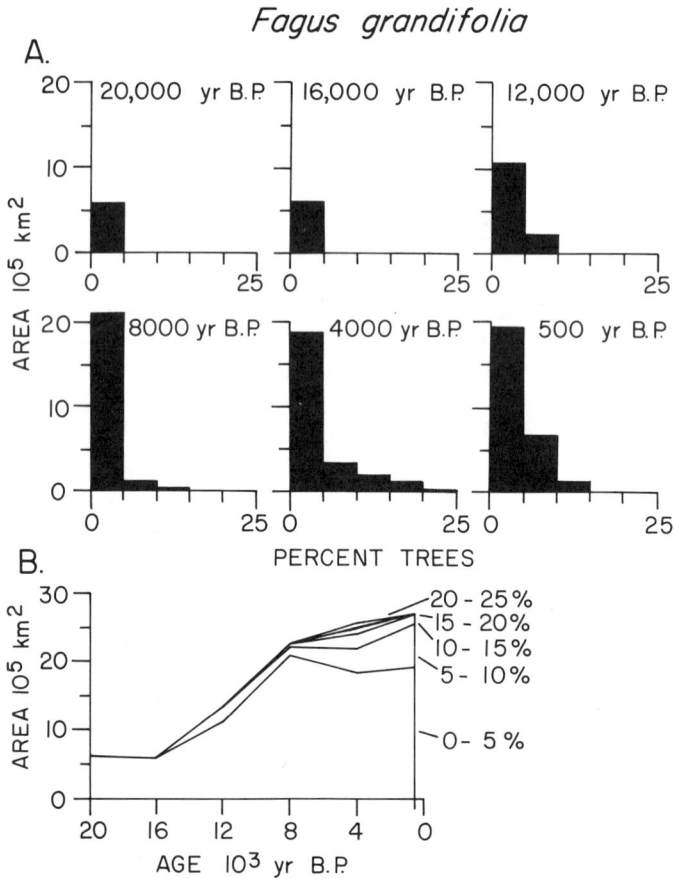

B.

Figure 5.15. Changes in area and dominance of beech (*Fagus grandifolia*). a. Histograms of area–dominance structure at 20,000, 16,000, 12,000, 8000, 4000, and 500 yr B.P. b. Cumulative plot from 20,000 yr B.P. to the present.

west through the central Great Lakes region. During the mid-Holocene interval, beech expanded north to 50°N and west nearly to Lake Superior. Beech also advanced to the northwest around Lake Huron into central Ontario and the Upper and Lower Peninsulas of Michigan. In addition, populations of beech spread around the southern shore of Lake Michigan and colonized southern Wisconsin. The western limit of beech generally extended to 90°W.

By 4000 yr B.P. (Fig. 5.14j), the primary population center for beech continued to expand in overall dominance values and to elongate as a band between 40°N and 46°N. Beech continued its migration northeast along the St. Lawrence River Valley and coastal Quebec. Elsewhere, the margins of its range remained relatively stationary. The maps for 2000 yr B.P. (Fig. 5.14k) and 500 yr B.P. (Fig. 5.14l) show a long-term decline in beech populations. Maximum dominance values dropped from a peak of 24% at 4000 yr B.P. to 18% at 2000 yr B.P. and

14% at 500 yr B.P. During the last 2000 years, beech has retreated along its northern margin but advanced to the northwest through the Upper Peninsula of Michigan as well as through Wisconsin. Key population centers persisted from Michigan east to southern New England, and northeast across New Brunswick, Nova Scotia, and Prince Edward Island.

The area–dominance histograms for beech (Fig. 5.15a,b) show a gradual increase in both area and number of dominance classes through time. The histograms for 20,000 yr B.P. and 16,000 yr B.P. are represented only by the 0% to 5% dominance class, with total mapped area remaining at $6.0 \cdot 10^5$ km^2. The total area occupied by beech increased rapidly during the late-glacial and early-Holocene intervals and then expanded more slowly from 8000 yr B.P. to the present, reaching $27.3 \cdot 10^5$ km^2 by 500 yr B.P. (Fig. 5.15b). Through the mid- and late-Holocene intervals, the number of dominance classes increased as beech increased its populations in the Great Lakes and New England regions. For example, at 4000 yr B.P., 26% of beech's distribution was dominated by $\geqslant$ 5% beech. By 500 yr B.P., 29% of the area of beech was $\geqslant$ 5% dominance.

Ash (*Fraxinus*)

The population trajectory for ash (Fig. 5.16a) illustrates very low mean dominance values (about 2%) from 20,000 yr B.P. to 17,000 yr B.P. In addition during that time interval, the variability about the mean values was low ($\pm$ 1% SD), with maximum reconstructed values from 3% to 5%. Dominance values for ash increased through the late-glacial interval, with highest reconstructed values for both mean and maximum dominance (8% and 51%, respectively) at 12,000 yr B.P. From 12,000 yr B.P. to 9000 yr B.P., dominance values for ash declined dramatically. Between 9000 yr B.P. and 6000 yr B.P., mean values were consistently about 3%, and maximum values were about 10%. After 6000 yr B.P., dominance of ash increased, with maximum values ranging between 17% and 26% in the interval from 5000 yr B.P. to 3000 yr B.P., and with mean values between 3% and 4%. During the last 3000 years, mean dominance values remained at 3%, although maximum values dropped to 9% at 500 yr B.P. and increased to 16% at 0 yr B.P.

The full-glacial paleo-dominance maps for ash (Figs. 5.16b,c) display ash populations located generally throughout the southern portion of unglaciated eastern North America. The paleo-dominance map for 20,000 yr B.P. (Fig. 5.16b) illustrates a northern boundary of ash distribution at about 37°N to 38°N, extending from northern North Carolina across central Kentucky and west to the continental interior. Although widely distributed through the Interior Low Plateaus and central to west portions of the Gulf Coastal Plain, ash typically constituted < 4% of overall forest composition. An irregular southeastern margin of ash distribution at both 20,000 yr B.P. and 18,000 yr B.P. extended from southern North Carolina west to northern Georgia and then south to northwestern Florida. By 18,000 yr B.P. (Fig. 5.16c), limited populations of ash reconstructed in northeastern Kansas (Gruger 1973) indicate a minor expansion of ash into the continental interior.

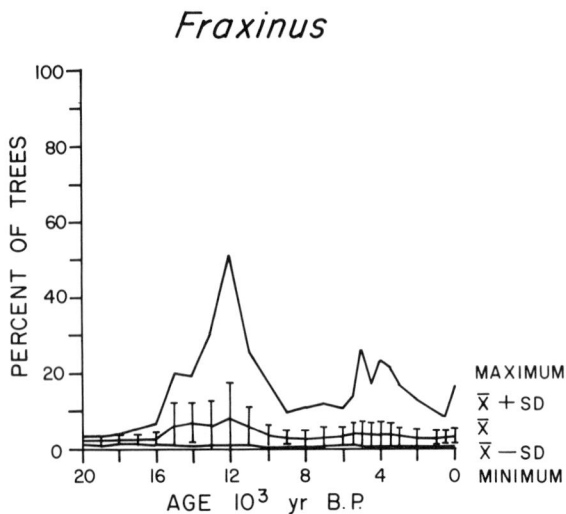

Figure 5.16. Ash (*Fraxinus*). a. Population trajectory, including maximum, minimum, and mean values of paleo-dominance as well as ± 1 SD of the mean, for the past 20,000 years.

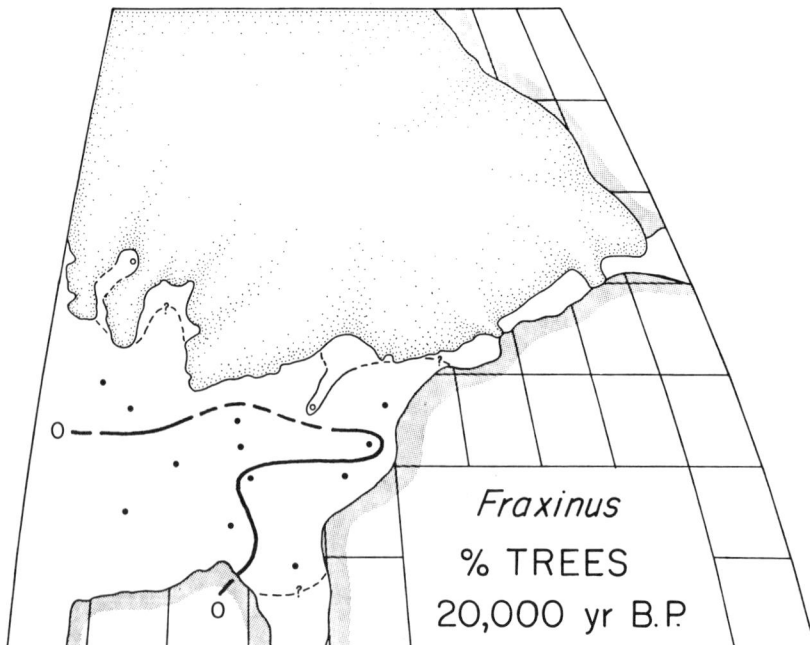

Figure 5.16. Ash (*Fraxinus*). b. Paleo-dominance map for 20,000 yr B.P.

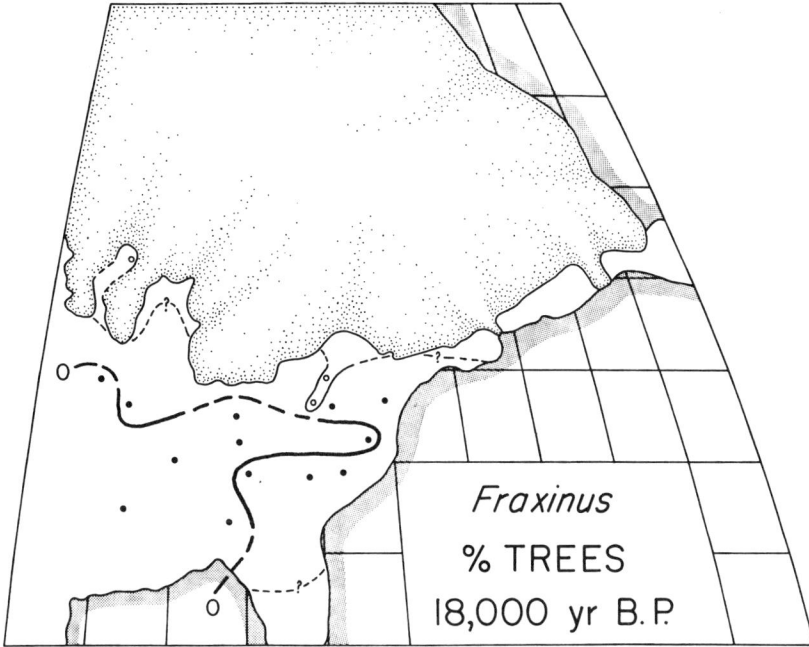

Figure 5.16. Ash (*Fraxinus*). c. Paleo-dominance map for 18,000 yr B.P.

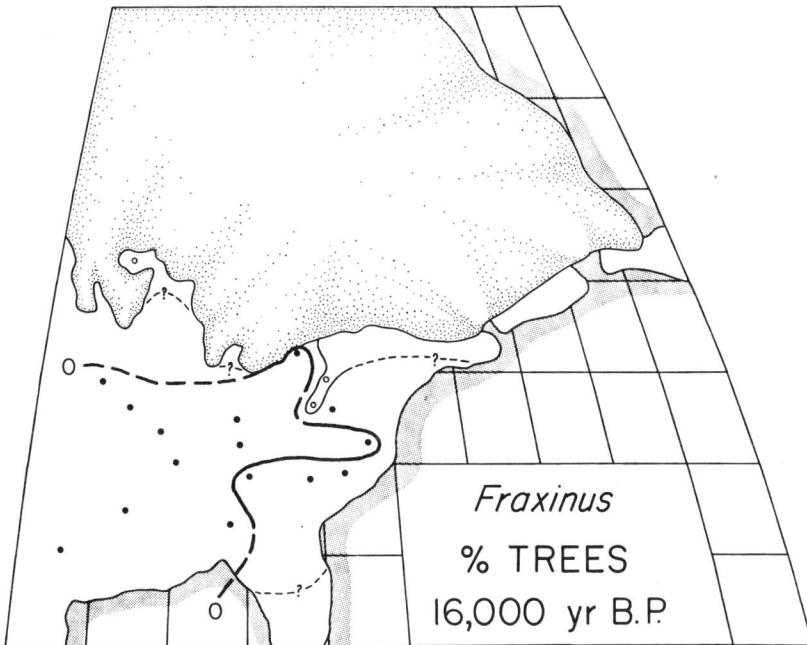

Figure 5.16. Ash (*Fraxinus*). d. Paleo-dominance map for 16,000 yr B.P.

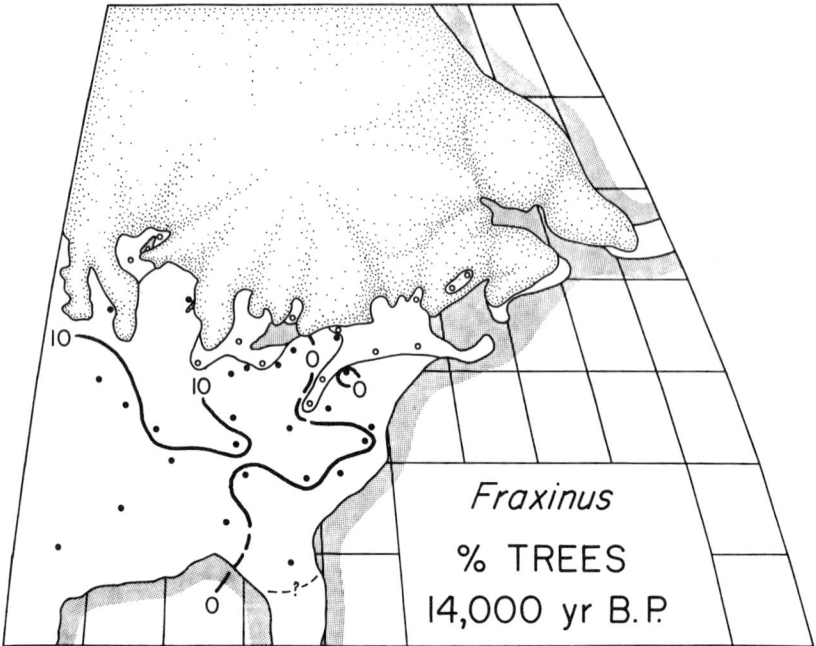

Figure 5.16. Ash (*Fraxinus*). e. Paleo-dominance map for 14,000 yr B.P.

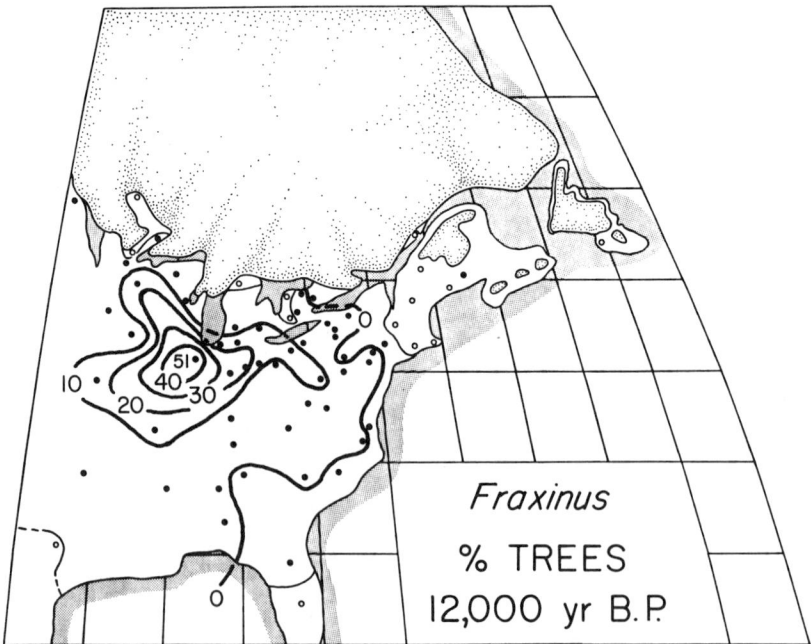

Figure 5.16. Ash (*Fraxinus*). f. Paleo-dominance map for 12,000 yr B.P.

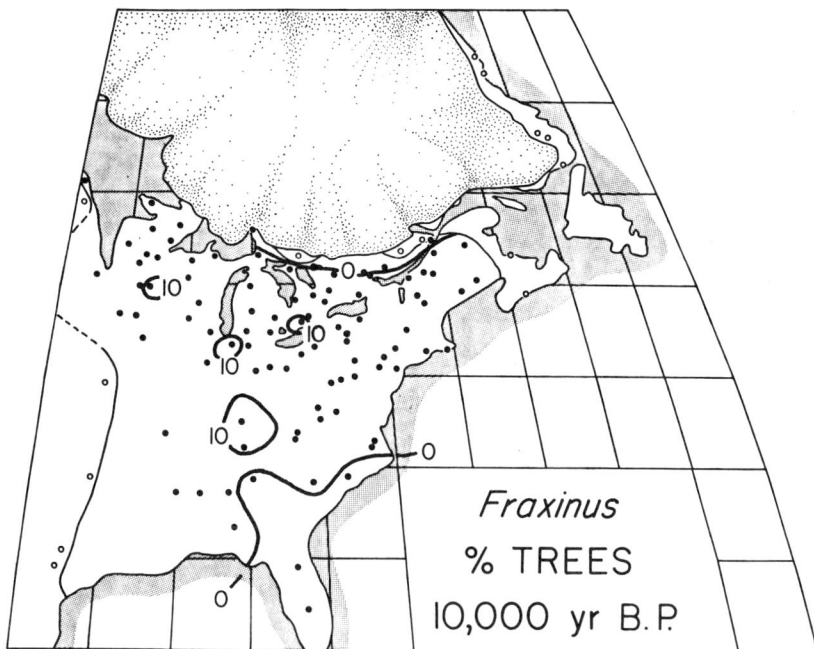

Figure 5.16. Ash (*Fraxinus*). g. Paleo-dominance map for 10,000 yr B.P.

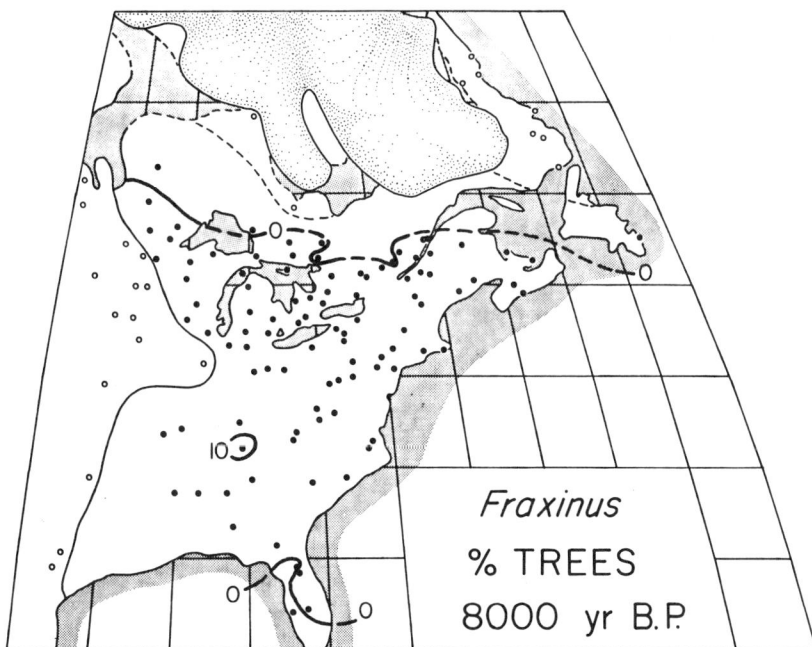

Figure 5.16. Ash (*Fraxinus*). h. Paleo-dominance map for 8000 yr B.P.

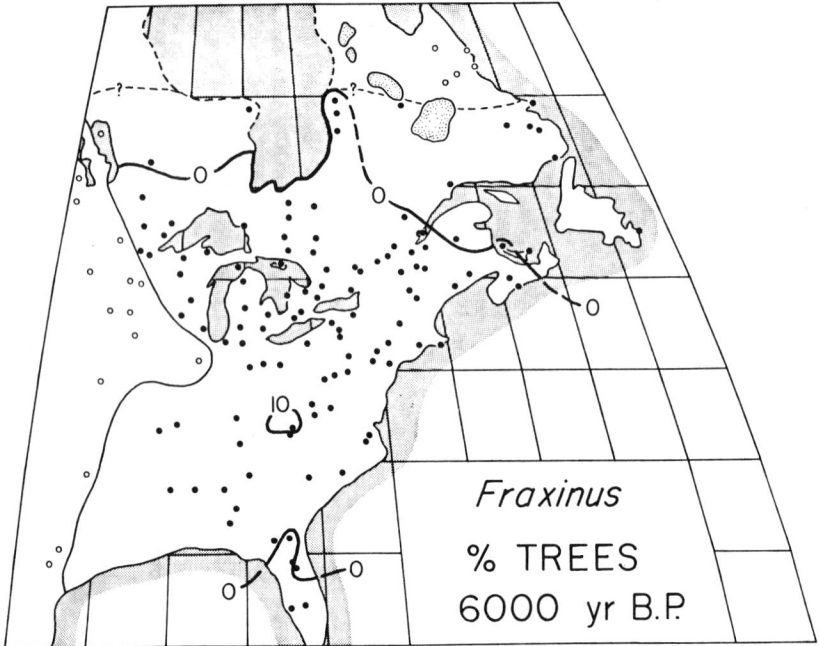

Figure 5.16. Ash (*Fraxinus*). i. Paleo-dominance map for 6000 yr B.P.

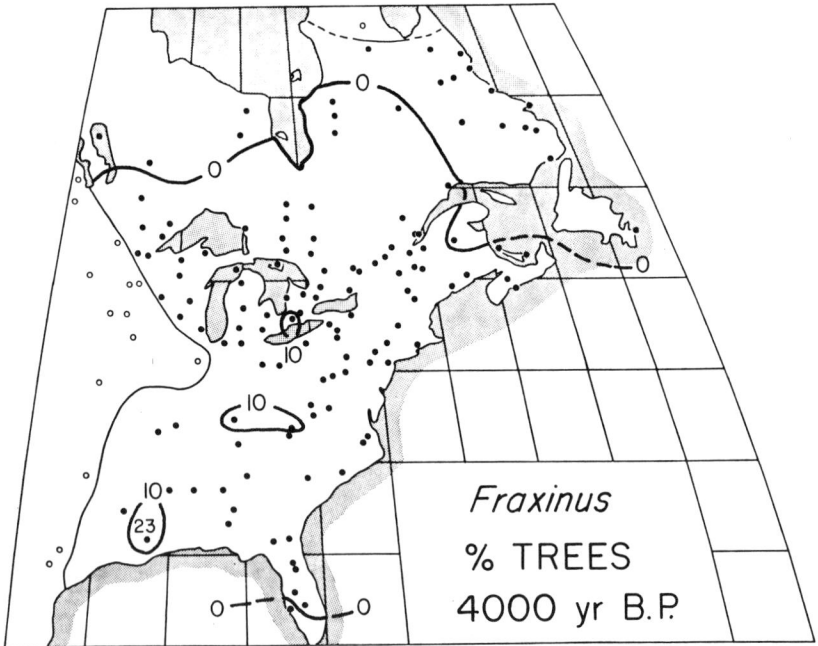

Figure 5.16. Ash (*Fraxinus*). j. Paleo-dominance map for 4000 yr B.P.

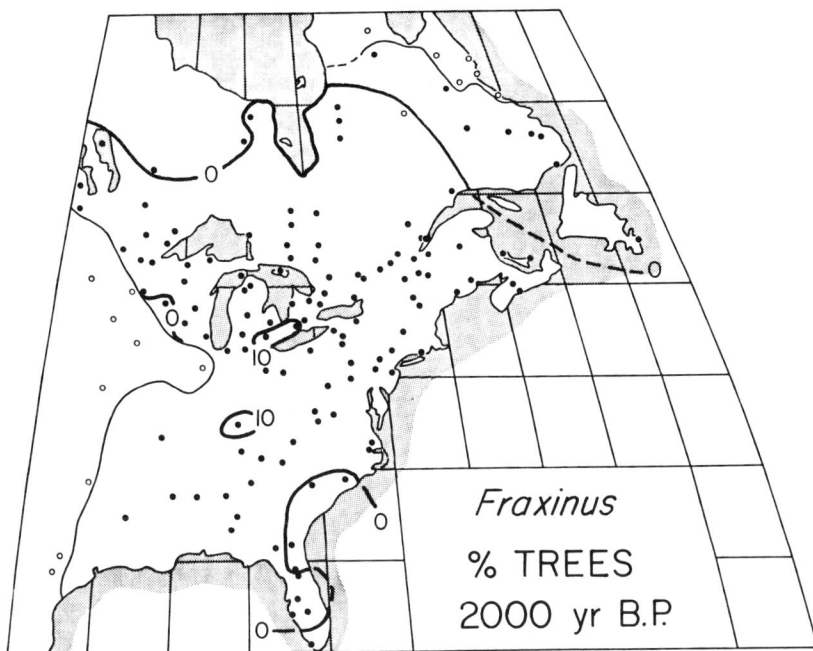

Figure 5.16. Ash (*Fraxinus*). k. Paleo-dominance map for 2000 yr B.P.

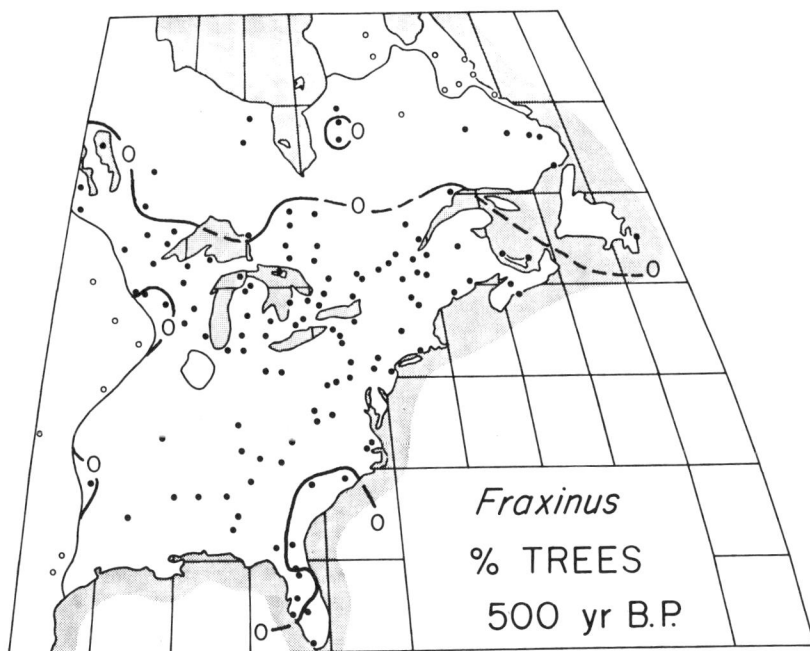

Figure 5.16. Ash (*Fraxinus*). l. Paleo-dominance map for 500 yr B.P.

By 16,000 yr B.P. (Fig. 5.16d), ash advanced northward along the central Appalachian Highlands and into eastern Ohio, and its northern boundary at that time is speculatively mapped at 40°N across central Indiana, Illinois, and northern Missouri. However, ash did not invade coastal-plain habitats east of the Appalachians (*e.g.*, through Virginia) in the early late-glacial interval. The map for 14,000 yr B.P. (Fig. 5.16e) displays a northward advance of ash through deglaciated terrain north to about 45°N in the western Great Lakes region. Dominance values of 10% to 19% ash were recorded across the region from Middle Tennessee and southeastern Missouri northwest through Iowa and into Wisconsin. East of the Appalachians, disjunct outliers of ash established in central Pennsylvania. The southeastern margin of ash distribution expanded to include central South Carolina. Local, late-glacial extinctions in the south-western range of ash were reflected by the decreasing dominance percentages across Louisiana and eastern Texas.

By 12,000 yr B.P. (Fig. 5.16f), ash grew throughout nearly all of the forested region in eastern North America. Along its northern border, ash had established populations throughout the early-Holocene forests nearly to the forest–tundra ecotone in the northeast, as well as to the margins of proglacial lakes and glacial ice in the north and northwest. Along the southern margin of distribution, ash was restricted to the north and west of Georgia and Florida. The expansion of prairie in eastern Texas also displaced the southwestern limit of ash to the east. At 12,000 yr B.P., ash was a dominant tree over a large region of the Midwest, reaching values from 10% to 51% across the area from Kansas to Maryland and from Missouri to Minnesota. The peak value of 51% reconstructed for Chatsworth Bog, Illinois (King 1981) is the highest value for ash in the time series of paleo-dominance maps (Fig. 5.16a).

During the early Holocene, between 12,000 yr B.P. and 10,000 yr B.P. (Fig. 5.16g), dominance values for ash decreased throughout the Midwest region. By 10,000 yr B.P., only four limited areas remained with dominance values for ash between 10% and 17%. These areas included (1) Middle Tennessee and central Kentucky; (2) northeastern Indiana; (3) south-central Minnesota; and (4) southern Ontario. Elsewhere, ash advanced to the forest limit, bordered by tundra in Nova Scotia, New Brunswick, and coastal Quebec, bounded by glacial ice in southern Quebec and Ontario, and limited by proglacial lakes including Agassiz and Superior. The western border of ash coincided with the eastern limit of prairie.

The 8000 yr B.P. map (Fig. 5.16h) illustrates that relative dominance of ash was typically < 10% throughout its range in the mid-Holocene interval. The northern irregular border of its distribution traced an arc from about 48°N in southeastern Canada to about 50°N in central Canada. Between 10,000 yr B.P. and 8000 yr B.P., ash extended southeast of its previous distribution on its southern margin, colonizing Georgia and as far south as central Florida. The western limit of ash retreated, particularly through Iowa and Illinois. By 6000 yr B.P. (Fig. 5.16i), ash extended northward throughout the coastal zone surrounding Hudson Bay, but remained relatively stationary along its western and southern margins, and was < 10% dominance throughout its range.

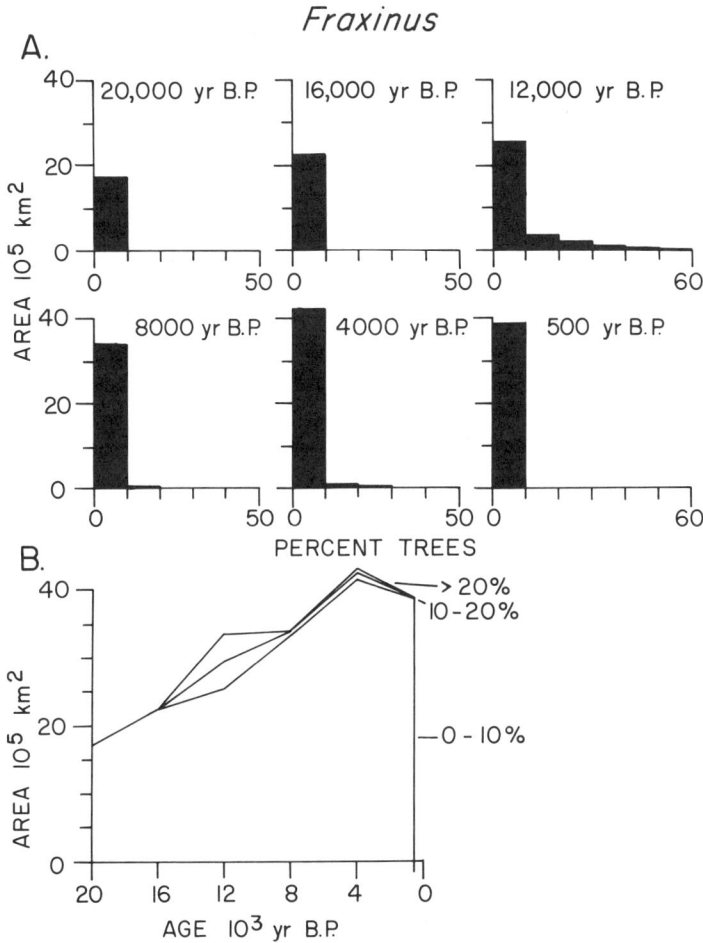

Figure 5.17. Changes in area and dominance of ash (*Fraxinus*). a. Histograms of area–dominance structure at 20,000, 16,000, 12,000, 8000, 4000, and 500 yr B.P. b. Cumulative plot from 20,000 yr B.P. to the present.

During the late Holocene, geographically limited population centers of ash developed by 4000 yr B.P. (Fig. 5.16j) in the Lower Mississippi Alluvial Valley (up to 23%), Kentucky and southwestern Virginia (10% to 18%), and southern Ontario (up to 12%). By 4000 yr B.P., ash invaded northeastward through central and northern Quebec to about 55°N. West of Hudson Bay, ash advanced to approximately 51°N. By 2000 yr B.P. (Fig. 5.16k), the northern limit of ash extended from the Gulf of St. Lawrence northwest across Quebec to about 55°N along the shore of Hudson Bay, then southwest through central Ontario into central Manitoba. As the prairie–forest ecotone was displaced westward, ash invaded southern Manitoba, western Minnesota, and Wisconsin. From 2000 yr B.P. to 500 yr B.P. (Fig. 5.16l), the northern limit of ash shifted south to

50°N in eastern Canada. The western border continued to expand as forest displaced prairie. By presettlement times, ash was geographically widespread, although as a genus it comprised < 10% of the forest composition of temperate and southern boreal forest.

The area–dominance histograms for ash (Fig. 5.17a) display an increase from 1 dominance class (0% to 10%) at 20,000 yr B.P. and 16,000 yr B.P. to a maximum number of 6 dominance classes at 12,000 yr B.P. Thereafter, the number of dominance classes declined to 2 classes at 8000 yr B.P., with 3 at 4000 yr B.P., and only 1 at 500 yr B.P. During the 8000-year span from 20,000 yr B.P. to 12,000 yr B.P., the area occupied by ash increased by 66%, from $17.2 \cdot 10^5$ km^2 to $28.5 \cdot 10^5$ km^2. Despite the general decrease in number of dominance classes throughout the Holocene, ash expanded in total distribution to an interglacial maximum area of $43.2 \cdot 10^5$ km^2 at 4000 yr B.P. (Fig. 5.17b). Although both area–dominance histograms for 20,000 yr B.P. and 500 yr B.P. exhibit comparable population structure, with only the lowest dominance class represented, the change in ash distribution represents an increase of 227% over the glacial–interglacial cycle (at 500 yr B.P., the area occupied was $39.0 \cdot 10^5$ km^2).

Walnut (*Juglans*)

Mean values of dominance reconstructed for walnut remained at levels of ≤ 1.5% throughout the interval from 20,000 yr B.P. to the present (Fig. 5.18a). Maximum dominance values were typically < 2% from 20,000 yr B.P. to 11,000 yr B.P. and from 9000 yr B.P. to 7000 yr B.P. Peak values of 5% were reached at 10,000 yr B.P., and maximum values ranged from 3% to 7% during the mid- and late-Holocene intervals (Fig. 5.18a).

On the paleo-dominance map for 20,000 yr B.P. (Fig. 5.18b), walnut is reconstructed as 0.3% of the forest composition in north-central Louisiana, based upon palynological evidence from Rayburn's Salt Dome (Kolb and Fredlund 1981). Walnut is not yet known from any other site in eastern North America at 20,000 yr B.P. At Rayburn's Salt Dome, walnut increased to 1.4% dominance at 19,000 yr B.P. but was not represented at 18,000 yr B.P. No paleoecological sites dating from 18,000 yr B.P., 17,000 yr B.P., or 16,000 yr B.P. contain pollen evidence for reconstruction of paleo-dominance of walnut. However, a macrofossil nut of black walnut (*Juglans nigra*) from the Nonconnah Creek site in southwestern Tennessee was radiocarbon-dated at 17,200 yr B.P. (Delcourt *et al.* 1980), documenting the occurrence of walnut during the full-glacial interval locally along the river bluffs east of the Lower Mississippi Alluvial Valley. At 14,000 yr B.P. (Fig. 5.18c), walnut appeared again at 1.4% dominance only at Rayburn's Salt Dome, indicating that its populations remained rare and widely dispersed throughout the full- and late-glacial intervals. By 13,000 yr B.P., however, small populations of walnut extended along the Mississippi Alluvial Valley from southeastern Louisiana through west-central Mississippi and was also present in Middle Tennessee.

During the early Holocene, major changes occurred in the distribution of

Juglans

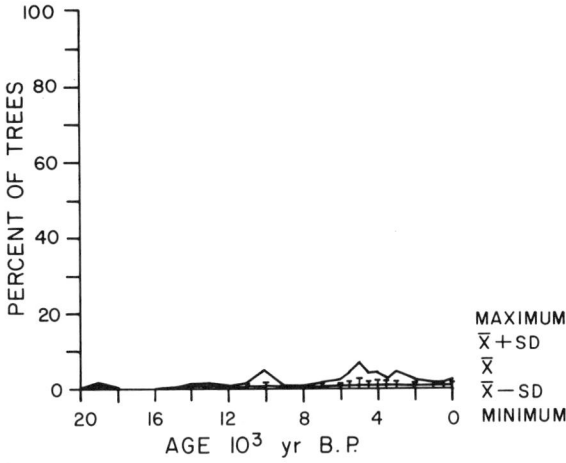

Figure 5.18. Walnut (*Juglans*). a. Population trajectory, including maximum, minimum, and mean values of paleo-dominance as well as ± 1 SD of the mean, for the past 20,000 years.

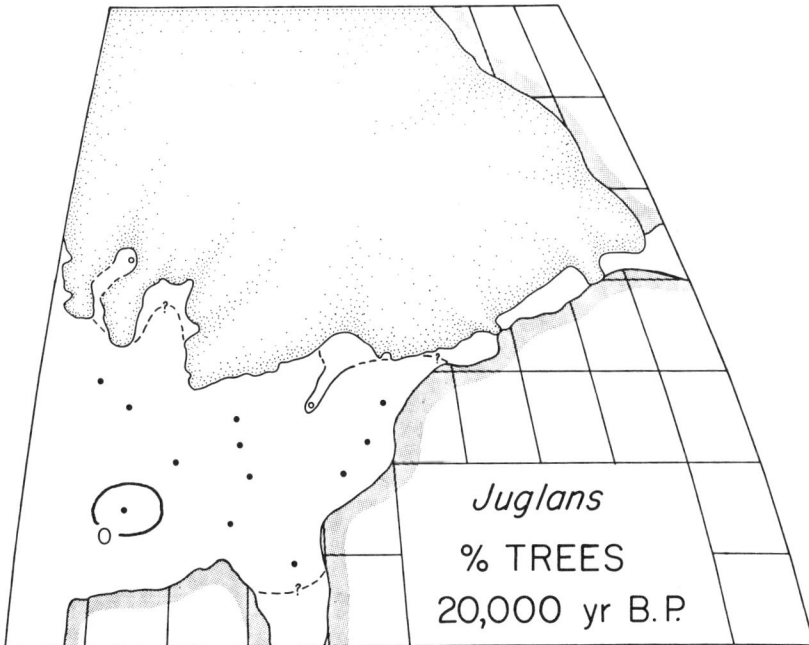

Figure 5.18. Walnut (*Juglans*). b. Paleo-dominance map for 20,000 yr B.P.

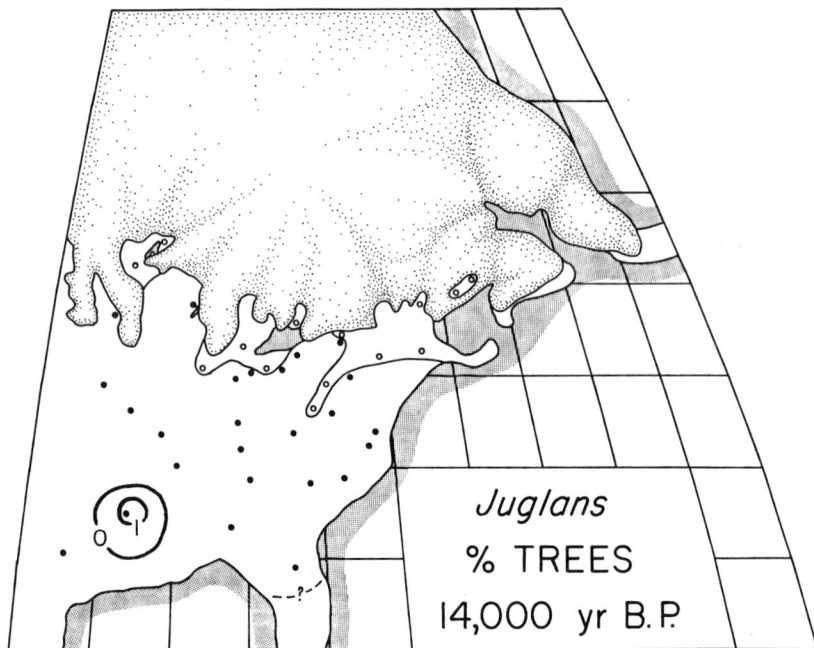

Figure 5.18. Walnut (*Juglans*). c. Paleo-dominance map for 14,000 yr B.P.

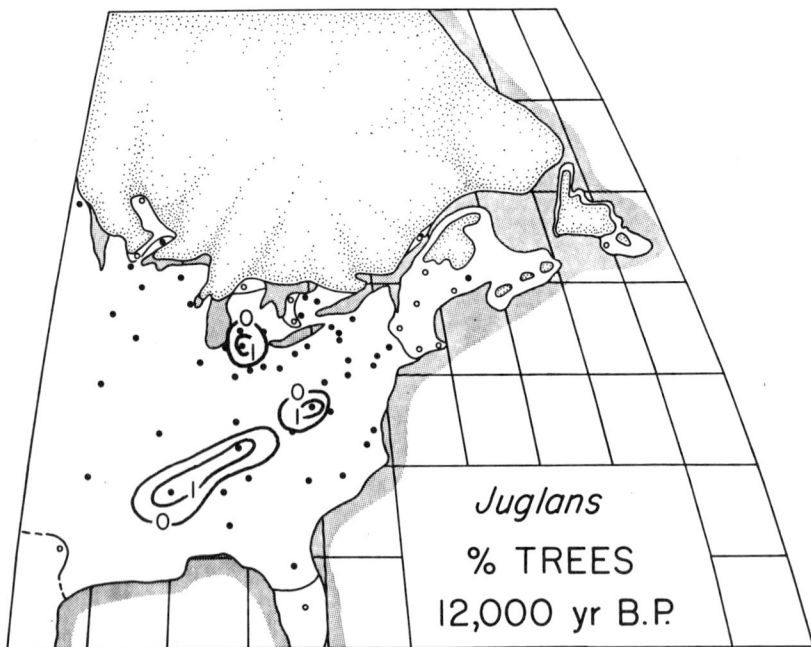

Figure 5.18. Walnut (*Juglans*). d. Paleo-dominance map for 12,000 yr B.P.

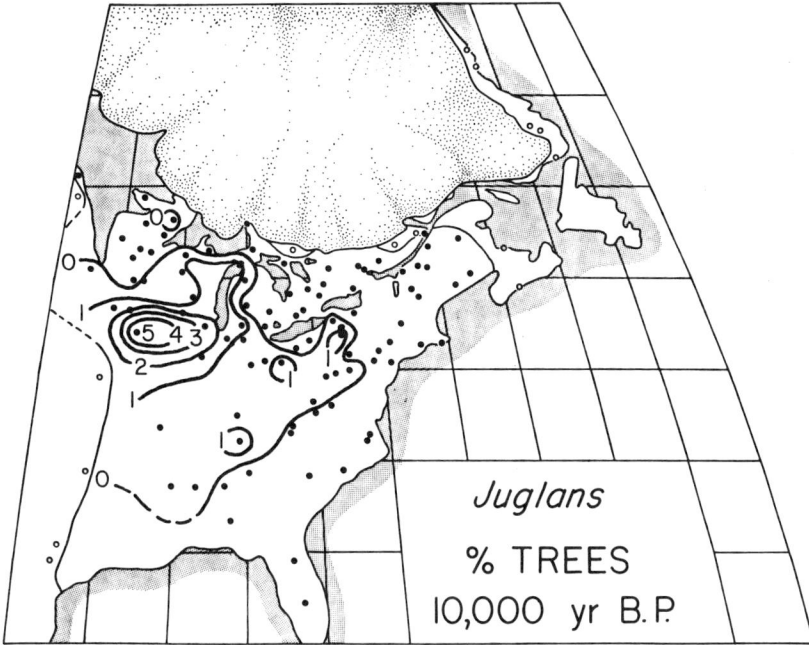

Figure 5.18. Walnut (*Juglans*). e. Paleo-dominance map for 10,000 yr B.P.

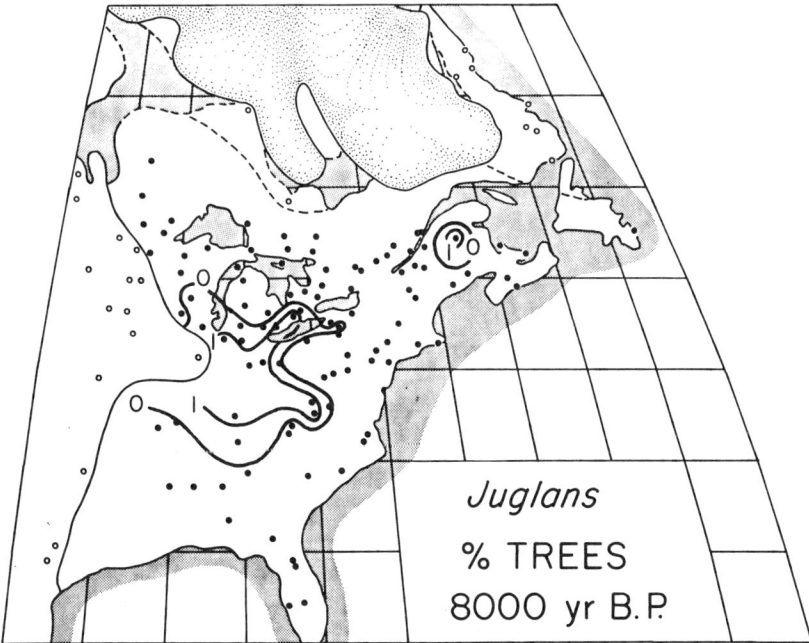

Figure 5.18. Walnut (*Juglans*). f. Paleo-dominance map for 8000 yr B.P.

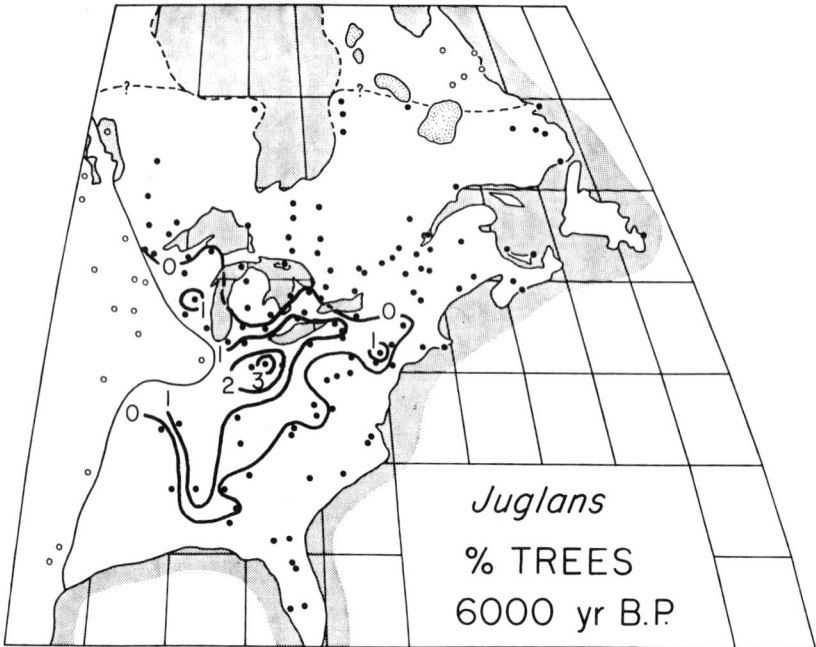

Figure 5.18. Walnut (*Juglans*). g. Paleo-dominance map for 6000 yr B.P.

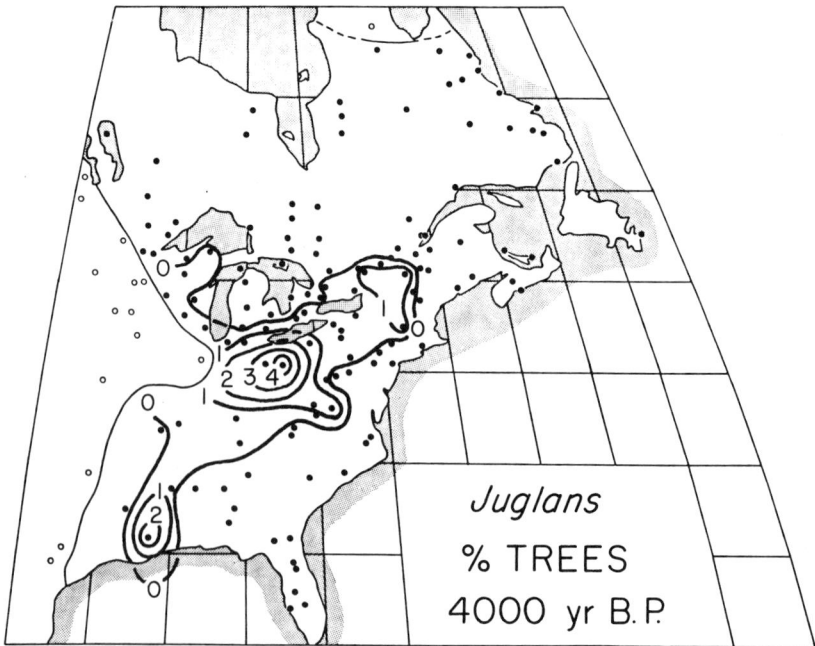

Figure 5.18. Walnut (*Juglans*). h. Paleo-dominance map for 4000 yr B.P.

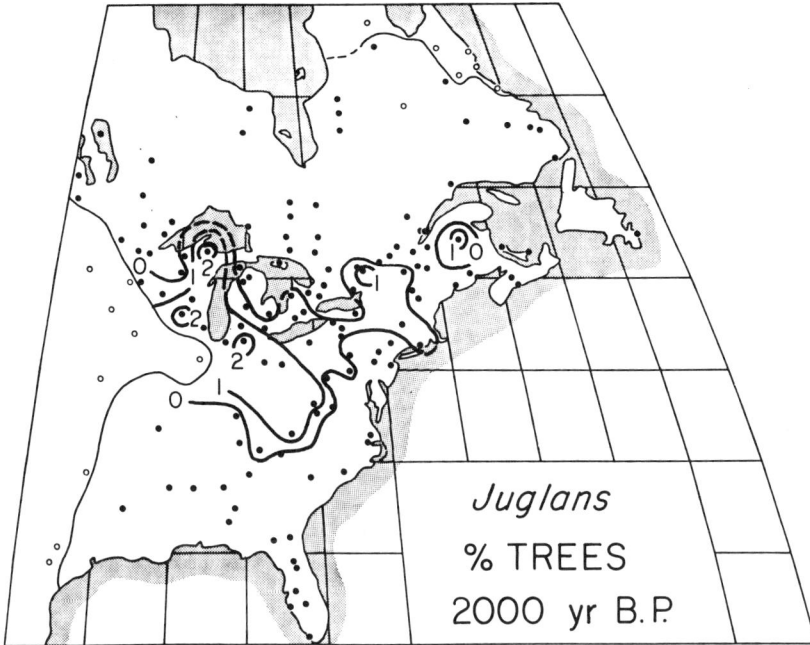

Figure 5.18. Walnut (*Juglans*). i. Paleo-dominance map for 2000 yr B.P.

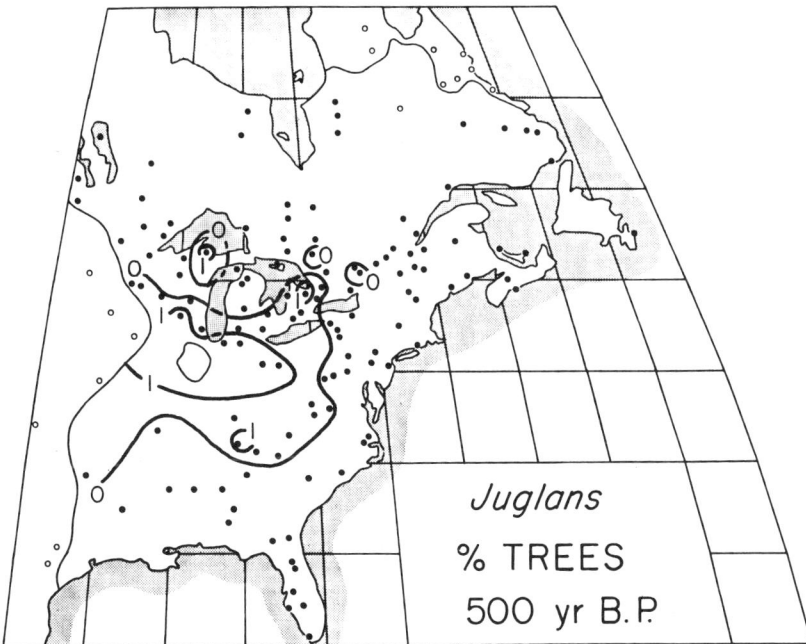

Figure 5.18. Walnut (*Juglans*). j. Paleo-dominance map for 500 yr B.P.

walnut. Limited populations of 1% extended from west-central Mississippi to Middle Tennessee by 12,000 yr B.P. (Fig. 5.18d). Additional outliers were established in central Maryland and northeastern Indiana (Fig. 5.18d). Between 12,000 yr B.P. and 10,000 yr B.P. (Fig. 5.18e), walnut expanded in distribution northeastward along the Appalachians to Pennsylvania and western New York. In addition, walnut extended north from the Lower Mississippi Alluvial Valley across the Interior Low Plateaus and Ozark Plateaus, reaching dominance values of 1% to 5% across the western Great Lakes region. The northern continuous limit of walnut's range at 10,000 yr B.P. extended from western Pennsylvania and northern Ohio through Michigan, northern Wisconsin, southern Minnesota, and eastern South Dakota (Fig. 5.18e). The western boundary of *Juglans* distribution coincided with the prairie–forest ecotone. Walnut occurred consistently at almost every fossil-pollen site throughout its range at 10,000 yr B.P., although it was generally represented at low values.

Between 10,000 yr B.P. and 8000 yr B.P., the area of distribution constricted on all borders, and at 8000 yr B.P. the center of dominance for walnut was located in the geographic center of its distribution, within Indiana, Ohio, and Kentucky (Fig. 5.18f). By 8000 yr B.P. (Fig. 5.18f), walnut populations extended east of the prairie–forest boundary from Minnesota and Illinois to the central Appalachians. At that time, however, walnut was confined to a latitudinal belt from 35°N to 44°N. The walnut paleo-dominance map for 6000 yr B.P. (Fig. 5.18g) illustrates a subsequent extension of the population center (1% to 3%) south into the central Gulf Coastal Plain of Mississippi, with the irregular margin of walnut distribution extending eastward into southern New England, northeast through southern Ontario, and northwest across northern Wisconsin and central Minnesota. The western limit bordered the prairie–forest ecotone throughout the mid- and late-Holocene intervals.

Walnut's paleo-dominance map for 4000 yr B.P. (Fig. 5.18h) documents the primary population center of 1% to 4% in Indiana and Ohio, with peninsular extensions into the central Appalachians and into the western Great Lakes as far as Wisconsin. Secondary population centers of walnut of 1% occurred in the northern Appalachians of New York and 1% to 2% in the Mississippi Alluvial Valley of northern Louisiana. The maps of walnut for 2000 yr B.P. (Fig. 5.18i) and 500 yr B.P. (Fig. 5.18j) illustrate the progressive fragmentation of walnut populations with a general retraction of continuous distributional limits on both their northern and southern borders. The primary center for walnut persisted in Ohio and Indiana and spread westward through Illinois. By 500 yr B.P., the area of continuous distribution of walnut corresponded generally with the region of eastern deciduous forest (Fig. 4.2a).

The area–dominance histograms for walnut (Fig. 5.19a) depict a minimum in distributional area and dominance classes during the full- and late-glacial intervals. The Holocene interval exhibits a marked increase in area, with as many as 5 dominance classes (from 0% to 5% in 1% increments) prominent on the maps for 10,000 yr B.P. and 4000 yr B.P. The histogram for 4000 yr B.P. shows a population structure with a nearly logarithmic decrease in area occupied by progressively higher dominance classes (Fig. 5.19a). Although widespread

Juglans

A.

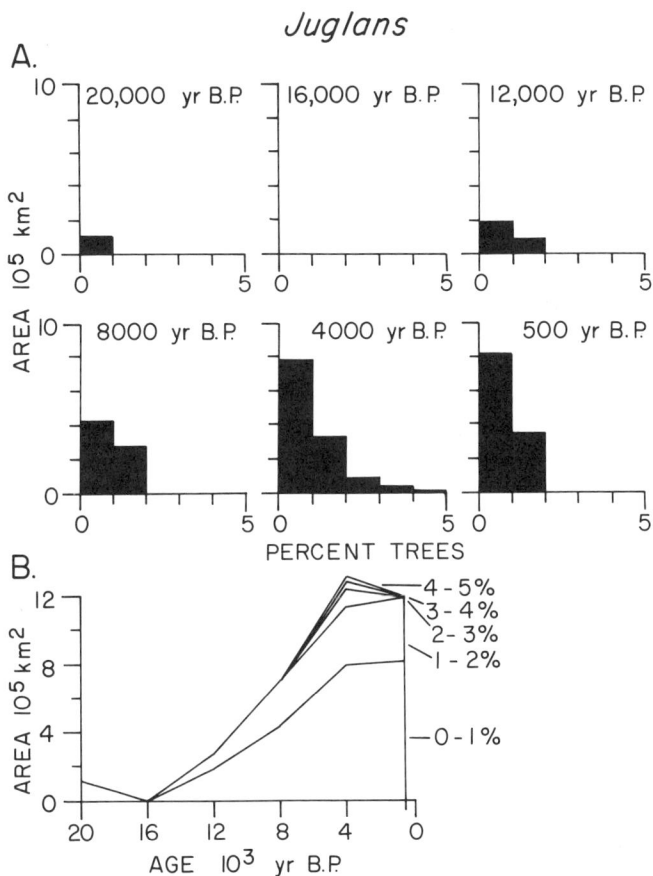

Figure 5.19. Changes in area and dominance of walnut (*Juglans*). a. Histograms of area–dominance structure at 20,000, 16,000, 12,000, 8000, 4000, and 500 yr B.P. b. Cumulative plot from 20,000 yr B.P. to the present.

in overall distribution, walnut constituted more than 1% of the late-Holocene forest composition only in a restricted geographic area. Walnut increased in area from $1.1 \cdot 10^5$ km^2 at 20,000 yr B.P. to $12.5 \cdot 10^5$ km^2 by 4000 yr B.P. (Fig. 5.19b).

Tamarack (*Larix*)

The population trajectory for tamarack (Fig. 5.20a) displays mean dominance values of < 2% from 20,000 yr B.P. to 17,000 yr B.P. The mean dominance value increased to 11% at 16,000 yr B.P. (however, this is based upon its occurrence at only one site). After 16,000 yr B.P., mean values were consistently between 2% and 4% for the remainder of the late Pleistocene and Holocene. Maximum dominance values for tamarack were typically low during the full-

Larix laricina

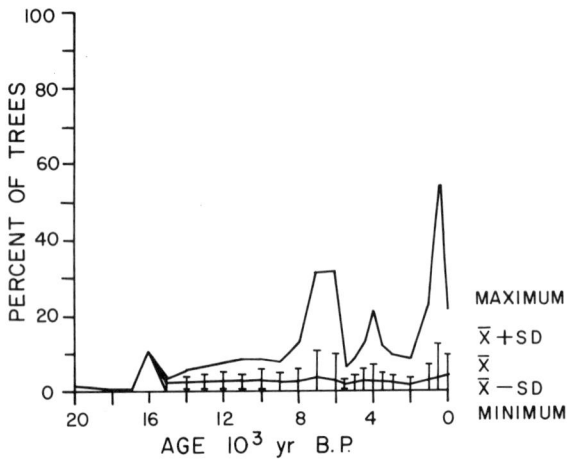

Figure 5.20. Tamarack (*Larix laricina*). a. Population trajectory, including maximum, minimum, and mean values of paleo-dominance as well as ± 1 SD of the mean, for the past 20,000 years.

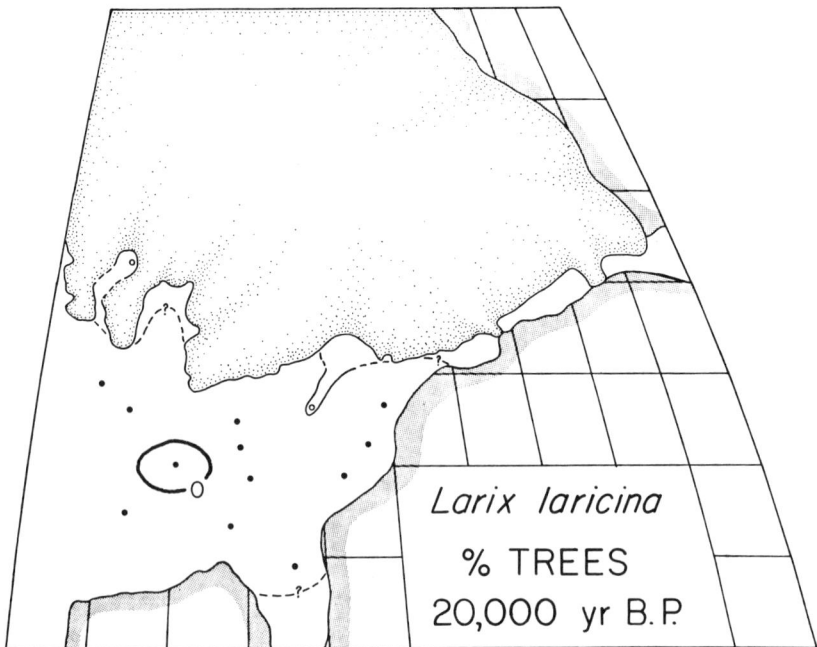

Figure 5.20. Tamarack (*Larix laricina*). b. Paleo-dominance map for 20,000 yr B.P.

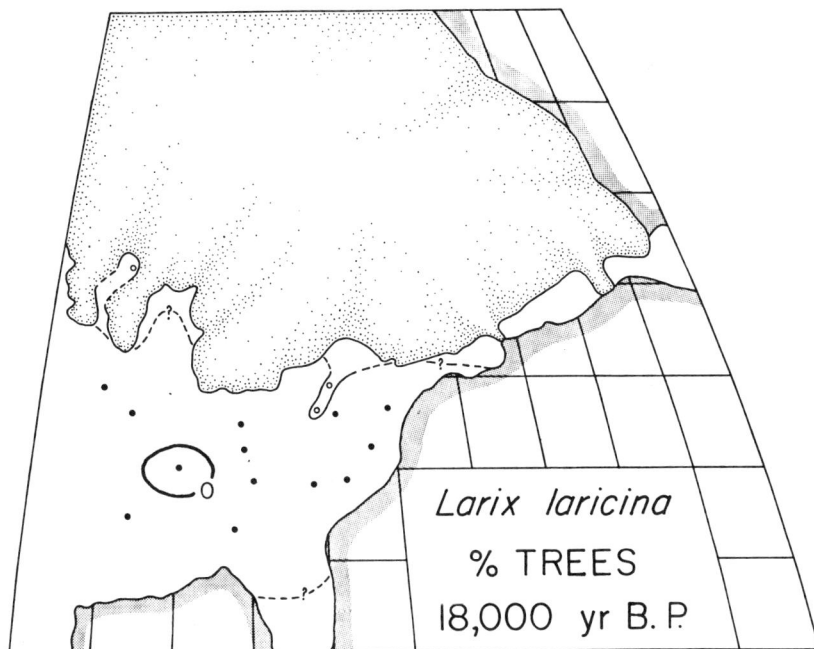

Figure 5.20. Tamarack (*Larix laricina*). c. Paleo-dominance map for 18,000 yr B.P.

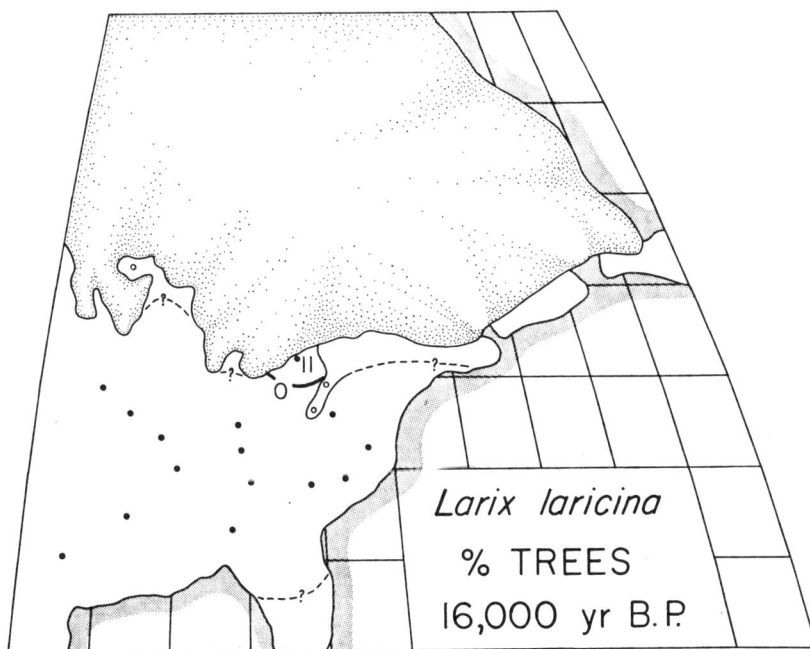

Figure 5.20. Tamarack (*Larix laricina*). d. Paleo-dominance map for 16,000 yr B.P.

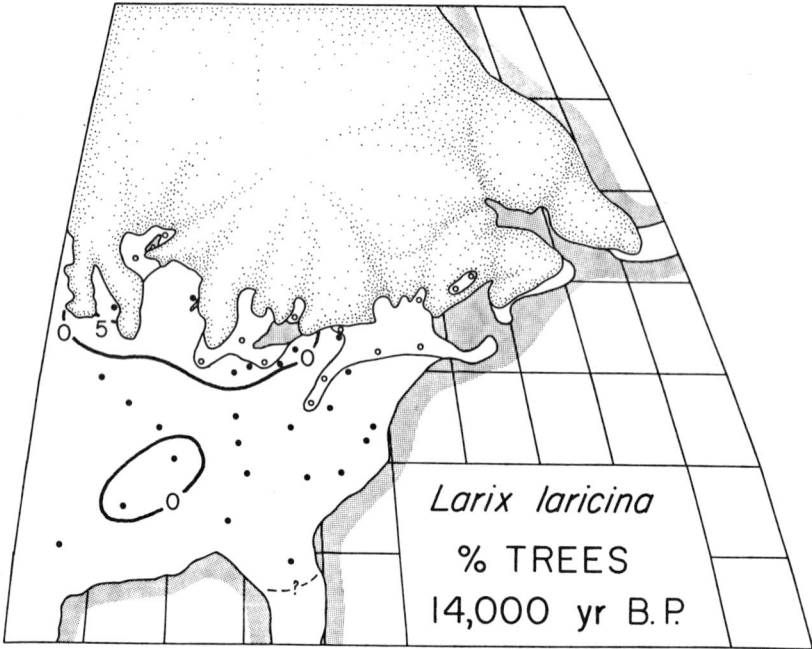

Figure 5.20. Tamarack (*Larix laricina*). e. Paleo-dominance map for 14,000 yr B.P.

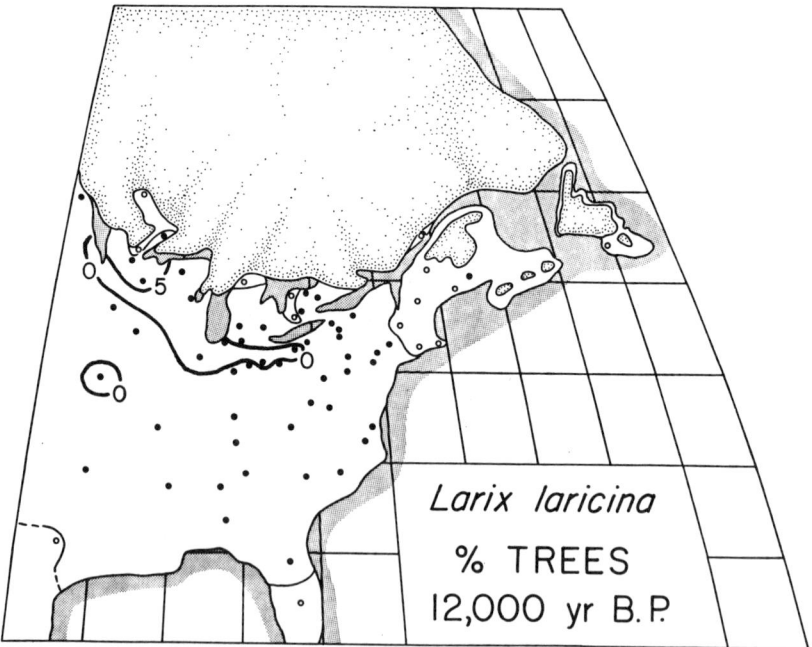

Figure 5.20. Tamarack (*Larix laricina*). f. Paleo-dominance map for 12,000 yr B.P.

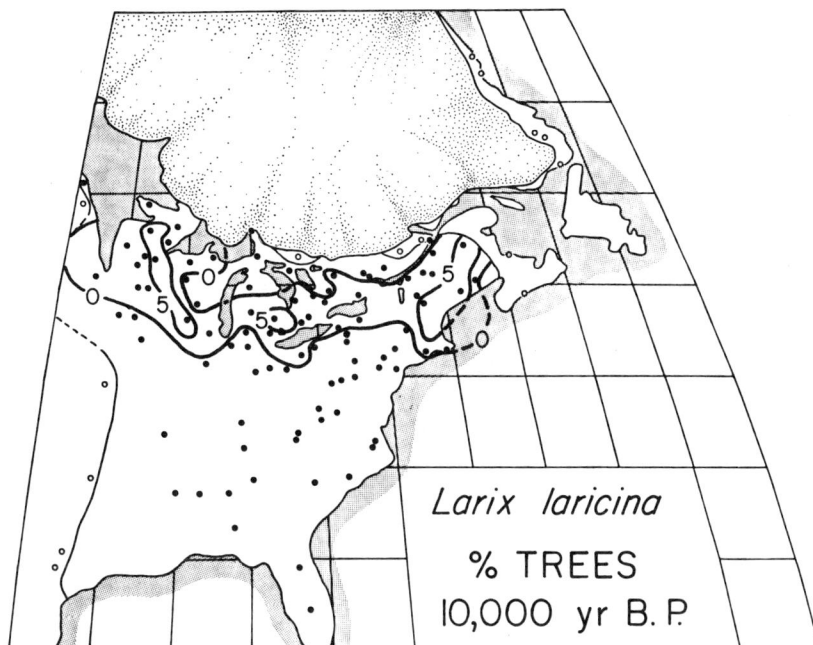

Figure 5.20. Tamarack (*Larix laricina*). g. Paleo-dominance map for 10,000 yr B.P.

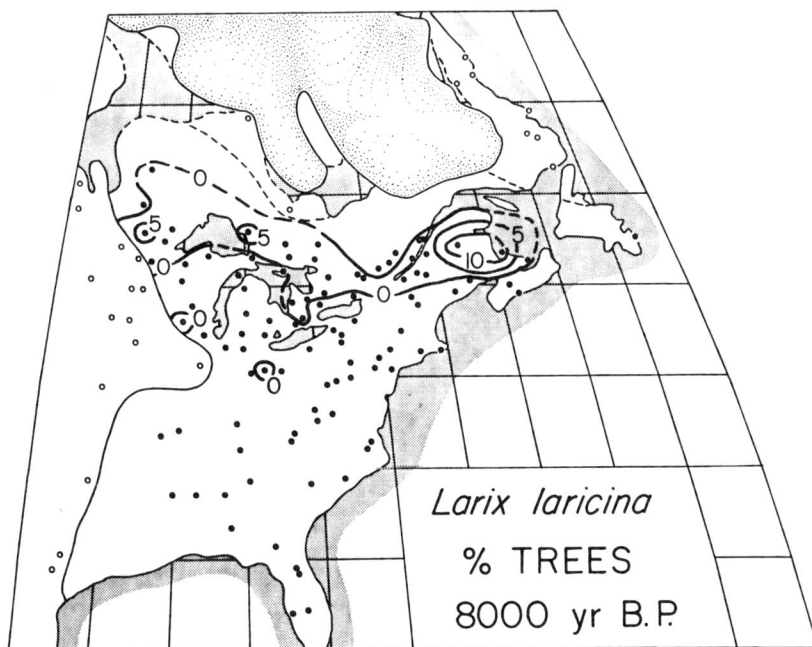

Figure 5.20. Tamarack (*Larix laricina*). h. Paleo-dominance map for 8000 yr B.P.

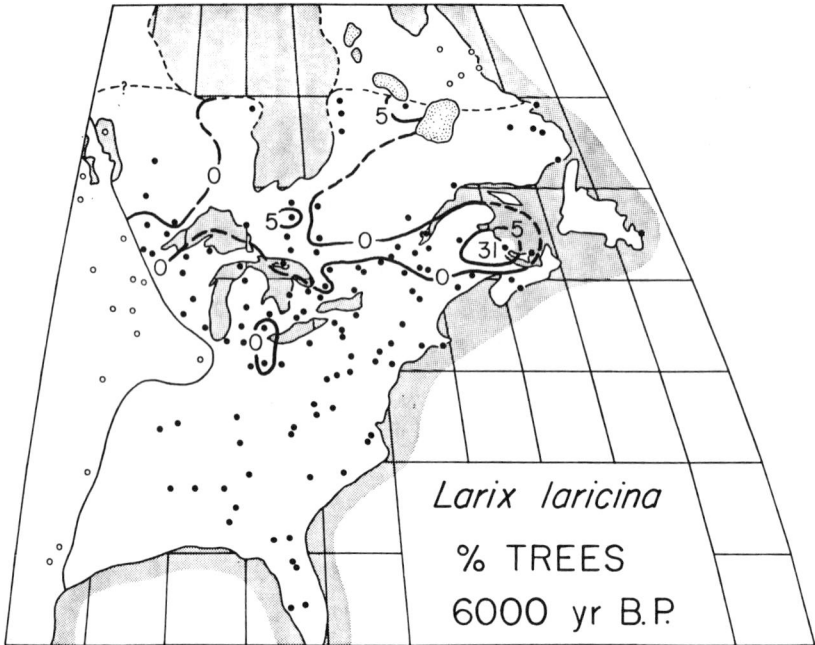

Figure 5.20. Tamarack (*Larix laricina*). i. Paleo-dominance map for 6000 yr B.P.

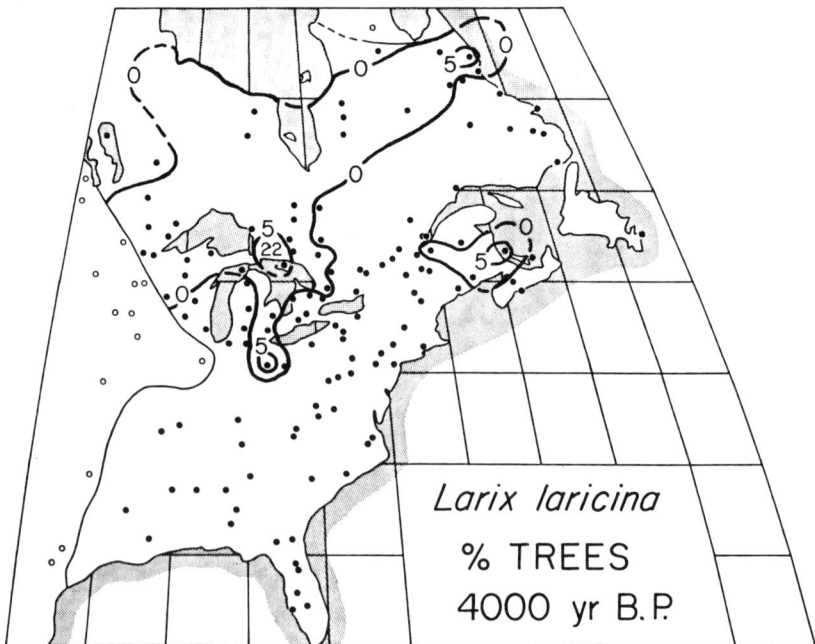

Figure 5.20. Tamarack (*Larix laricina*). j. Paleo-dominance map for 4000 yr B.P.

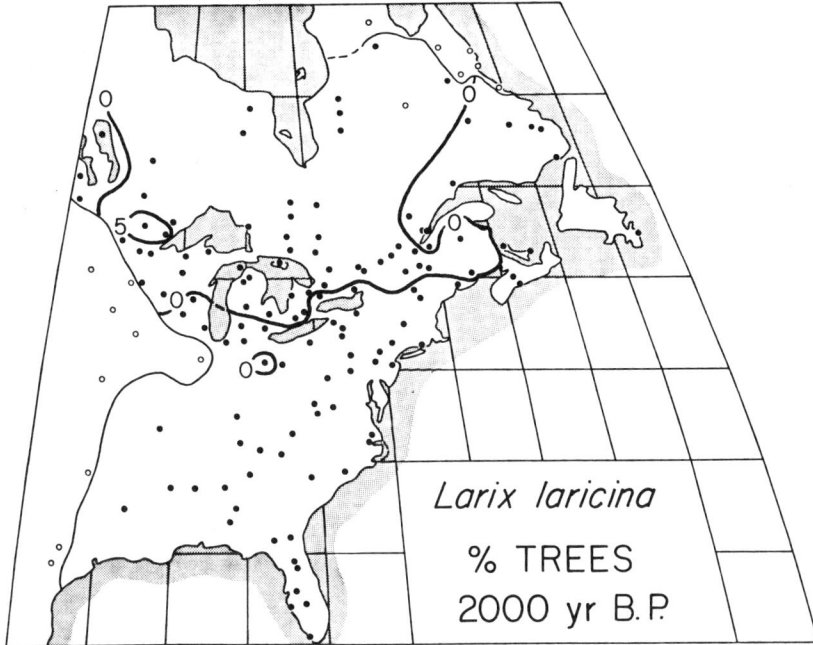

Figure 5.20. Tamarack (*Larix laricina*). k. Paleo-dominance map for 2000 yr B.P.

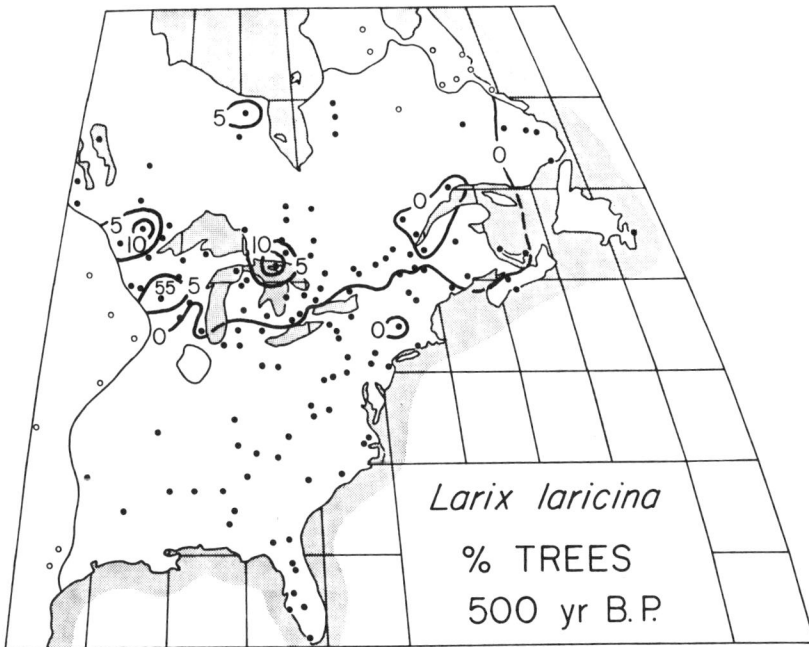

Figure 5.20. Tamarack (*Larix laricina*). l. Paleo-dominance map for 500 yr B.P.

glacial interval (< 2% from 20,000 yr B.P. to 17,000 yr B.P.), reached a peak of 11% at 16,000 yr B.P., and then diminished to between 4% and 8% from 15,000 yr B.P. to 9000 yr B.P. Maximum tamarack values increased sporadically through the mid-Holocene interval, reaching 13% at 8000 yr B.P., 31% at 7000 yr B.P. and 6000 yr B.P., and then oscillated between a low of 8% at 5000 yr B.P., 22% at 4000 yr B.P., and 8% at 2000 yr B.P. In the last 1000 years, tamarack increased from 22% at 1000 yr B.P. to 55% at 500 yr B.P., then decreased to 22% at 0 yr B.P. Through the last 20,000 years, values for ± 1 SD ranged between 2% and 7%, indicating that the highest peaks in dominance of tamarack at 7000 yr B.P., 6000 yr B.P., and 500 yr B.P. reflected the occurrence of *Larix* locally within swamp forests at only a few sites. For example, at 500 yr B.P., tamarack was quantitatively reconstructed at 40 sites; of these, 39 sites had values of 13% or less, and only one site, Tamarack Creek, Wisconsin (Davis 1977, 1979) had 55% dominance for *Larix* within a local tamarack swamp.

The maps of paleo-dominance for tamarack at 20,000 yr B.P. (Fig. 5.20b) and 18,000 yr B.P. (Fig. 5.20c) record < 2% dominance in the vicinity of Nonconnah Creek, Tennessee. This region adjacent the Lower Mississippi Alluvial Valley is the only known full-glacial refuge area for *Larix*. During the late-glacial interval, the paleo-dominance map for 16,000 yr B.P. (Fig. 5.20d) portrays the local occurrence of 11% tamarack near Crystal Lake, western Pennsylvania, which at that time was located in the northern boreal forest region adjacent the Laurentide Ice Sheet.

Between 16,000 yr B.P. and 14,000 yr B.P. (Fig. 5.20e), tamarack invaded a number of sites in deglaciated landscapes abandoned by the retreating ice margin. By 14,000 yr B.P., tamarack occupied sites in two areas (Fig. 4.3b): (1) the northern edge of the boreal forest region adjacent to either tundra or ice, extending from northeastern Ohio through central Indiana, northern Illinois, Wisconsin, and Iowa; and (2) near the southern fringe of the boreal forest, along the Mississippi Alluvial Valley from southwestern Tennessee to north-central Louisiana. The more northern populations reached up to 5% dominance (in Iowa), whereas the more southern populations were about 1% of the total forest composition.

By 12,000 yr B.P. (Fig. 5.20f), tamarack became locally extinct on the southern margin of its distribution. Tamarack populations persisted, however, west of the Appalachian Highlands in the central and western Great Lakes region, although its relative population center shifted northward from Iowa into Minnesota between 14,000 yr B.P. and 12,000 yr B.P. Sites in central Minnesota contained values of tamarack dominance of up to 5% to 7%. Traces of tamarack occurred as an outlier in northeastern Kansas, indicating that limited populations of *Larix* persisted throughout the western boreal forest region during the early Holocene. Between 12,000 yr B.P. and 10,000 yr B.P. (Fig. 5.20g), tamarack expanded eastward across the Great Lakes region and the northern Appalachians into New England. By 10,000 yr B.P., *Larix* occupied a latitudinal band between 41°N and 49°N and between 67°W and 100°W. Relative population centers of between 5% and 8% occurred in northeastern Minnesota, southwestern Wisconsin, southeastern Michigan, and across New England to Maine and New Brunswick.

During the early- and mid-Holocene intervals, tamarack continued to advance northward across deglaciated terrain. By 8000 yr B.P. (Fig. 5.20h), the latitudinal band occupied by tamarack shifted north and was located between 45°N and 50°N. With localized extinction and fragmentation of its populations, outliers of tamarack developed in Ontario and Wisconsin. Population centers tracked northward and by 8000 yr B.P. were situated in the maritime provinces of New Brunswick and Prince Edward Island (up to 13%), central Ontario (5%), and northern Minnesota (6%). At 10,000 yr B.P., tamarack was located on the leading edge of the advancing boreal forest region in Canada. However, by 8000 yr B.P., the northernmost advance of tamarack lagged behind the more rapid northward advance of the forest limit, delineated by the spread of aspen (Fig. 5.28h) and spruce (Fig. 5.24h). The paleo-dominance map for tamarack at 6000 yr B.P. (Fig. 5.20i) shows a latitudinal band between 45°N and 49°N extending from the primary population center of 31% in Prince Edward Island westward across southern Quebec and Ontario to northern Minnesota. By 6000 yr B.P., tamarack had preferentially expanded northward along the coastal regions of Hudson Bay as far north as the tundra–forest limit at about 55°N in northern Quebec and Ontario.

By 4000 yr B.P. (Fig. 5.20j), local extinctions of tamarack occurred along the St. Lawrence River Valley in Quebec. Populations of tamarack became separated into a primary center of dominance (up to 22%) in the central Great Lakes region and a large outlier region in northern New England. Tamarack expanded northeastward through northern Quebec to coastal Labrador as well as northwestward through Ontario into Manitoba. *Larix* also reexpanded southward and was distributed across northern Wisconsin and northern Lower Michigan.

From 4000 yr B.P. to 500 yr B.P., the western limit of tamarack was reached at the prairie–forest boundary. At 2000 yr B.P. (Fig. 5.20k), the southern boundary of tamarack distribution shifted from central Quebec southeast across the northern Appalachians and merged with the former outlier in northern New England. However, east of Hudson Bay on its northeastern border, tamarack retreated as tundra expanded southward, particularly along the coastal zone of northern Labrador. By 500 yr B.P. (Fig. 5.20l), the northeastern limit of tamarack shifted southward as the tundra–boreal forest ecotone continued to move. The southern margin of distribution generally paralleled 45°N in south-eastern Canada but was located at 43°N west of the Appalachians. In presettlement times, tamarack was widely distributed through the boreal forest region and the mixed conifer–northern hardwoods region (Fig. 4.2a). Typically, *Larix* was less than 5% of the forest composition, becoming dominant only locally within bogs and swamps. Along its southern and western margins, outlier populations developed within the last several thousand years. These southern outliers are advance disjunct colonies rather than relicts of Pleistocene distributions.

The area–dominance histograms (Fig. 5.21a,b) display changes in overall dominance structure of *Larix* populations that are consistent with the patchiness of its distributional history across the landscapes of eastern North America. The histogram for 20,000 yr B.P. illustrates that known refuge areas for *Larix* occupied a relatively small area, only $0.9 \cdot 10^5$ km^2, with values for dominance

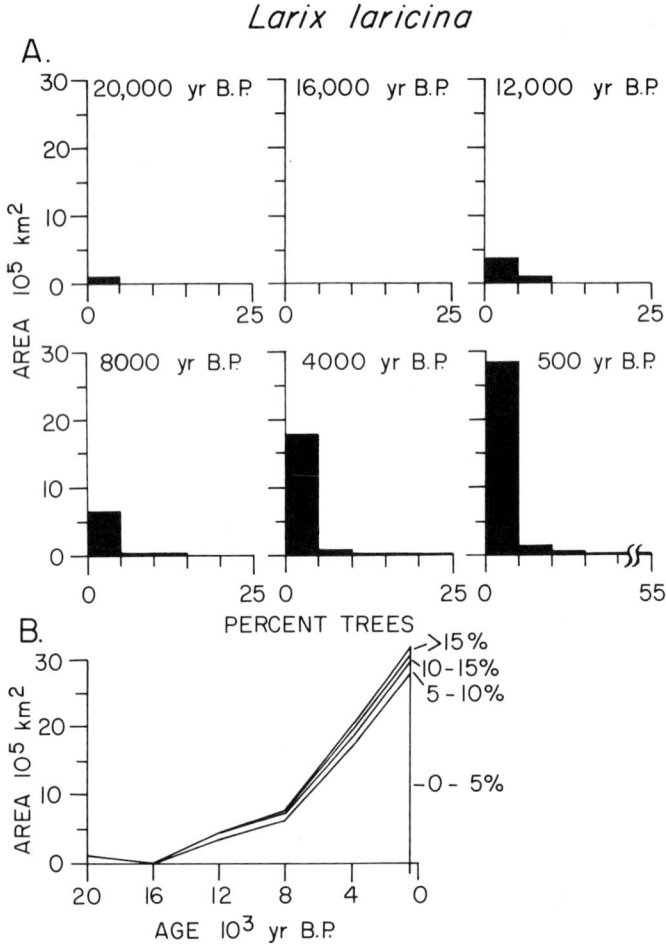

Figure 5.21. Changes in area and dominance of tamarack (*Larix laricina*). a. Histograms of area–dominance structure at 20,000, 16,000, 12,000, 8000, 4000, and 500 yr B.P. b. Cumulative plot from 20,000 yr B.P. to the present.

below 5%. The total area within the range occupied by tamarack increased gradually from 16,000 yr B.P. to 8000 yr B.P. and then increased rapidly during the last 8000 years. The interglacial maximum areal extent was $30.5 \cdot 10^5$ km^2 and occurred at 500 yr B.P. Overall, the area occupied by *Larix* increased by 3400% from 20,000 yr B.P. to 500 yr B.P. The cumulative plot of dominance structure (Fig. 5.21b) exhibits a change through time in the number of dominance classes, from 1 in full-glacial times, 2 in late-glacial and early-Holocene times, to between 3 and 5 classes for the mid- and late-Holocene intervals. However, even during the Holocene, the lowest dominance class represented > 90% of the total mapped area of tamarack distribution.

Tupelo (*Nyssa*)

The long-term population trajectory for tupelo (Fig. 5.22a) portrays three discrete modes of dynamics, punctuated by population shifts in trajectory completed within approximately one to two thousand years. From full-glacial through early-Holocene times, mean values for tupelo dominance were consistently low, between 5% and 7%, with limited variability about the mean (± 1 SD typically about 1% to 2%). Correspondingly, the maximum values for tupelo varied between 6% and 11% dominance for this 11,000-year interval from 20,000 yr B.P. to 9000 yr B.P. Between 9000 yr B.P. and 8000 yr B.P., the mean increased from 6% up to 10% and the peak values soared from 11% to 63%. The second mode of population dynamics persisted as a plateau of values from 8000 yr B.P. until 5000 yr B.P.; the means stabilized at the level of 10% to 11% (concomitant with increased population variability, ± 1 SD equaled approximately 11%). During this mid-Holocene interval, the maximum values of tupelo reached 63% at 8000 yr B.P., and then declined consecutively from 56% at 7000 yr B.P., to 43% at 6000 yr B.P., and 33% at 5000 yr B.P. The transition from the second to the third population mode occurred more gradually, from 5000 yr B.P. to 3000 yr B.P. The late-Holocene trajectory for tupelo populations reached a plateau with a higher mean of about 16%, accompanied by increased values for both variability (± 1 SD about 18%) and maximum values (fluctuating between 80% and 86%).

The paleo-dominance map for tupelo (Fig. 5.22b) identifies one refugial area at 20,000 yr B.P. in the Deep South. This refugium (bounded by the 0% contour line) is mapped in the region of the central and eastern Gulf Coastal Plain and the adjacent portion of the southern Atlantic Coastal Plain. The full-glacial map for 18,000 yr B.P. (Fig. 5.22c), as well as the subsequent late-glacial maps for 16,000 yr B.P. (Fig. 5.22d) and 14,000 yr B.P. (Fig. 5.22e), reveal a broader distribution for tupelo, generally situated south of about 33°N and spreading west from the Atlantic Coast of Georgia and northern Florida, across the Gulf Coastal Plain to eastern Texas.

During the early-Holocene interval, tupelo populations were curtailed along their southwestern margin with the northward spread of prairie from 12,000 yr B.P. (Fig. 5.22f) to 10,000 yr B.P. (Fig. 5.22g). However, during this interval, tupelo populations expanded continuously along their northwestern margin into the Ozark Plateaus of southeastern Missouri and northeast along the Atlantic Coastal Plain into the Carolinas and southeastern Virginia. At 10,000 yr B.P., the existence of three outlier stations in the southern and central Appalachians indicates that long-distance dispersal of *Nyssa* seeds occurred from coastal-plain populations to suitable montane sites. Transport of seeds by birds probably accounts for the sparse outlier pattern observed for this early-Holocene advance dispersal and establishment of such isolated tupelo populations in West Virginia, central Pennsylvania, and southeastern New York.

The 8000 yr B.P. map for tupelo populations (Fig. 5.22h) emphasizes the differential success of the populations situated along its northern border. West of the Appalachian Highlands, *Nyssa* populations advanced to 38°N in central Kentucky but retreated to about 35°N in Arkansas. Farther to the east, *Nyssa*

Nyssa

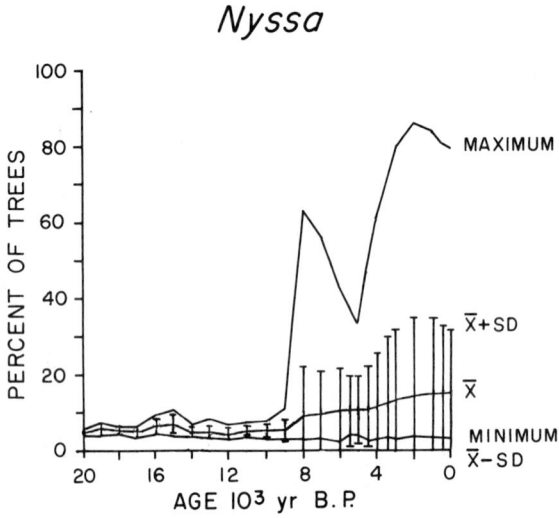

Figure 5.22. Tupelo (*Nyssa*). a. Population trajectory, including maximum, minimum, and mean values of paleo-dominance as well as ± 1 SD of the mean, for the past 20,000 years.

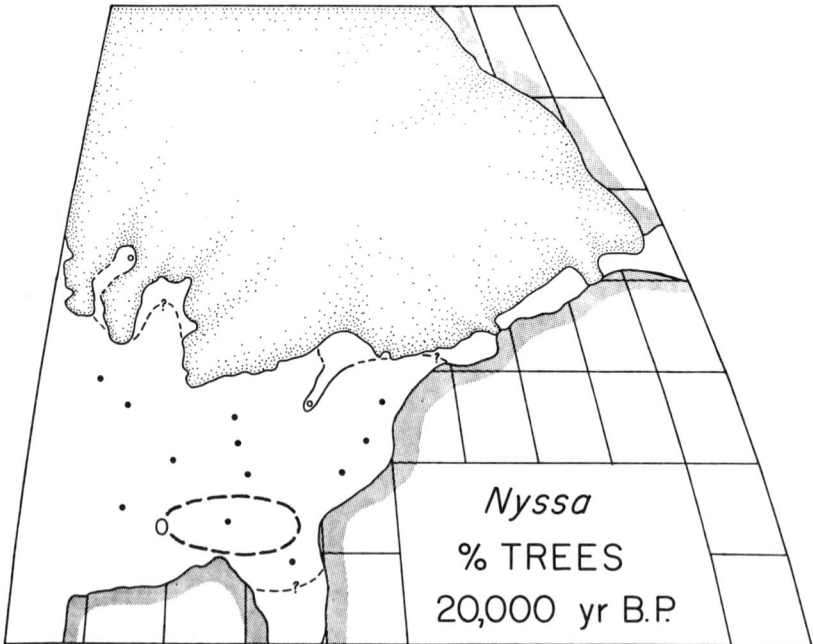

Figure 5.22. Tupelo (*Nyssa*). b. Paleo-dominance map for 20,000 yr B.P.

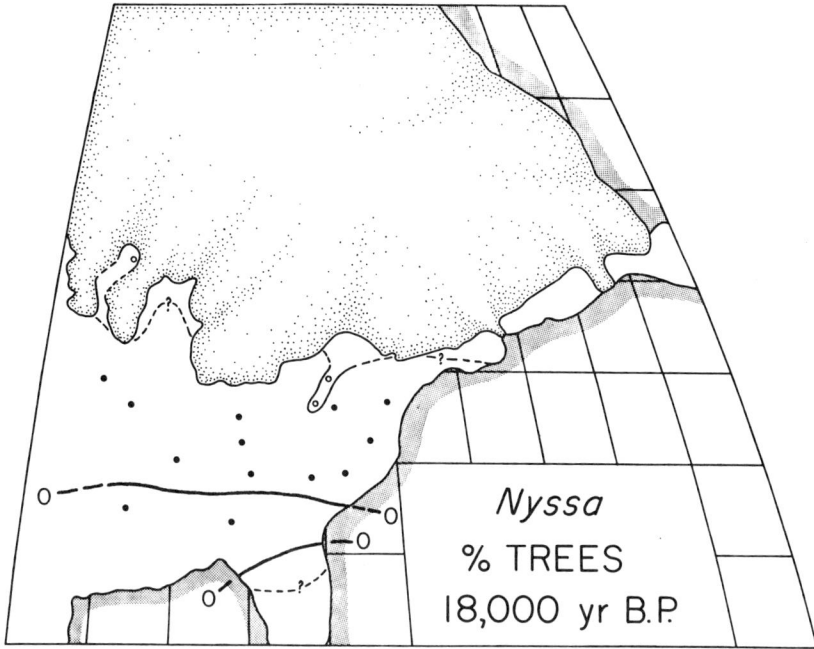

Figure 5.22. Tupelo (*Nyssa*). c. Paleo-dominance map for 18,000 yr B.P.

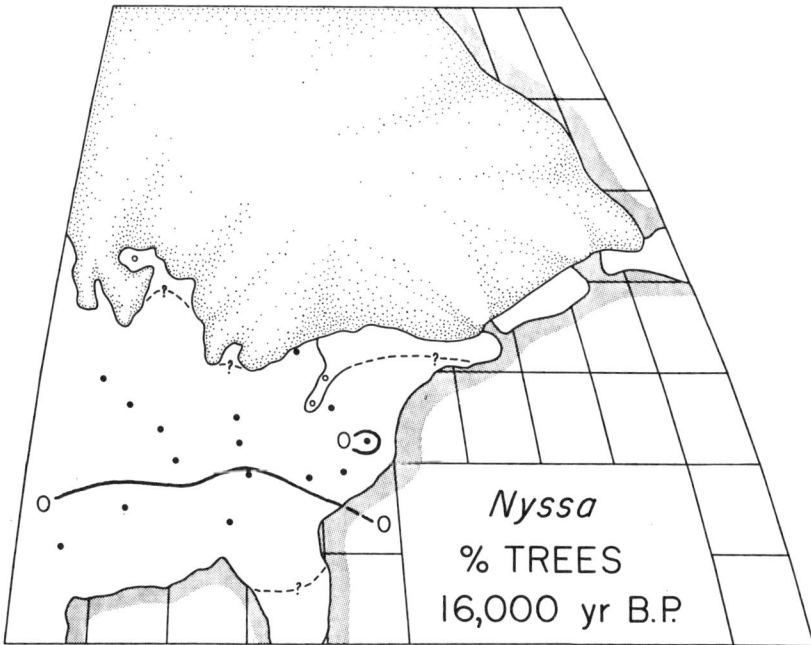

Figure 5.22. Tupelo (*Nyssa*). d. Paleo-dominance map for 16,000 yr B.P.

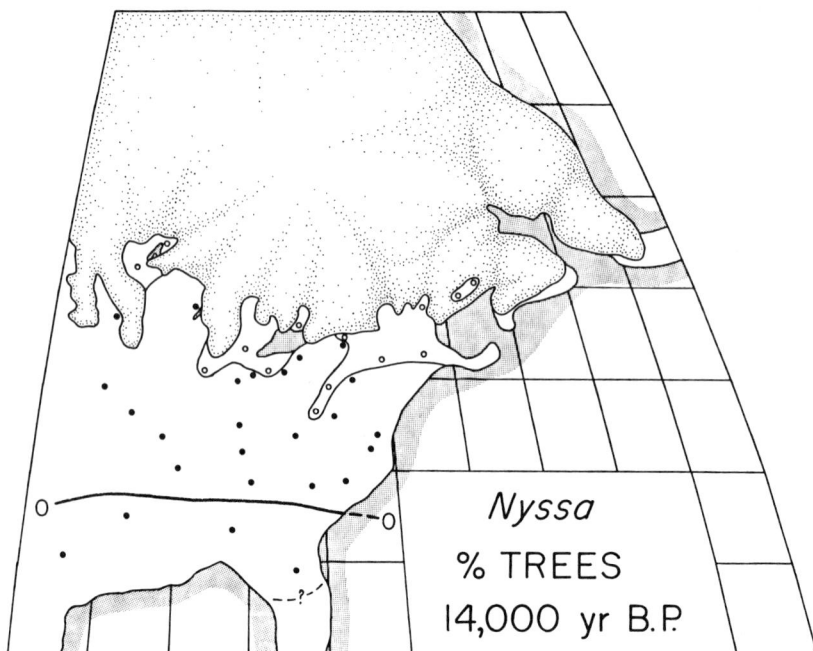

Figure 5.22. Tupelo (*Nyssa*). e. Paleo-dominance map for 14,000 yr B.P.

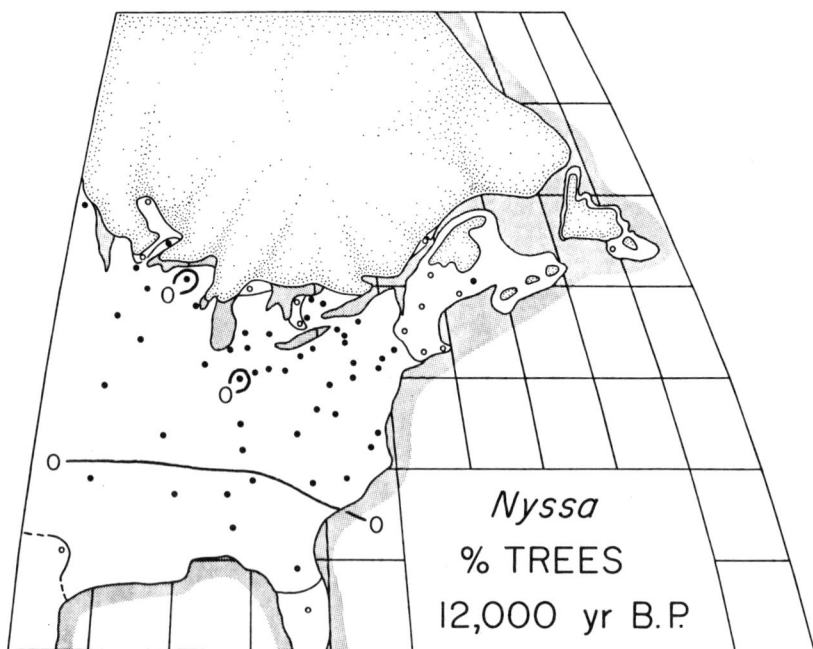

Figure 5.22. Tupelo (*Nyssa*). f. Paleo-dominance map for 12,000 yr B.P.

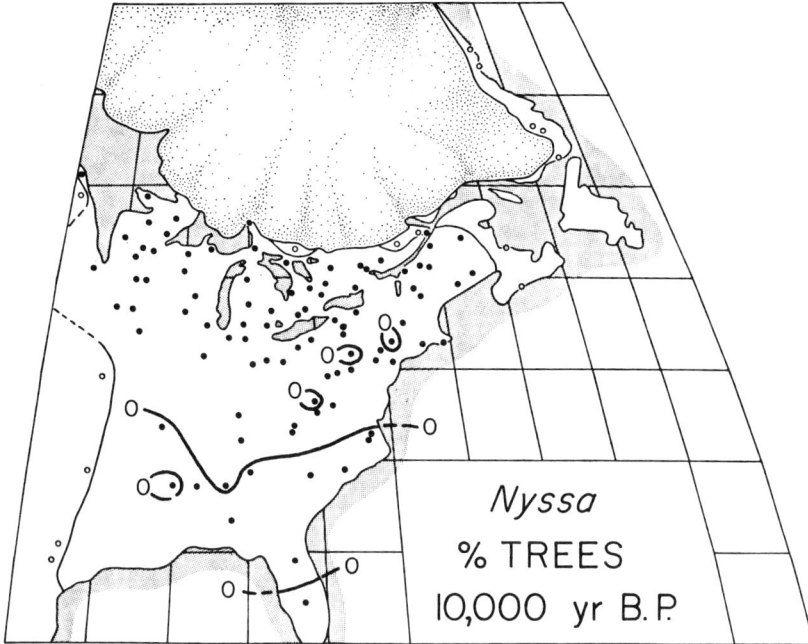

Figure 5.22. Tupelo (*Nyssa*). g. Paleo-dominance map for 10,000 yr B.P.

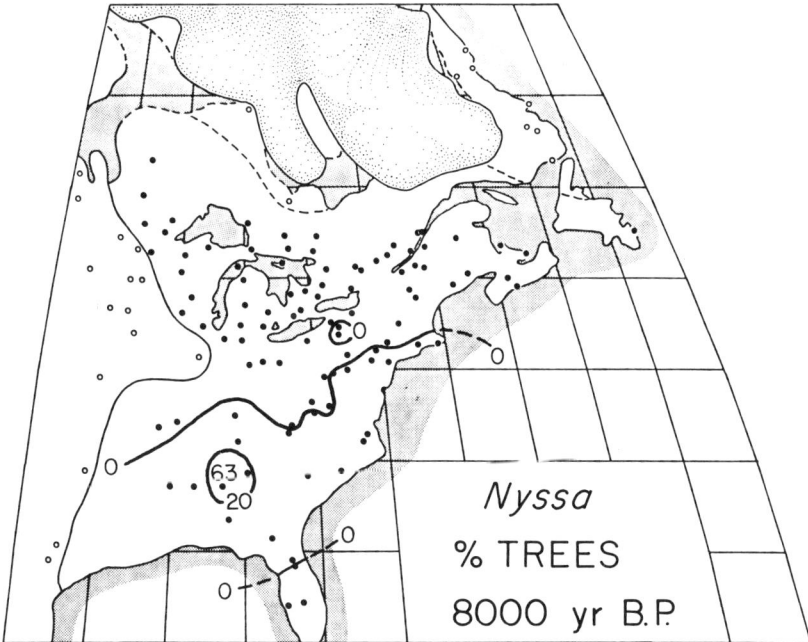

Figure 5.22. Tupelo (*Nyssa*). h. Paleo-dominance map for 8000 yr B.P.

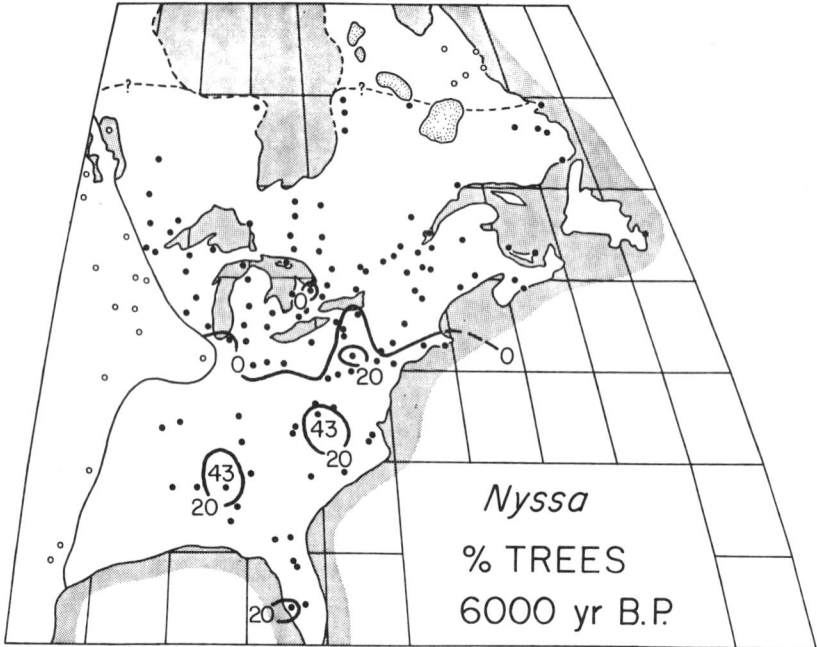

Figure 5.22. Tupelo (*Nyssa*). i. Paleo-dominance map for 6000 yr B.P.

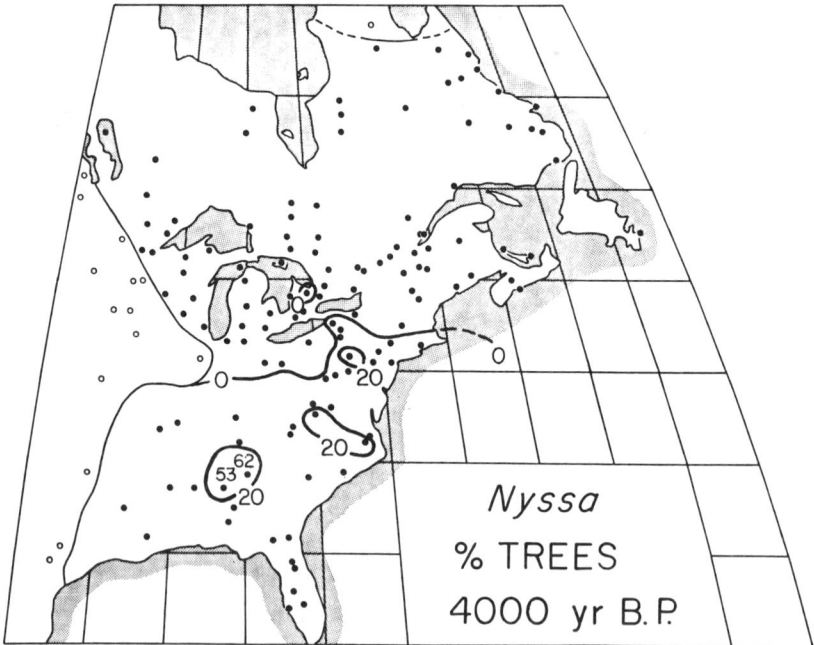

Figure 5.22. Tupelo (*Nyssa*). j. Paleo-dominance map for 4000 yr B.P.

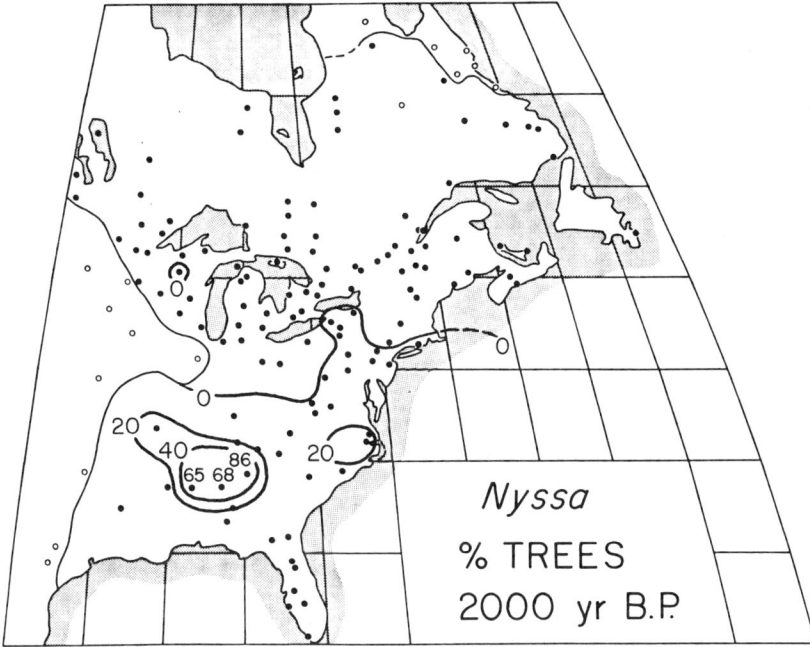

Figure 5.22. Tupelo (*Nyssa*). k. Paleo-dominance map for 2000 yr B.P.

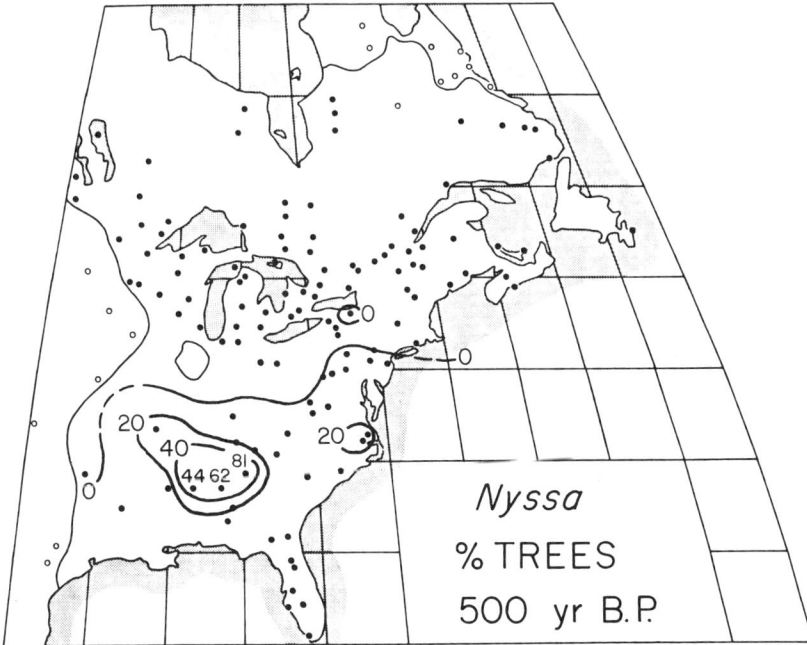

Figure 5.22. Tupelo (*Nyssa*). l. Paleo-dominance map for 500 yr B.P.

spread northeast into suitable habitats of the Appalachian Mountains, the Piedmont, and the Atlantic Coastal Plain as far as southern New England. An advance outlier occupied the Allegheny Plateau of north-central Pennsylvania and adjacent portions of western New York.

By 6000 yr B.P., mid-Holocene populations of tupelo (Fig. 5.22i) dispersed north and west in the region west of the Appalachians. The irregular northern margin of distribution swept from the prairie–forest border of Illinois, from the northeastern corner through central Indiana, central Ohio, then north through western New York, south across eastern Pennsylvania and New Jersey to coastal Massachusetts. The outlier population at 44°N in southern Ontario represented one of the northernmost colonies of *Nyssa* established during the Holocene. Along its southeastern border, tupelo secured new habitats as it migrated across central and southern Florida.

During the last stages of the mid- and late-Holocene intervals, the paleodominance maps for 6000 yr B.P. (Fig. 5.22i), 4000 yr B.P. (Fig. 5.22j), 2000 yr B.P. (Fig. 5.22k), and 500 yr B.P. (Fig. 5.22l) illustrate three trends. After reaching its maximum interglacial limit of migration 6000 years ago, the northern continuous range limit of *Nyssa* (the 0% contour) retracted southward along the Atlantic Coastal Plain to southern New York, within the northern Appalachians, and to about 38°N west of the Appalachian Highlands. The second pattern reflects the expansion of tupelo populations in poorly drained coastal habitats, inundated in part because of the mid- and late-Holocene rise in sea level. The increase in dominance of *Nyssa* populations to between 35% and 40% was observed (Figs. 5.22k,l) at the Great Dismal Swamp in eastern Virginia (Whitehead 1972). The third pattern is characterized by regionally high values (from 44% up to 86%) for tupelo within the Mid-South region of central Mississippi, Alabama, and Georgia. In this case, *Nyssa* occupied major alluvial swamps such as along the Tombigbee River in east-central Mississippi (Whitehead and Sheehan 1985), as well as occurring locally around the periphery of karst sinkhole lakes ("sag ponds") in the southernmost Ridge and Valley Province of central Alabama (Delcourt *et al.* 1983b) and northwestern Georgia (Watts 1970).

Species determinations for *Nyssa* can be made readily on fossil seeds; however, macrofossils of tupelo are rarely found in late-Quaternary sedimentary sequences. Seeds of black gum (*Nyssa sylvatica*) have been recovered throughout the last 9000 years from Cahaba Pond in central Alabama (Delcourt *et al.* 1983b) and in sediments dating 5300 yr B.P. in southeastern Louisiana (Delcourt and Delcourt 1977b). Seeds of water tupelo (*Nyssa aquatica*) found in sediments from Cupola Pond, southeastern Missouri (Smith 1984) confirm that tupelo migrated northward within the Lower Mississippi Valley during the Holocene and, during at least the last 1500 years, colonized upland ponds in the Ozark Plateaus.

The histograms of area and dominance structure (Fig. 5.23a,b) indicate an increase in areal extent of 535% from $2.3 \cdot 10^5$ km^2 at 20,000 yr B.P. up to $12.3 \cdot 10^5$ km^2 at 12,000 yr B.P.; during this peak-glacial through early-Holocene interval, populations of tupelo were distributed exclusively within the lowest

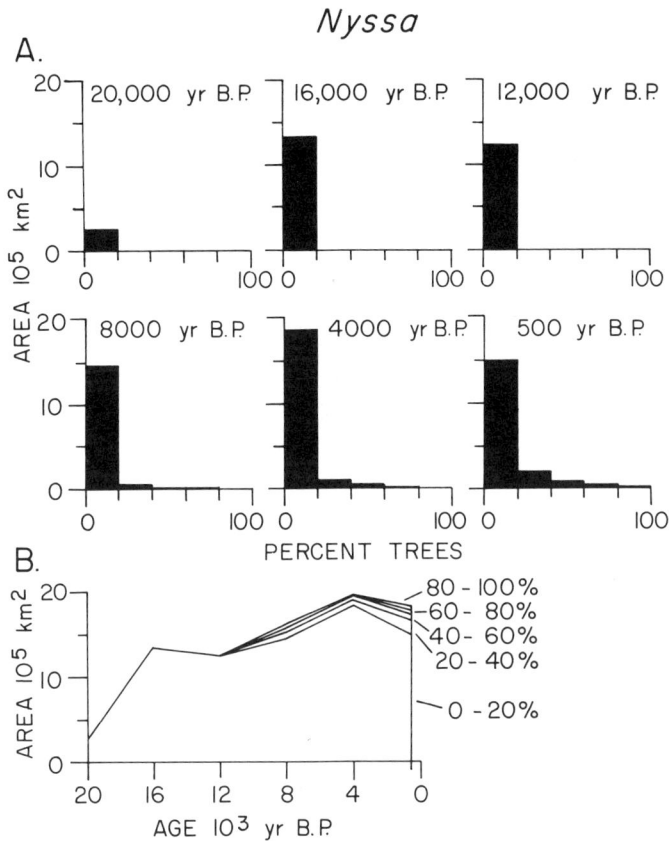

Figure 5.23. Changes in area and dominance of tupelo (*Nyssa*). a. Histograms of area–dominance structure at 20,000, 16,000, 12,000, 8000, 4000, and 500 yr B.P. b. Cumulative plot from 20,000 yr B.P. to the present.

dominance class measured (0% to 20%). The histograms for the last 8000 years illustrate the progressive increase in dominance classes represented (including the 80% to 100% class by 500 yr B.P.), although between 84% and 98% of the area portrayed in these mid- and late-Holocene histograms for *Nyssa* were still confined by the 0% to 20% contour lines. The cumulative plot for dominance structure (Fig. 5.23b) depicts the moderate 9% decrease in the distributional range of tupelo from its maximum interglacial extent ($19.7 \cdot 10^5$ km^2 at 4000 yr B.P.) to the present day ($17.8 \cdot 10^5$ km^2 at 500 yr B.P.).

Spruce (*Picea*)

In terms of the relative percent of forest composition, spruce has been a dominant genus in eastern North America for most of the past 20,000 years (Fig. 5.24a). Spruce percentages reached regionally widespread, maximum values of 80% to 100% both during the full-glacial and late-glacial intervals and during

Picea

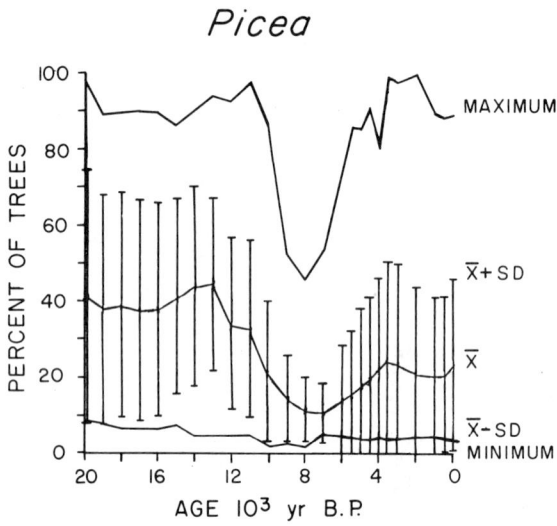

Figure 5.24. Spruce (*Picea*). a. Population trajectory, including maximum, minimum, and mean values of paleo-dominance as well as ± 1 SD of the mean, for the past 20,000 years.

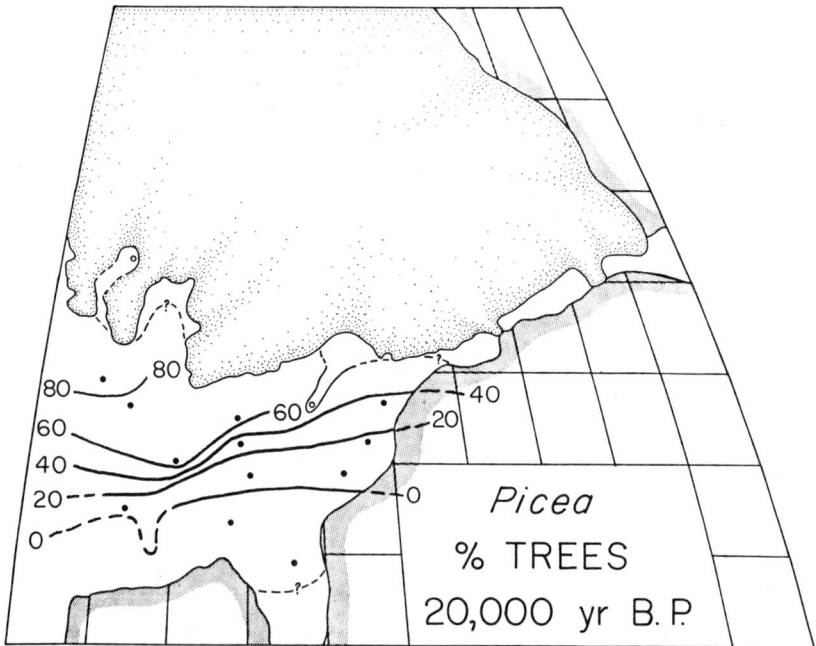

Figure 5.24. Spruce (*Picea*). b. Paleo-dominance map for 20,000 yr B.P.

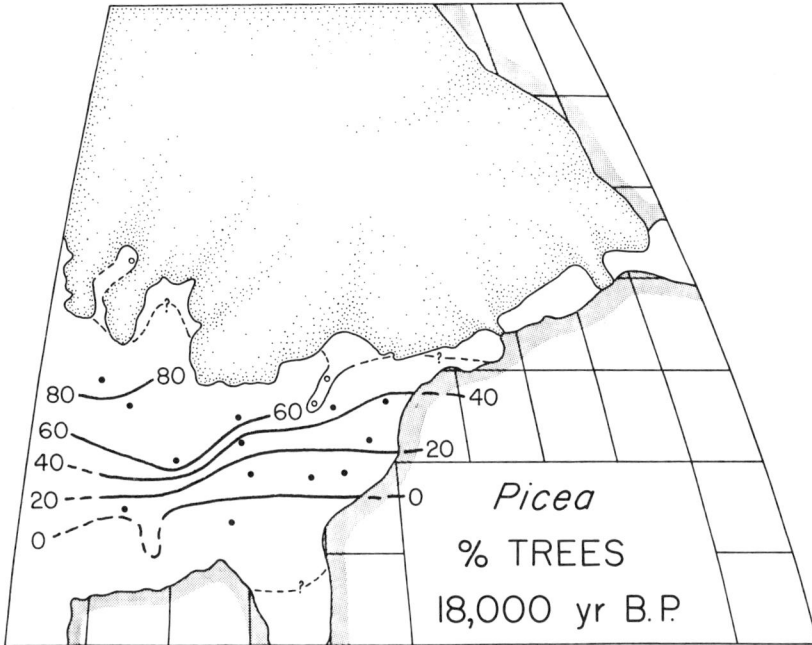

Figure 5.24. Spruce (*Picea*). c. Paleo-dominance map for 18,000 yr B.P.

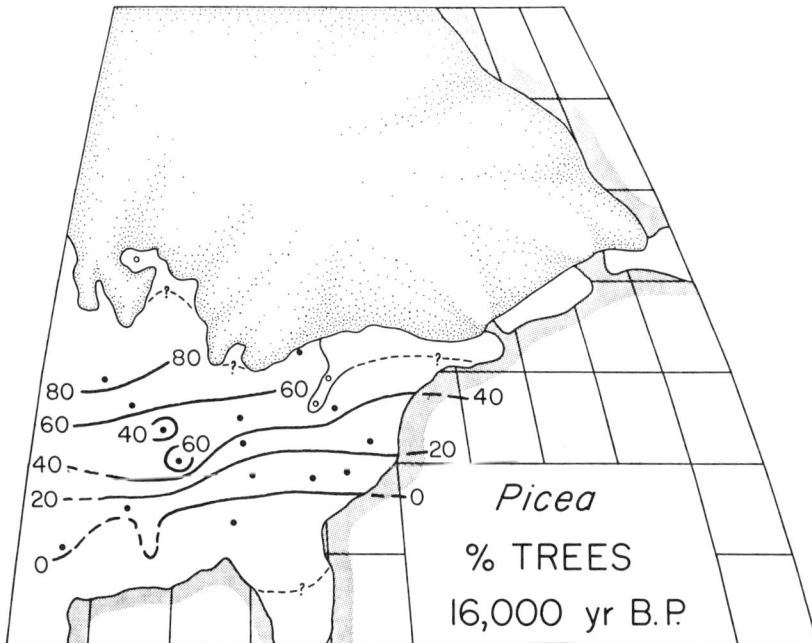

Figure 5.24. Spruce (*Picea*). d. Paleo-dominance map for 16,000 yr B.P.

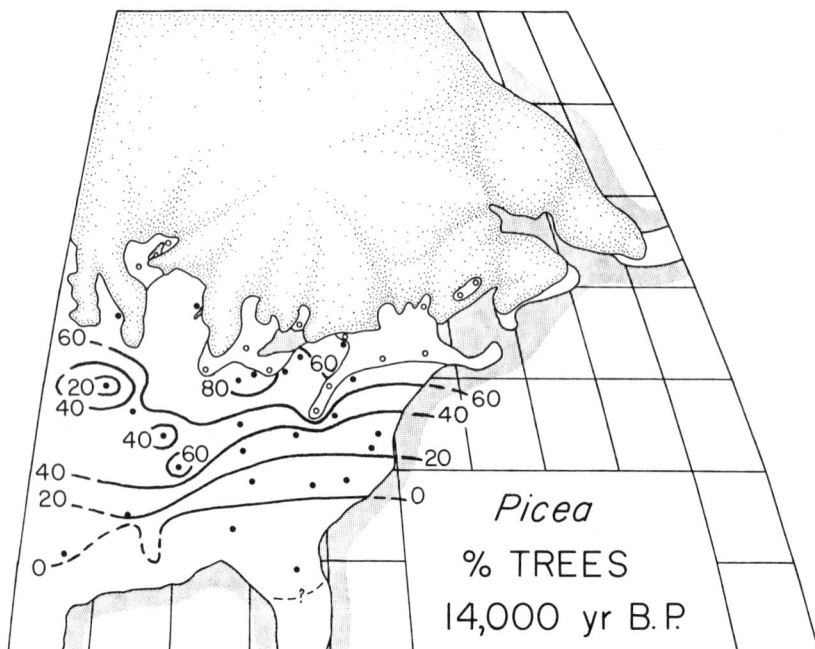

Figure 5.24. Spruce (*Picea*). e. Paleo-dominance map for 14,000 yr B.P.

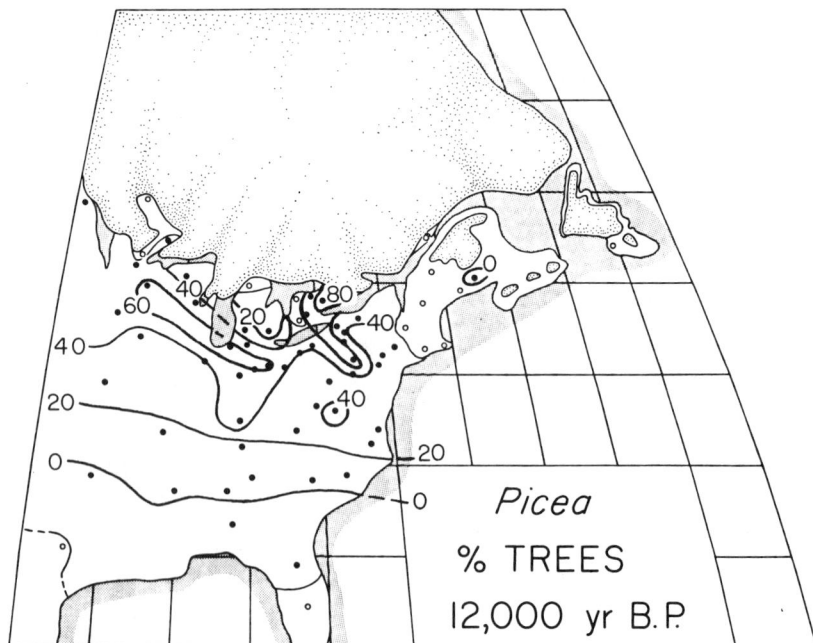

Figure 5.24. Spruce (*Picea*). f. Paleo-dominance map for 12,000 yr B.P.

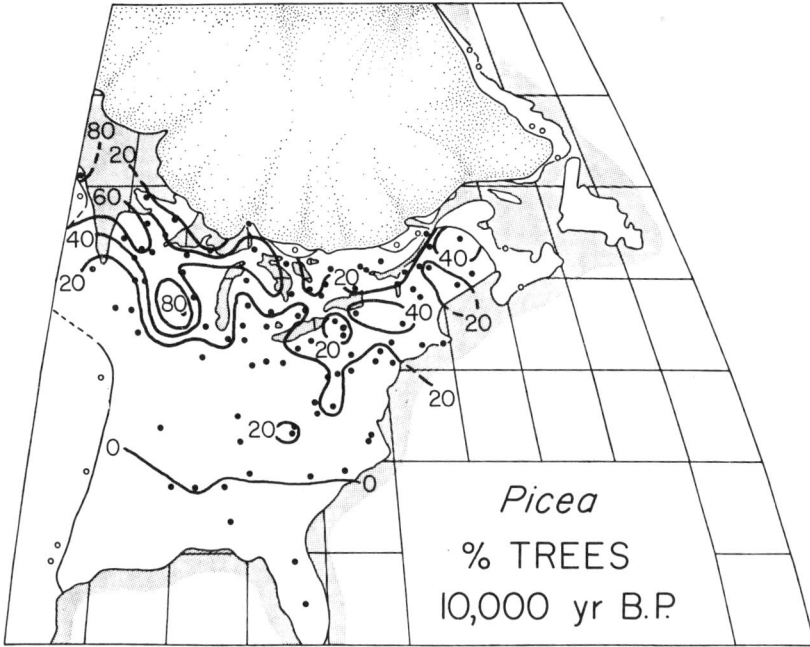

Figure 5.24. Spruce (*Picea*). g. Paleo-dominance map for 10,000 yr B.P.

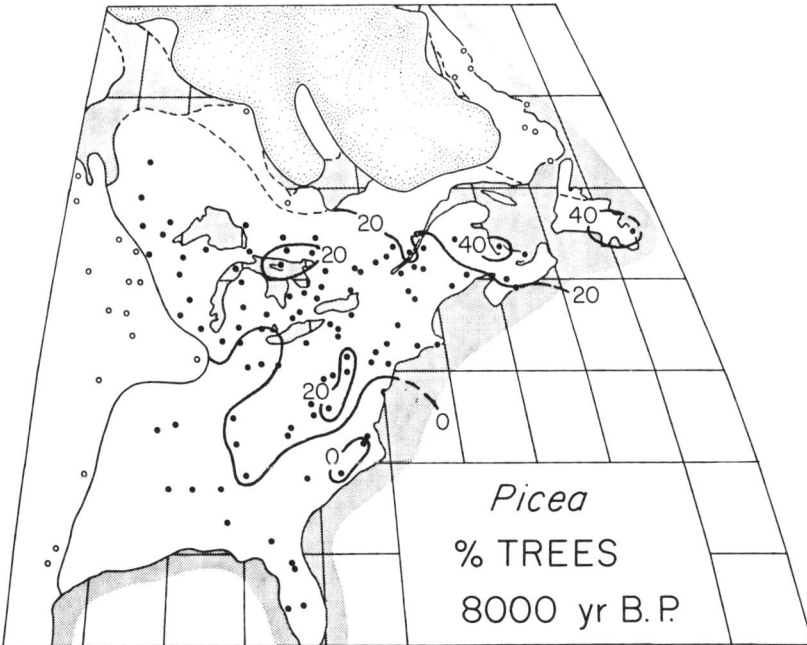

Figure 5.24. Spruce (*Picea*). h. Paleo-dominance map for 8000 yr B.P.

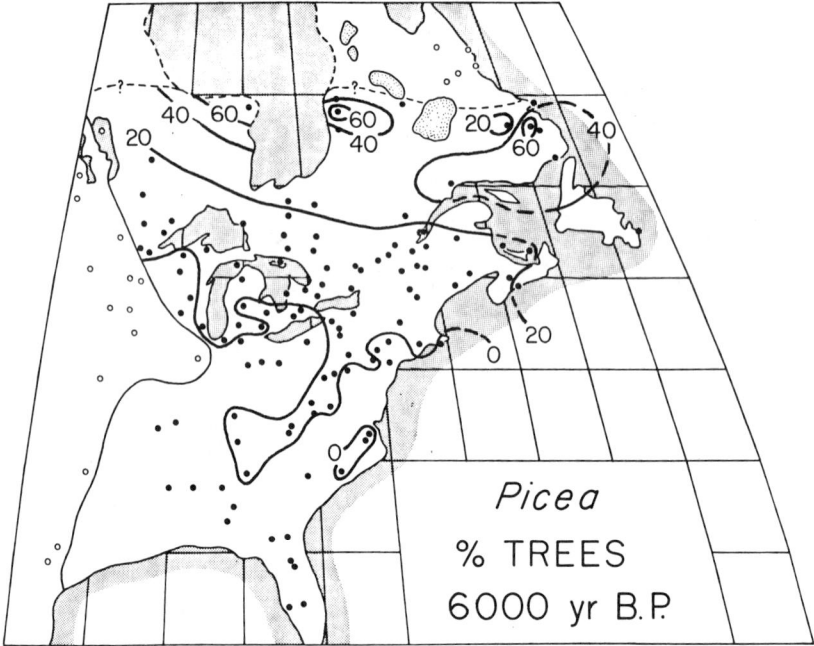

Figure 5.24. Spruce (*Picea*). i. Paleo-dominance map for 6000 yr B.P.

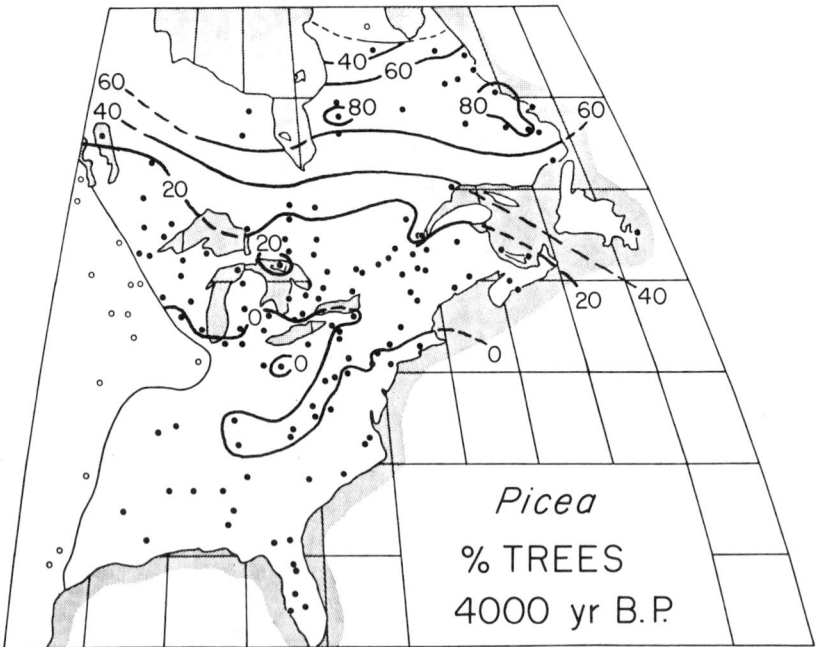

Figure 5.24. Spruce (*Picea*). j. Paleo-dominance map for 4000 yr B.P.

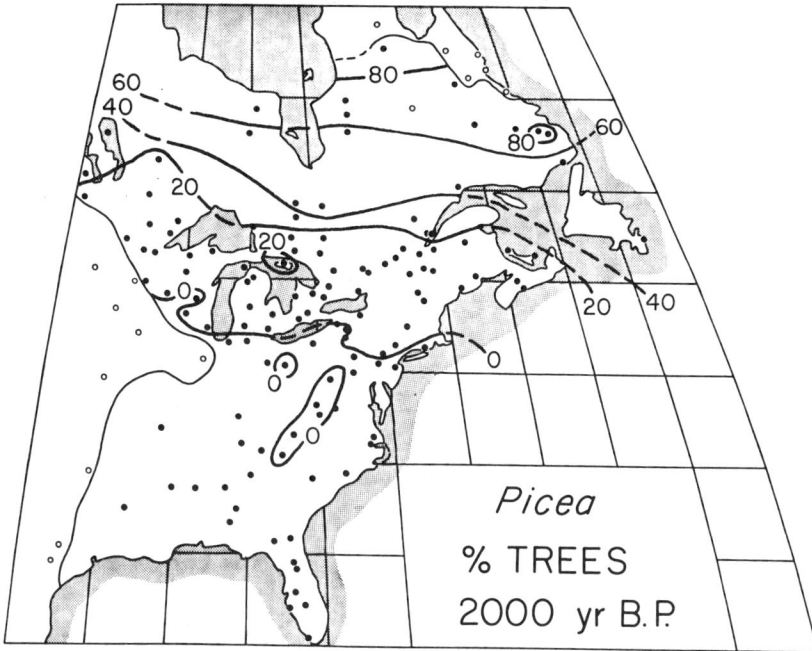

Figure 5.24. Spruce (*Picea*). k. Paleo-dominance map for 2000 yr B.P.

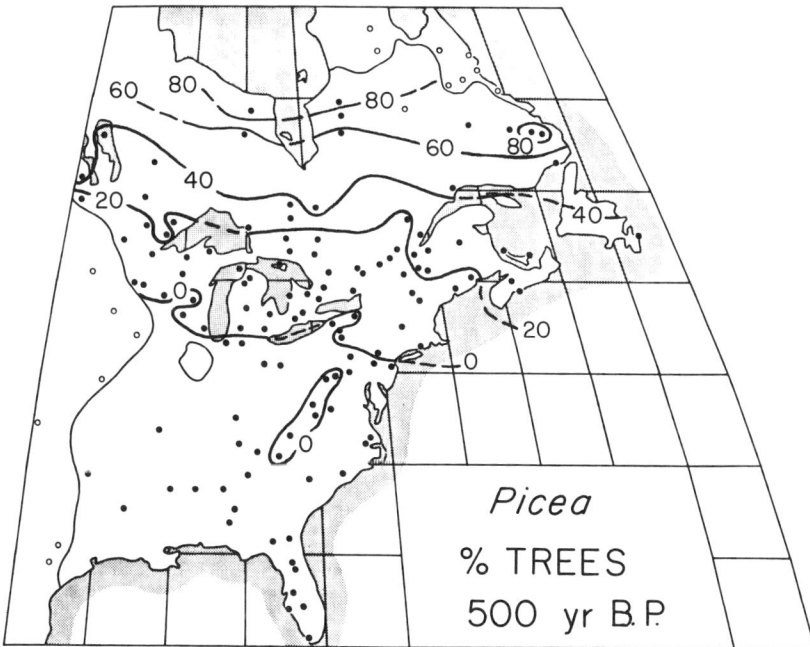

Figure 5.24. Spruce (*Picea*). l. Paleo-dominance map for 500 yr B.P.

the late Holocene. Between 10,000 yr B.P. and 6000 yr B.P., however, spruce percentages of dominance declined throughout its range. The early-Holocene regional decline in pollen percentages of *Picea* was first recognized in the isopollen map analysis of Bernabo and Webb (1977). By 8000 yr B.P., maximum values for spruce dominance were below 50% and its mean values had dropped from characteristic full-glacial levels of about 40% to about 11%. Mean dominance values for spruce rose only to between 20% and 25% in the late Holocene.

The contoured paleo-dominance map for 20,000 yr B.P. (Fig. 5.24b) illustrates the full-glacial distribution of spruce. Spruce population centers of over 60% forest composition were located north of 35°N and generally west of the Appalachian Mountains, increasing to more than 80% dominance in the northernmost forested region (Fig. 4.3a) northwest of the Mississippi Alluvial Valley. Spruce trees may have grown as far north as the glacial margin, particularly in the central Lower Midwest region. A strong gradient in dominance occurred south of 35°N, with spruce reaching its general southern latitudinal limits in distribution between 33°N and 34°N. An extension of spruce [with the species identification of white spruce (*Picea glauca*) based upon fossil cones from Nonconnah Creek in southwestern Tennessee, Delcourt *et al.* 1980] is speculatively mapped southward within the Lower Mississippi Alluvial Valley. Spruce populations typically comprised between 15% and 35% of the forest composition throughout the Carolinas, eastern parts of Virginia and Maryland, and Delaware. The relatively low spruce values correspond with the region of the central Atlantic Seaboard, which may have been situated within a rain shadow of the Appalachian Highlands (Delcourt and Delcourt 1986). Patterns for 18,000 yr B.P. (Fig. 5.24c) and for 16,000 yr B.P. (Fig. 5.24d) were similar to those at 20,000 yr B.P., indicating that spruce populations experienced at least a 4000-year period of relative stability during peak glacial times.

However, during the late-glacial time span from 16,000 yr B.P. to 13,000 yr B.P., mean dominance values for spruce rose progressively from 38% (± 1 SD of 28%) up to 44% (± 1 SD of 23%), the highest mean value of spruce populations reconstructed during the past 20,000 years (Fig. 5.24a). This late-glacial increase in mean population dominance, accompanied by an overall reduction in variability indicated by the reduced standard deviation, reflected a northward expansion of the population center as spruce trees successfully invaded tundra and parkland communities on deglaciated landscapes. The paleo-dominance map for 14,000 yr B.P. (Fig. 5.24e) strikingly emphasizes the north–south asymmetry of the spruce dominance pattern between the advancing northern border and the stationary southern boundary of its distributional limits. Although at 14,000 yr B.P. its range encompassed 15° latitude in eastern North America, the principal population centers for spruce (*e.g.*, regions with more than 60% spruce) were concentrated in the periglacial and deglaciated terrains north of about 39°N. At forested sites near the tundra–boreal forest boundary or adjacent to the glacial-ice margin, spruce constituted 65% to 90% of the forests. Farther south between 33°N and 37°N, the spruce isophytes or contours for 0%, 20%, and 40% remained in relatively stationary positions from 20,000 until 14,000 yr B.P.

The northern limit of spruce populations continued to track the margin of the rapidly retreating Laurentide during the late-Pleistocene to early-Holocene transition. By 12,000 yr B.P., the primary population center for spruce extended as a pronounced band along an arc from central Minnesota, through central Wisconsin, southern Michigan, northern Ohio, northwestern Pennsylvania, western New York and southeastern Ontario (Fig. 5.24f). The latitudinal asymmetry in spruce dominance intensified with the peak mode within 1° to 3° south of its northern limit and then the population values skewed to much lowered dominance southward. Near the glacial-ice margin, spruce dominance values were generally low, between 9% and 32%, attained peak values of 60% to 92% along the arc-shaped population center, and dropped progressively toward its southern range boundary at about 33°N.

Between 12,000 yr B.P. and 10,000 yr B.P. in the Great Lakes region, the prominent arc of spruce fragmented into several restricted population centers. By 10,000 yr B.P. (Fig. 5.24g), pockets of spruce forest (*i.e.*, > 60% dominance) persisted in western Wisconsin and near the climate-ameliorating influences of the margins of meltwater-fed proglacial lakes (Gordon 1985, personal communication), such as in southern Manitoba and northern Minnesota near Glacial Lake Agassiz and in northern Michigan and adjacent Ontario near Lake Superior. Two additional, minor centers of spruce (with values between 40% and 60%) occurred east of Lake Erie across central New York and directly south of the tundra border in western New Brunswick. With the notable exception of the central and southern Appalachians, spruce declined to < 20% dominance in eastern North America south of 41°N.

The paleo-dominance map at 10,000 yr B.P. (Fig. 5.24g) illustrates that the northward migration of spruce during the late-glacial and early-Holocene intervals was bounded by physical and biotic barriers. Along its northern leading edge, spruce advanced up to the margin of the Laurentide Ice Sheet. To the northwest, the extensive Glacial Lake Agassiz inundated vast areas, leaving no suitable terrestrial habitat to invade. During the early Holocene, the northern expansion of prairie through the continental interior resulted in a range constriction of spruce along its western boundary from 95°W to 100°W.

The spruce map for 8000 yr B.P. (Fig. 5.24h) demonstrates the virtual collapse of spruce-dominated forest during the mid-Holocene interval. Spruce trees represented between 20% and 40% of overall forest composition only in the northeastern sector of its range. Limited areas with 40% to 45% spruce occurred in Newfoundland and on the western portion of Prince Edward Island. From 10,000 yr B.P. to 8000 yr B.P., spruce populations advanced to the northeast along the coastal corridor of southeastern Quebec vacated by the glacial ice. Spruce invaded across New Brunswick and Nova Scotia, reaching the southern shoreline of the Gulf of St. Lawrence by 9000 yr B.P. and, with airborne seeds, jumping the water barrier [or perhaps blowing across winter ice (Glaser 1981)] to the island of Newfoundland by 8000 yr B.P. Proglacial lakes and tundra formed the northern and northwestern boundary with spruce. The western limit of spruce 8000 years ago continued to contract eastward, displaced by prairie. During the mid-Holocene interval, remnant spruce populations experienced local extinction generally in the region south of 40°N, within the Ozark Plateaus of Missouri, the Interior Low Plateaus, and the central Atlantic Coastal Plain. Spruce populations survived in the Appalachian Highlands along a peninsula-like southwestward extension from their primary distribution between 40°N and 50°N. Spruce was, however, typically less than 20% dominant along the southern boundary of its range.

The mean value for spruce populations of 11% dominance at 8000 yr B.P. dropped to 10% at 7000 yr B.P. and then rose to 14% at 6000 yr B.P. (Fig. 5.24a); correspondingly, the peak values during this 2000-year span rose from 45% to 53% and then to 74% at 6000 yr B.P. This dynamic transition from population collapse to renewed population expansion in northerly latitudes is clearly seen in the spruce map for 6000 yr B.P. (Fig. 5.24i). The position of the spruce population center (> 40%) had shifted from Prince Edward Island and Newfoundland at 8000 yr B.P. north to Newfoundland and easternmost Quebec at 7000 yr B.P., and then north from Newfoundland by 6000 yr B.P. to three centers with > 60% dominance in eastern Labrador, as well as spreading westward through central Quebec and northern Ontario. The region of 20% to 40% spruce at 6000 yr B.P. extended from 45°N to 55°N in the eastern maritime provinces of Canada, and tapered west of Hudson Bay to between about 52°N to 55°N in Ontario.

By 4000 yr B.P., in the early late-Holocene interval, both the spruce map (Fig. 5.24j) and its population trajectory (Fig. 5.24a) exhibited substantial changes reflecting increases in both the population mean (22%) and maximum (82%). By 4000 years ago, the continuous northern limit of spruce extended from about 53°N across Labrador and Quebec and, west of Hudson Bay, to north of 60°N into the Canadian Arctic. Spruce-dominated boreal forest (> 60% spruce) encompassed the region north of about 52°N and across nearly 30° longitude from 56°W to at least 85°W.

The paleo-dominance map for 2000 yr B.P. (Fig. 5.24k) differs from the earlier Holocene spruce maps in four key respects. The shape of the latitudinal gradient in spruce dominance was concave southward, reaching its highest values along its northernmost border (e.g., as much as 99% spruce in northern Quebec) and declined progressively to its 0% contour line on its southern border at about 42°N. The arctic tundra had shifted to the south to as far as 55°N in coastal Labrador. The northern spruce border, formerly the advancing population edge, was curtailed and retreated from its farthest northward extensions in Quebec and Labrador. The periphery of spruce populations had invaded prairie as forest reestablished in central and southern Manitoba and Minnesota. The southernmost peninsular configuration along the Appalachians had fragmented into disjunct populations of *Picea rubens* within high-elevation southern Appalachian habitats.

The spruce map for 500 yr B.P. (Fig. 5.24l) exhibits the late-Holocene pattern, with principal population centers distributed across the northern half of the boreal forest. During the last several thousand years, the tundra–forest boundary in eastern Canada has continued to be displaced southward. The position of the contour line for 20% spruce dominance has shifted as much as several de-

Picea

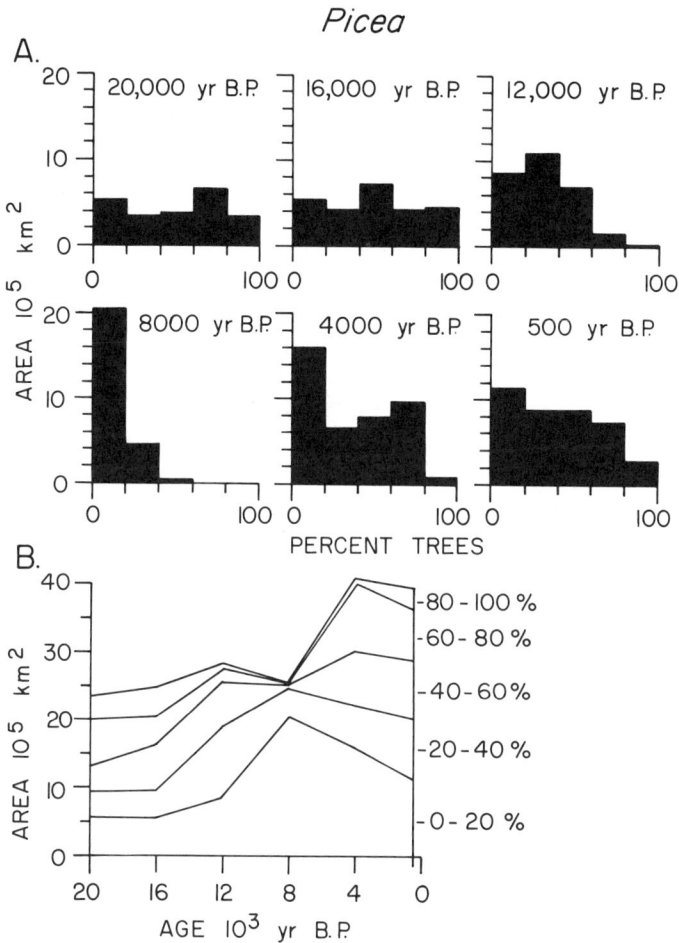

Figure 5.25. Changes in area and dominance of spruce (*Picea*). a. Histograms of area–dominance structure at 20,000, 16,000, 12,000, 8000, 4000, and 500 yr B.P. b. Cumulative plot from 20,000 yr B.P. to the present.

grees of latitude in New England, southern Quebec, southwestern Ontario, and southern Manitoba. The southern border of continuous spruce occurrence at 500 yr B.P. followed the "ecological tension zone" at about 42°N across the Great Lakes region and southern new England.

The area–dominance histograms for spruce (Fig. 5.25a) document for glacial-age times relatively even areal distribution across all dominance classes, from the 0% to 20% class up to the class of 80% to 100%. The histograms for peak-glacial (20,000 yr B.P.) and early late-glacial times (16,000 yr B.P.) illustrate that spruce tended to be the key constituent where its populations occurred in the forests. However, in the early-Holocene interval (12,000 yr B.P.), the areal extent corresponding to higher dominance classes (*e.g.*, the 60% to 80% class and 80% to 100% class) declined by more than 60% of the area occupied, while lower dominance classes (*e.g.*, 0% to 20% class and 20% to 40% class) nearly doubled in extent. During the mid-Holocene interval (8000 yr B.P.), the lowest dominance class (0% to 20%) coincided with more than 80% of the area for which spruce occurred. With no area mapped for a dominance class exceeding 60%, this interglacial dominance structure contrasts markedly with the glacial-age pattern. The suite of histograms for 8000 yr B.P., 4000 yr B.P., and 500 yr B.P. reveal an areal trend over eight millennia to build up the population levels for the higher dominance classes, accompanied by an areal reduction of 44% for the lowest dominance class. Figure 5.25b shows the cumulative areal changes in dominance structure over the past glacial–interglacial cycle for spruce. In total area of distribution, spruce populations occupied between $22.6 \cdot 10^5$ km^2 (20,000 yr B.P.) and $25.3 \cdot 10^5$ km^2 (16,000 yr B.P.) in glacial times, expanding up to $27.2 \cdot 10^5$ km^2 at 12,000 yr B.P. as they tracked the initial major glacial retreat. Although spruce occupied more area at 8000 yr B.P. ($25.2 \cdot 10^5$ km^2) than during the peak glacial at 20,000 yr B.P., the dominance structure was radically different, with virtually all of the populations comprising less than 20% of the forests in which they occurred, rather than becoming regional dominants with equal proportions distributed over all dominance classes. The mid-Holocene diminishment in spruce populations is an example of a fundamental ecological, if not genetic, crisis that probably affected all three species of eastern North American *Picea*. By 4000 yr B.P., spruce populations had dispersed across nearly 80% more land area than they occupied during peak-glacial times. However, the late-Holocene trend was toward a more even pattern of dominance across all five dominance classes. In the millennia following the initial migration of spruce into the modern boreal forest region, long-term shifts have occurred in its pattern of dominance across the landscape that indicate long-term changes in the competitive success of *Picea* relative to other boreal tree taxa.

Pine (*Pinus*)

The population trajectory for pine (Fig. 5.26a) is compiled for the entire genus, which includes 12 species of the subgenus Diploxylon and one species of the subgenus Haploxylon in eastern North America. Two species of Diploxylon *Pinus*, jack pine (*P. banksiana*) and red pine (*P. resinosa*), are today characteristic of mixed conifer–northern hardwoods and boreal forests. The Haploxylon eastern white pine (*P. strobus*) is distributed across the mixed conifer–northern hardwoods forest region as well as throughout the Appalachian Mountains. Several Diploxylon pine species are primarily distributed within the Appalachian Mountains, including Virginia pine (*P. virginiana*), pitch pine (*P. rigida*), and table mountain pine (*P. pungens*). The remainder of the Diploxylon pines are predominant across the Atlantic and Gulf Coastal Plains, and include shortleaf pine (*P. echinata*), longleaf pine (*P. palustris*), loblolly pine (*P. taeda*), spruce pine (*P. glabra*), scrub pine (*P. clausa*), slash pine (*P. elliottii*), and pond pine (*P. serotina*).

During the full-glacial interval, pine populations reached the highest mean

Pinus

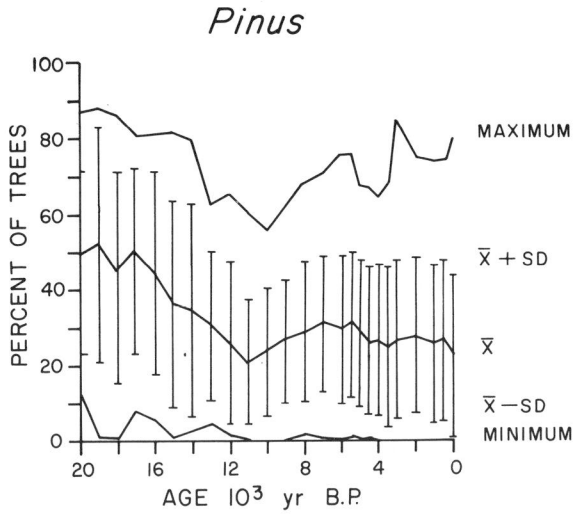

Figure 5.26. Pine (*Pinus*). a. Population trajectory, including maximum, minimum, and mean values of paleo-dominance as well as ± 1 SD of the mean, for the past 20,000 years.

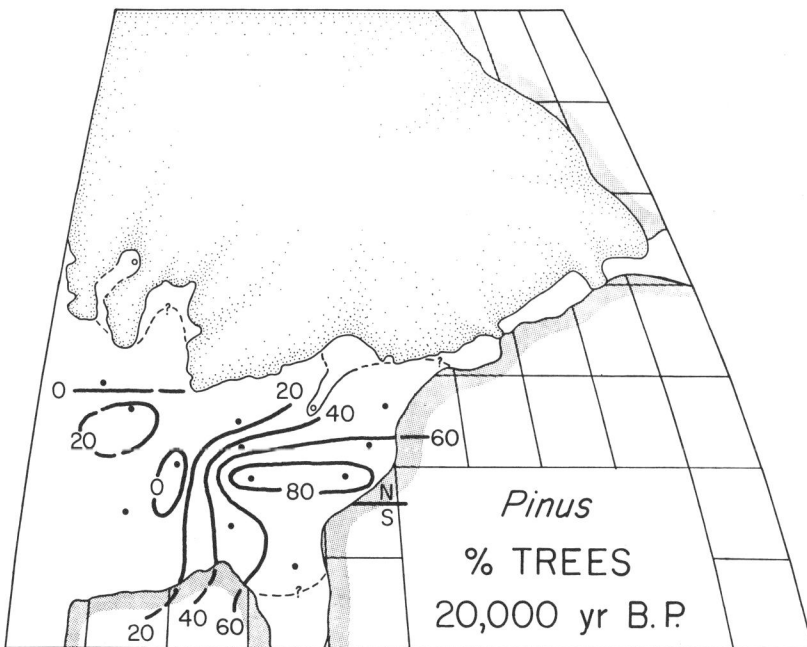

Figure 5.26. Pine (*Pinus*). b. Paleo-dominance map for 20,000 yr B.P.

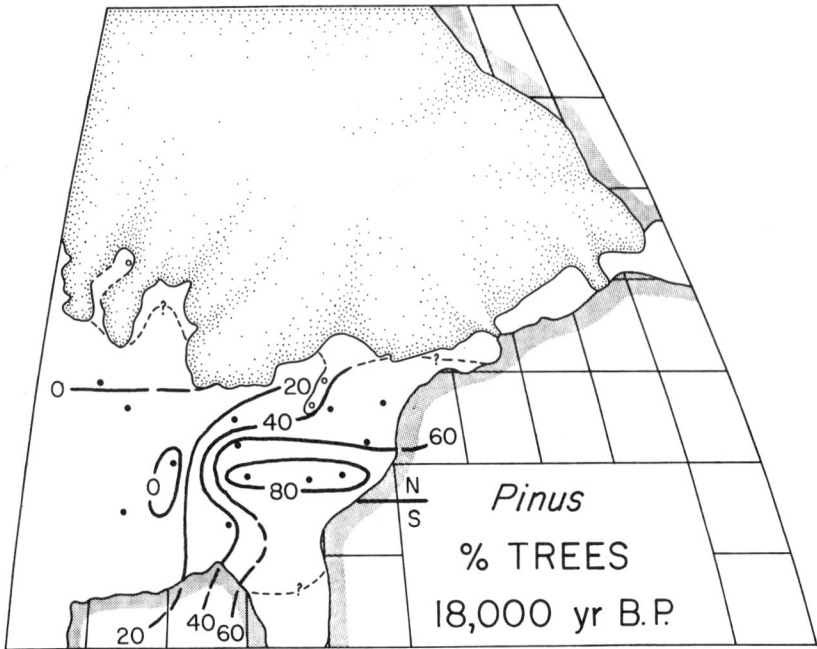

Figure 5.26. Pine (*Pinus*). c. Paleo-dominance map for 18,000 yr B.P.

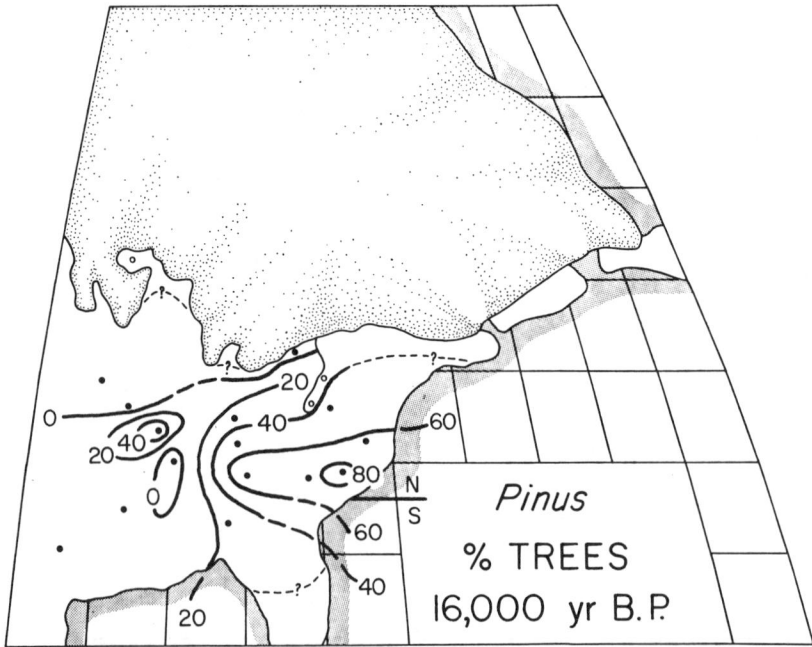

Figure 5.26. Pine (*Pinus*). d. Paleo-dominance map for 16,000 yr B.P.

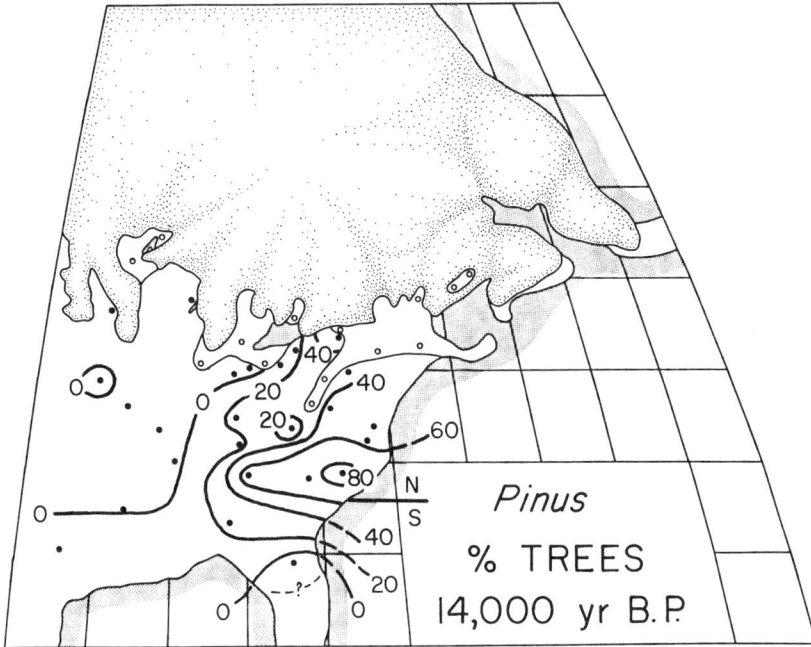

Figure 5.26. Pine (*Pinus*). e. Paleo-dominance map for 14,000 yr B.P.

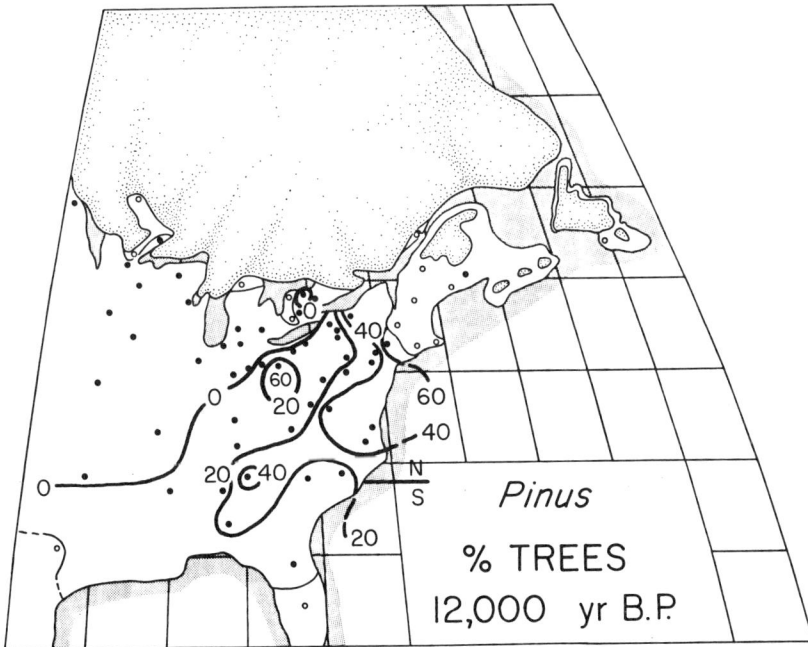

Figure 5.26. Pine (*Pinus*). f. Paleo-dominance map for 12,000 yr B.P.

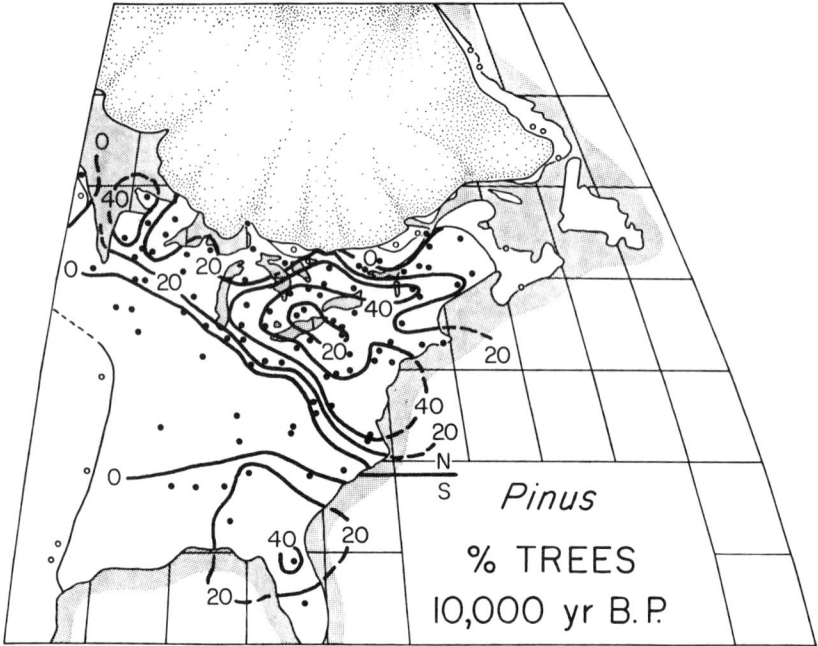

Figure 5.26. Pine (*Pinus*). g. Paleo-dominance map for 10,000 yr B.P.

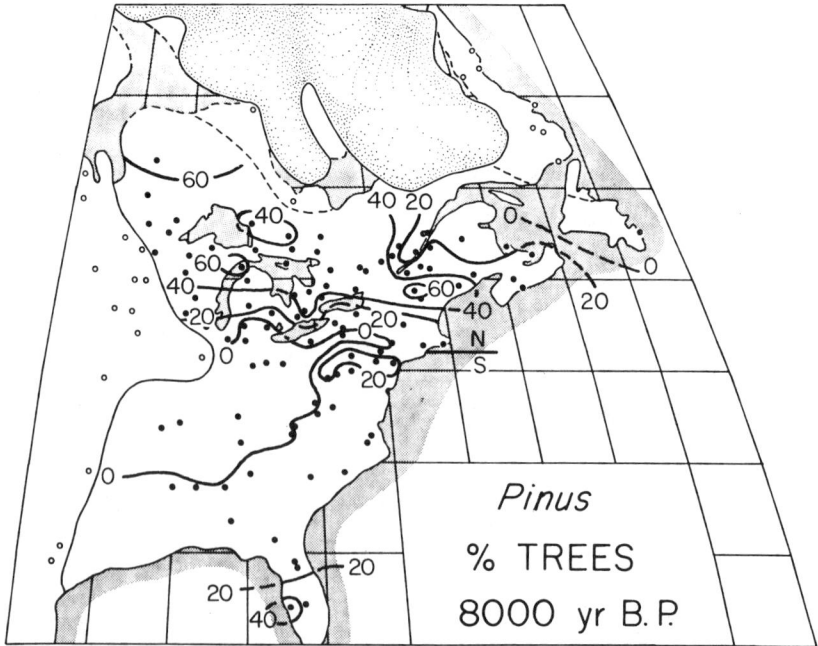

Figure 5.26. Pine (*Pinus*). h. Paleo-dominance map for 8000 yr B.P.

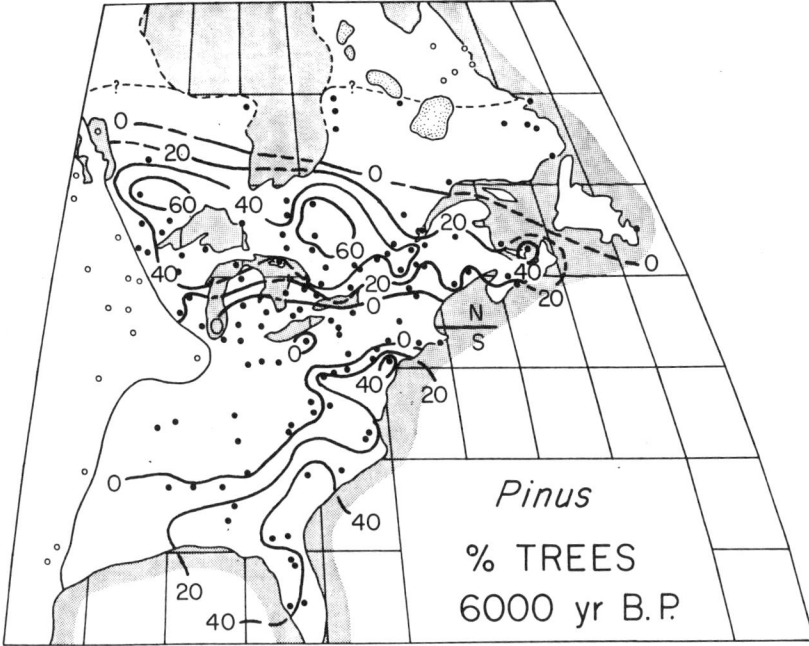

Figure 5.26. Pine (*Pinus*). i. Paleo-dominance map for 6000 yr B.P.

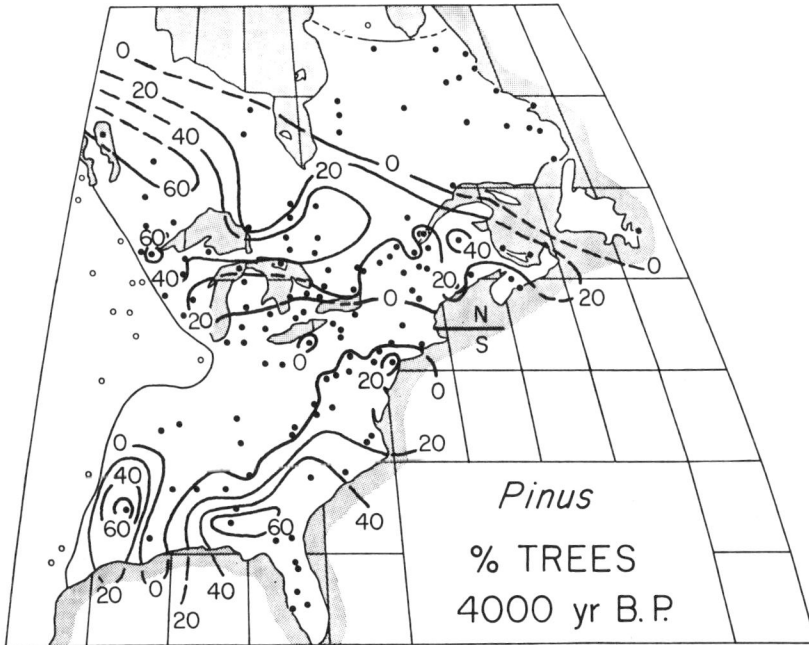

Figure 5.26. Pine (*Pinus*). j. Paleo-dominance map for 4000 yr B.P.

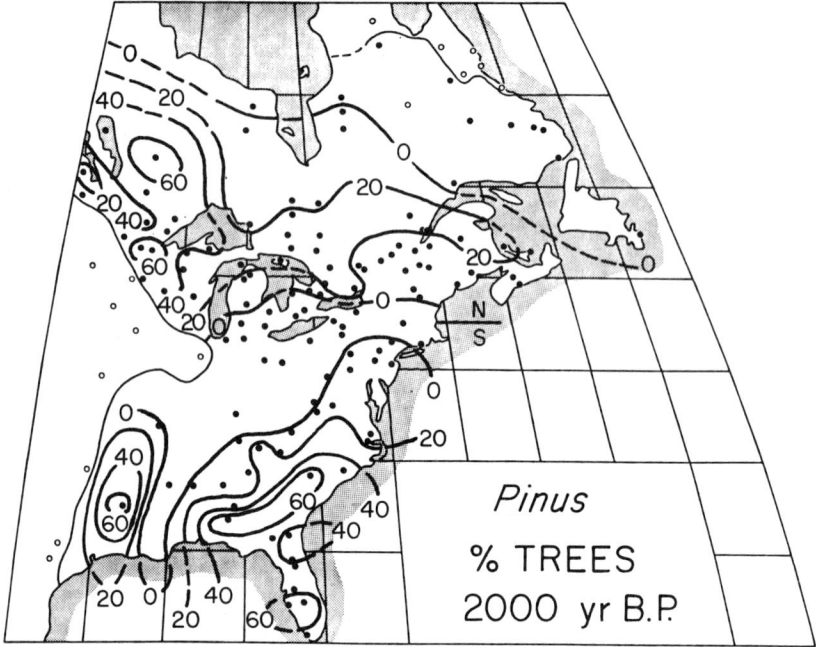

Figure 5.26. Pine (*Pinus*). k. Paleo-dominance map for 2000 yr B.P.

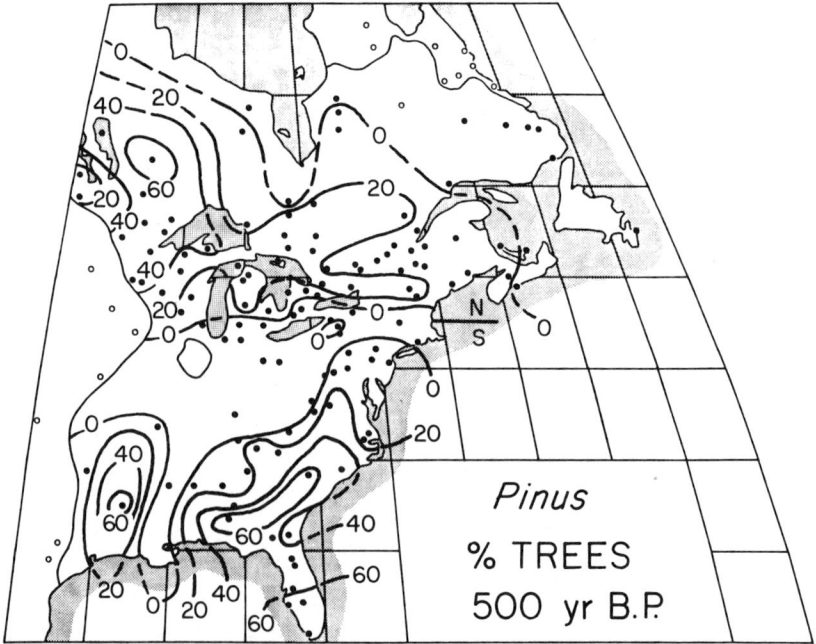

Figure 5.26. Pine (*Pinus*). l. Paleo-dominance map for 500 yr B.P.

values of the late Quaternary (46% to 52%). Variability about those mean values was ± 1 SD of 26% to 31%, and maximum reconstructed values were 81% to 88% (Fig. 5.26a). During the late-glacial interval, from 16,000 yr B.P. to 13,000 yr B.P., both means and standard deviations dropped from 45% to 31% and 26% to 19%, respectively. Late-glacial maximum values reached a plateau at about 81% from 16,000 yr B.P. to 14,000 yr B.P., and then dropped abruptly to 63% by 13,000 yr B.P. Between 12,000 yr B.P. and 9000 yr B.P., mean and maximum values for pine reached their lowest values in the late Quaternary. At 11,000 yr B.P., mean values were 21%; by 10,000 yr B.P., maximum values were only 56% (Fig. 5.26a). From 8000 yr B.P. to 5000 yr B.P., during the mid-Holocene interval, the population means rose to between 29% and 31%, with a reduced population variability of ± 1 SD of 18% to 20% about the mean values. Mid-Holocene populations of pine achieved maximum values between 67% and 76%. During the last 4000 years, maximum values for pine have fluctuated between a late-Holocene low of 64% at 4000 yr B.P. and a high of 84% at 3000 yr B.P. The mean values converged toward 26% (± 1 SD of 21%) in the late Holocene. Overall, the maximum values of pine during the full- and late-glacial intervals were comparable in magnitude to those reached during the late-Holocene interval. The full-glacial means, however, were about twice as high as those achieved in the mid- and late-Holocene intervals. During the interval from 12,000 yr B.P. to 9000 yr B.P., mean dominance values for pine were relatively low.

The paleo-dominance maps for pine at 20,000 yr B.P. (Fig. 5.26b), 18,000 yr B.P. (Fig. 5.26c), and 16,000 yr B.P. (Fig. 5.26d) display primary population centers of 40% to 87% in the region of the unglaciated Atlantic Coastal Plain, the central and eastern Gulf Coastal Plain (generally east of 88°W), and the northern half of the Florida peninsula. At 20,000 yr B.P. (Fig. 5.26b), dominance values for reconstructed pine populations reached 53% in the Delmarva Peninsula, increased southward to 80% to 87% in the Carolinas and northern Georgia, and then decreased to 65% in north-central Florida. Full-glacial populations of pine were typically < 20% along the north–south axis of the Lower Mississippi Alluvial Valley. They reached a secondary population center of between 20% and 40% in Missouri along the northwest portion of their full-glacial distribution in eastern North America. The maps for 20,000 yr B.P. and 18,000 yr B.P. (Fig. 5.26b,c) show the northern distributional limit for pine at approximately 40°N, reaching the forest–tundra ecotone within and east of the Appalachians, the glacial margin immediately west of the Appalachians, and confined to the boreal forest region on the northwestern portion of the range. The southeastern range limit of pine coincided with the limit of forest in central Florida. Where evidence is available based upon measurements of fossil-pollen grains of *Pinus* (*e.g.*, Whitehead 1964) or macrofossil needle remains (*e.g.*, Watts 1970), it appears that northern Diploxylon pines (*P. banksiana* and *P. resinosa*) and southern Diploxylon pines were distributed north and south of about 33°N, respectively, during the full-glacial interval. Some intermixing of populations may have occurred between 33°N and 34°N, the full-glacial ecotone between boreal and temperate forest, but the spatial resolution available from sites studied to date is insufficient to definitively document this possibility.

The paleo-dominance maps for 16,000 yr B.P. (Fig. 5.26d) and 14,000 yr B.P. (Fig. 5.26e) document the persistence of major pine populations in the northern and central Atlantic Coastal Plain as well as the local expansion of pine within and west of the Appalachians. However, major reductions in dominance of pine occurred in the northern half of the Florida peninsula and in the region west of the Mississippi Alluvial Valley and north of about 33°N. Between 16,000 yr B.P. and 14,000 yr B.P., jack pine populations that were formerly abundant within the Ozarks and continental interior became extinct; however, the boundary separating northern from southern groups of Diploxylon pines east of the Mississippi River persisted at 33°N.

Between 14,000 yr B.P. and 12,000 yr B.P., the coastal-plain population center for pine shifted northward from the Carolinas into southern New England, where it reached up to 65%. By 12,000 yr B.P. (Fig. 5.26f), northern Diploxylon pines advanced northward along the Appalachian Highlands and into the eastern Great Lakes region. Generally west of the Appalachians, paleo-dominance of northern Diploxylon pine diminished to < 20%, although locally attaining outlier values of 60% in central Ohio. By 12,000 yr B.P., the latitudinal ecotone between northern and southern pine groups shifted to 34°N. Southern Diploxylon pines decreased to < 40% of forest composition in the southern Atlantic and Gulf Coastal Plains. The expansion of prairie in eastern Texas resulted in diminishment of paleo-dominance of southern Diploxylon pine on its southwestern range margin.

Between 12,000 yr B.P. and 10,000 yr B.P. (Fig. 5.26f,g), population limits of northern and southern Diploxylon pine groups became geographically separated by 1° latitude in the Carolinas and by as much as 11° latitude along 95°W. At 10,000 yr B.P., northern pines, including eastern white pine, advanced through the eastern Great Lakes region and New England but remained southeast of the migrational barrier of the Champlain Sea. Primarily northern Diploxylon pines expanded northwestward across the central and western Great Lakes region. Populations of northern pines represented 40% to 56% of the forest composition in the arc extending from New Jersey and Virginia northwest to central Michigan, then east across southern Ontario and New York. A second center of population for northern pines established in northern Minnesota and southwestern Ontario. By 10,000 yr B.P., the southern Diploxylon pine group advanced northward into the southern Appalachians and extended southeastward throughout the Florida peninsula (Fig. 5.26g). The western limit coincided with the forest–prairie ecotone in eastern Texas. Maximum paleo-dominance values for southern pines were only 20% to 40%, located in southern Alabama, Georgia, and northern Florida.

By 8000 yr B.P. (Fig. 5.26h), northern pines advanced across New Brunswick, Prince Edward Island, and Nova Scotia, as well as north through central Quebec and into central and western Ontario. The western limit of northern pines occurred at the prairie–forest border in Minnesota and Wisconsin, and the southern boundary occurred along 42°N. Paleo-dominance values for northern pines attained maximum values of 40% to > 60% in the western portion of their ranges and decreased to the east, to generally about 40% in the central Great Lakes region and to < 40% in New England. The southern pines were distributed along the Appalachians and Atlantic Coastal Plain north to approximately 41°N by 8000 yr B.P. (Fig. 5.26h). Two population centers of > 20% occurred in Maryland, Pennsylvania, and New Jersey as well as the southern half of Florida. Throughout the majority of their ranges at 8000 yr B.P., however, southern Diploxylon pines did not achieve paleo-dominance values > 20%.

During the mid-Holocene interval, the northern pine group continued to spread northward, reaching about 50°N in central Quebec and 52°N in central Ontario by 6000 yr B.P. (Fig. 5.26i). Northern pines occupied a broad latitudinal band extending south to 44°N from the southeastern Canadian maritime provinces to the prairie–forest border. Areas of pine dominance > 60% were located across southwestern Quebec and southwestern Ontario. A latitudinal band of 20% to 60% pine extended from the St. Lawrence Valley at 71°W to 95°W in the western Great Lakes region. By 6000 yr B.P., the southern Diploxylon pine group occupied the Atlantic and Gulf Coastal Plains, Florida, and the central to southern Appalachians. Approximately 3° latitude separated the distributional limits of the northern and southern pine groups through the southern New England and Great Lakes regions. Southern pines expanded in paleo-dominance between 8000 and 6000 yr B.P., reaching > 20% throughout the eastern two-thirds of their distribution.

By 4000 yr B.P. (Fig. 5.26j), northern pines advanced northwest of Hudson Bay, with primary population centers of > 60% developed in west-central Ontario and central Manitoba. Bounded on the southwest by prairie and with a general distributional limit to the south at 43°N, northern pine populations exhibited a gradient in dominance with peak values in the continental interior, intermediate values between Hudson Bay and the central Great Lakes, and lowest values in southeastern Canada. South of 35°N, southern pines increased in dominance to > 40% throughout their ranges both east and west of the Lower Mississippi Alluvial Valley.

During the last 2000 years, minor changes occurred in both the distributional limits and dominance values for northern and southern pines. By 2000 yr B.P. (Fig. 5.26k), northern Diploxylon pines advanced into north-central Quebec east of Hudson Bay. Along their southwestern border, northern Diploxylon pines and eastern white pine spread into southern Manitoba, central Minnesota, and western Wisconsin as the prairie–forest ecotone shifted westward (Jacobson 1979). The southern pine group expanded along its northern border into East Tennessee as well as into southern Missouri. By 500 yr B.P. (Fig. 5.26l), southern pines advanced into eastern Oklahoma. During the last 2000 years, southern pines increased in dominance to > 60% of the forest composition throughout the uplands of the Gulf Coastal Plain and Florida.

At the genus level, the areal distribution for *Pinus* stayed relatively consistent at approximately $27 \cdot 10^5$ km^2 for the full- and late-glacial intervals, constricted to approximately $20 \cdot 10^5$ km^2 during the Pleistocene–Holocene transition, and then reexpanded by 68% of its minimum area to about $34 \cdot 10^5$ km^2 by the mid-

and late-Holocene intervals. The area–dominance histograms for *Pinus* are given in Fig. 5.27 for both the northern and southern groups. In the first case of the southern and Appalachian species of pine, the total distributional area has fluctuated between a minimum of $10.1 \cdot 10^5$ km^2 at 12,000 yr B.P. and a maximum of $15.0 \cdot 10^5$ km^2 at 500 yr B.P. (Fig. 5.27a,b). Although for much of the last 20,000 years the area occupied by southern pines has remained at generally about $12 \cdot 10^5$ km^2, the dominance structure of southern pine populations has changed markedly. The area–dominance histogram for southern pines reveals dual modes at 20,000 yr B.P.; the greatest area was occupied by < 20%, with a secondary peak in area between 60% and 80%. However, the area–dominance histograms for 16,000 yr B.P., 12,000 yr B.P., and 8000 yr B.P. exhibit a trend toward reduced number of dominance classes, such that the < 20% class rep-

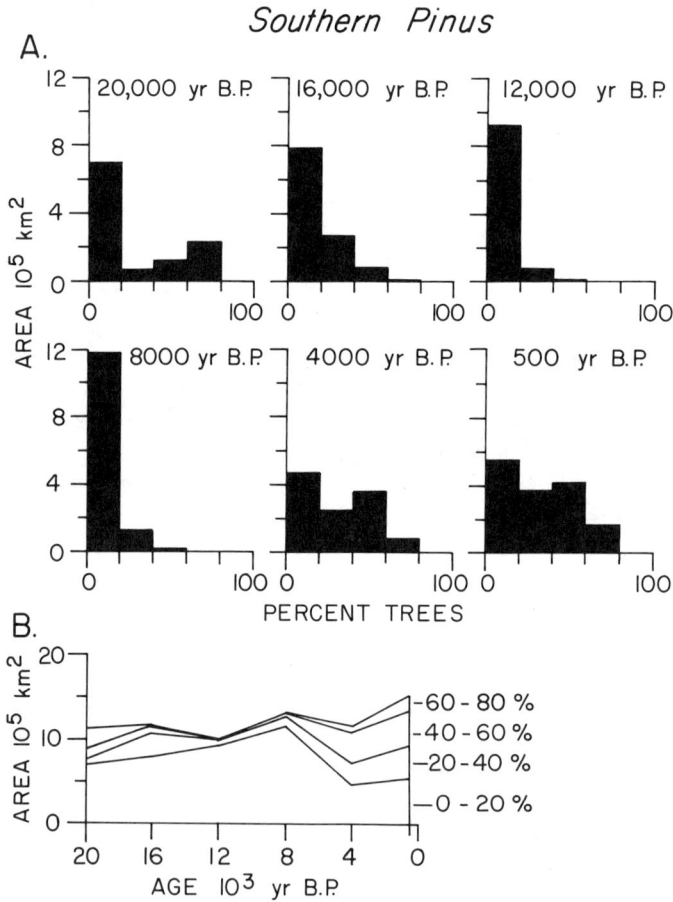

Figure 5.27. Changes in area and dominance of the southern group of pines: a. Histograms of area–dominance structure at 20,000, 16,000, 12,000, 8000, 4000, and 500 yr B.P.; b. Cumulative plot from 20,000 yr B.P. to the present.

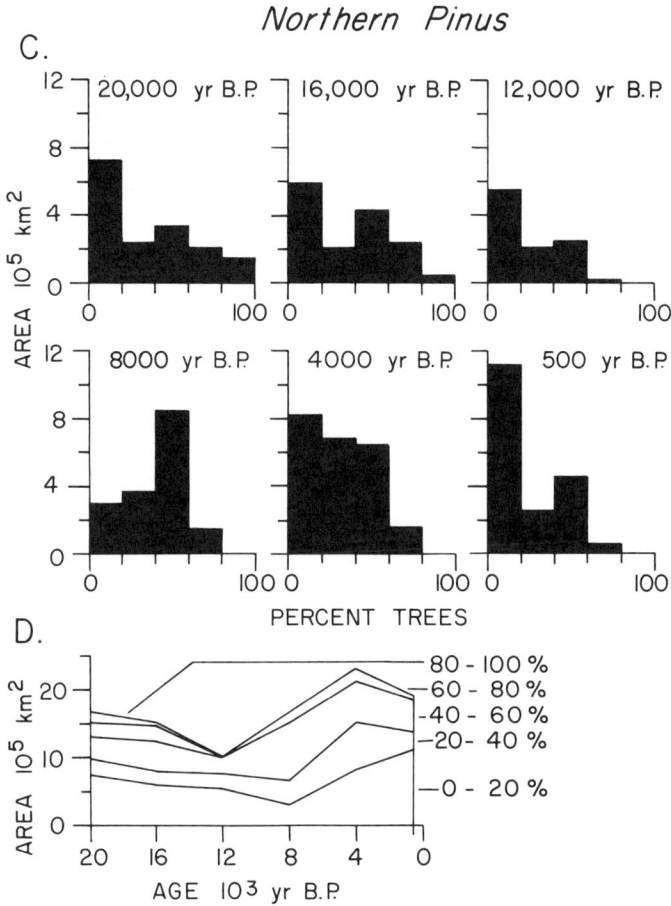

Figure 5.27. Changes in area and dominance of the northern group of pines: c. Histograms of area–dominance structure at 20,000, 16,000, 12,000, 8000, 4000, and 500 yr B.P.; d. Cumulative plot from 20,000 yr B.P. to the present.

resented 69%, 92%, and 89% of their spectra, respectively. During the last 8000 years, southern pines expanded both in number of dominance classes and more of the distributional area occupied by the higher dominance classes. Although there has been only a 14% increase in total area in the last 8000 years, there has been a 679% increase in the area dominated by 20% or more southern pines, from $1.4 \cdot 10^5$ km^2 at 8000 yr B.P. to 9.5 $\cdot$ 10^5 km^2 at 500 yr B.P. The pronounced increase in relative dominance of southern pines in the mid- and late-Holocene intervals has been referred to previously in the literature as the "southern pine rise" (Watts 1969).

The area–dominance histograms for the northern pine group (Fig. 5.27c,d) reflect changes in both the total distributional area and dominance structure within the group through the late-Quaternary interval. These histograms display 5 dominance classes for 20,000 yr B.P. and 16,000 yr B.P. and 4 dominance classes for the remainder of the time interval (Fig. 5.27c). Therefore, northern pines were dominant in part of their range throughout the entire time series. However, the area occupied by < 20% northern pine was greater than that of any other single dominance class in all histograms except that for 8000 yr B.P. The cumulative plot of dominance structure (Fig. 5.27d) displays a progressive decline in area from 16.6 · 10^5 km^2 at 20,000 yr B.P. to a minimum of 10.2 · 10^5 km^2 at 12,000 yr B.P., and then a reexpansion to a peak at 4000 yr B.P. of 23.0 · 10^5 km^2. This documents a substantial distributional range during the full-glacial interval, a diminishment of range during the late-glacial and early-Holocene intervals, and an expansion by more than 39% of full-glacial values by late-Holocene times. The histogram for 12,000 yr B.P. documents the loss in area for high and intermediate dominance classes that corresponded to the major reduction in overall area of distribution. In the mid-Holocene interval, northern pines both expanded in total distributional area and increased in the area occupied by the intermediate dominance classes. By 500 yr B.P., however, northern pines decreased in total distributional area to 19.0 · 10^5 km^2, at which time their lowest dominance class represented 59% of their dominance structure. The decrease in northern pines during the late Holocene thus contrasts with the late-Holocene expansion in southern pines.

Aspen or Cottonwood (*Populus*)

The trajectory plot for *Populus* (Fig. 5.28a) reveals major late-Quaternary oscillations in the contribution of aspen and cottonwood populations to the composition of late-Quaternary forests. Full-glacial populations of *Populus* represented minor forest components, with mean dominance values of about 5% and maximum values never exceeding 13% from 20,000 yr B.P. to 17,000 yr B.P. The late-glacial interval from 16,000 yr B.P. to 13,000 yr B.P. coincided with major population expansion and a tripling in both its mean and peak values (*i.e.*, from 7% up to 19% mean dominance and maximum values from 21% up to 61%). Aspen values reached a plateau of about 90% dominance in maximum values from 12,000 yr B.P. to 8000 yr B.P. Its mean values varied between 14% and 25% during this interval; relatively large standard deviations of approximately ± 18% accompanied these relatively high mean values. Aspen and cottonwood populations converged through the mid- and late-Holocene intervals toward a mean value of about 6%. Although sporadically fluctuating over the interval from 7000 yr B.P. to 500 yr B.P., maximum values of *Populus* generally dropped to about 30%. The modern mean and peak values for aspen have more than doubled since 500 yr B.P.; they reflect the substantial impact of Euro-American settlement, widespread forest clearance, and the historic regrowth of early-successional forests, particularly in the Great Lakes region.

The paleo-dominance maps for 20,000 yr B.P. (Fig. 5.28b) and 18,000 yr B.P. (Fig. 5.28c) indicate populations of aspen situated along a northwest–southeast trend from the Ozark Plateaus of central Missouri, across the Mississippi Embayment to western Tennessee, and to the Gulf Coastal Plain of south-central Alabama. This refugium is defined by paleoecological records from three forest

Populus

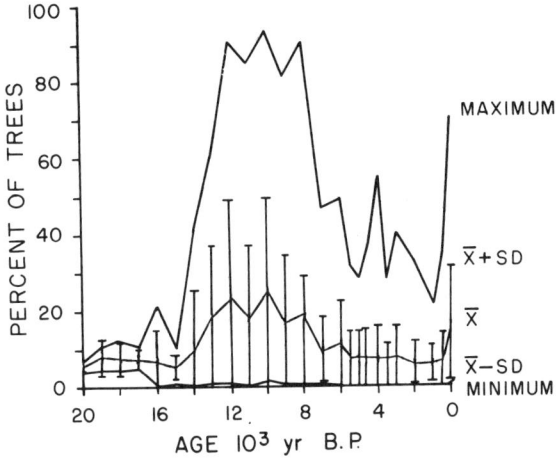

Figure 5.28. Aspen (*Populus*). a. Population trajectory, including maximum, minimum, and mean values of paleo-dominance as well as ± 1 SD of the mean, for the past 20,000 years.

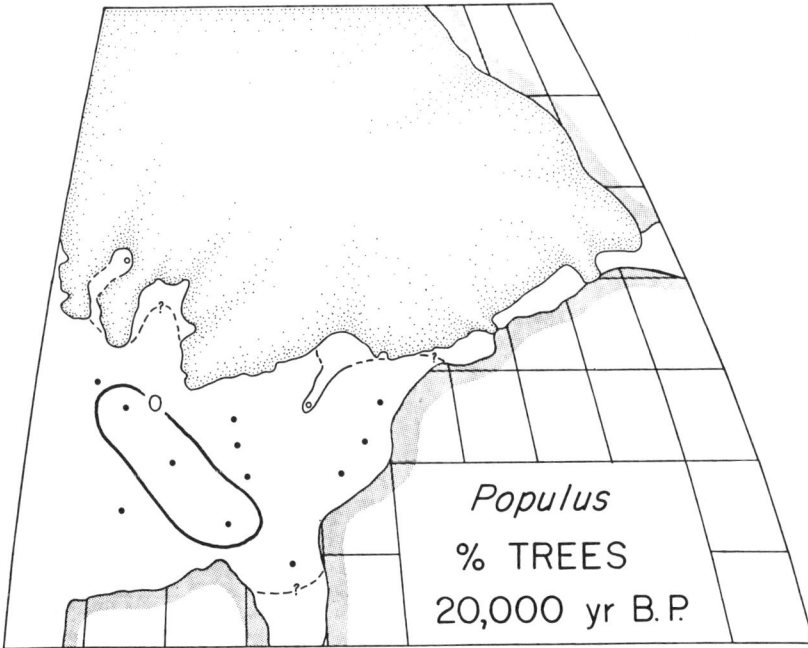

Figure 5.28. Aspen (*Populus*). b. Paleo-dominance map for 20,000 yr B.P.

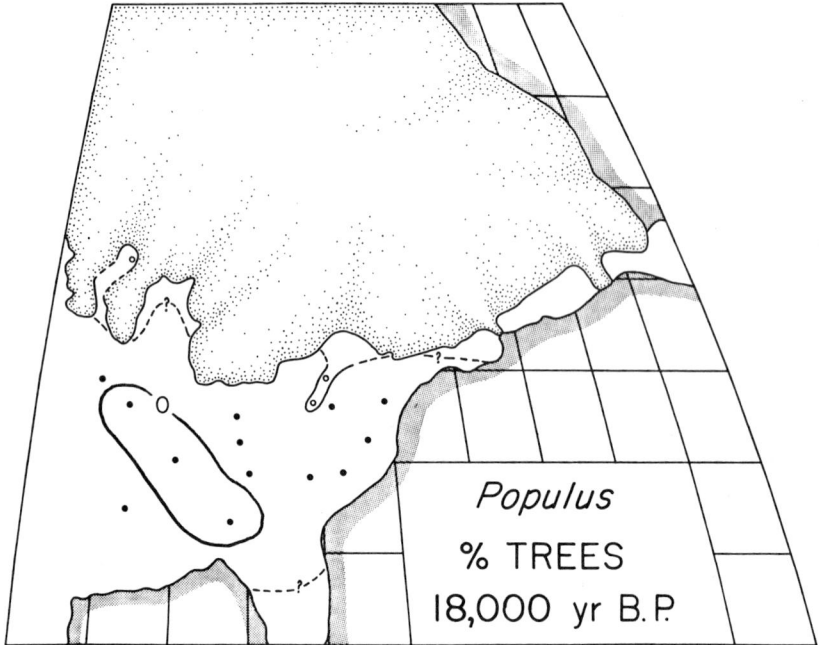

Figure 5.28. Aspen (*Populus*). c. Paleo-dominance map for 18,000 yr B.P.

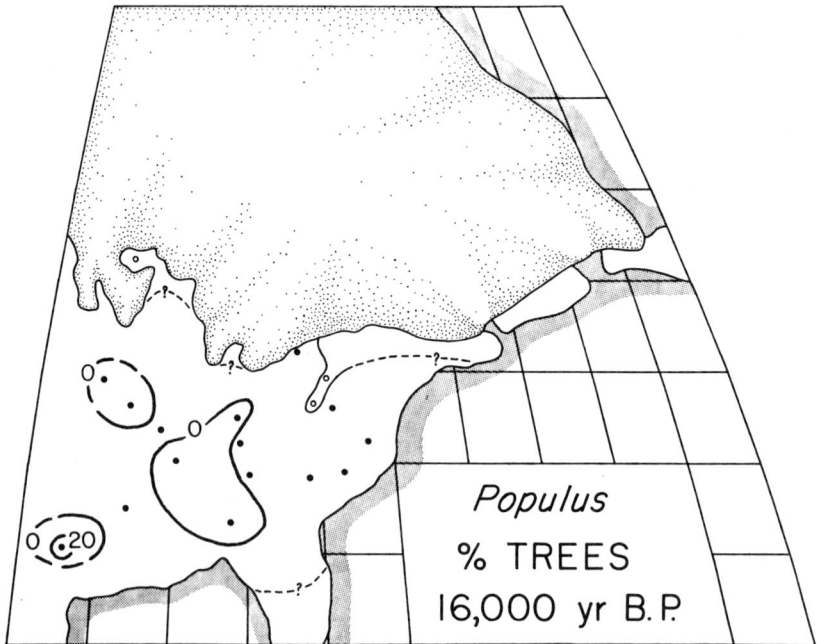

Figure 5.28. Aspen (*Populus*). d. Paleo-dominance map for 16,000 yr B.P.

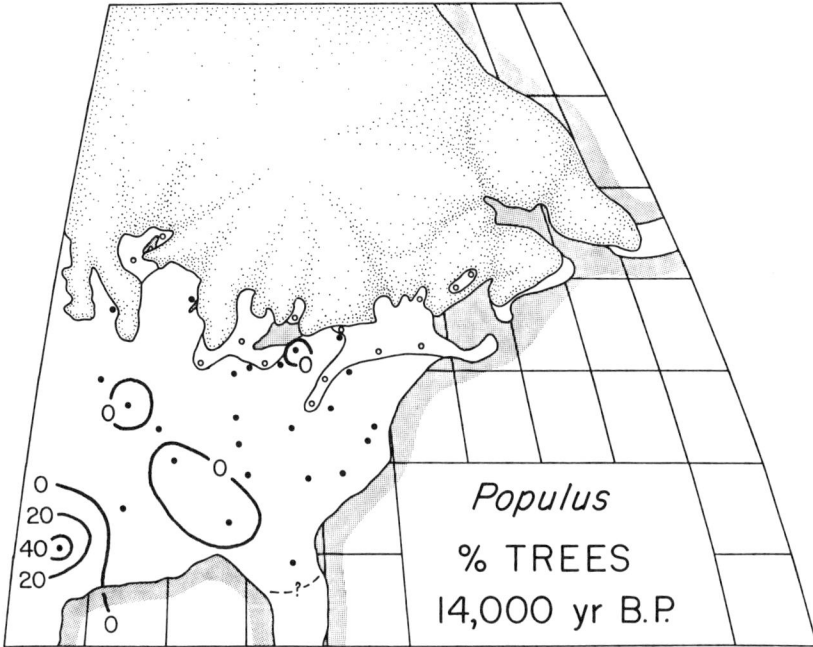

Figure 5.28. Aspen (*Populus*). e. Paleo-dominance map for 14,000 yr B.P.

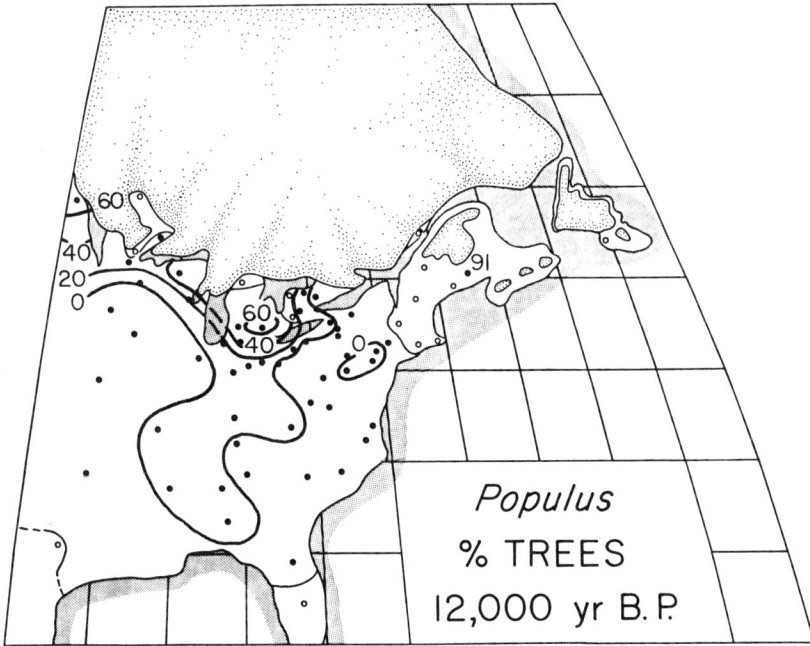

Figure 5.28. Aspen (*Populus*). f. Paleo-dominance map for 12,000 yr B.P.

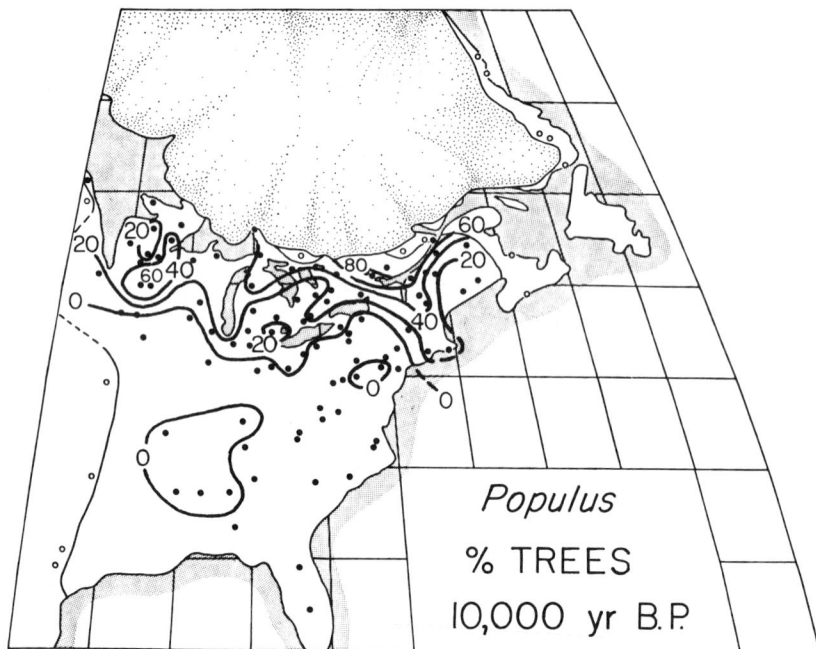

Figure 5.28. Aspen (*Populus*). g. Paleo-dominance map for 10,000 yr B.P.

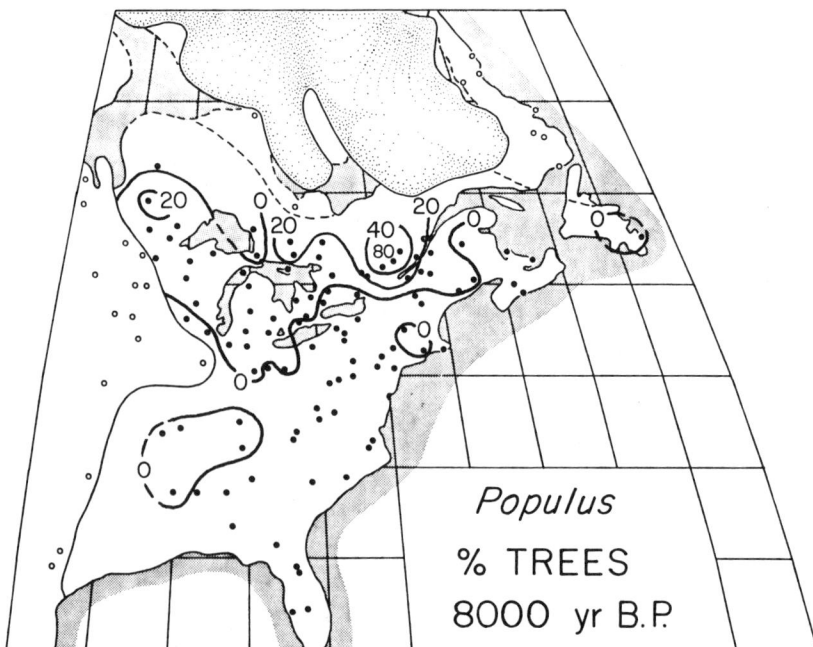

Figure 5.28. Aspen (*Populus*). h. Paleo-dominance map for 8000 yr B.P.

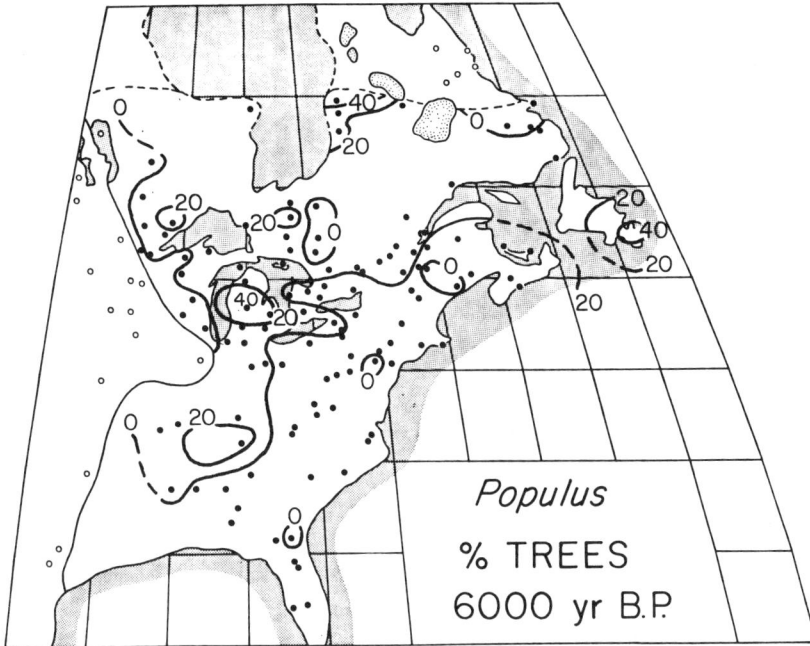

Figure 5.28. Aspen (*Populus*). i. Paleo-dominance map for 6000 yr B.P.

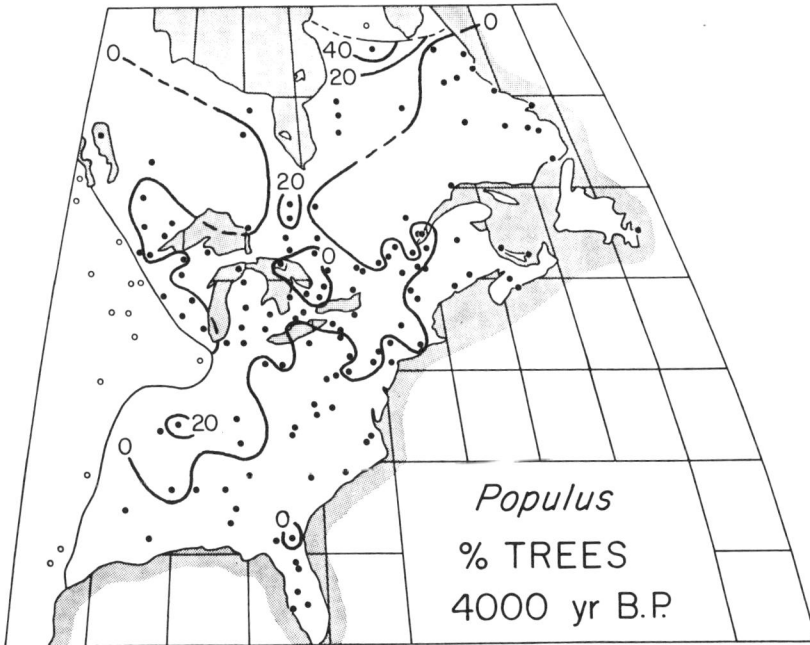

Figure 5.28. Aspen (*Populus*). j. Paleo-dominance map for 4000 yr B.P.

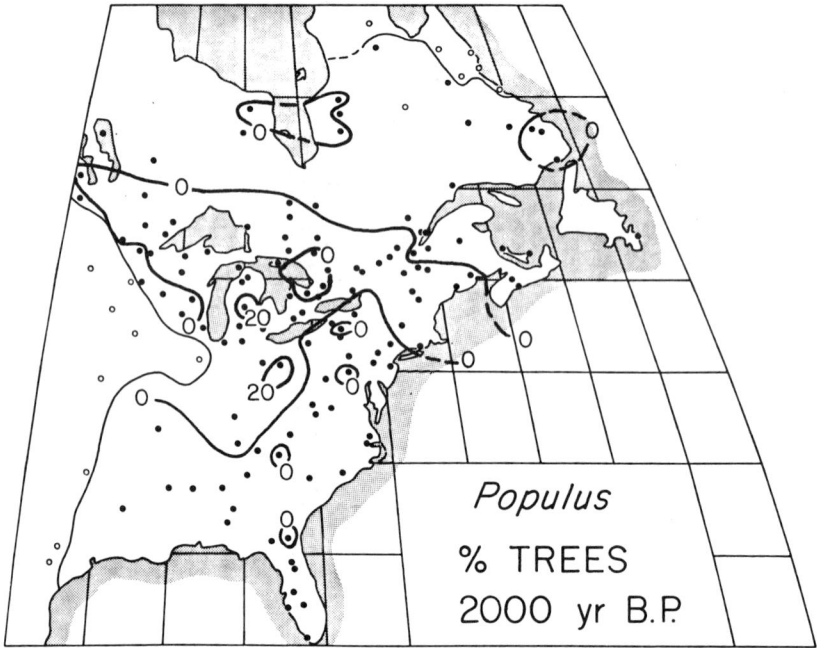

Figure 5.28. Aspen (*Populus*). k. Paleo-dominance map for 2000 yr B.P.

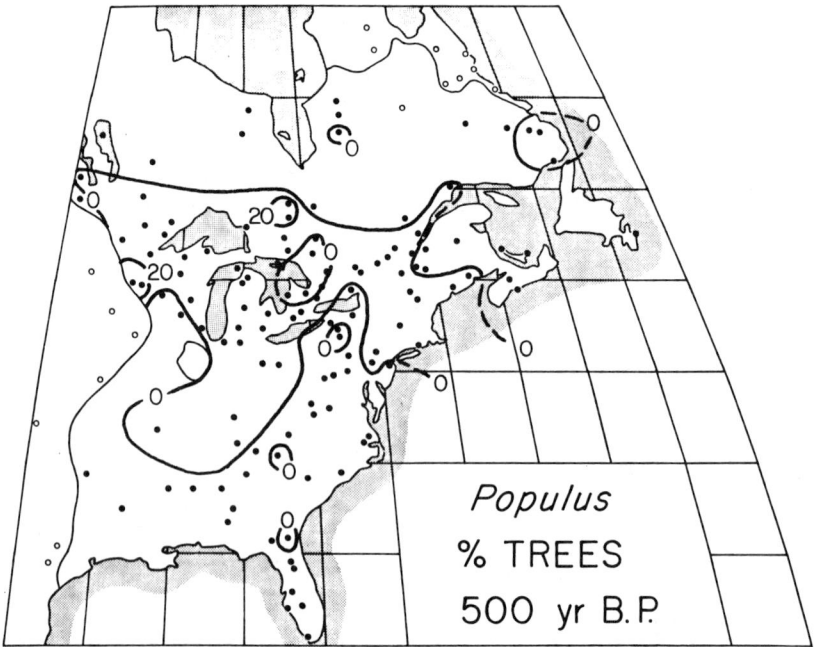

Figure 5.28. Aspen (*Populus*). l. Paleo-dominance map for 500 yr B.P.

sites and it traverses the full-glacial regions mapped for boreal forest, mixed conifer–northern hardwoods forest, and the southern admixture of southeastern evergreen forest and deciduous forest (Fig. 4.3a). The mapped area for aspen and cottonwood is bordered by the 0% dominance contour. Nonconnah Creek, Tennessee (Delcourt *et al*. 1980) has the highest values of dominance for *Populus*, with 7% and 13% at 20,000 yr B.P. and 18,000 yr B.P., respectively.

Populus trees grew at generally low population levels throughout the full-glacial and late-glacial intervals in both boreal and temperate forests. The time series of maps for 16,000 yr B.P. (Fig. 5.28d), 14,000 yr B.P. (Fig. 5.28e), and 12,000 yr B.P. (Fig. 5.28f) identifies two distinctive patterns of population dynamics for this genus. At 16,000 yr B.P., the dominance value of 21% in southeastern Texas reflects a locally important outlier population possibly represented by cottonwood trees (*Populus deltoides*). This southwestern outlier increased to 41% at 14,000 yr B.P., reached 61% at 13,000 yr B.P., and then disappeared by 12,000 yr B.P. as prairie displaced the forest at the site. This late-glacial rise in *Populus* in forests of the Southern High Plains and its subsequent replacement by prairie may indicate biotic dynamics along the forest–prairie ecotone responding to both warming and drying climatic conditions. In contrast, from 14,000 yr B.P. to 12,000 yr B.P., the second principal grouping of *Populus* extended north from the central Gulf Coastal Plain to near the glacial margin in the eastern Great Lakes region. By 13,000 yr B.P., aspen populations existed in two centers, one in northern Ohio and western New York and the second located in northern Wisconsin and northeastern Minnesota. By 12,000 years ago, substantial populations of aspen continuously occupied the northern perimeter of boreal forest from western New York, west across Ontario and Michigan to as far as southern Manitoba. To the north and northwest, the aspen limit was anchored by physical barriers such as the shoreline of proglacial lakes and the margin of the Laurentide Ice Sheet, or by the biotic boundary of the boreal forest–tundra ecotone. However, physical barriers did not bar aspen migrations to the northeast. A limited outlier of aspen trees in the Appalachian Mountains of central Pennsylvania apparently provided the seed source for long-distance dispersal of airborne propagules; the establishment of the advance colony of aspen, situated within the tundra landscape of southern New Brunswick, occurred at a distance of about 700 km beyond the closest aspen population mapped (see the 91% dominance value on Fig. 5.28f).

Between 12,000 yr B.P. (Fig. 5.28f) and 10,000 yr B.P. (Fig. 5.28g), glacial ice and major proglacial lakes continued to block appreciable aspen movement north and northwest. *Populus* populations fragmented into two groups, a cluster in the Mid-South region and another occupying the latitudinal band between 40°N and approximately 50°N. The more northerly band included substantial population centers (> 40% dominance) in Minnesota and along the ice or tundra border from northern Michigan, through southern Ontario, to the northern New England states. This pronounced band of extremely high dominance values (up to 93%) represented the invasion of the first principal tree taxon into tundra across this northeastern sector.

As earlier suggested by Davis and Jacobson (1985), the advance colonization of aspen probably met preferential success in displacing tundra. As it spread across New England, *Populus* was initially funneled through suitable habitats along the corridors of major river valleys and coastal lowlands. From 12,000 to 10,000 years ago, the Champlain Sea extended as a marine embayment southwest along the St. Lawrence Valley to southeastern Ontario and northern New York. At 10,000 years ago, aspen comprised 60% to 80% of the trees occupying this corridor.

By 8000 yr B.P., *Populus* had moved farther along the St. Lawrence Valley across coastal Quebec, with advance colonies established on Newfoundland (Fig. 5.28h). Although one major population center (up to 91% dominance) survived in southern Quebec, aspen declined by 40% to 60% dominance along the forest–tundra ecotone on central Quebec and central Ontario. In the western Great Lakes region, *Populus* comprised less than 20% of the forest composition in Minnesota, Wisconsin, and Michigan, northeast of the prairie–forest ecotone. Southeast of the Prairie Peninsula of Iowa and Illinois, limited *Populus* remained in the Ozarks and Interior Low Plateaus of the Mid-South region.

As the last major remnants of the Laurentide Ice Sheet stagnated and melted between 8000 yr B.P. and 6000 yr B.P. (Fig. 5.28h,i), aspen populations opportunistically invaded the freshly deglaciated terrain of northern Canada, effectively displacing the forest–tundra ecotone to about 55°N. Population centers with more than 40% aspen developed in central Lower Michigan, Newfoundland, and in west-central Quebec. By 4000 yr B.P. (Fig. 5.28j), *Populus* continuously extended across 27° latitude but were generally concentrated in three geographic areas. Mid-South populations were sustained across Missouri, Kentucky, Tennessee, and Mississippi, southeast of the Prairie Peninsula. A second area of continued dominance extended from about 40°N to 48°N, from the western limit of the forest–prairie border of Minnesota across the Great Lakes and southern New England regions, and along the St. Lawrence Valley. A third cluster of aspen populations nearly encircled the southern shore of Hudson Bay and stretched east along the forest periphery at 58°N. Although aspen populations were reconstructed at 57 forest sites for 4000 yr B.P., only at one site did they exceed the dominance value of 40%; that site was located in northern Quebec, with 55% near the forest–tundra ecotone.

During the last several thousand years, the northernmost and areally continuous aspen population around Hudson Bay has progressively fragmented into smaller, more isolated patches. By 2000 yr B.P. (Fig. 5.28k), the spatial continuity between Hudson Bay and Great Lakes populations of aspen was broken. The northern latitudinal boundary of continuous aspen occurrence (corresponding to the 0% contour) had moved south from 58°N in northern Quebec at 4000 yr B.P. to about 48°N in southern Quebec at 2000 yr B.P. From 2000 to 500 years ago (Fig. 5.28l), aspen populations were widely distributed but at low dominance levels (0% to 20%) along the southern fringe of the boreal forest, throughout the entire region of the mixed conifer–northern hardwoods forest and within the central core of the eastern deciduous forest region (Fig. 4.2a).

The area–dominance histograms for *Populus* (Fig. 5.29a) display an intriguing and temporally consistent pattern of dominance structure. For six representative

Populus

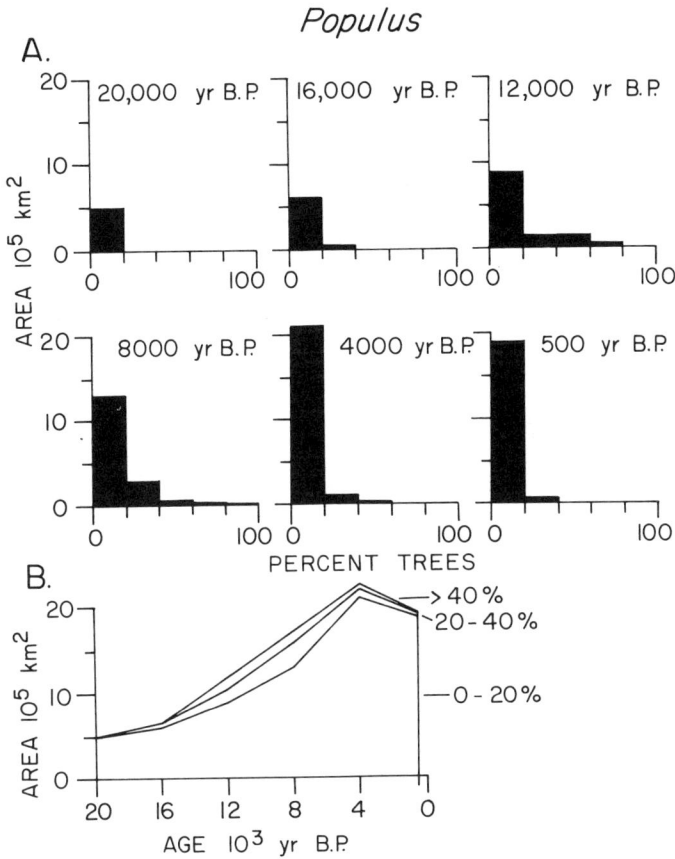

Figure 5.29. Changes in area and dominance of aspen (*Populus*). a. Histograms of area–dominance structure at 20,000, 16,000, 12,000, 8000, 4000, and 500 yr B.P. b. Cumulative plot from 20,000 yr B.P. to the present.

times during the past 20,000 years, the lowest dominance class (0% to 20%) visually dominates the spectrum of each aspen histogram, possessing from 73% (12,000 yr B.P.) to as much as 100% (20,000 yr B.P.) of the total aspen area measured for each time. Larger dominance classes of aspen increased in importance only during times of major climatic change, extensive ice stagnation, and retreat of continental glaciers (*e.g.*, 12,000 yr B.P. and 8000 yr B.P.) that provided newly deglaciated terrains suitable for aspen colonization. Shown on the cumulative plot of dominance structure (Fig. 5.29b), the mapped distribution of aspen equaled $4.8 \cdot 10^5$ km^2 during the full-glacial interval and expanded progressively through the late-glacial to early-Holocene transition (to $12.1 \cdot 10^5$ km^2 at 12,000 yr B.P.) and the mid-Holocene interval ($16.4 \cdot 10^5$ km^2 at 8000 yr B.P.). Aspen populations reached their peak in areal extent in the early part of the late-Holocene interval ($22.1 \cdot 10^5$ km^2 at 4000 yr B.P.) and then, by pre-settlement times (500 yr B.P.), diminished by about 12% to $19.1 \cdot 10^5$ km^2).

During glacial times (20,000 yr B.P.) aspen populations extended across 9° latitude. In contrast, 16,000 years later in the present interglacial, *Populus* had expanded 460% in terms of area occupied, and its distribution range tripled, continuously spanning more than 27° latitude in eastern North America.

Oak (*Quercus*)

The population trajectory for *Quercus* (Fig. 5.30a), shows mean values of approximately 9% from 20,000 yr B.P. to 17,000 yr B.P., an increase to between 10% and 15% from 16,000 yr B.P. to 10,000 yr B.P., and then a rise to between 19% and 25% from 9000 yr B.P. to the present. Variability about the mean was generally ± 1 SD of 10% to 14% from 20,000 yr B.P. to approximately 11,000 yr B.P. and increased to as much as ± 24% at 8000 yr B.P. before leveling off at about ± 20% for the remainder of the Holocene. Maximum values for oak rose from 29% at 20,000 yr B.P. to between 42% and 51% during the late-glacial interval; they increased dramatically to 96% at 10,000 yr B.P. before generally declining to values fluctuating between 62% and 91% throughout the remainder of the Holocene interval. Times of peak values reconstructed for oak populations were 10,000 yr B.P. (96%), 5000 yr B.P. (86%), and 0 yr B.P. (91%).

The map of paleo-dominance for oak at 20,000 yr B.P. (Fig. 5.30b) illustrates that limited oak populations persisted in the southernmost boreal forest region during the full-glacial interval. The population center of > 20% oak dominance was located south of 34°N. The full-glacial map for 18,000 yr B.P. (Fig. 5.30c) reflects this strong latitudinal pattern, with dominance of oak increasing southward to as much as 32% across the Gulf Coastal Plain.

By 16,000 yr B.P. (Fig. 5.30d), oak populations increased across the southern portion of the distribution of *Quercus*, reaching 46% of the forest composition in central Alabama and 31% dominance in north-central Louisiana. To the west, oak diminished to 2% in eastern Texas, but to the north *Quercus* expanded generally to 39°N across the Interior Low Plateaus of Tennessee and into the western Ozarks. The northern range limit of oak at 16,000 yr B.P. is uncertain across much of the Midwest, but one or more species may have colonized in limited populations as far as 41°N in northern Ohio. East of the Appalachians, oak was confined to the south of about 37°N.

By 14,000 yr B.P. (Fig. 5.30e), many sites document the northward shift in oak distribution west of the Appalachian Mountains. The northern limit of oak at 14,000 yr B.P. extended nearly to the ice front, with reconstructed values consistently between 2% and 14% throughout the northern portion of its late-glacial range (Fig. 5.30e). Oak expanded in dominance throughout the region between 40°N and 35°N, reaching paleo-dominance values consistently greater than 20% across the region from Tennessee to Missouri and Kansas. In the region south of 35°N, oak maintained values of 23% to 42% across the Gulf Coastal Plain from eastern Texas to northern Florida.

Between 14,000 yr B.P. and 12,000 yr B.P. (Fig. 5.30f), the northern range limit continued to extend northward, reaching between 46°N and 49°N in the northwestern part of its range across Wisconsin, Minnesota, and southern

Quercus

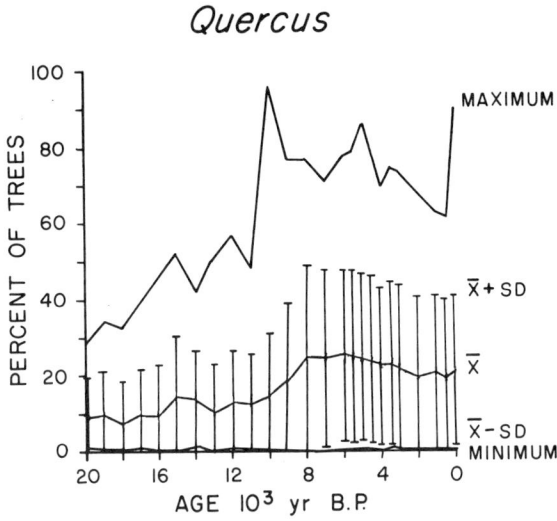

Figure 5.30. Oak (*Quercus*). a. Population trajectory, including maximum, minimum, and mean values of paleo-dominance as well as ± 1 SD of the mean, for the past 20,000 years.

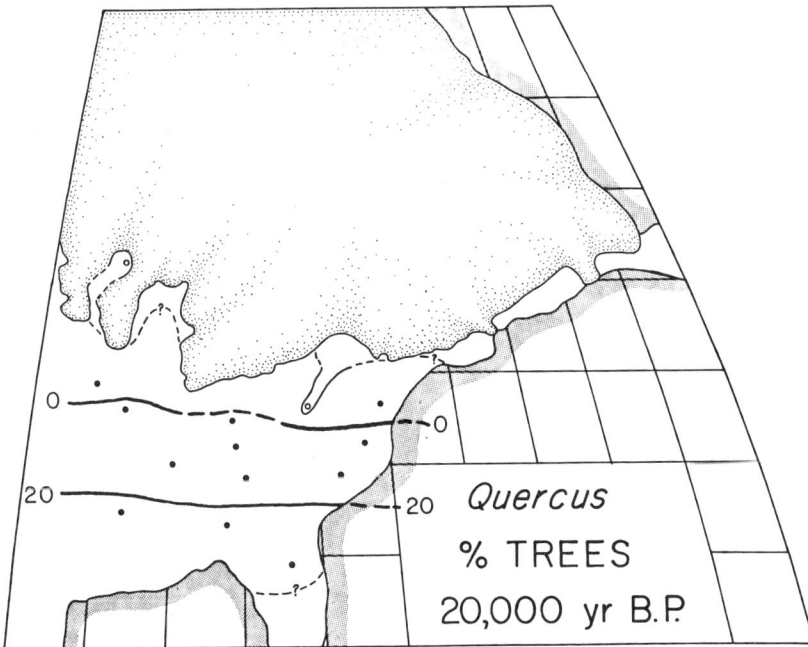

Figure 5.30. Oak (*Quercus*). b. Paleo-dominance map for 20,000 yr B.P.

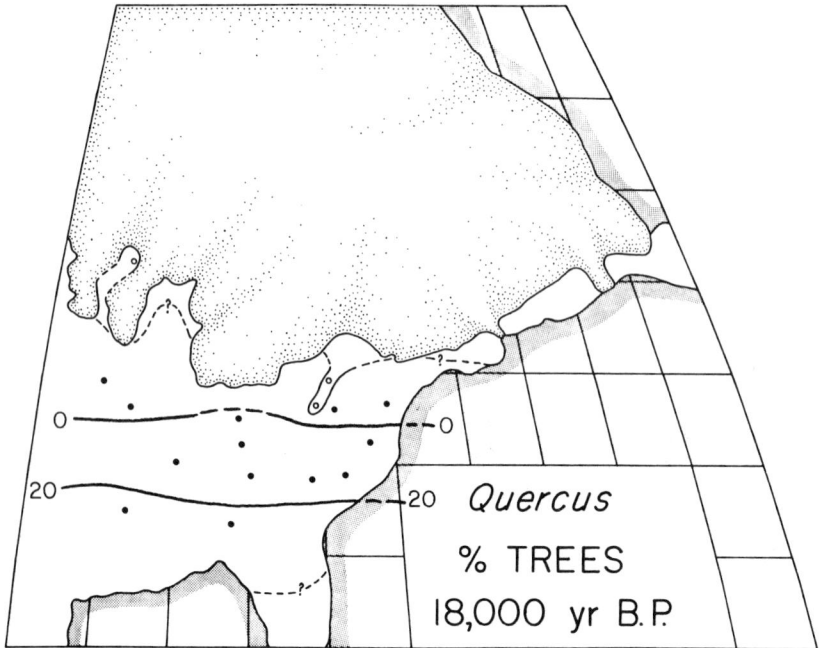

Figure 5.30. Oak (*Quercus*). c. Paleo-dominance map for 18,000 yr B.P.

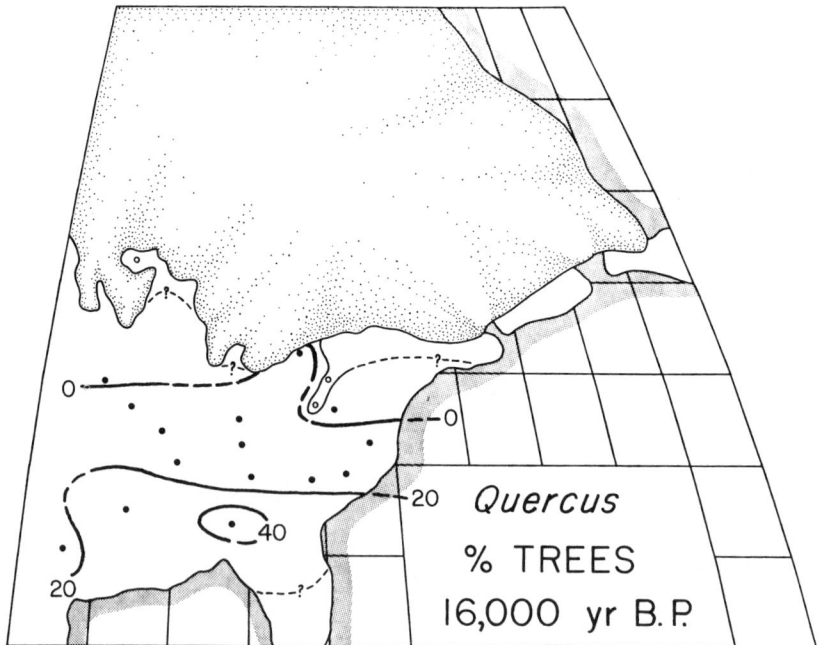

Figure 5.30. Oak (*Quercus*). d. Paleo-dominance map for 16,000 yr B.P.

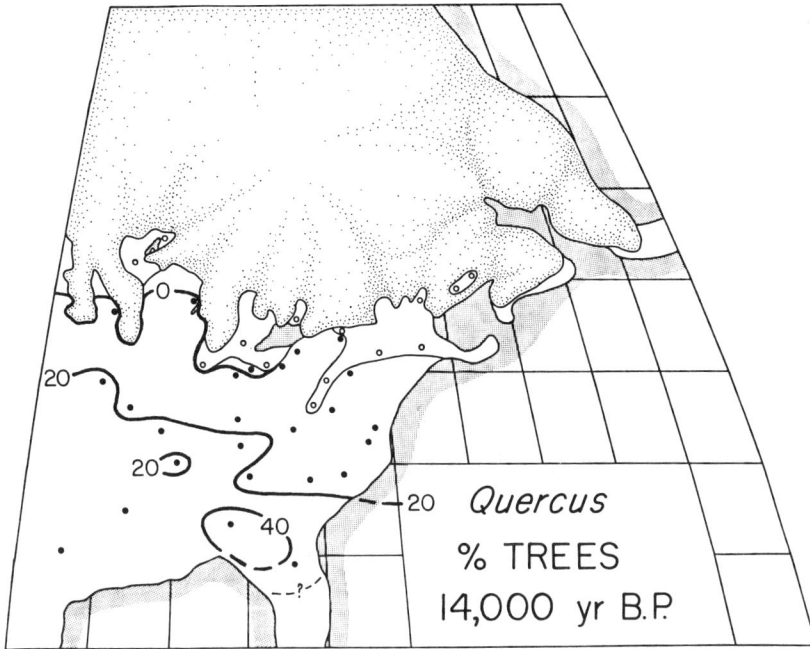

Figure 5.30. Oak (*Quercus*). e. Paleo-dominance map for 14,000 yr B.P.

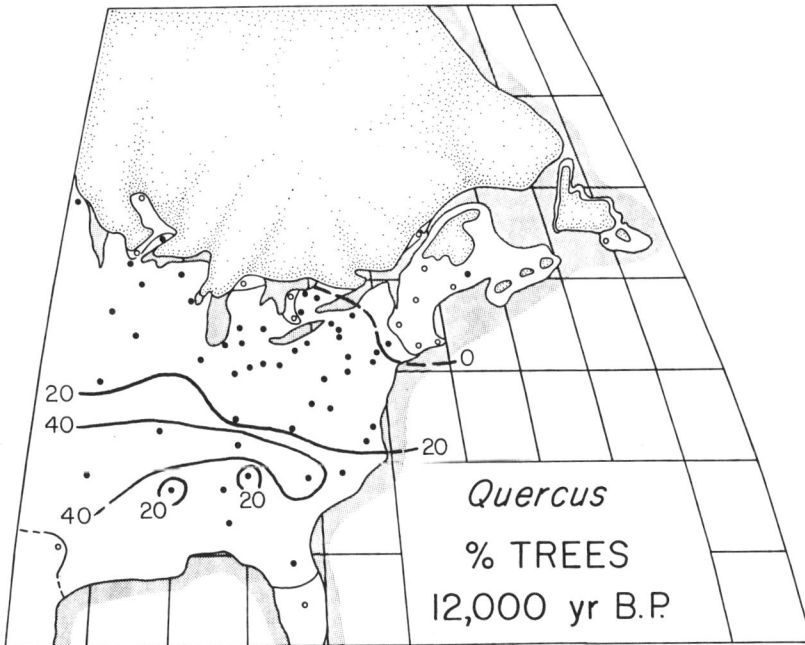

Figure 5.30. Oak (*Quercus*). f. Paleo-dominance map for 12,000 yr B.P.

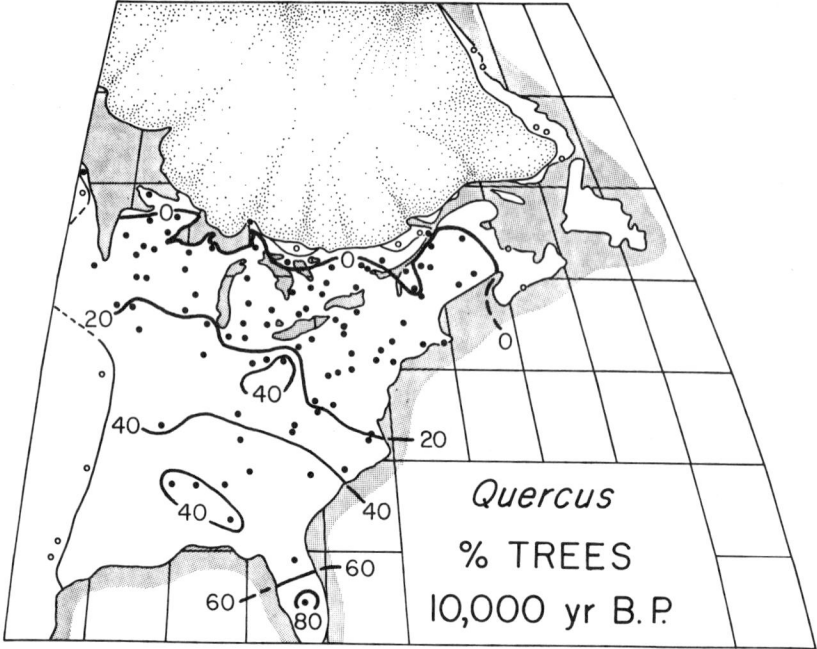

Figure 5.30. Oak (*Quercus*). g. Paleo-dominance map for 10,000 yr B.P.

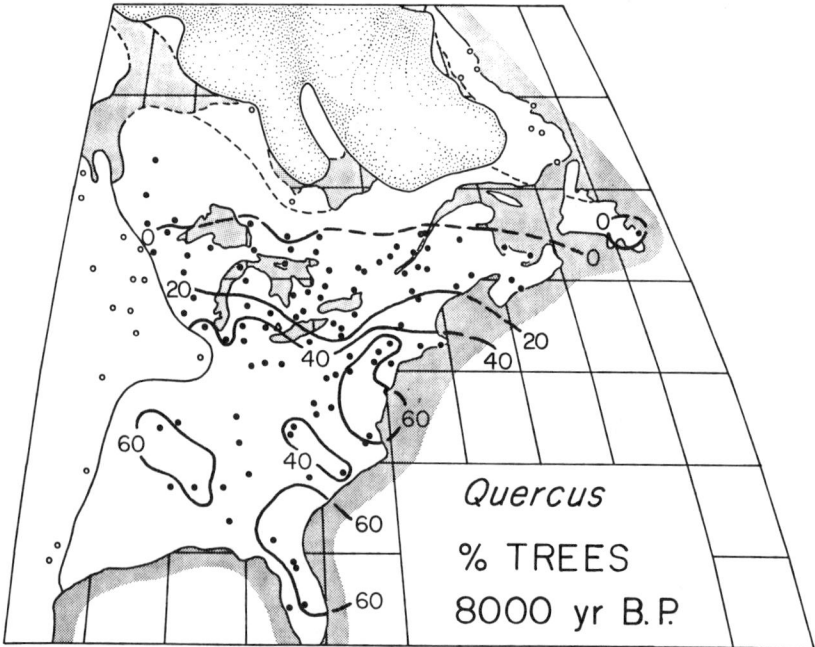

Figure 5.30. Oak (*Quercus*). h. Paleo-dominance map for 8000 yr B.P.

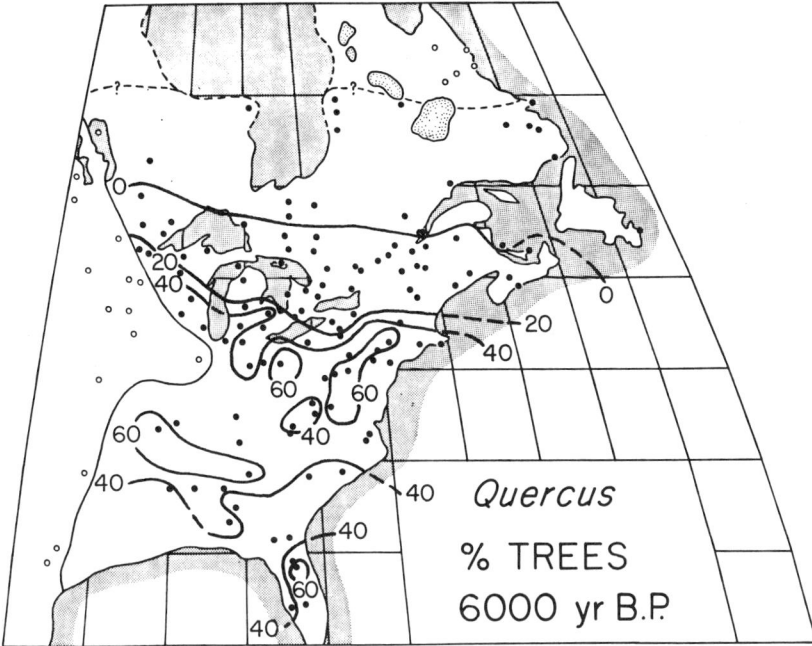

Figure 5.30. Oak (*Quercus*). i. Paleo-dominance map for 6000 yr B.P.

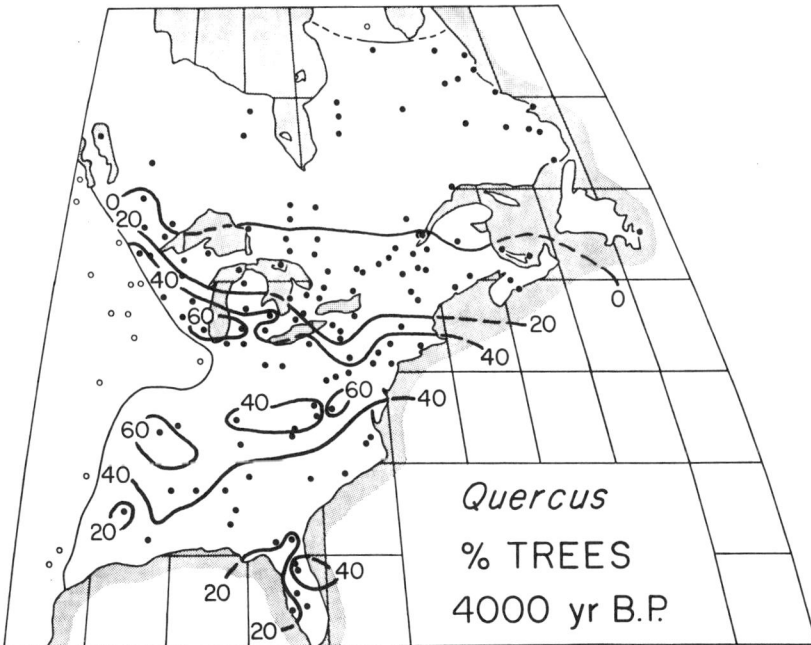

Figure 5.30. Oak (*Quercus*). j. Paleo-dominance map for 4000 yr B.P.

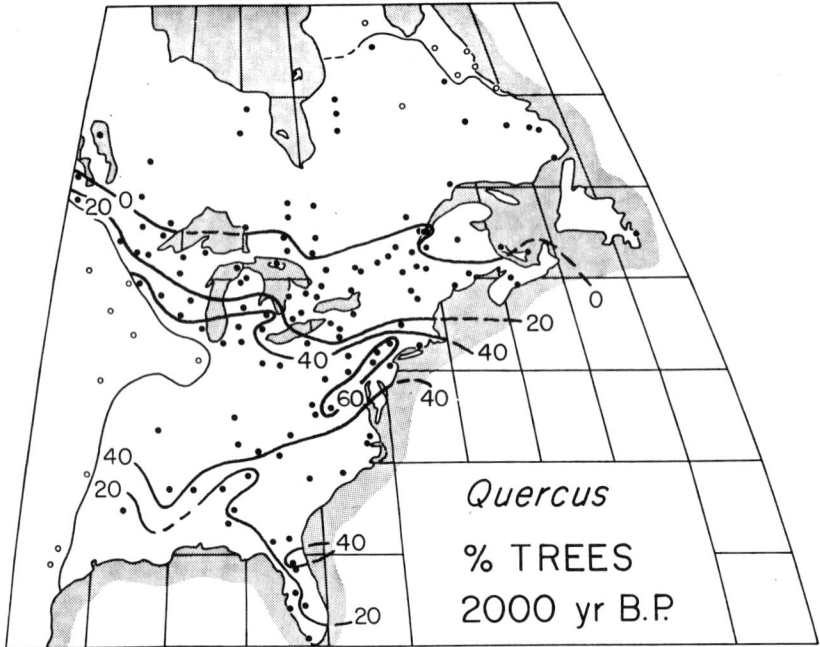

Figure 5.30. Oak (*Quercus*). k. Paleo-dominance map for 2000 yr B.P.

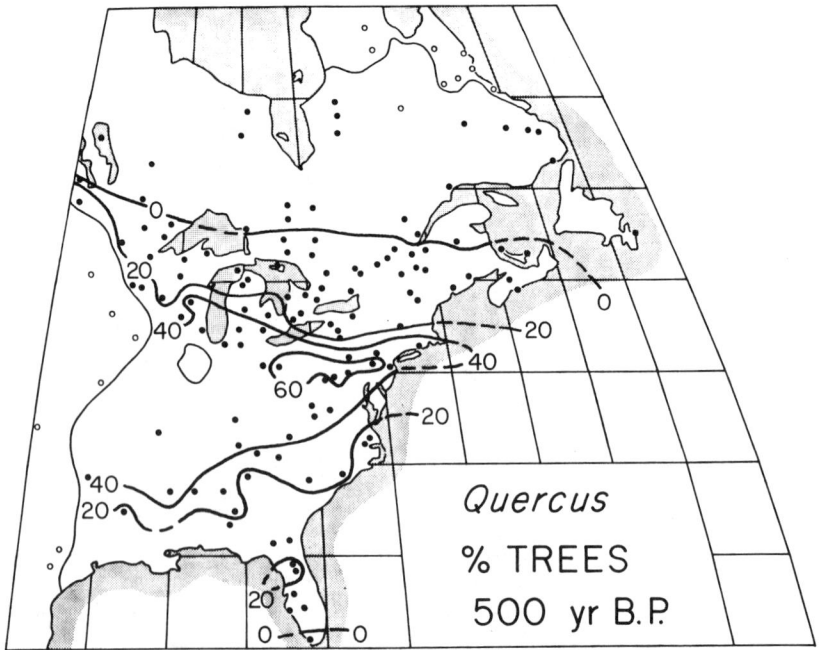

Figure 5.30. Oak (*Quercus*). l. Paleo-dominance map for 500 yr B.P.

Manitoba. However, oak only advanced to between 42°N and 44°N in the eastern part of its range through southern Ontario, New York, and Pennsylvania. A broad zone of relatively moderate paleo-dominance for oak (between 5% and 20%) was distributed in a latitudinal band extending southward from its northern range limits to southeastern Kansas, Kentucky, and Virginia. The center of dominance of > 40% oak shifted northward from the southern coastal plains to a zone between 33°N and 37°N, extending from eastern Oklahoma and southern Missouri across Tennessee and into the Carolinas. Across the Gulf and southern Atlantic Coastal Plains, oak populations diminished in abundance but still comprised between 20% and 40% of the forests. Retaining latitudinally elongate patterns, oak developed an asymmetrical longitudinal distribution, attaining a primary center along 35°N and decreasing along steep compositional gradients to 0% at its range limits in both the north and southeast.

By 10,000 yr B.P. (Fig. 5.30g), the northern distributional limit for *Quercus* reached the northern Great Lakes region near the glacial ice limit, extending from 48°N in Minnesota east of Glacial Lake Agassiz, along the the southern shore of Lake Superior, and eastward across central Ontario and southern Quebec to Maine and New Brunswick. Paleo-dominance values for oak were consistently between about 1% and 3% in Minnesota, central Wisconsin, Michigan, and across New York and Maine. However, oak populations increased southward across a steep gradient to regional values over 20% from southern Wisconsin and Iowa to northern Illinois, Ohio, and then southeastward to southwestern Virginia and North Carolina. Paleo-dominance values typically between 20% and 40% were mapped extending southward to southeastern Missouri, Kentucky, and South Carolina. One population center > 40% oak was located in Ohio (up to 53%). South of about 37°N on the western range limit and 35°N to 34°N in the east, oak was a major dominant of forests at 10,000 yr B.P. Values of > 40% oak were attained generally across the region south from Tennessee to the Gulf of Mexico. By 10,000 yr B.P., oak colonized central and southern Florida, with oak populations reaching up to 96% dominance in the reconstructed forests. The western oak limit was bounded by the developing prairie–forest ecotone.

Between 10,000 yr B.P. and 8000 yr B.P. (Fig. 5.30h), the western range margin of oak shifted eastward with the Prairie Peninsula, particularly between 39°N and 43°N. The northern range limit extended to about 48°N, from about 98°W in northern Minnesota, across south-central Ontario, southern Quebec, New Brunswick, and Nova Scotia. Values for *Quercus* dominance were between 1% and 20% throughout the Great Lakes region and across southern Ontario to the northern New England states. However, a strong gradient in oak paleo-dominance from 20% to 40% occurred through central Wisconsin, central Michigan, northern Ohio, and central New England. Values for oak represented 40% to 77% of the forests across nearly the entire region of eastern North America south of about 42°N. Population centers of > 60% were located in three areas: (1) the central Atlantic states from eastern Pennsylvania to Maryland and coastal Virginia; (2) the Ozark Plateaus in southeastern Missouri and the central Gulf Coastal Plain to central Alabama; and (3) the southeastern coastal plain from central Georgia to southern Florida.

From 8000 yr B.P. to 6000 yr B.P., minor changes in dominance of oak occurred across the northern portion of its range (Fig. 5.30i). The northern range limit was relatively stationary at about 48°N east of Lake Superior but advanced to 50°N in western Ontario. The strong gradient from minor ($<$ 20%) to major ($>$ 40%) oak paleo-dominance remained stationary in its location across central portions of Minnesota, Wisconsin, Michigan, and eastward at about 42°N through the southern New England region. Dominance of oak was $>$ 40% throughout the central portion of its distributional range, with values $>$ 60% reached in the central Appalachian Mountains, the Ozarks and southern Interior Low Plateaus, the Lower Midwest of Ohio, and the Florida Peninsula. Across the Gulf and southern Atlantic Coastal Plains, however, oak populations diminished between 8000 and 6000 yr B.P. to typically $<$ 40% of the total forest composition.

By 4000 yr B.P. (Fig. 5.30j), the distributional range limits for oak were constrained to the west by prairie and to the south and east by ocean. Along its northern periphery, oak remained relatively stationary at about 48°N through the mid- and late-Holocene intervals. Oak populations declined in abundance in the southern Atlantic and Gulf Coastal Plains, progressively displaced by southern pines on well-drained, fire-prone upland sites. The mid- and late-Holocene region of oak dominance (from 40% to 70%) corresponded with the wedge-shaped central portion of the eastern deciduous forest region (Fig. 4.4b). Three oak population centers of $>$ 60% were situated at 4000 yr B.P. within the central Appalachian Mountains, Ozark Plateaus, and along the prairie–forest ecotone in southeastern Wisconsin.

The late-Holocene, paleo-dominance maps for 2000 yr B.P. (Fig. 5.30k) and 500 yr B.P. (Fig. 5.30l) display the progressive reduction of oak populations to less than 20% in the southeastern coastal plains. By presettlement times, the oak population center ($>$ 40%) was located in the geographic center of its range, with respect to latitude. During the last 2000 years, oak constituted more than 60% of the forest composition within the central Appalachian Mountains. The population center $>$ 40% was coincident with the eastern deciduous forest region and was delimited both to the north and to the south by steep gradients in dominance of oak (*i.e.*, between the 20% and 40% contours). Thus, by 500 yr B.P., oak represented $<$ 20% of both the mixed conifer–northern hardwoods and the southeastern evergreen forest regions (Fig. 4.2a). With late-Holocene retreat of the prairie–forest ecotone between 87°W and 100°W, oak populations expanded toward the continental interior along the western margin of distribution.

The dominance–structure histograms for oak (Fig. 5.31a) show that the number of dominance classes (in 20% increments) increased from 2 classes at 20,000 yr B.P. to 3 classes at 16,000 yr B.P. and 12,000 yr B.P. and to 4 classes during the last 8000 years. The area–dominance histogram for 20,000 yr B.P. displays similar areas of about $12 \cdot 10^5$ km^2 for both the 0% to 20% class and the 20% to 40% class. The histograms for 16,000 yr B.P. and 12,000 yr B.P. show progressive increases in area occupied by the 0% to 20% dominance class as well as by the 40% to 60% dominance class, with the area occupied by the 20% to

Quercus

A.

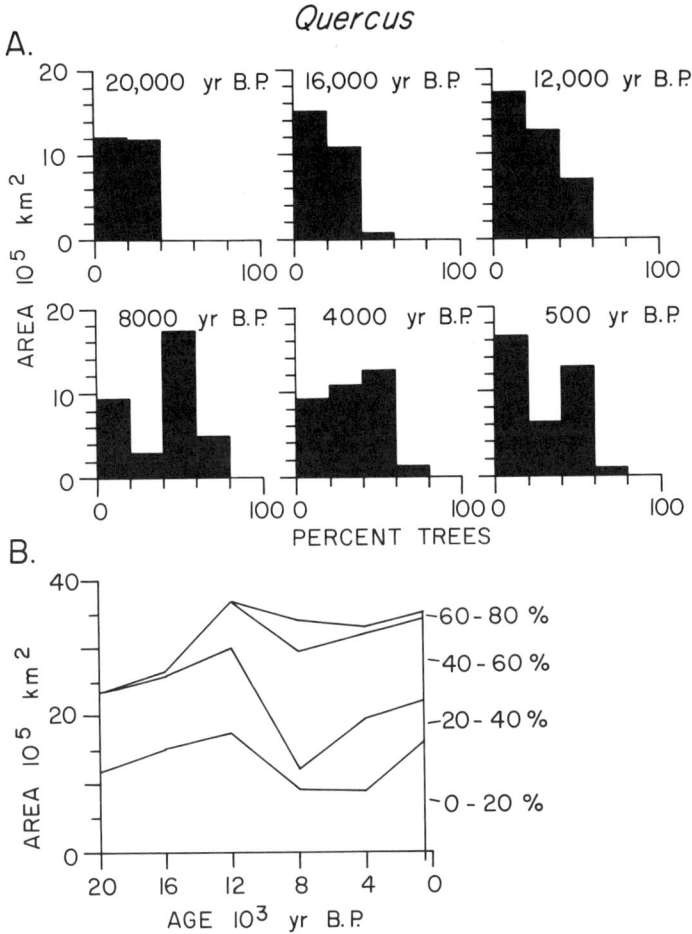

Figure 5.31. Changes in area and dominance of oak (*Quercus*). a. Histograms of area–dominance structure at 20,000, 16,000, 12,000, 8000, 4000, and 500 yr B.P. b. Cumulative plot from 20,000 yr B.P. to the present.

40% class remaining nearly constant in area. These changes reflected both the northward expansion of the range of oak and the development of a large population center in the southern portion of its range. By 8000 yr B.P., a major change in dominance structure occurred, with the greatest area represented in the 0% to 20% and the 40% to 60% dominance classes, and with a diminishment in the area occupied by the 20% to 40% dominance class. This reduction was the result of the development of a steepened gradient in oak dominance through the Great Lakes and New England regions. The histogram for 4000 yr B.P. illustrates a reciprocal decrease in area occupied by the 40% to 60% dominance class and an increase in the area covered by the 20% to 40% class, reflecting the late-Holocene decrease in oak within the southern coastal plains. By 500 yr B.P., the histogram for oak shows two dominance classes (the 0% to 20% class and the 40% to 60% class) that comprised 81% of the spectrum. The loss in area of the 20% to 40% dominance class between 4000 yr B.P. and 500 yr B.P. illustrates the results of development of the steepened gradient across the southern portion of the range of oak. The cumulative plot of dominance structure for oak (Fig. 5.31b) illustrates a 54% increase in area from $23.9 \cdot 10^5$ km^2 at 20,000 yr B.P. to $36.8 \cdot 10^5$ km^2 at 12,000 yr B.P. The total distribution of oak decreased through the mid-Holocene interval to $33.7 \cdot 10^5$ km^2 at 4000 yr B.P. in response to the expansion of prairie. Oak reexpanded to $36 \cdot 10^5$ km^2 by 500 yr B.P. as oak reinvaded prairie along its western margin of distribution.

Willow (*Salix*)

The population trajectory for *Salix* (Fig. 5.32a) shows mean values of 1% at 17,000 yr B.P., with no sites for which willow is reconstructed between 20,000 yr B.P. and 18,000 yr B.P. The mean values for willow increased to 4% by 16,000 yr B.P. and then fluctuated between 1% and 3% from 15,000 yr B.P. to 6000 yr B.P. Mean values for *Salix* rose to between 5% and 6% from 5000 yr B.P. to 4000 yr B.P. and then declined to about 1% during the last 1000 years. The maximum reconstructed value for willow at 17,000 yr B.P. was 1%, and willow dominance fluctuated between 2% and 11% between 16,000 yr B.P. and 10,000 yr B.P. From 9000 yr B.P. to 7000 yr B.P., willow maxima ranged between 11% and 22% and then rose to between 42% and 54% from 6000 yr B.P. to 4000 yr B.P. Late-Holocene willow maxima oscillated between 4% and 12% during the last 2000 years. Variability about mean values was typically ± 1 SD < 4% from 17,000 yr B.P. to 8000 yr B.P. and from 3000 yr B.P. to present; however, this variability increased up to 13% from 6000 yr B.P. to 4000 yr B.P.

During the full-glacial interval, from 20,000 yr B.P. to 17,000 yr B.P., tree *Salix* was not an important constituent of the forests of eastern North America, as reconstructed from the array of sites available south of the glacial boundary. Although no sites contain willow dating from 20,000 yr B.P. to 18,000 yr B.P., trace values were reconstructed for 17,000 yr B.P. at Muscotah Marsh, Kansas (Gruger 1973) and Cupola Pond, Missouri (Smith 1984), both located west of the Mississippi Alluvial Valley. By 16,000 yr B.P. (Fig. 5.32b), the range of willow extended south from Kansas to eastern Texas, grading from 1% of the forest composition at 40°N to 7% at 30°N.

During the late-glacial interval, between 16,000 yr B.P. and 14,000 yr B.P. (Fig. 5.32c), willow populations expanded to 11% in the western Gulf Coastal Plain. Along the northern and eastern border, willow dispersed across Missouri, Kentucky, and into western Pennsylvania in a broad band. The northern distributional limit of willow trees was located between 39°N and 42°N at 14,000 yr B.P. (Fig. 5.32c).

By 12,000 yr B.P. (Fig. 5.32d), the northern range limit of willow shifted as far north as 50°N in southern Manitoba. Willow invaded the poorly drained Interior Low Plateaus region, the recently deglaciated landscape across the Great Lakes region, and the formerly periglacially modified landscapes of the central

Salix

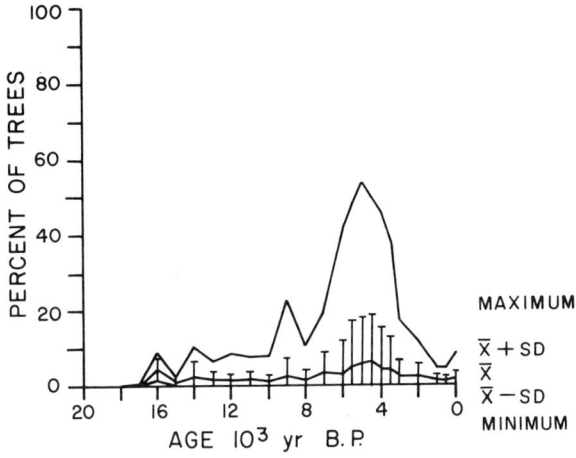

Figure 5.32. Willow (*Salix*). a. Population trajectory, including maximum, minimum, and mean values of paleo-dominance as well as ± 1 SD of the mean, for the past 20,000 years.

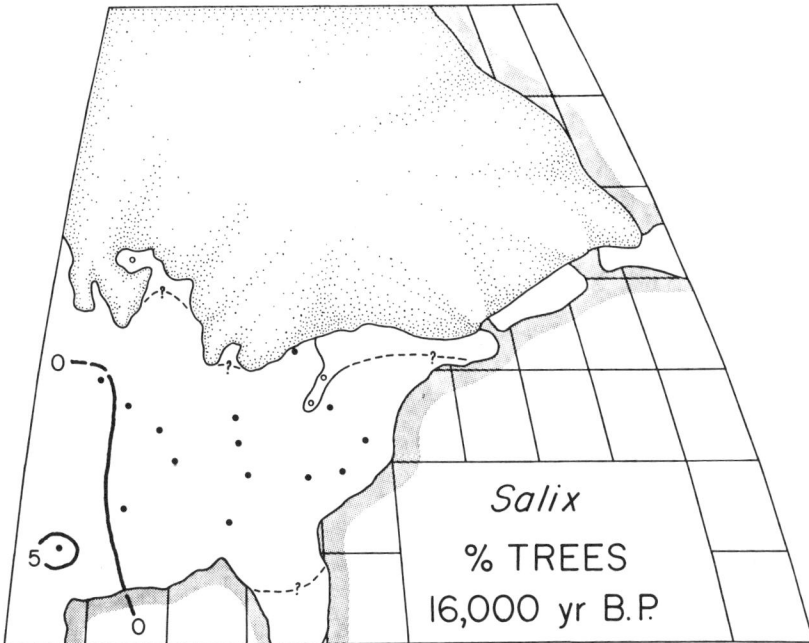

Figure 5.32. Willow (*Salix*). b. Paleo-dominance map for 16,000 yr B.P.

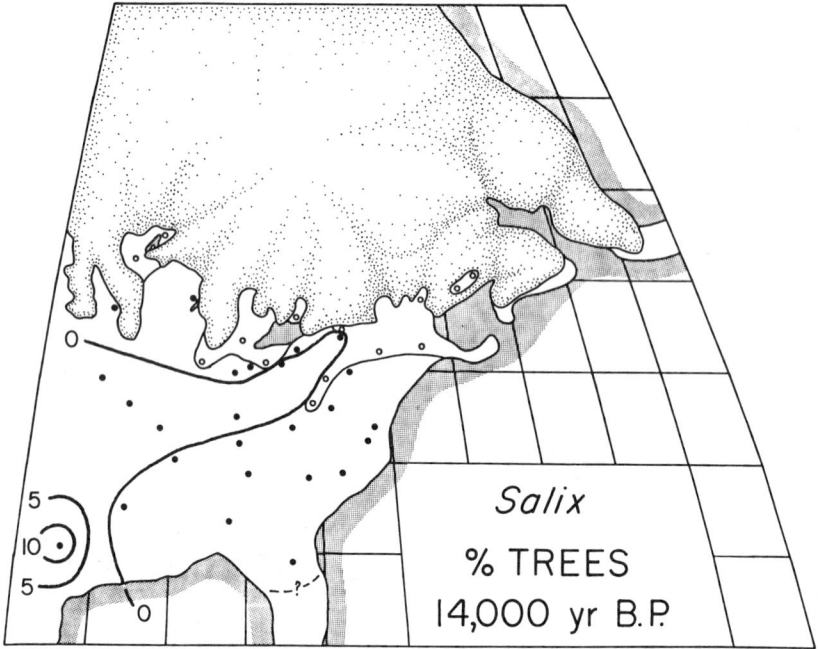

Figure 5.32. Willow (*Salix*). c. Paleo-dominance map for 14,000 yr B.P.

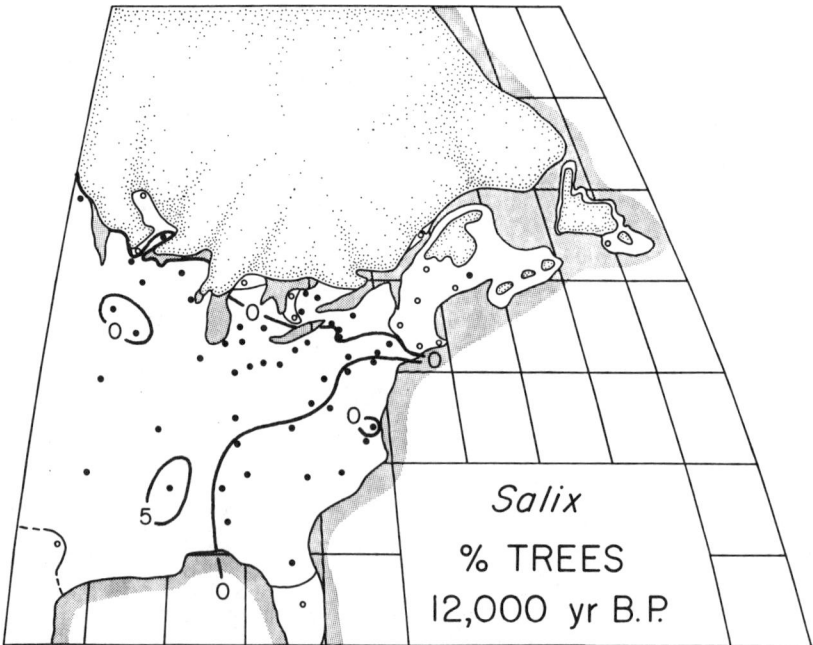

Figure 5.32. Willow (*Salix*). d. Paleo-dominance map for 12,000 yr B.P.

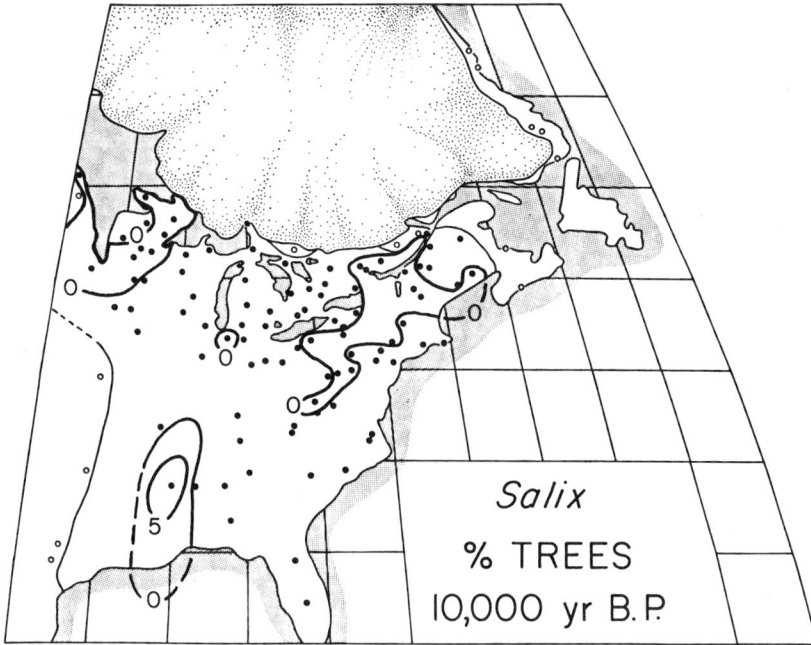

Figure 5.32. Willow (*Salix*). e. Paleo-dominance map for 10,000 yr B.P.

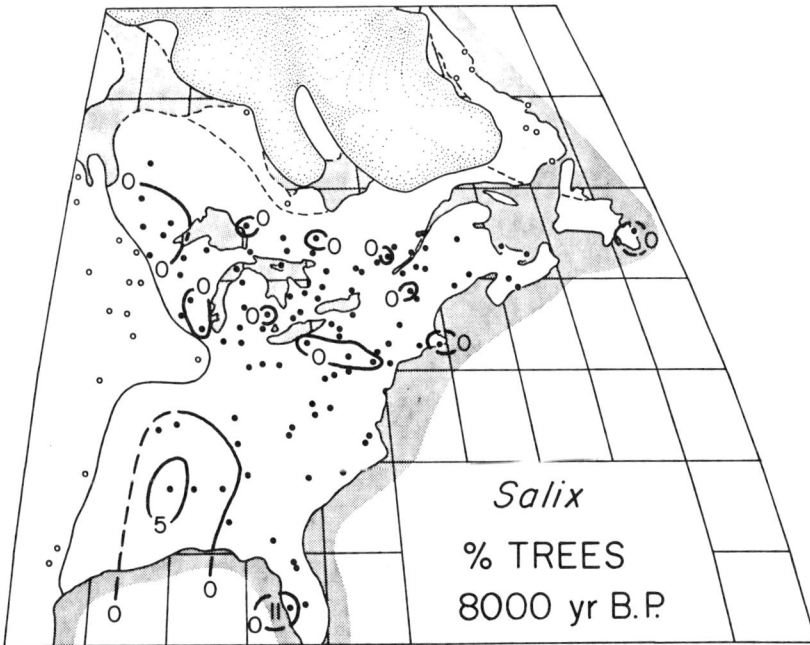

Figure 5.32. Willow (*Salix*). f. Paleo-dominance map for 8000 yr B.P.

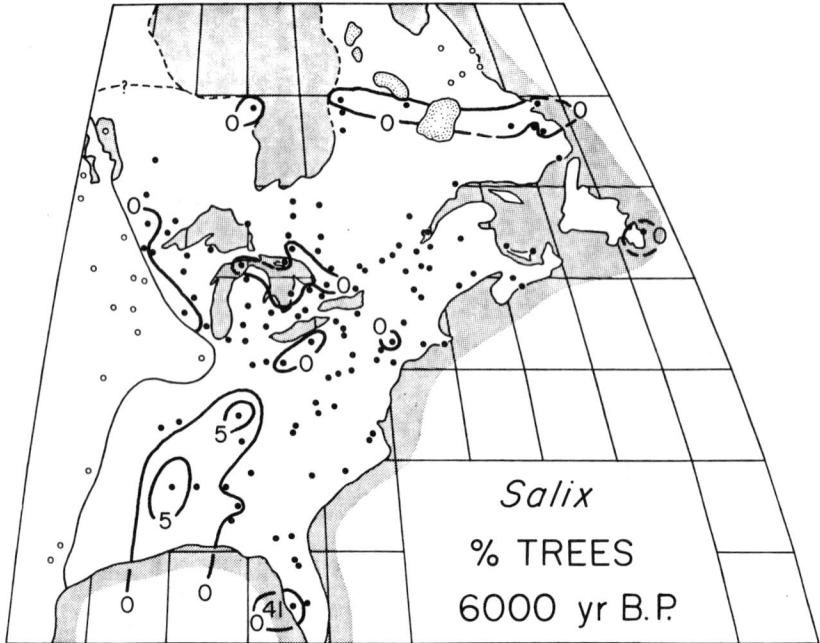

Figure 5.32. Willow (*Salix*). g. Paleo-dominance map for 6000 yr B.P.

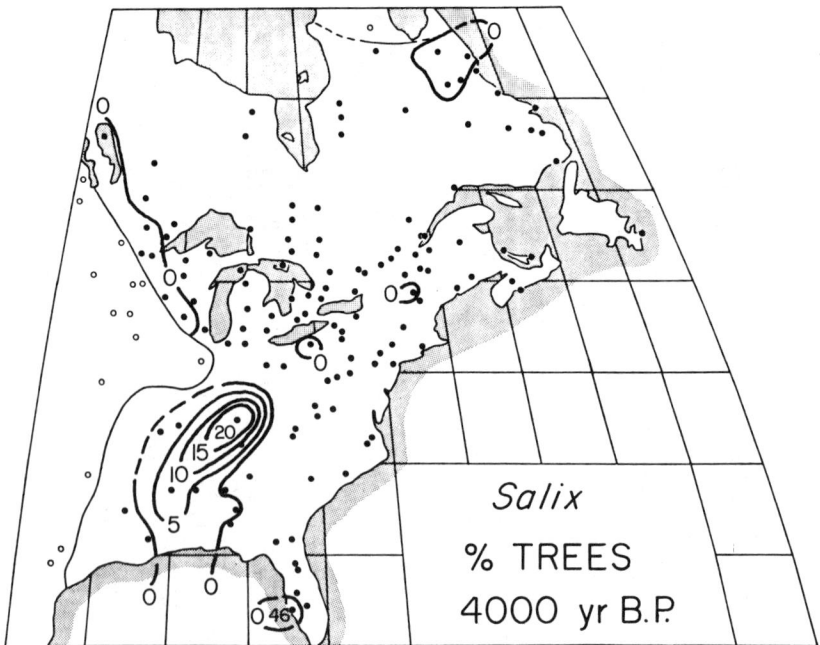

Figure 5.32. Willow (*Salix*). h. Paleo-dominance map for 4000 yr B.P.

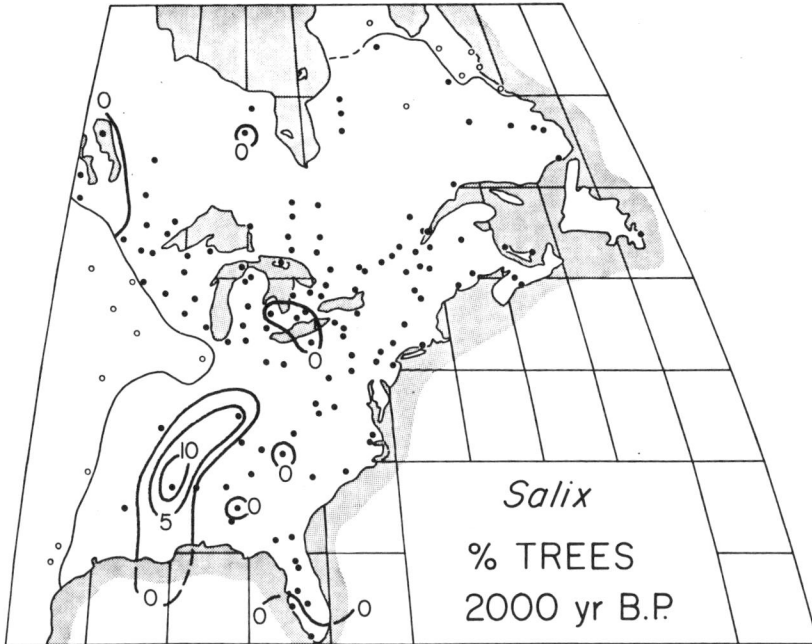

Figure 5.32. Willow (*Salix*). i. Paleo-dominance map for 2000 yr B.P.

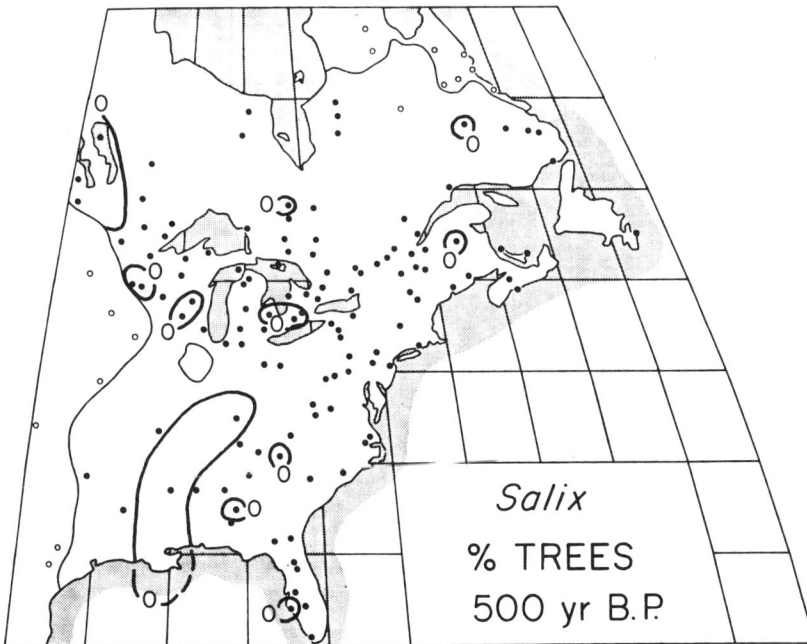

Figure 5.32. Willow (*Salix*). j. Paleo-dominance map for 500 yr B.P.

Appalachians. By 12,000 yr B.P., the formerly large population of willow in eastern Texas declined as prairie expanded eastward. With the abandonment of braided streams in the Lower Mississippi Alluvial Valley, the shift to a meandering fluvial regime favored the establishment of riparian forests containing as much as 8% willow, for example in the Yazoo Basin of west-central Mississippi (Holloway and Valastro 1983).

By 10,000 yr B.P. (Fig. 5.32e), the apparent range of willow was fragmented across the middle latitudes from 37°N to 43°N, with continuous populations of willow mapped primarily in the Lower Mississippi Alluvial Valley (up to 8%). Within the northern and central Appalachian Mountains, and extending northeastward along the St. Lawrence River Valley, willow remained up to 2% of the reconstructed forests at 10,000 yr B.P. A third pocket of up to 2% willow grew between the western shore of Lake Superior and the southern shore of Glacial Lake Agassiz.

By 8000 yr B.P. (Fig. 5.32f), willow was locally prominent (up to 9%) in the Lower Mississippi Alluvial Valley and adjacent portions of the central Gulf Coastal Plain. In south-central Florida, one site (Little Salt Spring; Brown and Cohen 1985) recorded a local population of willow of 11%. Elsewhere, willow was a minor forest component, representing < 3% dominance in fragmented populations distributed between 40°N and 50°N, generally across the Great Lakes region. Between 8000 yr B.P. and 6000 yr B.P. (Fig. 5.32g), *Salix* persisted through the Lower Mississippi Alluvial Valley and expanded into the Interior Low Plateaus of central Kentucky (Wilkins 1985). The outlier population in Florida increased locally to 41%. Patches of willow were recorded in the central and western Great Lakes region, particularly along the prairie–forest border in Minnesota and Wisconsin, as well as in sites around Lake Huron and Lake Erie. The map for 6000 yr B.P. (Fig. 5.32g) also shows a latitudinal band of willow between 53°N and 55°N; however, those populations occurred along the tundra–forest ecotone and may have been species of shrub willows rather than tree willows.

The paleo-dominance map for *Salix* at 4000 yr B.P. (Fig. 5.32h) highlights the primary population center along the Mississippi Valley, central Gulf Coastal Plain, and adjacent portions of the Interior Low Plateaus. High outlier values for willow occurred in central Kentucky (20%) and west-central Florida (46%). A second band of < 2% willow extended from 43°N to 53°N along the prairie–forest ecotone. Additional restricted occurrences of willow were located in the eastern Great Lakes region, the northern Appalachian Mountains, and along the tundra–forest ecotone in northern Labrador and Quebec (Fig. 5.32h).

The maps for 2000 yr B.P. (Fig. 5.32i) and 500 yr B.P. (Fig. 5.32j) depict a broad population center for willow within the Lower Mississippi Alluvial Valley, scattered populations of *Salix* across both the Southeast and the Great Lakes region, and the development of < 4% willow along the prairie–forest margin in southern and central Manitoba, Minnesota, and central Wisconsin. By 500 yr B.P., willow was widely distributed in geographically dispersed populations across both boreal and temperate forested regions, although locally concentrated along major river systems and at the margins of lakes and ponds.

Salix

A.

B.

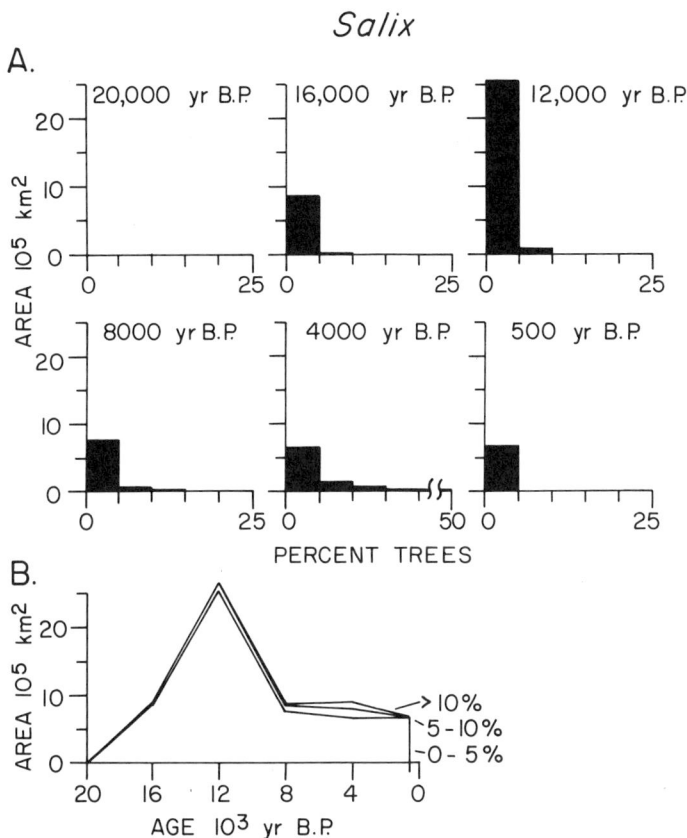

Figure 5.33. Changes in area and dominance of willow (*Salix*). a. Histograms of area–dominance structure at 20,000, 16,000, 12,000, 8000, 4000, and 500 yr B.P. b. Cumulative plot from 20,000 yr B.P. to the present.

The area–dominance histograms for *Salix* (Fig. 5.33a) display a consistent pattern in that the largest portion of areal distribution is represented by the class of 0% to 5% dominance. For example, at 12,000 yr B.P., 4000 yr B.P., and 500 yr B.P., respectively, the lowest dominance class was 97%, 73%, and 100% of the total area occupied by willow. However, the absolute area within the lowest dominance class has changed markedly through the late Quaternary, from a maximum at 12,000 yr B.P. of $25.7 \cdot 10^5$ km^2 to a late-Holocene minimum of $6.6 \cdot 10^5$ km^2 at 4000 yr B.P. The number of dominance classes increased from 0 recorded at 20,000 yr B.P., to 2 at 12,000 yr B.P. and to 10 at 4000 yr B.P.; this probably reflects the development of locally prominent willow populations in bottomland habitats at some of the sites investigated. The distribution of willow expanded in area by 300% from $8.8 \cdot 10^5$ km^2 at 16,000 yr B.P. to a maximum of $26.5 \cdot 10^5$ km^2 at 12,000 yr B.P., and then declined during the Holocene to $6.7 \cdot 10^5$ km^2 at 500 yr B.P. (Fig. 5.33b).

Basswood (*Tilia*)

The population trajectory for basswood (Fig. 5.34a) displays full-glacial mean values between 3% and 5% from 20,000 yr B.P. to 17,000 yr B.P., a spike increase to 12% at 16,000 yr B.P., and subsequent reduction to about 2% from 15,000 yr B.P. to 13,000 yr B.P. Holocene means for *Tilia* ranged between 4% and 5% from 12,000 yr B.P. to the present. Maximum values for basswood were typically 3% to 5% from 20,000 yr B.P. to 17,000 yr B.P., 21% at 16,000 yr B.P., and about 3% through the late-glacial interval until 13,000 yr B.P. Maximum values increased from 8% at 12,000 yr B.P. to a peak in the interglacial of 27% and 29% at 10,000 yr B.P. and 9000 yr B.P., respectively. Mid-Holocene maximum values oscillated between 9% and 20%, and late-Holocene values ranged between 9% and 14%. Variability about the mean values was generally ± 1 SD < 3%, except for 16,000 yr B.P. (± 1 SD of 12%) and the interval from 10,000 yr B.P. to 9000 yr B.P. (± 1 SD of 6%).

The full-glacial paleo-dominance maps for *Tilia* (Figs. 5.34b,c) identify one refugium in south-central Alabama (Goshen Springs; Delcourt 1980). At this site, paleo-dominance values reconstructed for basswood exceeded 5% at 20,000 yr B.P. (Fig. 5.34b) and 4% at 18,000 yr B.P. (Fig. 5.34c). The late-glacial basswood maps (16,000 yr B.P., Fig. 5.34d; 14,000 yr B.P., Fig. 5.34e) also illustrate the persistence of basswood in the central Gulf Coastal Plain. At 16,000 yr B.P., a second population of basswood was located in eastern Texas, reaching over 20% of the forest composition there, and accounting for the high maximum value, mean, and apparent variability about the mean for basswood indicated in the population trajectory (Fig. 5.34a).

Between 14,000 yr B.P. and 12,000 yr B.P. (Fig. 5.34f), basswood expanded northward and northeastward along mesic habitats with fertile soils. By 13,000 yr B.P., *Tilia* had advanced from the central coastal plain into the loess-mantled uplands of western Tennessee (Nonconnah Creek; Delcourt *et al.* 1980). Basswood also colonized the Appalachian Highlands northeastward from Alabama into East Tennessee and southwestern Virginia. At 12,000 yr B.P. (Fig. 5.34f), basswood was distributed along the Upper Mississippi Valley, including portions of Missouri, Iowa, and southern Wisconsin. During the early Holocene, basswood also had colonized lower elevations of the central and northern Appalachian Mountains from Maryland to New York (Fig. 5.34f). In Pennsylvania and New York, paleo-dominance values reconstructed for basswood reached up to 8% by 12,000 yr B.P.

By 10,000 yr B.P. (Fig. 5.34g), basswood populations had expanded across the Great Lakes region, advancing to 47°N. A major population center of > 5% dominance developed from western Pennsylvania and Maryland west across Ohio, Indiana, and Illinois, reaching 18% to 27% in central Iowa. The western range limit was reached at the prairie–forest ecotone. *Tilia* generally occupied a broad zone within the forested region west of the Appalachians and north of about 35°N. Populations of basswood persisted in the central Gulf Coastal Plain of Alabama and Georgia but became locally extinct in Mississippi and Louisiana by 10,000 yr B.P.

Tilia

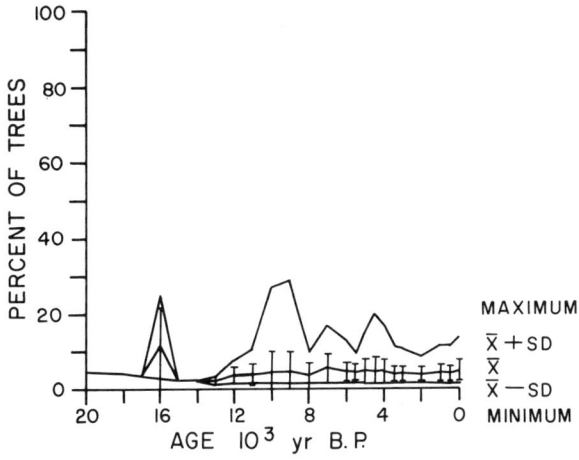

Figure 5.34. Basswood (*Tilia*). a. Population trajectory, including maximum, minimum, and mean values of paleo-dominance as well as ± 1 SD of the mean, for the past 20,000 years.

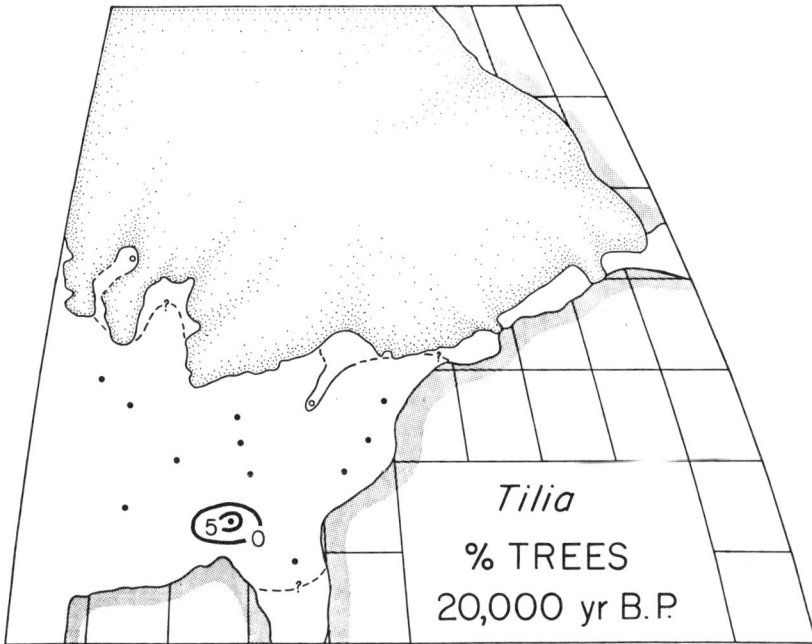

Figure 5.34. Basswood (*Tilia*). b. Paleo-dominance map for 20,000 yr B.P.

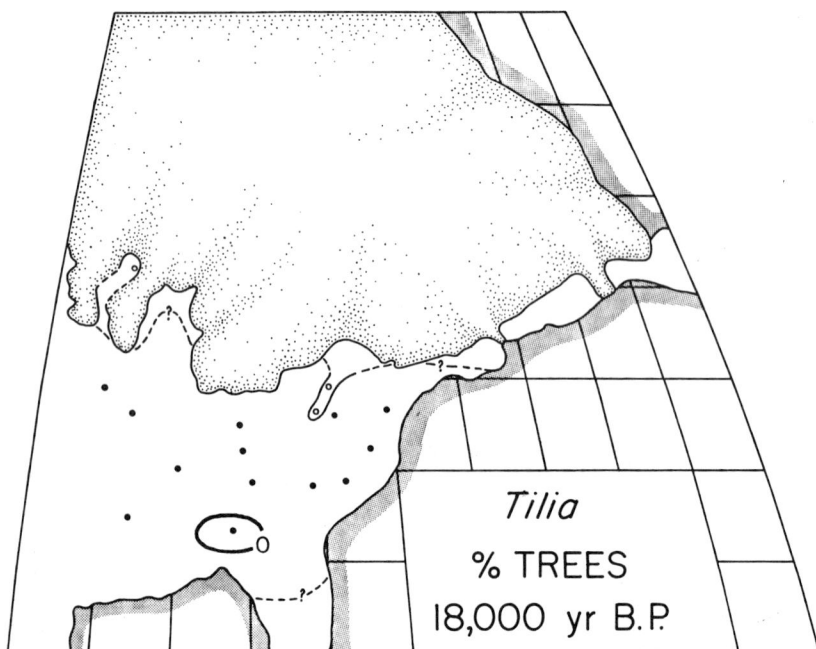

Figure 5.34. Basswood (*Tilia*). c. Paleo-dominance map for 18,000 yr B.P.

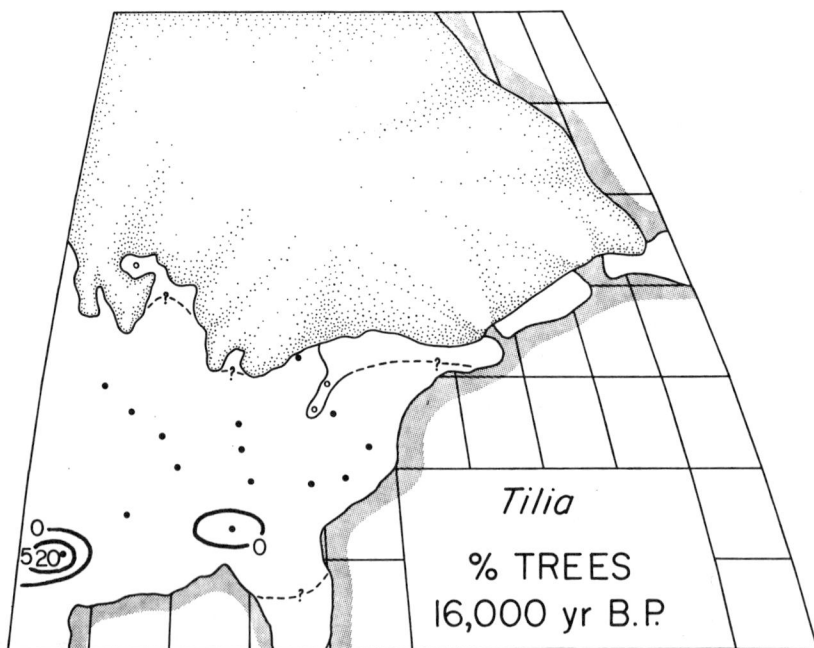

Figure 5.34. Basswood (*Tilia*). d. Paleo-dominance map for 16,000 yr B.P.

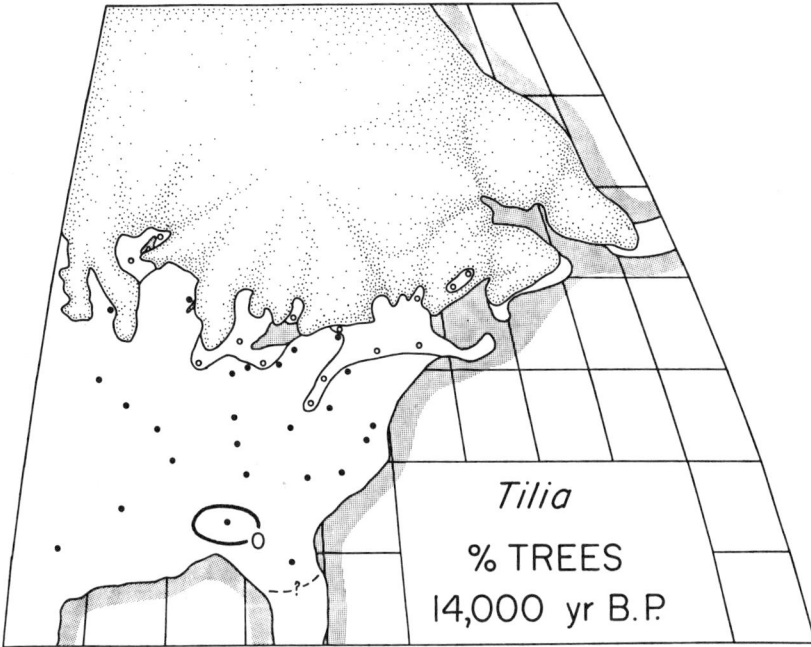

Figure 5.34. Basswood (*Tilia*). e. Paleo-dominance map for 14,000 yr B.P.

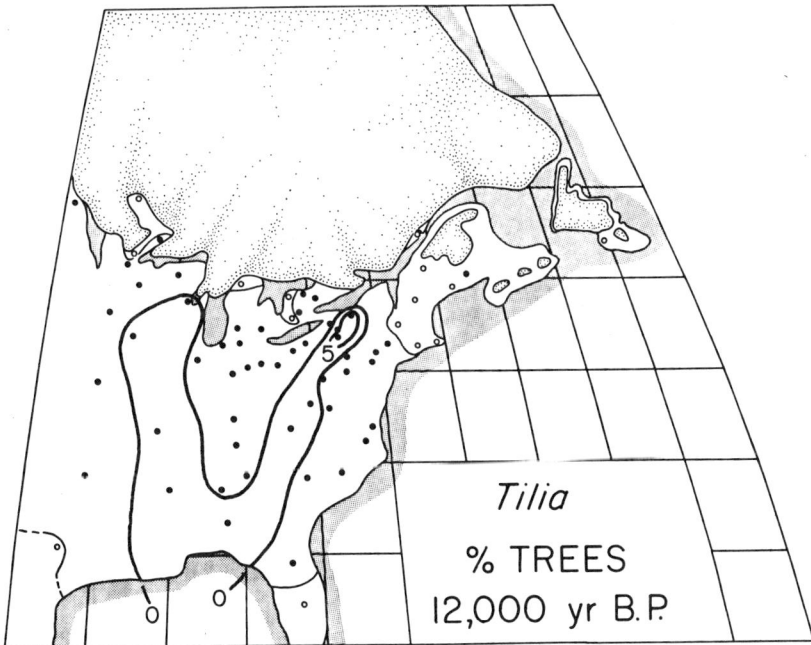

Figure 5.34. Basswood (*Tilia*). f. Paleo-dominance map for 12,000 yr B.P.

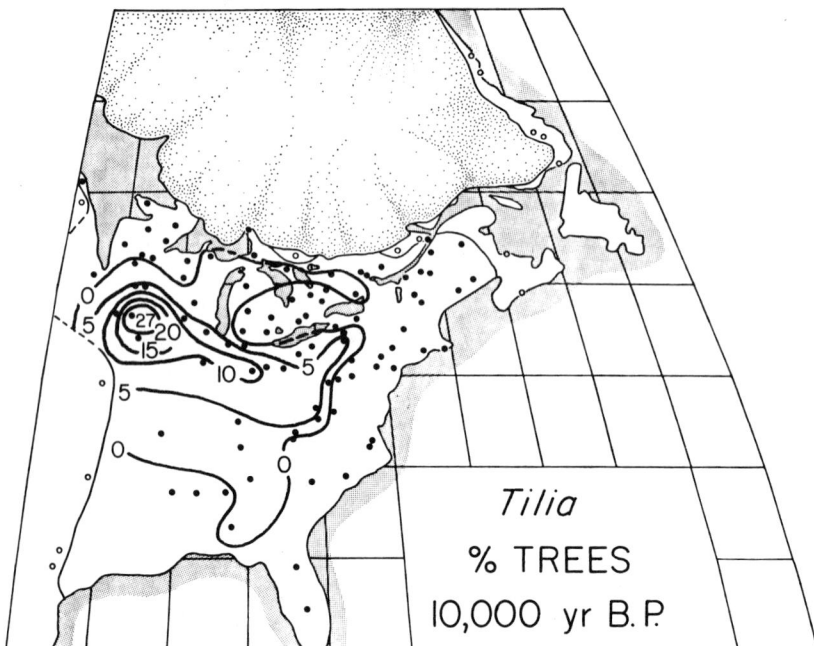

Figure 5.34. Basswood (*Tilia*). g. Paleo-dominance map for 10,000 yr B.P.

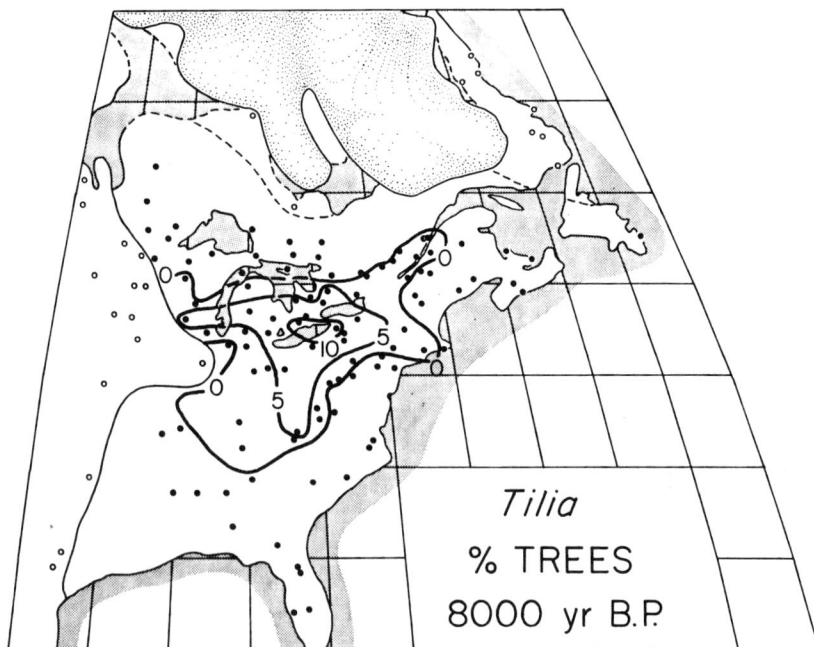

Figure 5.34. Basswood (*Tilia*). h. Paleo-dominance map for 8000 yr B.P.

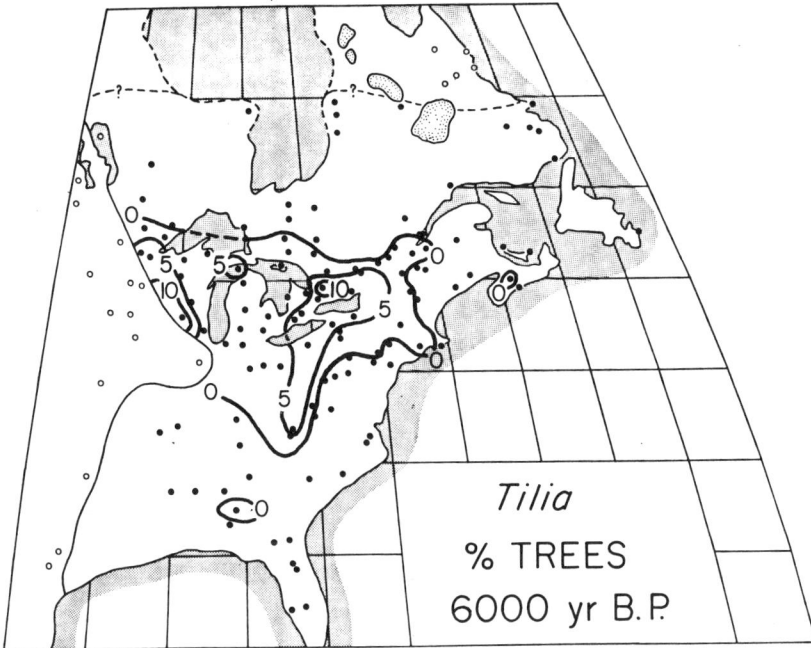

Figure 5.34. Basswood (*Tilia*). i. Paleo-dominance map for 6000 yr B.P.

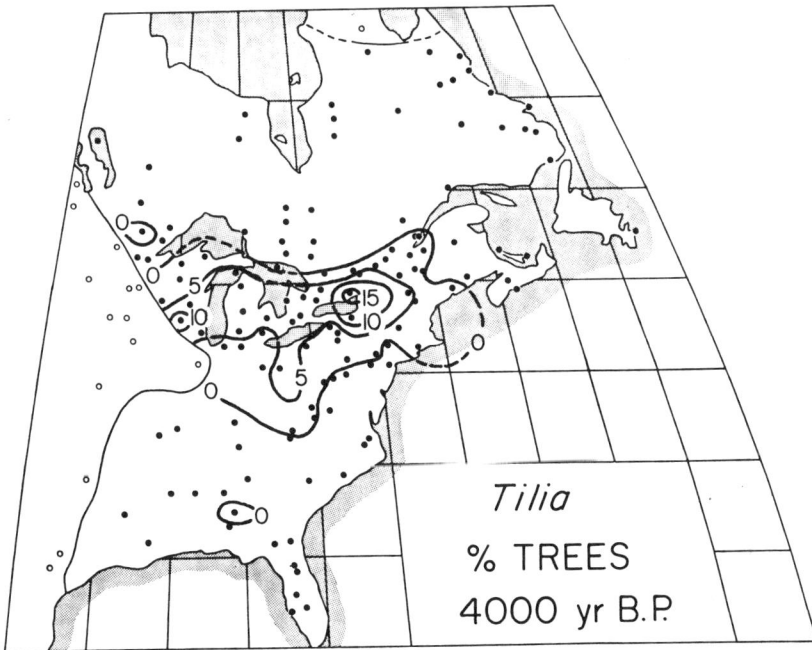

Figure 5.34. Basswood (*Tilia*). j. Paleo-dominance map for 4000 yr B.P.

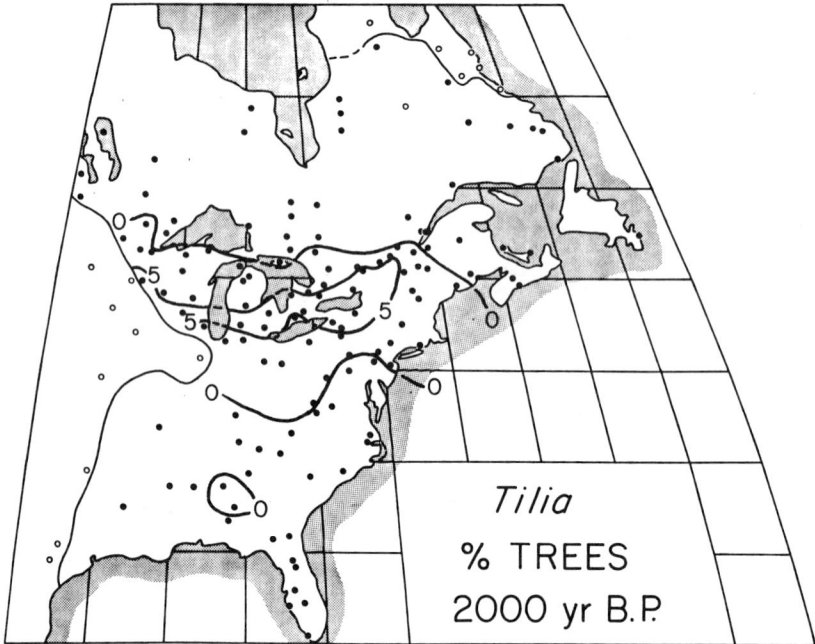

Figure 5.34. Basswood (*Tilia*). k. Paleo-dominance map for 2000 yr B.P.

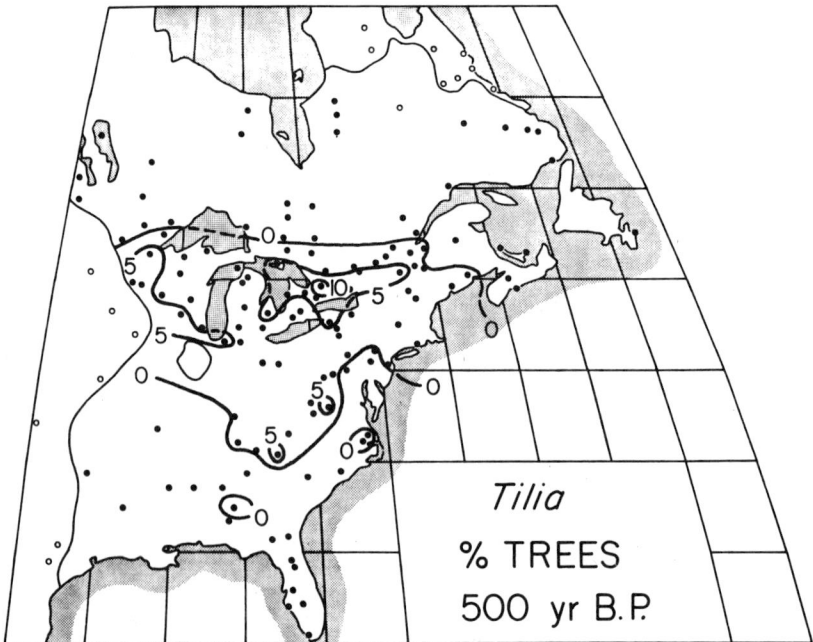

Figure 5.34. Basswood (*Tilia*). l. Paleo-dominance map for 500 yr B.P.

By 8000 yr B.P. (Fig. 5.34h), the primary population center for basswood (5% to 10%) was located within the central and eastern Great Lakes region. *Tilia* became locally extinct south of 38°N and west of 91°W near the prairie–forest margin. The northern range limit of basswood extended generally along 46°N from Wisconsin through Michigan and Ontario and northeastward to 47°N along the St. Lawrence River Valley of southeastern Quebec. By 8000 yr B.P., basswood migrated east from the northern Appalachians across southern New England.

The paleo-dominance map for basswood at 6000 yr B.P. (Fig. 5.34i) shows that the former population center had fragmented, with three areas remaining where basswood was more than 5% of the forest composition: (1) the eastern Great Lakes and Appalachian Mountains (up to 11% in southern Ontario); (2) the western Great Lakes region along the prairie–forest border from Minnesota to Wisconsin (up to 13%); and (3) islands in northern Lake Michigan (up to

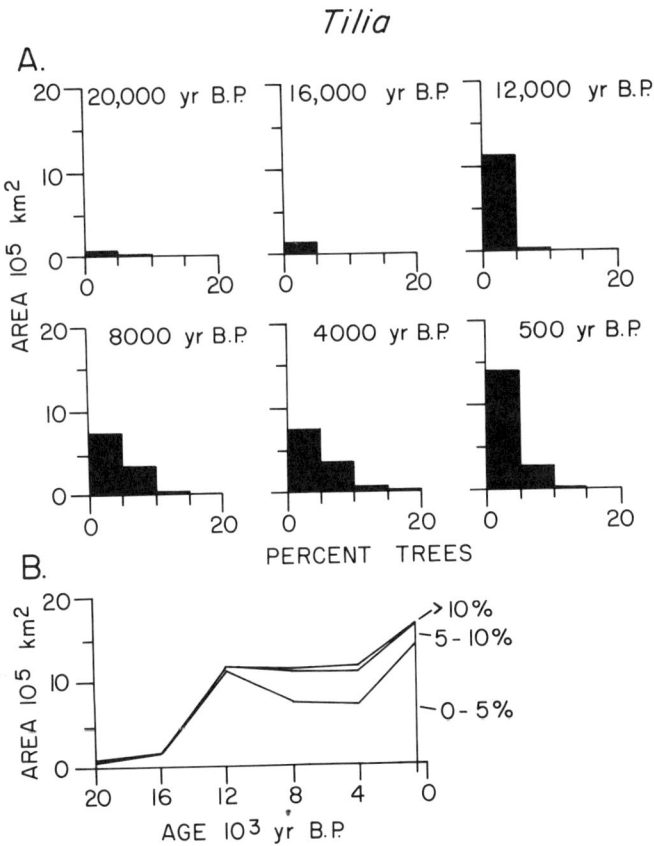

Figure 5.35. Changes in area and dominance of basswood (*Tilia*). a. Histograms of area–dominance structure at 20,000, 16,000, 12,000, 8000, 4000, and 500 yr B.P. b. Cumulative plot from 20,000 yr B.P. to the present.

6%). Outlier populations of basswood colonized Nova Scotia and persisted in Alabama. By 6000 yr B.P., basswood advanced along its northern range limit to 47°N in central Ontario and 48°N into northern Minnesota (Fig. 5.34i).

Between 6000 yr B.P. and 4000 yr B.P. (Fig. 5.34j), both the northern and southern range margins of basswood constricted, with its distribution at 4000 yr B.P. restricted to generally between 37°N and 47°N. Basswood increased in dominance to > 5% across the Great Lakes region in the latitudinal band extending from 42°N to 46°N. Population centers were located in southeastern Ontario and New York (10% to 16%) and in southern Wisconsin (up to 13%). By 2000 yr B.P. (Fig. 5.34k), basswood dropped below 10% dominance in both of its previous population centers. A narrow band of 5% to 10% dominance remained from southern Wisconsin across Lower Michigan and southern Ontario into New York and southern Quebec. From 4000 yr B.P. to 2000 yr B.P., *Tilia* spread into the northern half of New England. The paleo-dominance map for 500 yr B.P. (Fig. 5.34l) reflects a fragmentation of the latitudinally elongate basswood population center within the Great Lakes region, which produced one remnant along the prairie–forest border of Minnesota and Wisconsin and a second one across the eastern Great Lakes region. Limited populations of > 5% basswood established within the central and southern Appalachian Mountains during the interval from 2000 yr B.P. to 500 yr B.P. Minor reexpansions in range limits occurred along the Appalachians to 35°N, westward along the prairie–forest margin, and north to 47°N.

The paleo-dominance histograms for basswood (Fig. 5.35a,b) illustrate that full-glacial and late-glacial populations in the lower dominance classes occupied relatively small areas. The total area of basswood distribution increased by 200% from $0.7 \cdot 10^5$ km^2 at 20,000 yr B.P. to $1.4 \cdot 10^5$ km^2 at 16,000 yr B.P., and again by 829% from 16,000 yr B.P. to $11.6 \cdot 10^5$ km^2 at 12,000 yr B.P. However, between 12,000 yr B.P. and 4000 yr B.P., *Tilia*'s areal extent remained consistent at $11.6 \cdot 10^5$ km^2. Basswood populations then expanded by 42%, reaching $16.5 \cdot 10^5$ km^2 at 500 yr B.P. From 12,000 yr B.P. to 8000 yr B.P., dominance classes > 5% increased from 2% to 33% of the total area of distribution of basswood. Even though the area remained nearly constant, the relative dominance of basswood in the 5% to 10% dominance class increased progressively through the Holocene.

Hemlock (*Tsuga*)

The population trajectory for hemlock (Fig. 5.36a) portrays population means < 2% from 20,000 yr B.P. to 14,000 yr B.P., a late-glacial to early-Holocene rise to 5% to 6% from 13,000 yr B.P. to 11,000 yr B.P., and a prominent early-Holocene decline to 3% by 10,000 yr B.P. During the Holocene, hemlock mean values rose from 5% at 9000 yr B.P. to an interglacial maximum of 8% at 5000 yr B.P. and then dropped progressively to 6% at 4500 yr B.P., 5% at 4000 yr B.P., and 4% at 3500 yr B.P. During the last 3000 years, mean values for hemlock regained earlier levels, fluctuating between 5% and 6%. The maximum paleo-dominance values reconstructed for hemlock varied between 1% and 3% from

Tsuga

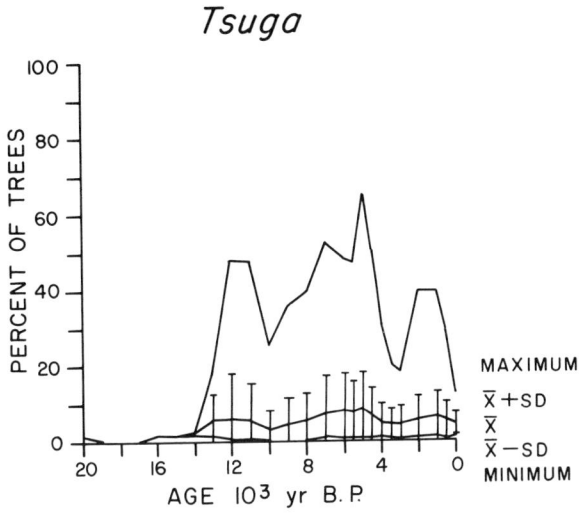

Figure 5.36. Hemlock (*Tsuga*). a. Population trajectory, including maximum, minimum, and mean values of paleo-dominance as well as ± 1 SD of the mean, for the past 20,000 years.

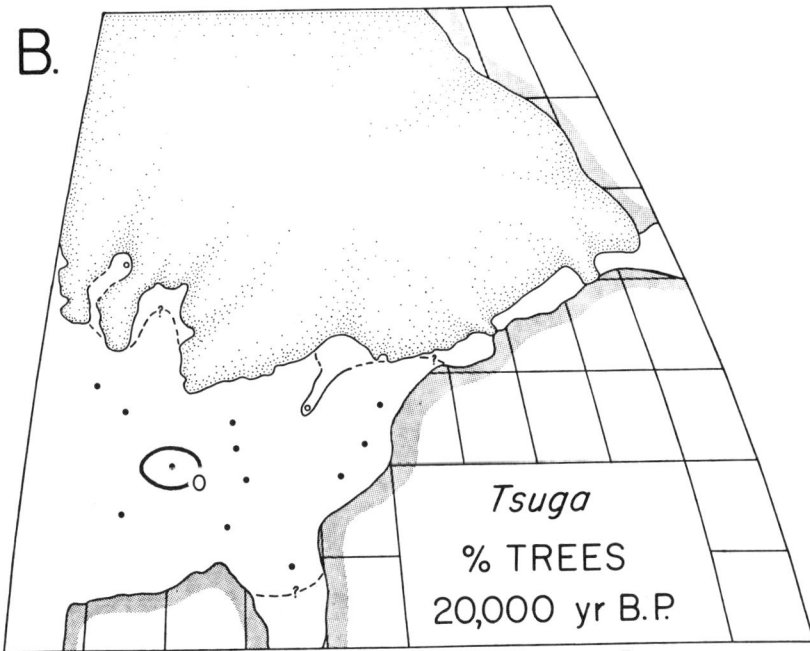

Figure 5.36. Hemlock (*Tsuga*). b. Paleo-dominance map for 20,000 yr B.P.

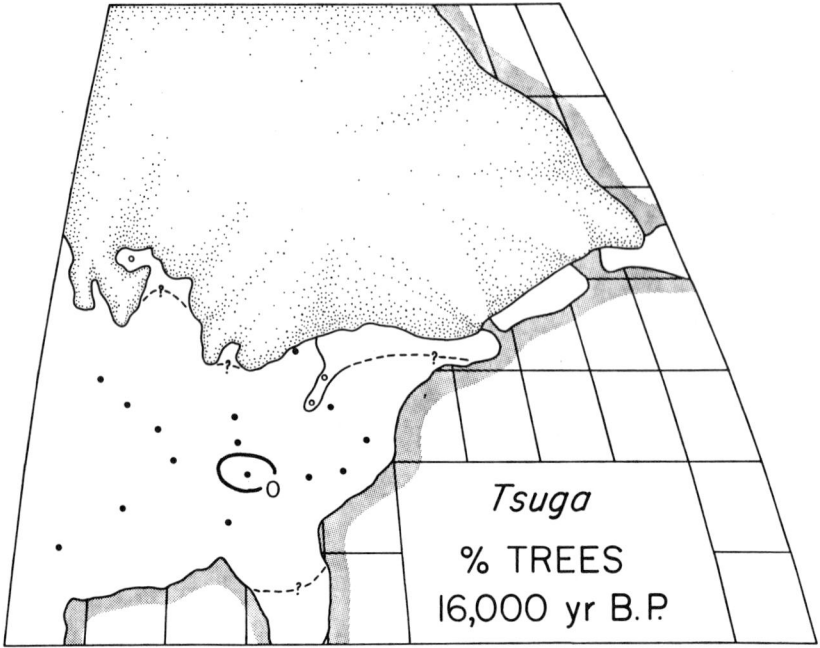

Figure 5.36. Hemlock (*Tsuga*). c. Paleo-dominance map for 16,000 yr B.P.

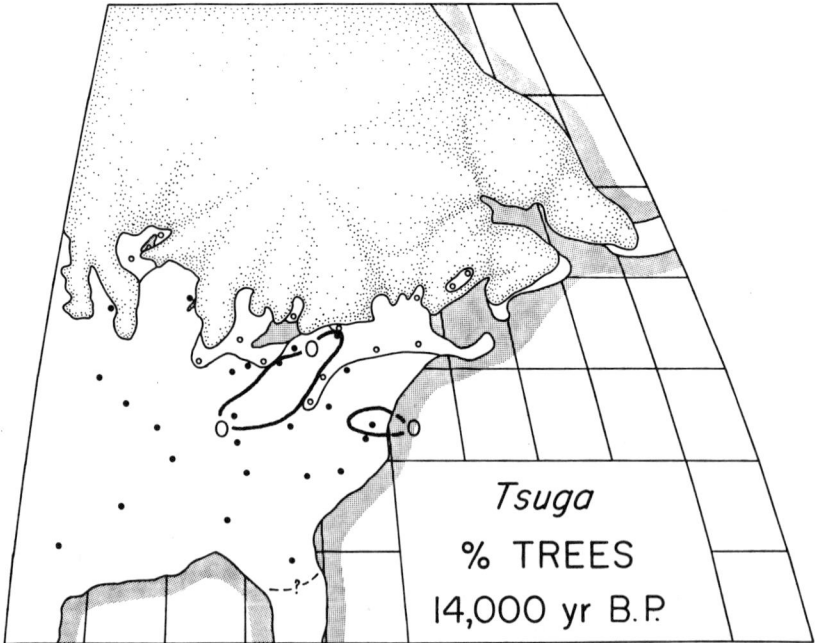

Figure 5.36. Hemlock (*Tsuga*). d. Paleo-dominance map for 14,000 yr B.P.

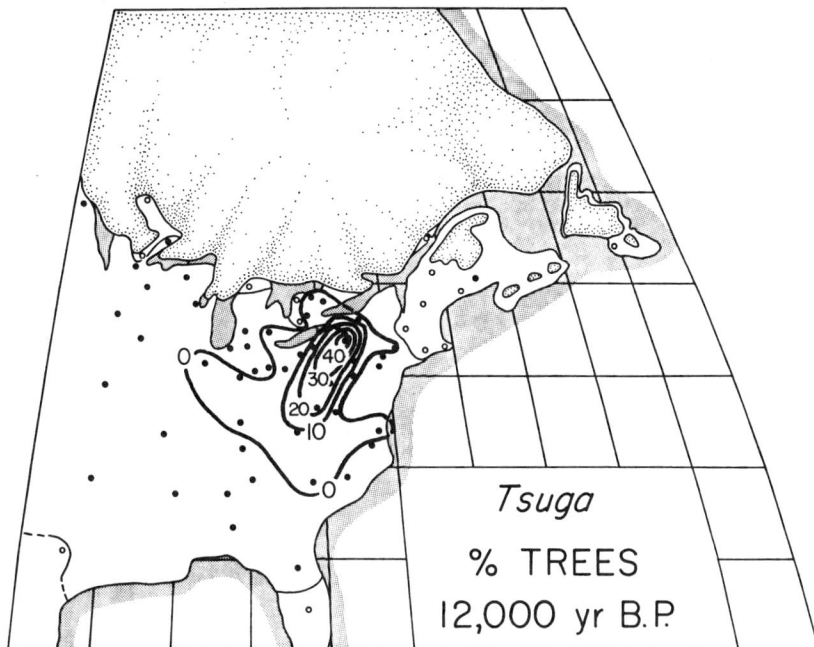

Figure 5.36. Hemlock (*Tsuga*). e. Paleo-dominance map for 12,000 yr B.P.

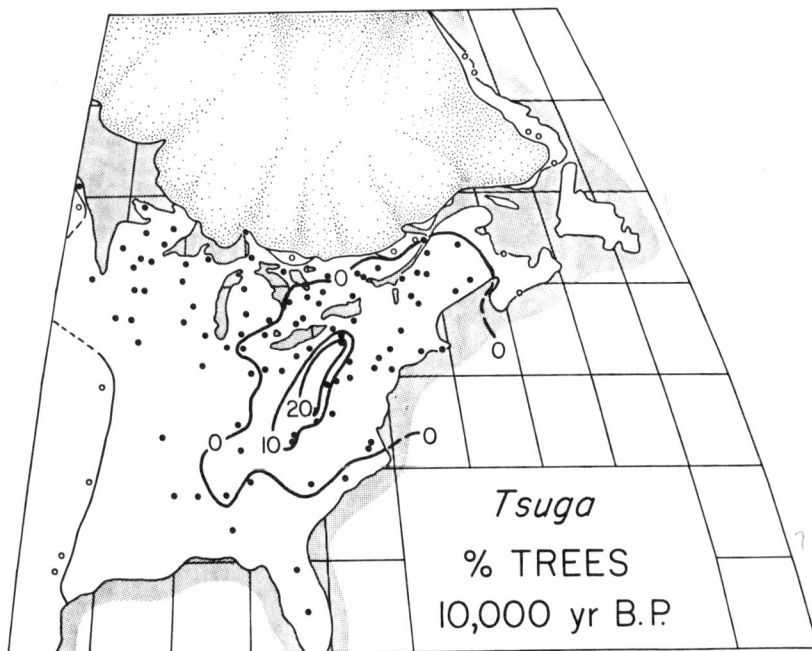

Figure 5.36. Hemlock (*Tsuga*). f. Paleo-dominance map for 10,000 yr B.P.

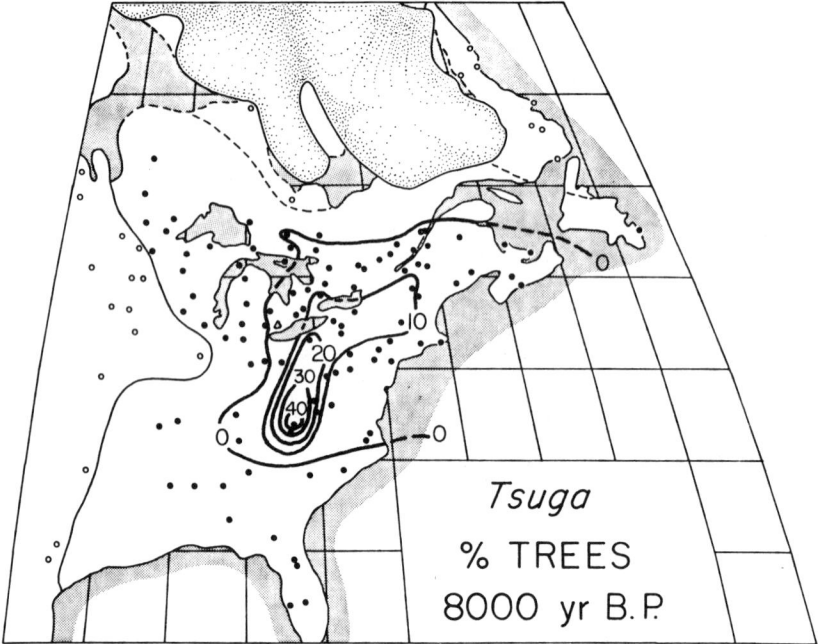

Figure 5.36. Hemlock (*Tsuga*). g. Paleo-dominance map for 8000 yr B.P.

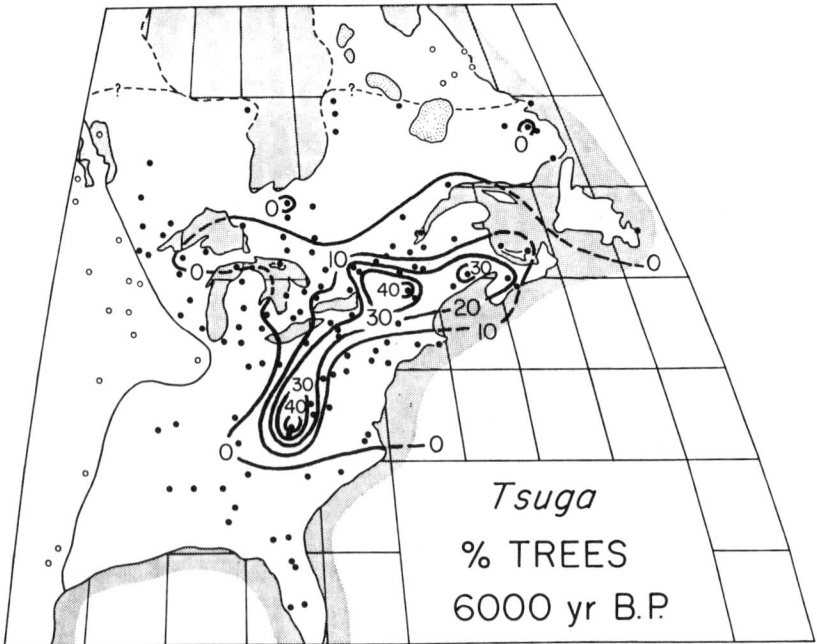

Figure 5.36. Hemlock (*Tsuga*). h. Paleo-dominance map for 6000 yr B.P.

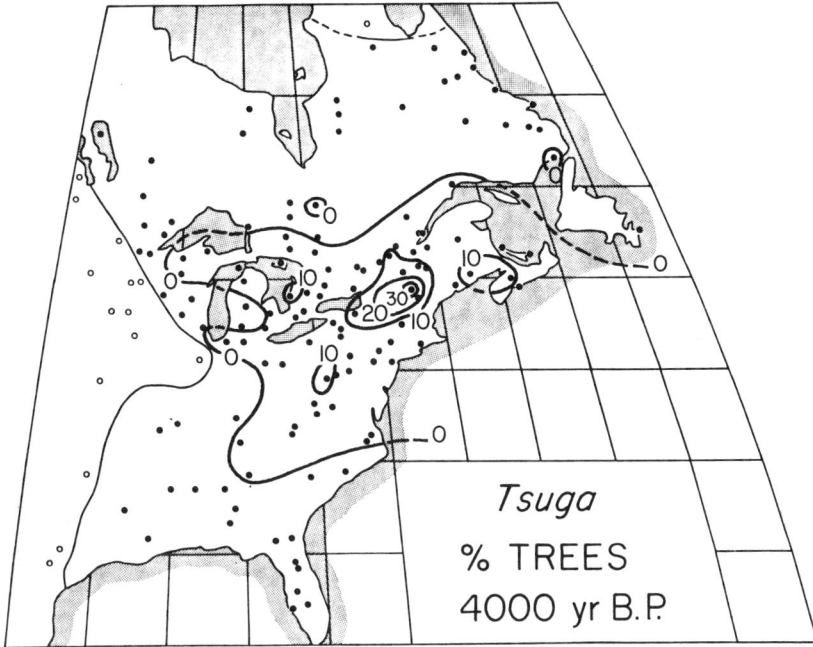

Figure 5.36. Hemlock (*Tsuga*). i. Paleo-dominance map for 4000 yr B.P.

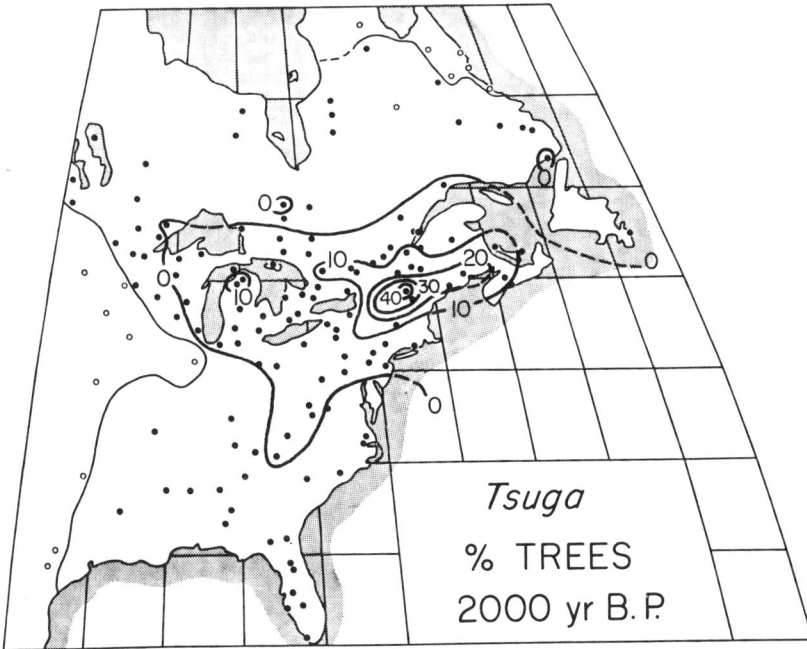

Figure 5.36. Hemlock (*Tsuga*). j. Paleo-dominance map for 2000 yr B.P.

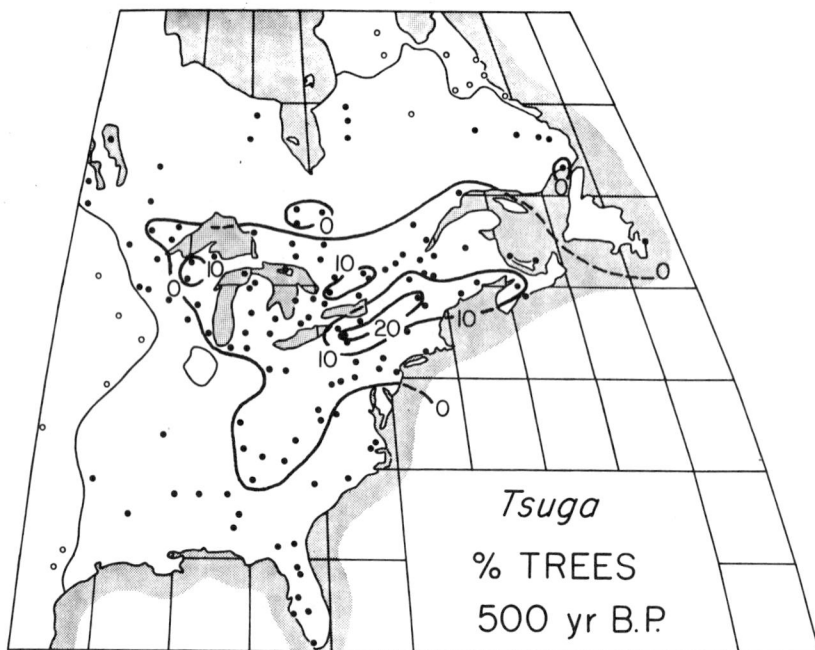

Figure 5.36. Hemlock (*Tsuga*). k. Paleo-dominance map for 500 yr B.P.

20,000 yr B.P. to 14,000 yr B.P., increased to 48% by 12,000 yr B.P., declined to 25% by 10,000 yr B.P., and then progressively increased to 65% by 5000 yr B.P. Hemlock population maxima declined to 51% by 4500 yr B.P., 30% at 4000 yr B.P., 20% by 3500 yr B.P., and 18% at 3000 yr B.P. Maximum values of paleo-dominance for hemlock rose to 40% at 2000 yr B.P. but declined markedly after 1000 yr B.P. Variability in paleo-dominance estimates about mean values was ± 1 SD of 1% from 20,000 yr B.P. to 14,000 yr B.P. but increased to the level of ± 3% to 12% through the last 13,000 years. Times of maximum population variability (± 1 SD > 10%) occurred at 12,000 yr B.P., 11,000 yr B.P., and 5,000 yr B.P.

The paleo-dominance map for hemlock at 20,000 yr B.P. (Fig. 5.36b) identifies one full-glacial refugium with 1% *Tsuga* located at Nonconnah Creek, Tennessee (Delcourt *et al.* 1980). Hemlock was not reconstructed at any site available in the array for 19,000 yr B.P., 18,000 yr B.P., or 17,000 yr B.P. The paleo-dominance map for 16,000 yr B.P. (Fig. 5.36c) portrays a minor hemlock population (2% dominance) at Quicksand Pond, northwestern Georgia (Watts 1970). It is plausible that restricted populations of hemlock occurred during the full-glacial and early late-glacial intervals within the narrow latitudinal band of mixed conifer–northern hardwoods forest mapped between more northern boreal forests and more southern temperate forests (Fig. 4.3a). Between 16,000 yr B.P. and 14,000 yr B.P. (Fig. 5.36d), hemlock dispersed along the Appalachian Mountains and locally colonized portions of the central Atlantic Seaboard. The paleo-dominance map for 14,000 yr B.P. illustrates that up to 2% hemlock occurred across the region from central Kentucky to eastern Ohio and western Pennsylvania. A second population of hemlock occupied coastal Virginia (Fig. 5.36d).

By 12,000 yr B.P. (Fig. 5.36e), hemlock colonized the eastern Great Lakes region, extending into southern Ontario and eastern Lower Michigan. Populations established northwestward across Indiana and Illinois, northeastward into New York, and southeastward along the Atlantic Coastal Plain into South Carolina. The primary population center of hemlock was situated along the central Appalachians from West Virginia (21% at Cranberry Glades; Watts 1979) northeast to north-central Pennsylvania (48% at Rose Lake; Cotter and Crowl 1981).

Between 12,000 yr B.P. and 10,000 yr B.P., hemlock expanded north to the shores of Lake Huron and northeast across southern Ontario and southeastern Quebec, along the St. Lawrence River Valley, and across New England to New Brunswick (Fig. 5.36f). By 10,000 yr B.P., hemlock's western margin extended from southern Michigan south to Middle Tennessee and northern Mississippi. Along its southern margin, hemlock advanced southward to approximately 33°N. The primary population center remained in the central Appalachian Mountains, with dominance levels of 24% in West Virginia and 25% in western Pennsylvania. Plant macrofossils of *Tsuga canadensis* have been recovered from sediments dating as old as 10,000 yr B.P. along the Allegheny Plateau in northern West Virginia (Larabee 1986). The 10% contour of paleo-dominance extended into the southern Appalachians as far south as Shady Valley in northeastern Tennessee (Barclay 1957).

Between 10,000 yr B.P. and 8000 yr B.P. (Fig. 5.36g), hemlock migrated north to 47°N in eastern Ontario and to 48°N in both southeastern Quebec and Manitoba. Hemlock also advanced across Prince Edward Island and Nova Scotia. By 8000 yr B.P., the western limit of hemlock extended from the eastern shores of Lake Huron across southeastern Michigan and central Ohio and into the Interior Low Plateaus of central Kentucky and Middle Tennessee. Hemlock's southern border extended along 35°N and its eastern margin was delimited by the shoreline of the Atlantic Ocean. By 8000 yr B.P., hemlock's primary population center of > 10% dominance had expanded from the central Appalachians across the northern Appalachians. At that time, populations of > 20% hemlock occurred from western Pennsylvania (29% at Crystal Lake), to southwestern Virginia (40% at Saltville Valley; Delcourt and Delcourt 1986) and northeastern Tennessee (23% at Shady Valley; Barclay 1957).

By 6000 yr B.P. (Fig. 5.36h), the primary population center of > 10% hemlock extended from the central Appalachians to the northern Appalachians and across the New England states into southeastern Canada. Areas of ≥ 30% hemlock paleo-dominance occurred in southwestern Virginia (45%), Vermont (48%), southern Quebec (30%), and southern New Brunswick (33%). By 6000 yr B.P., the northern distributional limit of hemlock extended northeastward to about 50°N in coastal Quebec. Hemlock populations migrated north and west of Lake Huron, advancing to the shores of Lake Superior in Ontario and the Upper Peninsula of Michigan by 6000 yr B.P. By 5000 yr B.P., hemlock populations in southwestern Virginia had declined to 18%. In contrast, northern Appalachian populations increased to a maximum of 65% within Vermont. The geographic region with > 20% hemlock extended across the central New England region and into Nova Scotia. Hemlock's migration halted along its northern boundary by 5000 yr B.P. but continued to extend westward through Ohio, northeastern Indiana, and Lower Michigan.

Between 5000 yr B.P. and 4000 yr B.P. (Fig. 5.36i), hemlock populations declined in dominance, possibly due to pathogen attack, throughout all of its former population centers (Davis 1981b; Allison *et al.* 1986). For example, in the northern Appalachians, peak values declined from 65% to 30% within the 1000 years between 5000 yr B.P. and 4000 yr B.P. Populations within both the central Appalachians and the eastern Great Lakes region typically dropped below 10%. Despite the loss in overall dominance, limited populations of hemlock continued to advance westward through northern Indiana and northeastern Illinois, as well as northern Wisconsin.

Between 4000 yr B.P. and 2000 yr B.P. (Fig. 5.36j), hemlock regained moderate dominance values (10% to 40%) in the northern Appalachian Mountains from New York to New Brunswick. Populations of hemlock continued to advance across the central portions of Lower Michigan and Wisconsin and into northeastern Minnesota, but they retracted eastward along their southwestern distributional limit. By 2000 yr B.P. (Fig. 5.36j), relict hemlock populations were extirpated from coastal-plain habitats of the central Atlantic Seaboard.

During the last 2000 years, hemlock has persisted with moderate population centers (10% to 29%) in the region from 42°N to 46°N and east of 79°W. One minor population center (up to 16%) established in northern Wisconsin by 500 yr B.P. (Fig. 5.36k). The presettlement distribution of hemlock (Fig. 5.36k) extended from 34°N in the southern Appalachians continuously north to about 50°N in coastal Quebec. By 500 yr B.P., hemlock occupied a latitudinal band from the New England region west across the Great Lakes into western Wisconsin and northern Minnesota.

The area–dominance histograms for hemlock (Fig. 5.37a) illustrate major changes in dominance structure throughout the late Quaternary. There was only 1 dominance class (0% to 10%) for 20,000 yr B.P. and 16,000 yr B.P., an increase to 5 dominance classes (in 10% increments, up to 50%) by 12,000 yr B.P., followed by a reduction to 3 classes at 10,000 yr B.P., then a reexpansion to 5 classes at 8000 yr B.P. and 6000 yr B.P. (Fig. 5.37b). The number of dominance classes increased to 7 at 5000 yr B.P. and then dropped progressively, to 6 at 4500 yr B.P. and 4 at 4000 yr B.P. Following that mid-Holocene decline, there was a reexpansion to 5 classes at 2000 yr B.P. and then a reduction to only 3 classes by 500 yr B.P. In all histograms shown, the lowest dominance class represented the greatest amount of area for any class. The cumulative plot of dominance structure for hemlock (Fig. 5.37b) shows that hemlock maintained a total area of distribution of $0.6 \cdot 10^5$ km^2 from 20,000 yr B.P. to 16,000 yr B.P. The area occupied by hemlock increased by 1183% to $7.1 \cdot 10^5$ km^2 by 12,000 yr B.P., and it continued to increase through the early- and mid-Holocene intervals to a maximum extent of $14.5 \cdot 10^5$ km^2 at 8000 yr B.P. Early postglacial expansion in range was accompanied by an areal expansion in all dominance

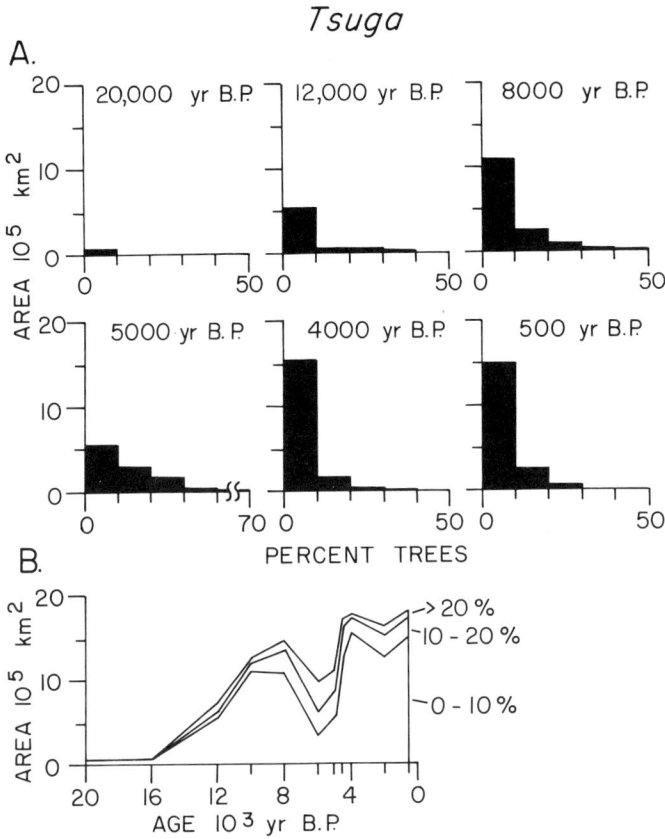

Figure 5.37. Changes in area and dominance of hemlock (*Tsuga*). a. Histograms of area–dominance structure at 20,000, 12,000, 8000, 5000, 4000, and 500 yr B.P. b. Cumulative plot from 20,000 yr B.P. to the present.

classes. However, during the mid-Holocene interval, the total area of hemlock fluctuated markedly, decreasing to $9.6 \cdot 10^5$ km^2 at 6000 yr B.P., then expanding to $11.2 \cdot 10^5$ km^2 at 5000 yr B.P., $17.1 \cdot 10^5$ km^2 at 4500 yr B.P., and $17.5 \cdot 10^5$ km^2 at 4000 yr B.P. Hemlock's area oscillated through the late Holocene between $16.1 \cdot 10^5$ km^2 at 2000 yr B.P. and $17.9 \cdot 10^5$ km^2 at 500 yr B.P.

Although the hemlock population trajectory (Fig. 5.36a) shows the greatest collapse in hemlock's maximum values occurring between 5000 yr B.P. and 3000 yr B.P., the total area for hemlock expanded by more than 150% during that interval. A marked decline in hemlock populations in the northern portion of its range at approximately 4800 yr B.P. has been attributed to an outbreak of pathogens (Davis 1981b; Allison *et al.* 1986). The total area occupied by dominance classes $> 10\%$ decreased by 65% from $5.7 \cdot 10^5$ km^2 at 5000 yr B.P., to $4.3 \cdot 10^5$ km^2 at 4500 yr B.P., and then to $2.0 \cdot 10^5$ km^2 at 4000 yr B.P. However, the area of the lowest dominance class (0% to 10%) increased by nearly 280% during that time interval, from $5.6 \cdot 10^5$ km^2 at 5000 yr B.P., to

$12.8 \cdot 10^5$ km^2 at 4500 yr B.P., and then to $15.5 \cdot 10^5$ km^2 at 4000 yr B.P. Thus, although dominance of major population centers collapsed in the mid-Holocene interval, hemlock continued its dramatic areal expansion in distribution, advancing along both its northern and western distributional limits (Fig. 5.36i,j).

Elm (*Ulmus*)

The population trajectory for elm (Fig. 5.38a) shows that, during the full-glacial and early late-glacial intervals from 20,000 yr B.P. to 16,000 yr B.P., mean values of paleo-dominance were low, generally about 1% of the reconstructed forest composition. During the late-glacial and early-Holocene intervals, from 15,000 yr B.P. to 11,000 yr B.P., mean values rose to between 2% and 3%, crested at 6% from 10,000 yr B.P. to 9000 yr B.P., and fluctuated between 3% and 5% during the last 8000 years. The maximum reconstructed paleo-dominance values for elm varied between 2% and 3% from 20,000 yr B.P. to 16,000 yr B.P. and rose to between 5% and 19% from 15,000 yr B.P. to 12,000 yr B.P. Elm increased in maximum dominance from 36% at 11,000 yr B.P. to the late-Quaternary maximum of 51% at 9000 yr B.P., before dropping to 20% by 8000 yr B.P. During the last 7000 years, maximum reconstructed values for elm fluctuated between the high of 26% at 7000 yr B.P. and the late-Holocene low of 13% at 500 yr B.P. Variability about mean values ($\pm$ 1 SD) was typically 1% for the full-glacial interval, ranged between 1% and 4% for the late-glacial interval, and fluctuated between 3% and 9% during the Holocene.

The paleo-dominance map for *Ulmus* at 20,000 yr B.P. (Fig. 5.38b) displays a widespread distribution at low percentages (up to 3%) between 31°N and 39°N. Elm extended from the coastal plain of North Carolina and Virginia west across the southern Appalachian Mountains to at least central Missouri. Elm was distributed generally throughout temperate forested regions as well as extending through much of the boreal forest region during the full-glacial interval. At 18,000 yr B.P. (Fig. 5.38c), the pattern of distribution for elm was comparable to that of 20,000 yr B.P., with primary distribution within and west of the Appalachians between 32°N and 39°N.

The map for 16,000 yr B.P. (Fig. 5.38d) shows a limited northeastward advance of elm into eastern Ohio. Paleo-dominance values remained $\leq$ 3% throughout the range of elm in the early late-glacial interval. However, between 16,000 yr B.P. and 14,000 yr B.P. (Fig. 5.38e), elm populations dispersed north to the glacial front or to the tundra–forest limit. By 14,000 yr B.P., the northern perimeter of elm extended from western Pennsylvania across Ohio and Indiana and northwestward through Wisconsin and Iowa (Fig. 5.38e). Elm was distributed continuously from 32°N (its southern border) to as far north as about 46°N in the continental interior. By 14,000 yr B.P., a primary population center of > 5% had developed in the western portion of the range of elm, with 5% elm dominance reconstructed at Boney Springs, west-central Missouri (King 1973) and up to 17% elm dominance reconstructed at Muscotah Marsh, Kansas (Gruger 1973).

By 12,000 yr B.P. (Fig. 5.38f), elm had established populations in deglaciated

Ulmus

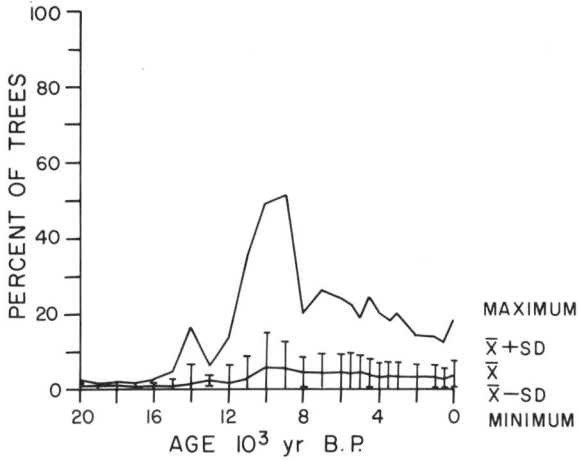

Figure 5.38. Elm (*Ulmus*). a. Population trajectory, including maximum, minimum, and mean values of paleo-dominance as well as ± 1 SD of the mean, for the past 20,000 years.

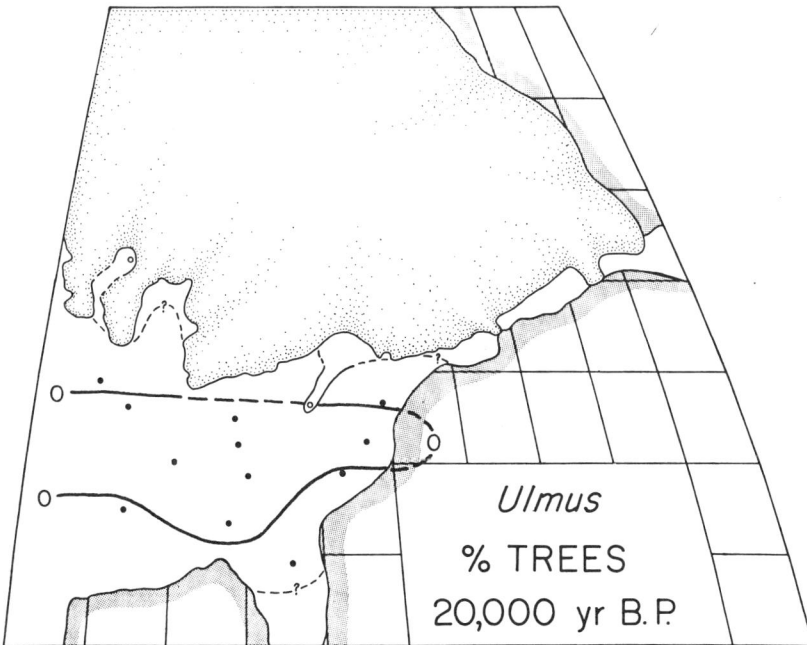

Figure 5.38. Elm (*Ulmus*). b. Paleo-dominance map for 20,000 yr B.P.

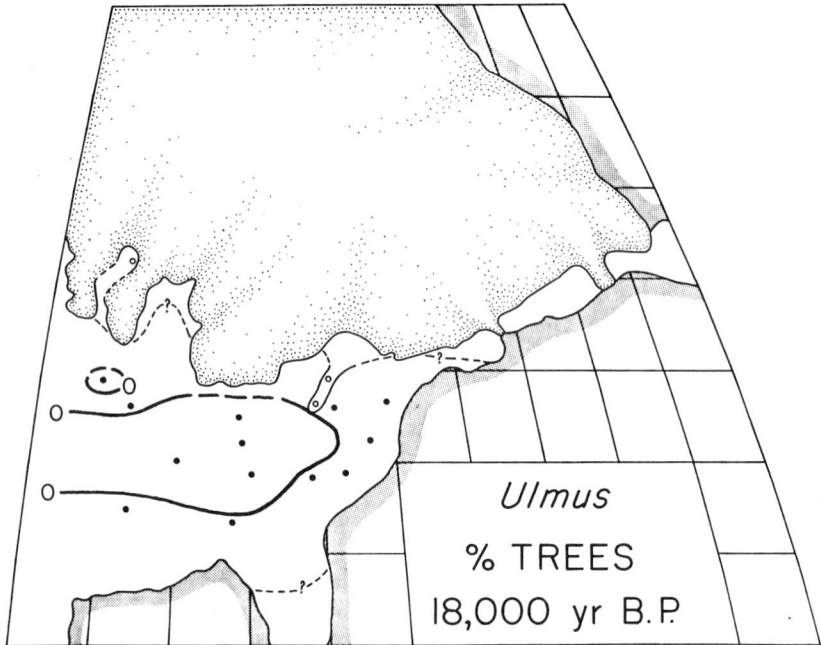

Figure 5.38. Elm (*Ulmus*). c. Paleo-dominance map for 18,000 yr B.P.

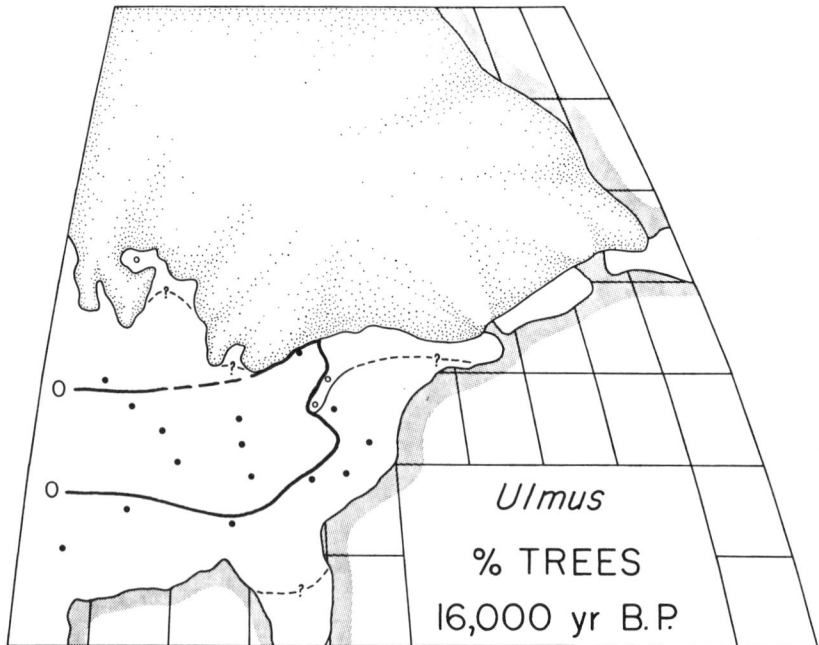

Figure 5.38. Elm (*Ulmus*). d. Paleo-dominance map for 16,000 yr B.P.

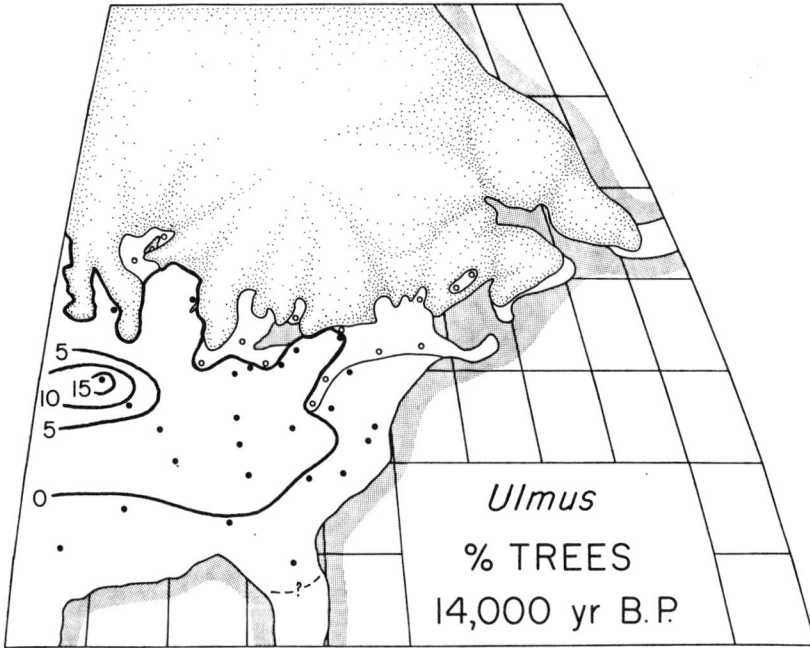

Figure 5.38. Elm (*Ulmus*). e. Paleo-dominance map for 14,000 yr B.P.

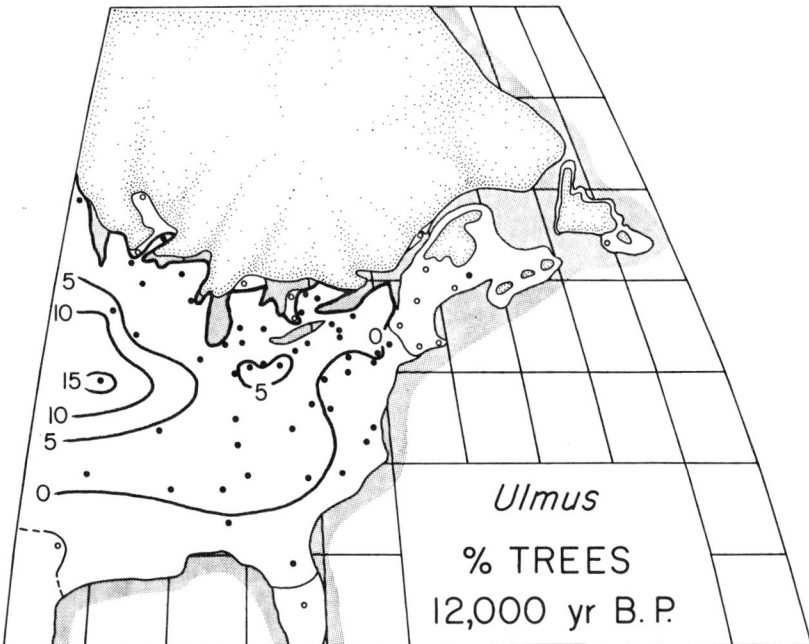

Figure 5.38. Elm (*Ulmus*). f. Paleo-dominance map for 12,000 yr B.P.

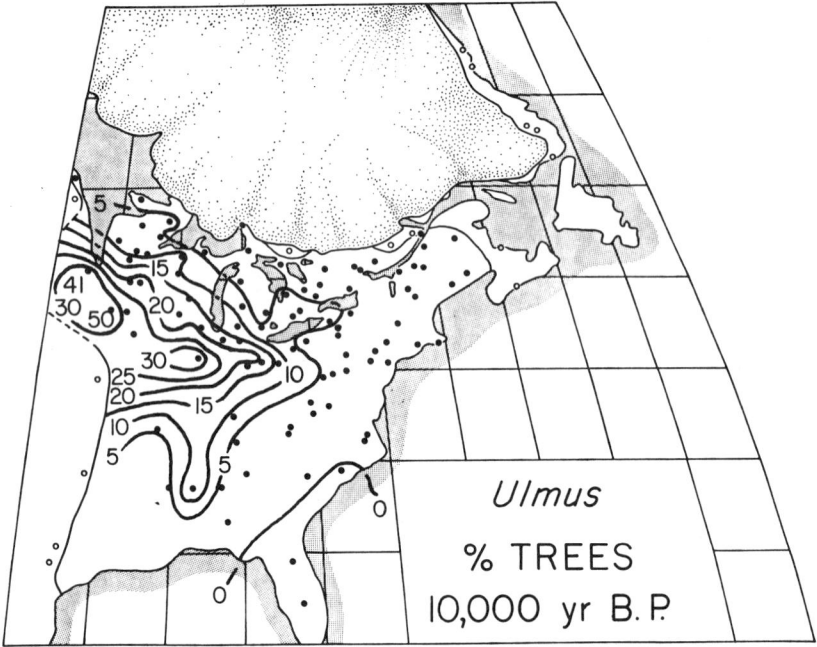

Figure 5.38. Elm (*Ulmus*). g. Paleo-dominance map for 10,000 yr B.P.

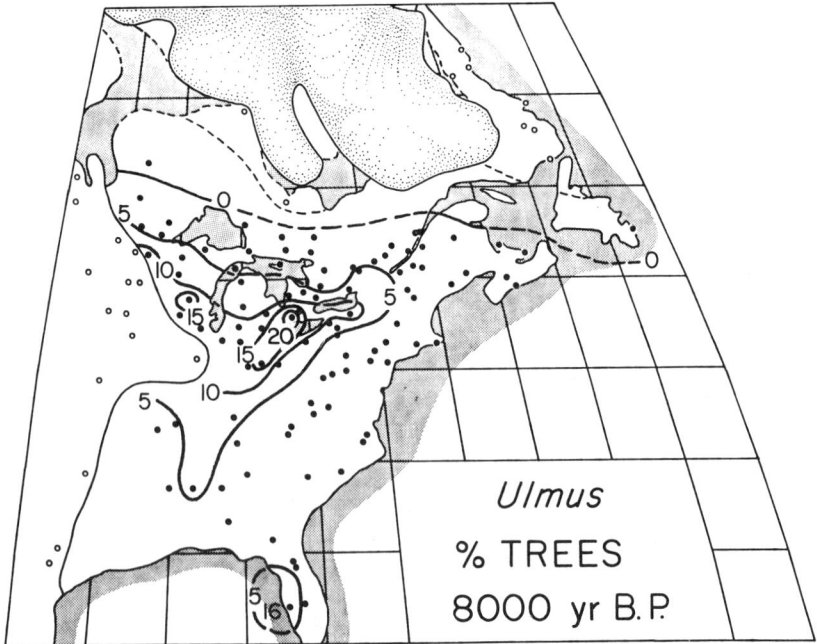

Figure 5.38. Elm (*Ulmus*). h. Paleo-dominance map for 8000 yr B.P.

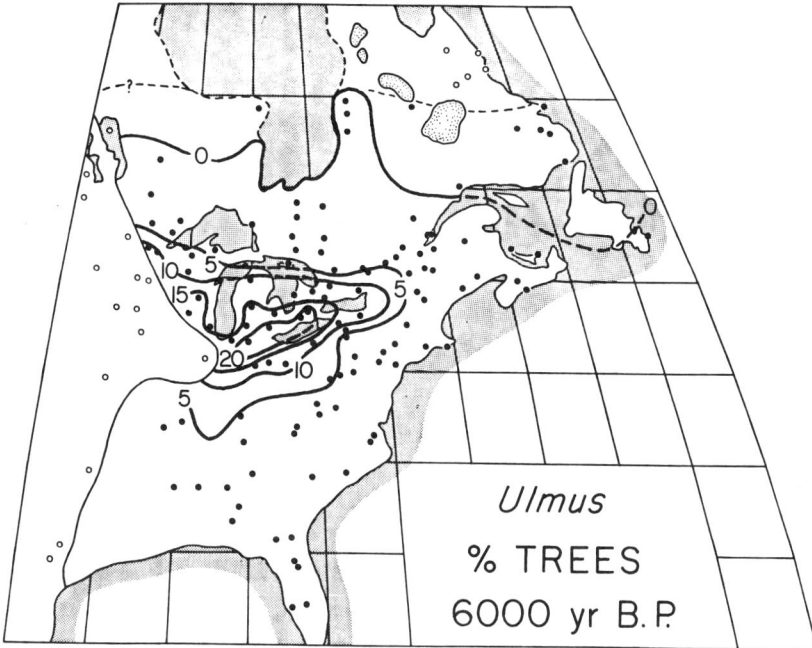

Figure 5.38. Elm (*Ulmus*). i. Paleo-dominance map for 6000 yr B.P.

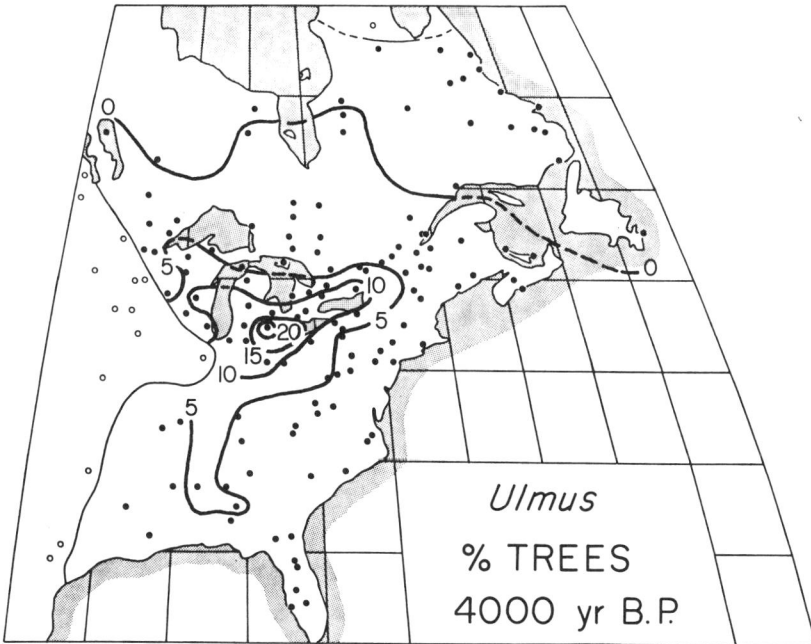

Figure 5.38. Elm (*Ulmus*). j. Paleo-dominance map for 4000 yr B.P.

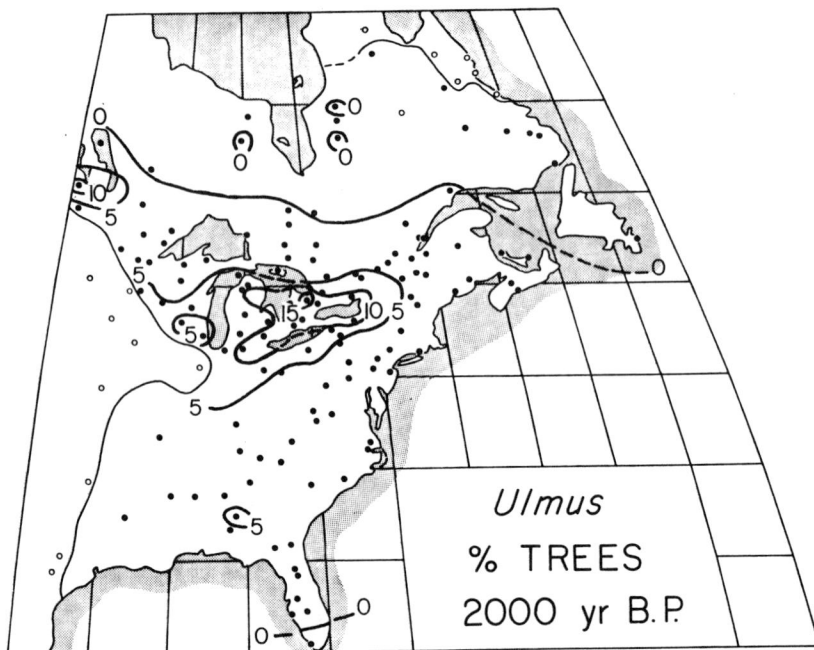

Figure 5.38. Elm (*Ulmus*). k. Paleo-dominance map for 2000 yr B.P.

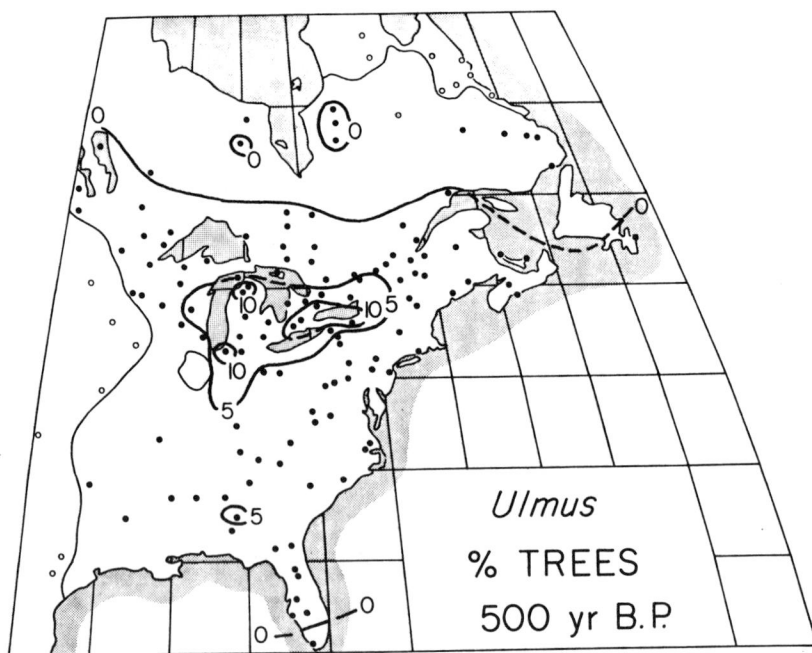

Figure 5.38. Elm (*Ulmus*). l. Paleo-dominance map for 500 yr B.P.

terrain of the northern Appalachian Mountains as well as across the Great Lakes region. The northern continuous range limit for elm extended to near the margin of the Laurentide Ice Sheet and was located generally between 45°N and 50°N. The areas of > 5% elm dominance expanded northward through the continental interior, with populations of 19% persisting in Kansas and 9% in western Iowa (Fig. 5.38f). A second population center, with values of elm dominance up to 7%, established in Ohio by 12,000 yr B.P.

Between 12,000 yr B.P. and 10,000 yr B.P. (Fig. 5.38g), *Ulmus* spread outward on all margins except its western range limit. By 10,000 yr B.P., elm extended from Manitoba to New Brunswick and occupied coastal-plain habitats from New Brunswick south to Georgia and west to eastern Texas. Elm was eliminated from much of the continental interior by the eastward expansion of the Prairie Peninsula. The area of elm dominance > 5% was situated west of the Appalachians, generally between 37°N and 49°N. Locally, elm reached > 10% within the central Gulf Coastal Plain. Major population centers of > 30% were located in central Illinois (37%) as well as throughout western Iowa (50%) and into eastern South Dakota (41%).

From 10,000 yr B.P. to 8000 yr B.P. (Fig. 5.38h), the northeastward expansion of the Prairie Peninsula eliminated the population centers of elm within the Lower Midwest region. At 8000 yr B.P., two population centers for elm were reconstructed: (1) the region including the Great Lakes and extending to the Interior Low Plateaus (reaching 20% in southern Ontario); and (2) a restricted population of 16% elm located in the western Florida Peninsula. By 8000 yr B.P., the northern border of elm extended from about 50°N in western Ontario to approximately 48°N in New Brunswick and adjacent portions of Nova Scotia. The western border of the range of *Ulmus* extended to the prairie–forest ecotone. During the early- and mid-Holocene intervals, both the areal extent and magnitude of dominance of population centers for elm diminished greatly. The region of > 10% elm extended along an arc from New York through southern Ontario and Lower Michigan through Ohio, Indiana, Wisconsin, and into eastern Minnesota.

At 6000 yr B.P. (Fig. 5.38i), the overall distribution of elm was widespread throughout eastern North America, extending from the Gulf Coastal Plain and Peninsular Florida northward to north-central Manitoba, around the shores of Hudson Bay in northeastern Ontario and northwestern Quebec, and then southeastward to Nova Scotia. During the mid-Holocene interval, elm occupied all of forested eastern North America except the northernmost fringe of boreal forest. The primary population center of > 5% elm extended east of the prairie–forest boundary from Minnesota to western New York. The region of > 10% elm generally extended from 40°N to 45°N west of the Appalachian Mountains. Maximum values of > 20% occurred from northern Indiana across southeastern Michigan and into southern Ontario.

The map of elm paleo-dominance for 4000 yr B.P. (Fig. 5.38j) shows that the elm population center persisted throughout the Great Lakes region, with a peninsular extension southward into Mississippi and Alabama. The northern range margin continued to expand northwestward through central Manitoba and northward along the retreating shores of Hudson Bay. During the late Holocene, the paleo-dominance maps for 2000 yr B.P. (Fig. 5.38k) and 500 yr B.P. (Fig. 5.38l) illustrate the fragmentation of elm populations in the vicinity of Hudson Bay and the general southward retreat of the northern range limit to approximately 50°N. The population center for elm persisted through the eastern and central Great Lakes region through the late Holocene, although maximum values of elm dominance dropped from 21% at 4000 yr B.P. to 15% at 2000 yr B.P. and then to 13% by 500 yr B.P. During the last several thousand years, as the prairie–forest border shifted westward, elm extended westward in gallery forests distributed from southern Manitoba through Minnesota, Iowa, Missouri, and eastern Oklahoma.

The area–dominance histograms for *Ulmus* (Fig. 5.39a) illustrate that from 20,000 yr B.P. to 16,000 yr B.P. only the 0% to 5% dominance class was represented throughout the full-glacial and late-glacial range of the genus. During the Holocene, the number of dominance classes increased to between 3 and 5 (in 5% increments up to 25%). The cumulative plot of dominance structure (Fig. 5.39b) shows that the total area for elm was $14.8 \cdot 10^5$ km^2 at 20,000 yr B.P. Elm's total distributional area declined to $13.2 \cdot 10^5$ km^2 at 16,000 yr B.P.

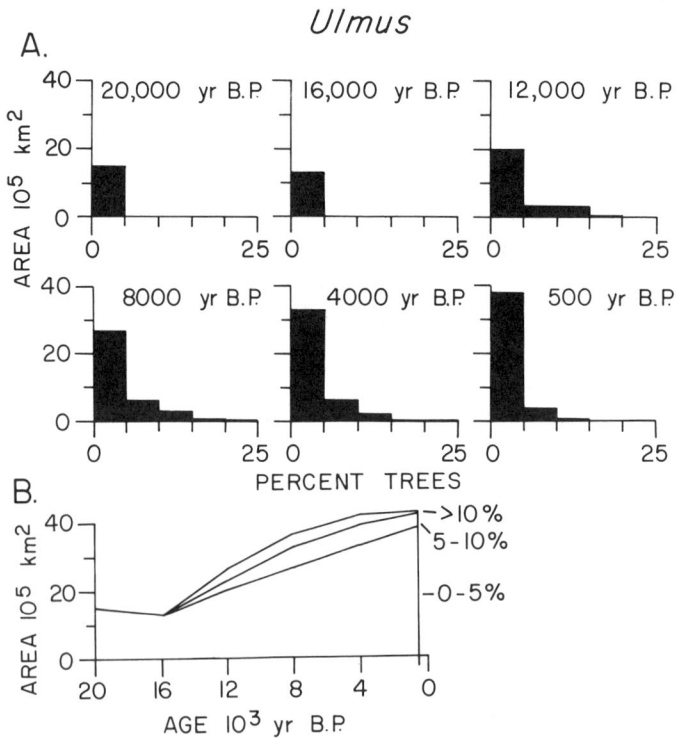

Figure 5.39. Changes in area and dominance of elm (*Ulmus*). a. Histograms of area–dominance structure at 20,000, 16,000, 12,000, 8000, 4000, and 500 yr B.P. b. Cumulative plot from 20,000 yr B.P. to the present.

During the interval from 16,000 yr B.P. to 500 yr B.P., elm's area increased more than 320% to an interglacial maximum of $42.4 \cdot 10^5$ km^2. The area–dominance histograms illustrate that the lowest dominance class represented 73% to 90% of the total distributional area throughout the Holocene. Thus, although elm was geographically widespread, its population centers were substantially restricted in overall areal extent.

Conclusions

1. The combination of population trajectories, time series of paleo-dominance maps, and area–dominance histograms of dominance structure provide complementary, quantitative summaries of long-term dynamics of the populations of important tree taxa of eastern North America.
2. Paleo-dominance maps prepared for the last 20,000 years in eastern North America graphically demonstrate that the locations of modern population centers and distributional limits for most tree taxa do not reflect their past distributional histories. Rather, dynamic and individualistic shifts have occurred in not only the locations, but also in the total area and relative dominance of all major temperate and boreal tree taxa during the late Quaternary.

6. Late-Quaternary Migrational Strategies of Tree Species

The widely held notion that modern distributions of taxa can be used to interpret their past history underestimates the complexity of shifts in range margins, centers of dominance, and changes in community composition through the late-Quaternary interval. Modern patterns of distribution of species along environmental gradients are often assumed to be in equilibrium with prevailing climatic conditions and disturbance regimes (*e.g.*, Whittaker 1975). In a conceptual model representing a three-dimensional map view of the pattern of relative dominance of a taxon in equilibrium (Fig. 6.1), the population center coincides with the geographic center of distribution. The population center is located in the optimal portion of the environmental gradient, and the dominance of the taxon decreases symmetrically away from its optimum in all directions. This assumption is incorporated within forest-stand simulation models such as FORET and JABOWA (Shugart 1984). In cross-section view, percent dominance of the taxon along an environmental gradient appears as a symmetrical bell-shaped or Gaussian curve.

Quantitative data sets covering the entire ranges of distribution of arboreal taxa (*e.g.*, Delcourt *et al.* 1983a, 1984) show that only rarely do the observed dominance distributions of species conform to this idealization. Instead, centers of dominance tend to be displaced from the geographic center of distribution of the taxon. Even where the geographic center of distribution coincides with the center of dominance, *e.g.*, in eastern North American *Quercus*, even the presettlement map pattern (Fig. 5.30l) is of little value in interpreting the history

CROSS-SECTION VIEW:

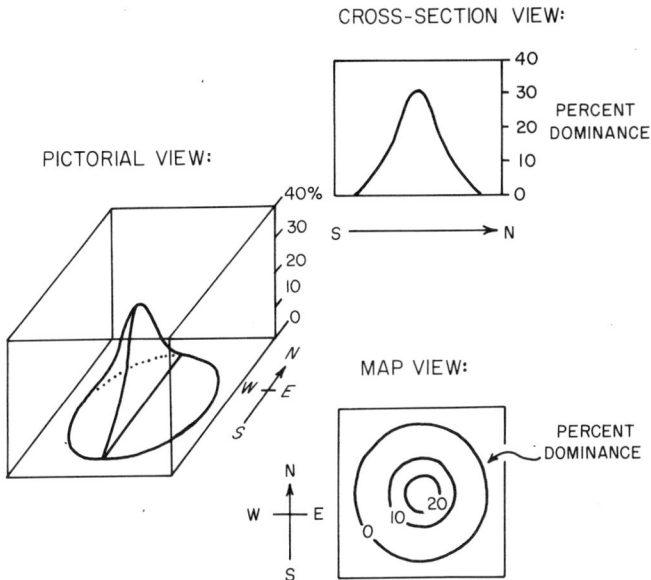

Figure 6.1. Hypothetical model of a population distribution in equilibrium with the environment.

of changes in location and importance of oaks in the vegetation over the past 20,000 years (Fig. 5.30).

During times of environmental change, populations of arboreal taxa are in disequilibrium, and the evidence presented in Chapter 5 indicates that their responses are individualistic in terms of changes in location and relative dominance of their populations. Whether these biotic responses are predictable is uncertain, and there is evidence from the fossil record of northwestern Europe (West 1970; Davis 1976b) that both the sequence of migration and changes in relative abundances of temperate tree taxa were different during each glacial–interglacial cycle. Forest-stand simulation models indicate that the responses of temperate tree taxa to climatic change are hysteretic; that is, with a given magnitude of temperature change, changes in community composition differ depending on whether climate is warming or cooling (Shugart *et al.* 1980). By measuring the rates of response and the quantitative shifts in dominance distributions across broad regions through the late Quaternary, we can see how the life-history strategies of tree taxa have influenced their overall patterns of response to environmental changes.

Ecological areography

On a broad geographic scale such as addressed with our Temperate-Zone Quaternary plant-ecological data set (Chapters 4 and 5), the behavior of tree taxa can be evaluated on a macro-scale intermediate between that studied in plant-successional time (micro-scale) and that of evolutionary time (mega-scale) (Delcourt *et al.* 1982). Among the processes relevant to the macro-scale of resolution are the migrations of individual taxa in response to broad environmental changes such as late-glacial and Holocene climatic change. We define **migration** as the continuing process through time involving both the dispersal of propagules and successful establishment of reproducing populations extending beyond the previous distributional limit of the taxon.

Time series of three-dimensional displays such as the paleo-dominance maps in Chapter 5 are appropriate as an aid in interpreting the migrational histories of the taxa through analysis of areographic dynamics, that is, evaluation of changes in the shape of perimeter, size in area, and morphology or shape of dominance structure across the overall population distribution (Rapoport 1982). We hypothesize that quantitative, four-dimensional changes in the distribution of taxa during migration events reflect ecophysiological and successional relationships, genetically controlled life-history characteristics, and the influence of physical factors such as migrational corridors or barriers. These map series of paleo-dominance distributions thus can be used to interpret the ecological properties of populations through time. These **migrational strategies** are analogous to life-history strategies in determining the roles of the taxa in plant succession. Migrational strategies, however, are an outcome of both life-history characteristics and the response of the taxa to differential possibilities afforded by changes in environment through space and time. For trees, the time scale of thousands of years is required for such analysis because it is the scale over which their invasions and major changes in population dominance have occurred.

Measurements of perimeter/$\sqrt{\text{area}}$

The ratio of the perimeter (p) of the continuous distributional limit of a taxon to the square root of the area of its distribution ($\sqrt{a}$) gives an index of the degree to which the environment acts as a filter or barrier to the spread of the taxon (Rapoport 1982). The relationship of $p/\sqrt{a}$ is a constant for geometric shapes irrespective of size; for example, the $p/\sqrt{a}$ of a square is always 4.0. The value of this ratio is smallest for a circle ($p/\sqrt{a} = 3.54$). The more convoluted the margins of the perimeter, the higher the value for $p/\sqrt{a}$. The perimeter of the distributional range of a taxon may be convoluted for any of a number of reasons. On a subcontinental map scale, physical barriers to migration such as high-elevation mountain ridges may result in an elongate margin of population distribution wrapping around the mountain range at low elevations. Great Lakes with complex shorelines may add convolutions to the range margins of certain taxa. On a finer map scale, presence of dendritic drainage patterns of rivers may result in highly convoluted range margins for riparian species. The mosaic of soil types across a region may also lead to irregularities in range boundaries for arboreal taxa.

To test Rapoport's (1982) hypothesis concerning the application of the p/√a̅ relationship in order to examine environmental or biological resistance during times of climatic change and species migrations, we measured the extent of contiguous outer perimeter of distribution as well as the total area of distribution within that perimeter at 20,000 yr B.P., 16,000 yr B.P., 12,000 yr B.P., 8000 yr B.P., 4000 yr B.P., and 500 yr B.P. (the same times used for the construction of paleo-dominance histograms in Chapter 5) for the calibrated tree taxa available across eastern North America (Table 6.1). For this analysis, we calculated perimeters and areas separately for northern and southern groups of *Pinus* as well as the two groups of northern Cupressaceae and of southern Cupressaceae/Taxodiaceae. We also calculated the perimeter and square root of the area occupied by all forest at these six times in order to compare trends in overall forest area with the individual taxa (Table 6.1). To our knowledge, this represents the first use of this areographic method involving a time series of quantitative paleoecological data.

Ecological areography of eastern North American tree taxa

In the late Quaternary, values for p/√a̅ for eastern North American tree taxa (Table 6.1) ranged from 3.7 (*Juglans* and *Larix* at 20,000 yr B.P.) to 15.8 (*Populus* at 4000 yr B.P.). These values are in the general range of those calculated by Rapoport (1982) for families of North American mammals and indicates that our measurements are realistic within the constraints of the map scale used.

A general trend toward an increase in the p/√a̅ of total forest occurred across eastern North America through the late Quaternary (Table 6.1). This trend was probably primarily a result of the progressive increase in suitable land mass because of climatic warming and retreat of the Laurentide Ice Sheet, in addition to development of major Great Lakes and increasing intricacy of shorelines depicted on the paleogeographic maps. The observed increase in p/√a̅ for all forests paralleled the increases in p/√a̅ for individual tree taxa. This correlation was statistically significant even for taxa such as *Populus* that increased in p/√a̅ from 20,000 yr B.P. to 4000 yr B.P. and that decreased thereafter. This trend was also reinforced by the increasing density of available paleoecological sites through time (Fig. 5.1), increasing the potential for spatial detail of the distributional margins portrayed on the maps. The areographic trends through time, therefore, for any given tree taxon cannot be evaluated independently of this overall context of the total forested landscape.

For any given time plane, however, the degree of convolution of the range margins of the taxa can be compared meaningfully with each other. At 20,000 yr B.P., for example, five taxa (*Acer*, *Juglans*, *Larix*, *Tilia*, and *Tsuga*) had p/√a̅ estimates ≤ 4.0, reflecting their known occurrences in the full-glacial interval in small, geographically isolated refugia (*Celtis* and *Salix* were absent in the reconstructions for this time). In contrast, two taxa had p/√a̅ relationships ≥ 6.0, approximating or exceeding that of the total forested area. When normalized against the p/√a̅ for the total forested area (ratio normalized to 1.00), these values were (1) northern Cupressaceae (1.00) and (2) *Picea* (1.08). These taxa were widespread throughout the full-glacial boreal forest region. Inter-

Table 6.1. Perimeter/square root of area relationships for tree taxa in eastern North America. For a square, $p/\sqrt{a} = 4.0$; for a circle, $p/\sqrt{a} = 3.54$ (the minimum value). The values underlined in the table represent taxon measures of perimeter/square root of area that are greater than corresponding measures for all forest

Age (in yr B.P.)	$p/\sqrt{a}$ (tree $(p/\sqrt{a})$/forest $(p/\sqrt{a}))$*					
	20,000	16,000	12,000	8000	4000	500
All Forest	6.0 (1.00)*	6.1 (1.00)	6.3 (1.00)	8.5 (1.00)	9.3 (1.00)	9.2 (1.00)
Abies	5.7 (0.95)	5.4 (0.89)	5.4 (0.86)	8.7 (1.02)	10.6 (1.14)	10.7 (1.16)
Acer	4.0 (0.67)	4.3 (0.70)	6.1 (0.97)	7.5 (0.88)	9.1 (0.98)	9.9 (1.08)
Betula	4.4 (0.73)	6.0 (0.98)	7.4 (1.17)	6.8 (0.80)	11.1 (1.19)	10.5 (1.14)
Carya	5.3 (0.88)	5.3 (0.87)	5.1 (0.81)	6.5 (0.76)	6.7 (0.72)	6.6 (0.72)
Celtis	—	4.3 (0.70)	5.0 (0.79)	5.0 (0.59)	4.9 (0.53)	5.2 (0.57)
Cupressaceae (Northern group)	6.0 (1.00)	5.7 (0.93)	11.8 (1.87)	7.2 (0.85)	10.4 (1.12)	11.4 (1.24)
Cupressaceae/Taxodiaceae (Southern group)	5.1 (0.85)	4.7 (0.77)	5.1 (0.81)	8.4 (0.99)	8.6 (0.92)	8.0 (0.87)
Fagus	4.6 (0.77)	5.0 (0.82)	5.9 (0.94)	7.3 (0.86)	9.4 (1.01)	8.1 (0.88)
Fraxinus	5.4 (0.90)	5.3 (0.87)	7.2 (1.14)	7.8 (0.92)	7.4 (0.80)	9.6 (1.04)
Juglans	3.7 (0.62)	—	6.7 (1.06)	7.6 (0.89)	7.6 (0.82)	7.8 (0.85)
Larix	3.7 (0.62)	4.2 (0.69)	8.1 (1.29)	11.4 (1.34)	9.7 (1.04)	9.4 (1.02)
Nyssa	4.1 (0.68)	5.5 (0.90)	5.5 (0.87)	6.4 (0.75)	6.3 (0.68)	6.9 (0.75)
Picea	6.5 (1.08)	6.6 (1.08)	6.6 (1.05)	9.9 (1.16)	11.4 (1.23)	11.2 (1.22)
Pinus (Northern group)	5.9 (0.98)	6.4 (1.05)	5.3 (0.84)	9.3 (1.09)	8.5 (0.91)	9.8 (1.07)
Pinus (Southern group)	5.1 (0.85)	5.6 (0.92)	5.5 (0.87)	6.5 (0.76)	8.8 (0.95)	8.2 (0.89)
Populus	4.3 (0.72)	6.6 (1.08)	9.6 (1.52)	11.4 (1.34)	15.8 (1.70)	9.7 (1.05)
Quercus	4.8 (0.80)	4.9 (0.80)	6.7 (1.06)	7.1 (0.84)	7.4 (0.80)	7.9 (0.86)
Salix	—	4.8 (0.79)	6.7 (1.06)	10.2 (1.20)	9.3 (1.00)	12.8 (1.39)
Tilia	3.8 (0.63)	5.3 (0.87)	7.0 (1.11)	7.8 (0.92)	8.8 (0.95)	7.1 (0.77)
Tsuga	3.9 (0.65)	3.7 (0.61)	6.2 (0.98)	6.5 (0.76)	11.1 (1.19)	9.8 (1.07)
Ulmus	4.8 (0.80)	4.6 (0.75)	7.0 (1.11)	6.9 (0.81)	8.4 (0.90)	8.8 (0.96)
Number of Taxa with $(p/\sqrt{a}) >$ forest $(p/\sqrt{a})$	1	3	11	6	8	11

*The value within parentheses is the taxon measure normalized to the forest measure.

mediate values relative to the value for total forest for the remainder of the taxa at 20,000 yr B.P. (Table 6.1) reflect their generally restricted occurrence within the latitudinal zone between about 30°N and 35°N.

The number of taxa for which the value of $p/\sqrt{a}$ exceeded that of the total forest increased from 1 at 20,000 yr B.P. to 3 at 16,000 yr B.P. and to 11 at 12,000 yr B.P. (Table 6.1). The timing of this increase corresponded to the interval of greatest environmental changes in the late Quaternary. The number of taxa with a ratio of $p/\sqrt{a}$ greater than that of the total forest decreased to 6 at 8000 yr B.P. and then increased to 8 at 4000 yr B.P., and to 11 at 500 yr B.P. (Table 6.1). The relatively large number of taxa with convoluted range margins during the Pleistocene–Holocene transition presumably reflected increased physical and biological resistance to their invasion during a time of active northward dispersal and establishment of both boreal and cool-temperate tree populations. The late-Holocene increase in number of taxa with convoluted range margins may reflect progressive adjustments in their ranges in response to intensified competition rather than prolonged environmental change.

Migrational strategies of tree taxa

Several considerations are relevant in evaluating the spread of a taxon during times of active migration. The vectors and rates of dispersal of propagules (seeds of gymnosperms and seeds or fruits of angiosperms) are often considered to limit the rate of advance of the leading edge of a population (Van der Pijl 1969; Bennett 1985). Wind-dispersal of winged seeds is usually considered to be more effective than mammal or bird dispersal of heavy nuts in terms of the average distance the propagule is carried, although cases of long-distance dispersal (*e.g.*, beech (*Fagus*) nuts by blue jays) are documented (Johnson and Adkisson 1985; Fenner 1985). Not only must the propagule be dispersed, but conditions where it falls must be favorable for germination and growth of the seedling. More importantly, a dispersal "event" must occur, particularly for plants with a dioecious breeding system in which, because male flowers are located on one plant and female flowers are on another, more than one seedling must germinate and survive in order to successfully establish a new, reproducing population of the species.

Changes in effectiveness of dispersal may occur through time because of loss of suitable animal vectors through extinction; this may have been a particularly important factor during the Pleistocene–Holocene transition because of the mass extinctions of megafauna (Martin and Klein 1984). In eastern North America, these animal species included the mastodon (*Mammut americanum*) and horse (*Equus conversidens*; Kurten and Anderson 1980), which may have been primary dispersers of propagules from fleshy-fruited or heavy-seeded trees, such as Osage orange (*Maclura pomifera*), Kentucky coffee tree (*Gymnocladus dioicus*), walnut (*Juglans nigra*), and butternut (*Juglans cinerea*) (Janzen and Martin 1982).

Together, dispersal and establishment may begin as a diffusion process with populations spreading outward from the leading edge of their distributions (Pielou 1979; Bennett 1985). The diffusion process involves the dispersal from areas of relatively high concentration toward peripheral regions of lower concentration (Rapoport 1982). Successful migration tends, however, to proceed in a way more analogous to osmosis than to diffusion (Rapoport 1982). Osmosis reflects the differential filter of the environment that limits the rate of dispersal and the availability of appropriate "target" sites for colonization. Populations tend to colonize favorable habitats that are patchily distributed on the landscape, such as along riverine corridors, rather than to spread uniformly across the terrain (Elton 1958). The degree of heterogeneity of the environment with respect to its suitability for colonization by a taxon has been termed "**environmental resistance**" by Rapoport (1982). The geographic distribution of both corridors and physical barriers in part determines the routes and rates of migration of a taxon. Inasmuch as their tolerance thresholds, adaptations to environmental conditions, and success rates for dispersal and establishment differ, the migrational responses of taxa to changes in climate and disturbance regimes are individualistic (Smith 1965). Establishment of a new population may also be slowed because of competition with another previously established species; this would be a form of biological "**inertia**" (Pearsall 1959) or "**biological resistance**" to invasion into an established plant community (Davis 1976b).

In previous studies of postglacial migrations of tree taxa (Davis 1976b, 1981a, 1983; Huntley and Birks 1983; Bennett 1985), attention has focused primarily on the leading, actively migrating edges of distribution. Additional important processes, including progressive fragmentation and local extinction of populations, occur on the retreating portions of the range margins. Changes in the overall shape, length of perimeter, and degree of convolution of the perimeter of a taxon's distributional range occur through time as portions of its range retract while other margins stabilize or expand beyond previous limits. Not all taxa have retreating range margins; some may simply expand to occupy increasing areas during intervals of favorable environmental conditions.

The rate of population increase after initial establishment may be slowed from its inherent or intrinsic rate either by unfavorable environmental conditions (including limiting factors imposed by climate, soil nutrients, soil moisture, or disturbance regime) or by competition (Diamond 1986; see discussion in Chapter 7). The shape of the initial "wave front" of colonization as well as the subsequent population increase may be interpreted in terms of the migrational "strategy" of the taxon.

The r-migration (r_m) strategist

The "r-strategist" has been described in the classical ecological literature (MacArthur and Wilson 1967; Harper 1977; Grime 1979; Rapoport 1982) as a relatively short-lived, early-successional taxon that is shade-intolerant and can germinate and grow in nutrient-poor soils. The r-strategist reaches reproductive maturity rapidly and invests relatively little of its energy in producing aboveground phytomass, but rather produces abundant propagules (seeds in gym-

CROSS-SECTION VIEW:

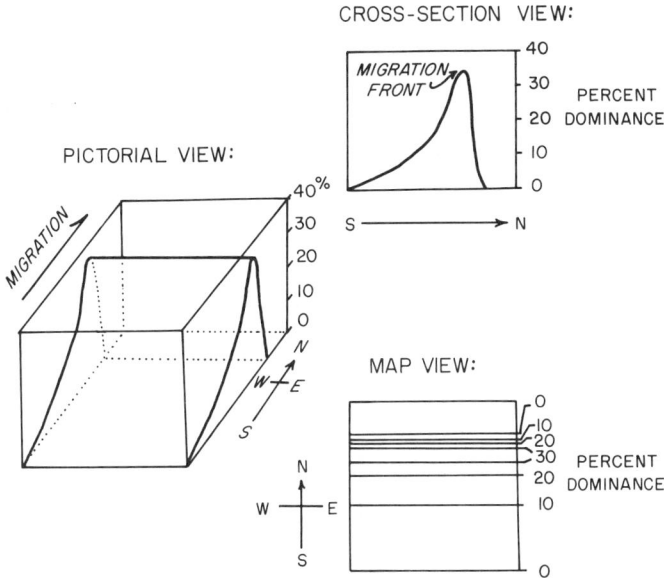

Figure 6.2. Hypothetical model of a population exhibiting an r-migration (r_m) strategy during a time of environmental change.

nosperms and seeds or fruits in angiosperms) that are dispersed widely, tending to saturate all available area.

We suggest that, during times of active migration, r-strategists would tend to exhibit characteristic map patterns (represented in three dimensions by the pictorial view, Fig. 6.2). For such r_m strategists, closely spaced contours of percent dominance, on a subcontinent scale occurring within as much as several hundred kilometers behind the advancing distributional limit, are interpretable as a steep and pronounced increase in population dominance along a migrational front. Contours would be more widely spaced with distance behind the front, diminishing with distance as the short-lived cohorts became locally extinct after the migration wave passed and were replaced by other, mid- to late-successional taxa. In cross-section view along a spatial transect, the migration front would appear as an asymmetrical curve, skewed toward the leading edge of migration (N in Fig. 5.2). Through time, the **"migration wave front"** should be traceable along the principal routes of migration, wherever competition or physical barriers did not limit the intrinsic rate of advance for the r_m strategist.

The K-migration (K_m) strategist

In contrast with the r-strategist, the "K-strategist" has been defined in successional time as a long-lived, late-successional taxon that tends to colonize and successfully maintain its populations on more nutrient-rich soils (MacArthur and Wilson 1967). The K-strategist is shade-tolerant, invests more energy in

CROSS-SECTION VIEW:

PICTORIAL VIEW:

PERCENT DOMINANCE

MIGRATION

MAP VIEW:

PERCENT DOMINANCE

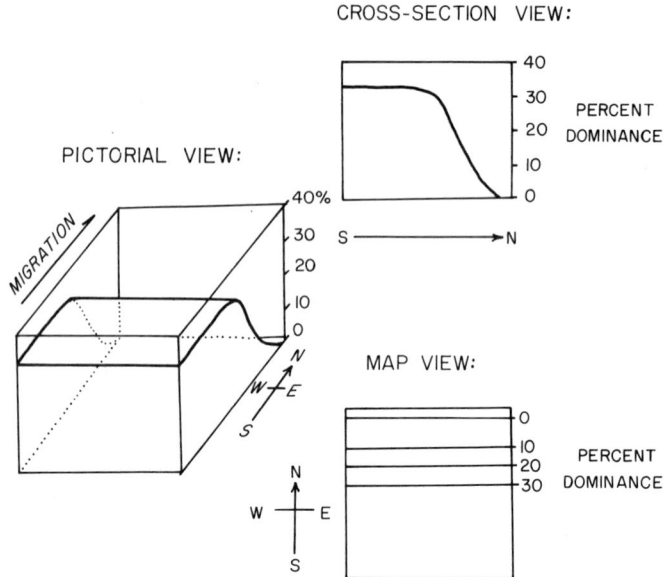

Figure 6.3. Hypothetical model of a population exhibiting a K-migration (K_m) strategy during a time of environmental change.

biomass, and produces fewer but larger propagules that can remain viable for longer periods of time in the soil seed bank (Fenner 1985). Its dispersal depends upon gravity or the activities of animal vectors such as birds or small mammals that may limit the rate of spread of the taxon (MacArthur and Wilson 1967; Harper 1977; Grime 1979; Rapoport 1982).

In the paleoecological record, we hypothesize that K-migration (K_m) strategists can be recognized in map or pictorial view (Fig. 6.3) by contours of percent dominance that are more widely spaced near the advancing range limit than for the r_m strategist. The sigmoid-shaped cross-section of dominance along spatial transects reflects the relatively slow initial rates of establishment and of population increase for taxa with this migrational strategy. When mapped on a subcontinent scale, the **migration plateau** reflects dominance values that tend to plateau many hundreds of kilometers behind the migrating edge of a late-successional tree taxon's range, however, as this taxon would tend to maintain high population levels as a forest dominant or subdominant (Fig. 6.3).

The fugitive-migration (f_m) strategist

A third possible pattern to be observed on the paleo-dominance maps is that reflecting the "fugitive"-migration (f_m) strategy. Many tree taxa may be rare, occurring only infrequently throughout their ranges; they may be locally important only in special habitats such as rock outcrops, mesic lower slopes, poorly drained alluvial or paludal habitats, or in temporally ephemeral landscape

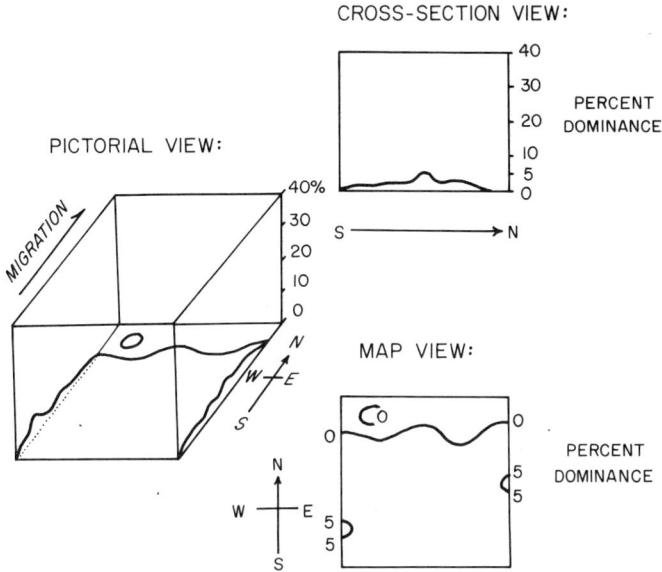

Figure 6.4. Hypothetical model of a population exhibiting a fugitive-migration (f_m) strategy during a time of environmental change.

patches disturbed by fire, debris avalanches, or wind throw (Harper 1977; Grime 1979; Rapoport 1982).

We hypothesize that presence of such taxa at any given time may be expressed on a broad-scale map (Fig. 6.4) either by an apparently generalized and uniform distribution with very low dominance values, or by a suite of geographically restricted populations that shift in their location through time. During times of environmental change, migration of such taxa would appear as a patchlike and ephemeral establishment of advance outlier populations "hopping" from one temporarily favorable "island" of habitat to another (Fig. 6.4). The fugitive strategy may be exhibited by late-successional taxa as well as by early- or mid-successional arboreal taxa.

Stress tactics

Rapoport (1982) has proposed an alternative to the notion that taxa exhibit rigid "strategies" in plant succession and species migration. Instead, the same taxon may behave differently in different portions of its range, exhibiting temporary and local "tactics" for its survival. Such behavior has been termed a **"stress response"** (Grime 1979) or an **"SOS tactic"** (Rapoport 1982). Stress may be placed on a population by unfavorable climatic conditions, geomorphic or other disturbances, competition from other species, or outbreaks of pathogen infestations. In map view, the results of such stresses would be reflected as departures from the hypothesized r_m, K_m, or f_m strategy models during times of active migration. Secondarily, stress may be reflected by a major decline in dominance within one or more population centers after migrational limits have been achieved.

An example of stress responses to environmental constraints occurs at tree line where *Abies* or *Picea* may persist for hundreds to thousands of years as stunted krummholz with minimal sexual reproduction. With climatic warming, the tree line may advance due to reinitiation of growth and production of pollen and seeds by formerly stressed individuals (Short and Nichols 1977). Stress response to biological factors has been observed in the case of the mid-Holocene decline in populations of hemlock (*Tsuga canadendis*) between 6000 yr B.P. and 4000 yr B.P. (Fig. 5.36a,h,i). This major, simultaneous decline in hemlock populations throughout much of its range in eastern North America has been attributed to an outbreak of a pathogen (Davis 1981b; Allison *et al.* 1986).

Late-Quaternary patterns and processes of migration

In the paleoecological record, where a number of the taxa mapped include several species or genera, differential responses in different portions of their distributional ranges can be expected and must be interpreted appropriately. In addition, because individualistic responses to environmental changes are to be expected, the mapped forms of the responses (Figs. 6.2–6.4) include a broad spectrum of possibilities that lie on a gradient for which the three hypothesized "strategies" or patterns of migration are nodes.

We used the time series of paleo-dominance maps for each of the 19 taxa (presented in Chapter 5) to prepare isochrone maps for both their northern and southern limits of continuous distribution. For each taxon, the position of the northern, continuous range limit (the 0% paleo-dominance contour) was delineated on an isochrone map at 2000-year intervals from 20,000 yr B.P. to 2000 yr B.P. and at 500 yr B.P., in order to map progressive shifts in the northern border of distribution. Comparable isochrone maps were drawn for the spatial displacements of each taxon's southern distributional limit during the past 20,000 years.

For both sets of isochrone maps, starting points for migrational Tracks 1 to 5 were located generally equidistant from each other, typically where the northern (or southern) 0% contour line for 20,000 yr B.P. intersected 75°W, 80°W, 85°W, 90°W, and 95°W. Equidistant points were identified on subsequent distribution limits (*e.g.*, 18,000 yr B.P.) and lines were drawn that connected these points along five separate tracks. The tracks thus documented representative migration paths for population advance (or retreat) across five geographic regions: (1) the continental interior, along 95°W; (2) the longitudinal axis of the Mississippi Alluvial Valley; (3) the migration route from the central Gulf Coastal Plain north to Hudson Bay; (4) the southwest–northeast trending axis of the Appalachian Mountains; and (5) the route traced along the Atlantic Coastal Plain into the maritime provinces of eastern Canada.

The distance between distributional limits was measured for successive (2000-year) time intervals along each migration track with an electronic digitizer (Model 1224, Numonics Corp., Lansdale, Pa.). Replicate measurements for known distances were accurate to ± 1%. The IBM-PC-compatible STAT-GRAPH software package (Version 1.1, STATGRAPHICS, Statistical Graphics Corporation 1985) was used on an IBM-AT microcomputer to calculate, statistically characterize, and graphically portray geographic and temporal patterns for the late-Quaternary rates of displacement of northern and southern borders of distribution for the 19 calibrated tree taxa.

Examples of r-migration (r_m) strategists

Aspen or Cottonwood (*Populus*)

Aspen (*Populus*; Fig. 5.28) exemplifies the ultimate r_m strategist (Fig. 6.2), termed an **"advanced colonizer"** by Davis and Jacobson (1985). From 20,000 yr B.P. to 14,000 yr B.P., the distribution of *Populus* reconstructed in eastern North America was confined to the southwestern quadrant of the study area, with the northern edge of distribution oriented northwest to southeast from approximately 38°N at 95°W to 32°N at about 84°W (Fig. 6.5a). The isochrone map for *Populus* (Fig. 6.5a) shows that northward advance of aspen began between 14,000 yr B.P. and 12,000 yr B.P. By 12,000 yr B.P., the leading edge of migration not only had spread into the Great Lakes region (Fig. 6.5a), but the centers of dominance of aspen (40% to 60% of forest composition) were located near the northern range margin (Fig. 5.28f). By 12,000 yr B.P., outlier populations of aspen had established in Maine in a region of predominantly tundra vegetation. This is an example of "jump dispersal" (Pielou 1979) as much as 700 km beyond the continuous range margin of *Populus*. The rate of advance of aspen was greatest during the time interval from 14,000 yr B.P. to 12,000 yr B.P., averaging 544 m/yr (Fig. 6.5b; Table 6.2).

After 12,000 yr B.P., aspen continued to advance northward, but at a slower rate averaging 212 m/yr from 12,000 yr B.P. to 10,000 yr B.P., reaching a relative minimum of 145 m/yr from 10,000 yr B.P. to 8000 yr B.P. During the early-Holocene interval, the rate of advance of aspen was largely limited by the rate at which the Laurentide Ice Sheet retreated. Aspen's population center remained immediately to the south of the leading edge of its distribution, but aspen populations rapidly became locally extinct to the south of about 40°N (*e.g.*, the *Populus* map at 10,000 yr B.P.; Fig. 5.28g).

The rate of advance of *Populus* increased to 295 m/yr from 8000 yr B.P. to 6000 yr B.P. (Fig. 6.5b; Table 6.2). During the mid-Holocene interval, aspen expanded into the area between 50°N and 55°N around Hudson Bay and across Labrador, following the rapidly diminishing remnants of the Laurentide Ice Sheet. The rate of advance then slowed to 185 m/yr between 6000 yr B.P. and 4000 yr B.P., as *Populus* reached its northernmost extent at the shores of Hudson Bay and in northern Labrador (Fig. 5.28j; Fig. 6.5b). After 4000 yr B.P., the northern range margin of aspen retracted rapidly southward at an average

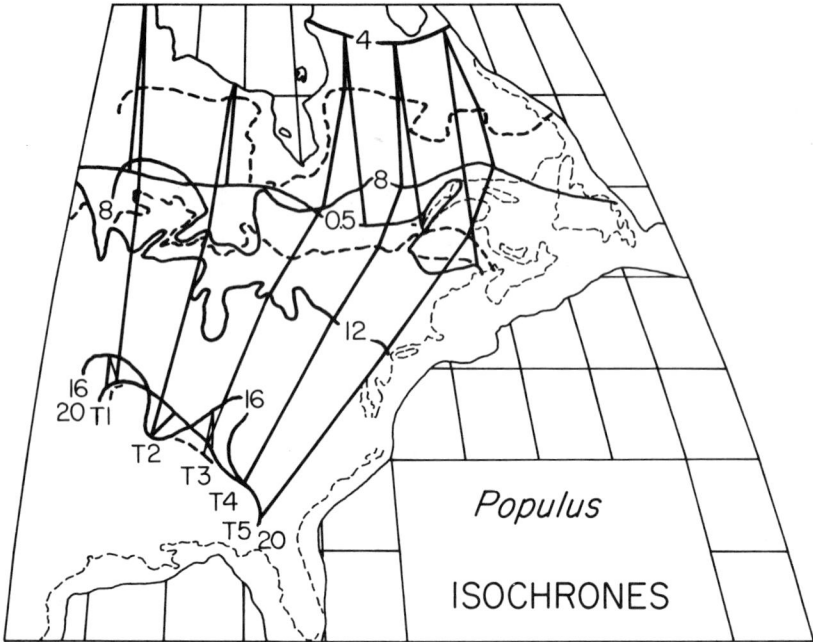

Figure 6.5. a. Isochrone map for *Populus*, showing its northern (leading) edge of migration (the 0% paleo-dominance contour) at 2000-year intervals from 20,000 yr B.P. to 2000 yr B.P. and at 500 yr B.P.

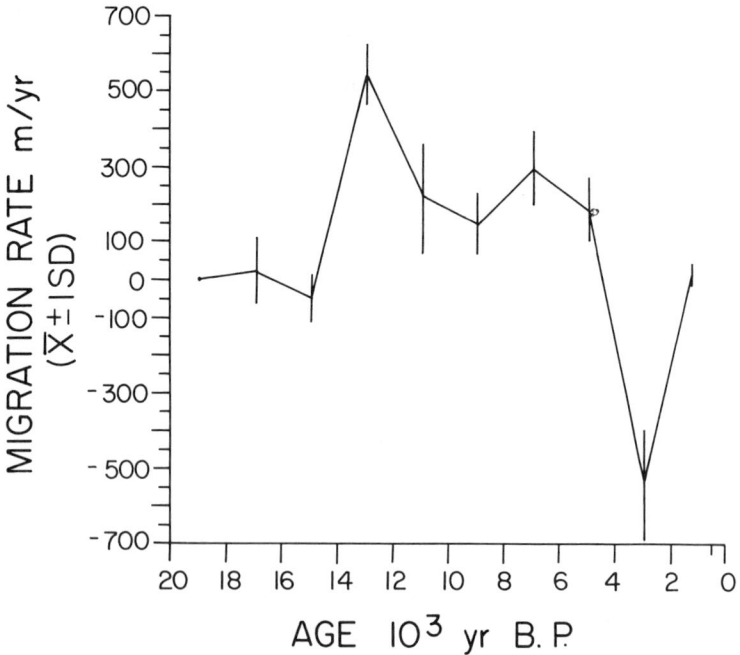

Figure 6.5. b. Rates of migration of *Populus* from 20,000 yr B.P. to 500 yr B.P., with mean ± 1 SD (vertical line) for migration rates averaged for all five tracks and calculated for successive (2000-year) time intervals.

Table 6.2. *Populus* migration rates (based on the northern edge of distribution)

Age (yr B.P.)	Time Span (yr)	Track 1		Track 2		Track 3		Track 4		Track 5		Mean Rate ($\overline{X} \pm 1$ SD) (m/yr)
		Distance (km)	Rate (m/yr)	Distance (km)	Rate (m/yr)	Distance (km)	Rate (m/yr)	Distance (km)	Rate (m/yr)	Distance (km)	Rate (m/yr)	
20,000	2000	0	0	0	0	0	0	0	0	0	0	0 ± 0
18,000	2000	187	94	-206	-103	240	120	0	0	0	0	22 ± 89
16,000	2000	-183	-92	0	0	-274	-137	0	0	0	0	-46 ± 65
14,000	2000	816	408	1141	571	1049	525	1160	580	1273	637	544 ± 86
12,000	2000	382	191	83	42	267	134	516	258	867	434	212 ± 147
10,000	2000	170	85	80	40	342	171	397	199	455	228	145 ± 79
8000	2000	462	231	828	414	753	377	541	271	360	180	295 ± 98
6000	2000	507	254	97	49	348	174	353	177	545	273	185 ± 88
4000	2000	-1028	-514	-634	-317	-1174	-587	-1153	-577	-1447	-724	-544 ± 148
2000	1500	0	0	0	0	0	0	100	67	0	0	13 ± 30
500												
Net Migration												
Distance		2347 km		2229 km		2759 km		2967 km		3500 km		
Interval (yr B.P.)		18,000 to 4000		14,000 to 4000		14,000 to 4000		14,000 to 4000		14,000 to 4000		
Total Time		14,000 yr		10,000 yr		10,000 yr		10,000 yr		10,000 yr		
Overall Rate		167 m/yr		223 m/yr		276 m/yr		297 m/yr		350 m/yr		263 m/yr
Minimum Rate		-514 m/yr		-317 m/yr		-587 m/yr		-577 m/yr		-724 m/yr		-724 m/yr
Maximum Rate		408 m/yr		571 m/yr		525 m/yr		580 m/yr		637 m/yr		637 m/yr

of 544 m/yr as aspen became restricted to the southern boreal forest region by 2000 yr B.P. The northern range limit of aspen generally stabilized between 2000 yr B.P. and 500 yr B.P.

For the times of active northward migration, mean rates of advance of aspen populations displayed differential geographic patterns. Along Track 1, through the continental interior, the mean rate of advance was 167 m/yr over the time interval from 18,000 yr B.P. to 4000 yr B.P. However, there was an increase in mean rate of migration from west to east, from 223 m/yr along Track 2 increasing to the maximum average value of 350 m/yr along Track 5 in the eastern coastal zone. Averaged for all tracks, aspen advanced northward at a mean rate of 263 m/yr (Table 6.2).

The data for *Populus* demonstrate that the late-glacial and mid-Holocene intervals were times when the taxon moved most rapidly across the subcontinent of eastern North America. Wind dispersal of its airborne propagules was apparently not a limiting factor to its rate of population spread, as distances of dispersal up to 637 m/yr were attained along the northern continuous margin of distribution. In map view (Figs. 5.28b to 5.28l), the migration of aspen approximated that depicted in the conceptual model for the r_m strategist (Fig. 6.2) through the time intervals from 12,000 yr B.P. to 10,000 yr B.P. and from 8000 yr B.P. to 6000 yr B.P. The loss of the migration wave front in the northwestern part of aspen's range between 10,000 yr B.P. and 8000 yr B.P. may have in part reflected biological stress imposed by the successful establishment of > 60% pine along the boreal forest–tundra ecotone. However, rates of advance were not uniform across the entire northern range margin from 12,000 yr B.P. to 4000 yr B.P., indicating that environmental conditions in part ultimately limited the expansion of *Populus* throughout much of the Holocene. As long as boreal forest actively invaded northward into terrain previously occupied by tundra, the aspen wave front represented the leading edge along that ecotone. However, as the tundra–forest ecotone stabilized and then shifted southward during the last 4000 years, the aspen wave front collapsed. Rather than opportunistically continuing to invade large, open areas, aspen's role in forest succession changed in the late Holocene to regeneration in smaller patches after disturbances such as fire within landscapes of predominantly closed boreal forest. Instead of a fundamental change in the biology of aspen, the change in its role through time was an outcome of new opportunities presented by the changing degree of openness of the landscape.

Spruce (*Picea*)

The northern (leading) edge of migration. At 20,000 yr B.P., the northern continuous limit for spruce extended to the southern margin of the Laurentide Ice Sheet in southern New England, wrapped southward around the alpine tundra zone of the central Appalachians, and extended west along 40°N in the Midwest to 45°N across the continental interior (Figs. 5.24b, 6.6a). From 20,000 yr B.P. to 14,000 yr B.P., the northern margin of spruce remained stationary east of the Appalachians but fluctuated by several hundred kilometers across the Great Lakes region in response to minor advances and retreats of glacial ice lobes (Fig. 6.6a).

Between 14,000 yr B.P. and 12,000 yr B.P., the mean rate of advance of spruce populations increased to 122 m/yr, averaged across all measured tracks (Table 6.3). Between 12,000 yr B.P. and 10,000 yr B.P., the advance of spruce diminished to 42 m/yr in the western Great Lakes region, but increased to 368 m/yr through the New England region. With major ice thinning and glacial retreat between 10,000 yr B.P. and 8000 yr B.P., spruce reached its highest mean rate of advance for all tracks, averaging 242 m/yr across central and eastern Canada. Spruce populations continued to advance across northern Canada during the mid-Holocene interval, with mean rates of 130 m/yr and 181 m/yr from 8000 yr B.P. to 6000 yr B.P. and from 6000 yr B.P. to 4000 yr B.P., respectively. During the last 4000 years, spruce advanced north of 60°N in the region west of Hudson Bay. However, east of Hudson Bay, the expansion of arctic tundra within northern Quebec and coastal regions of northern Labrador displaced the distributional limit of spruce southward at rates up to 174 m/yr (Fig. 6.6a; Table 6.3). The net migration rate of spruce averaged across all five measured migration tracks from 20,000 yr B.P. to the present was 141 m/yr. The overall average rate of migration was highest in the continental interior along Track 1 (146 m/yr) and along the easternmost Track 5 (221 m/yr). The overall slower rate of migration along Tracks 2 to 4 reflected the persistence of the Laurentide Ice Sheet as a physical barrier to migration of spruce throughout much of the late Quaternary.

From 20,000 yr B.P. to 14,000 yr B.P., *Picea* was dominant throughout the northern portion of the boreal forest region, reaching > 80% of the forest composition adjacent to the Laurentide Ice Sheet and along the tundra–forest ecotone (Fig. 5.24b–e). Sustained glacial retreat between 14,000 yr B.P. and 10,000 yr B.P. was followed by invasion of spruce populations across the freshly deglaciated terrain. With late-glacial and early-Holocene warming, progressive extinction of spruce populations occurred in the southern two-thirds of its range, sharpening the developing migrational wave front that was situated within 5° latitude of the glacial margin (Fig. 5.24e–g). The latitudinal band of the wave front of spruce migration was fragmented into isolated population centers between 10,000 yr B.P. and 8000 yr B.P. (Fig. 5.24g,h). By 6000 yr B.P., the wave front redeveloped between 50°N and 55°N as spruce invaded former tundra habitats during the mid-Holocene interval (Fig. 5.24i). The wave front extended northward to its interglacial limit by 4000 yr B.P. (Fig. 5.24j). With subsequent southward displacement of the northern limit of spruce, its pattern of dominance changed to an asymmetrical distribution, with highest values along its northern margin and a progressive decline in dominance southward (Fig. 5.24k,l).

Spruce thus exhibited map patterns interpretable as an r_m strategy of migration during the time intervals from about 12,000 yr B.P. to 10,000 yr B.P. and from 6000 yr B.P. to 4000 yr B.P. During these time intervals the average rate of migration of *Picea* was > 165 m/yr (Fig. 6.6b; Table 6.3). In the interval from

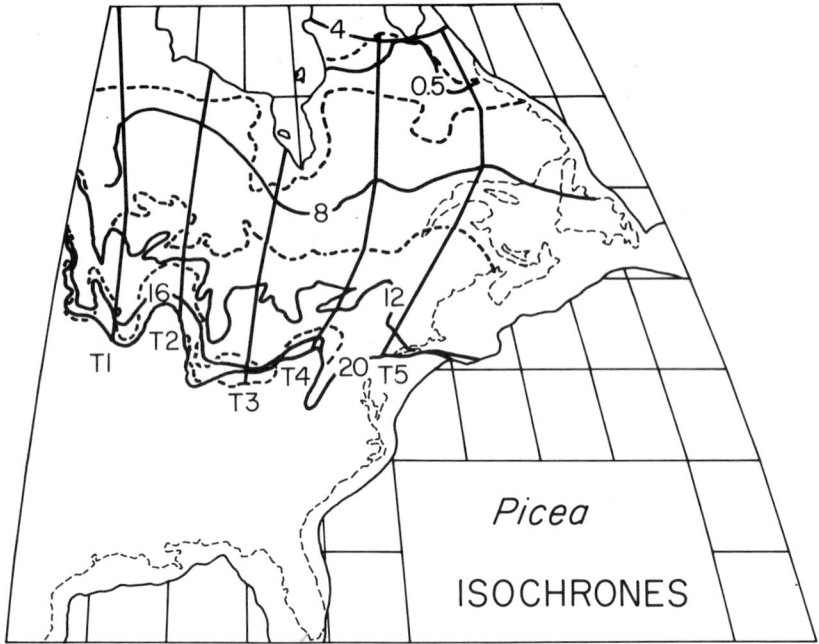

Figure 6.6. a. Isochrone map for *Picea*, showing its northern (leading) edge of migration (the 0% paleo-dominance contour) at 2000-year intervals from 20,000 yr B.P. to 2000 yr B.P. and at 500 yr B.P.

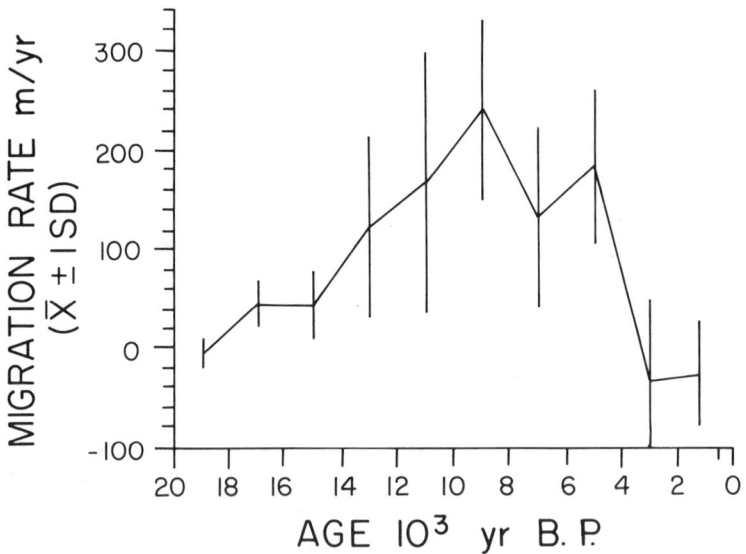

Figure 6.6. b. Rates of migration of *Picea* from 20,000 yr B.P. to 500 yr B.P., with mean ± 1 SD (vertical line) for migration rates averaged for all five tracks and calculated for successive (2000-year) time intervals.

Table 6.3. *Picea* migration rates (based on the northern edge of distribution)

Age (yr B.P.)	Time Span (yr)	Track 1 Distance (km)	Track 1 Rate (m/yr)	Track 2 Distance (km)	Track 2 Rate (m/yr)	Track 3 Distance (km)	Track 3 Rate (m/yr)	Track 4 Distance (km)	Track 4 Rate (m/yr)	Track 5 Distance (km)	Track 5 Rate (m/yr)	Mean Rate ($\overline{X} \pm 1$ SD) (m/yr)
20,000	2000	0	0	0	0	−68	−34	0	0	0	0	−7 ± 15
18,000	2000	126	63	125	63	87	44	75	38	0	0	42 ± 26
16,000	2000	88	44	198	99	46	23	92	46	0	0	42 ± 37
14,000	2000	404	202	49	25	465	233	132	66	166	83	122 ± 91
12,000	2000	157	79	84	42	269	135	406	203	735	368	165 ± 129
10,000	2000	560	280	751	376	272	136	418	209	418	209	242 ± 91
8000	2000	208	104	154	77	60	30	529	265	352	176	130 ± 92
6000	2000	496	248	185	93	259	130	330	165	539	270	181 ± 76
4000	2000	0	0	0	0	0	0	38	19	−348	−174	−31 ± 80
2000	2000	0	0	0	0	0	0	−182	−121	−29	−19	−28 ± 53
500	1500											
Net Migration												
Distance		2039 km		1546 km		1458 km		2020 km		2210 km		
Interval (yr B.P.)		18,000 to 4000		18,000 to 4000		18,000 to 4000		18,000 to 2000		14,000 to 4000		
Total Time		14,000 yr		14,000 yr		14,000 yr		16,000 yr		10,000 yr		
Overall Rate		146 m/yr		110 m/yr		104 m/yr		126 m/yr		221 m/yr		141 m/yr
Minimum Rate		0 m/yr		0 m/yr		−34 m/yr		−121 m/yr		−174 m/yr		−174 m/yr
Maximum Rate		280 m/yr		376 m/yr		233 m/yr		265 m/yr		368 m/yr		376 m/yr

10,000 yr B.P. to 6000 yr B.P., however, spruce populations became stressed throughout its distribution. During that time interval, its northward migration was stalled out against the physical barrier of the southern margin of the Laurentide Ice Sheet, and the centers of dominance for spruce diminished because of climatic warming that allowed replacement of spruce by temperate tree taxa. Widespread extinction of peripheral populations occurred along the western and southern limits because climatic warming there exceeded the tolerance limits of spruce.

The overall rate of late-glacial and Holocene migration of spruce (141 m/yr) was slower than that of aspen (263 m/yr). The invasion of spruce tracked after the initial colonization of deglaciated terrain by aspen. The wave front for spruce was less steep than that of aspen, and the latitudinal extent of *Picea*'s ridge of dominance behind the front was broader than that of *Populus*. Another difference in migrational pattern between these genera was that after the wave front of aspen passed through an area, its populations diminished to zero. In the case of spruce, after the passage of the wave front, its paleo-dominance values typically remained up to 40%. In the late-Holocene interval, aspen diminished in importance throughout the boreal forest region, whereas spruce remained an important dominant throughout much of central and eastern Canada.

The southern (receding) edge of migration. As *Picea* migrated northward during the late-glacial and early-Holocene intervals, its northern range limit advanced but its southern range limit receded. The local extinction of spruce populations near their southern distributional limits occurred generally after 14,000 yr B.P. in the western interior. The major pulses of change occurred between 12,000 yr B.P. and 8000 yr B.P. west of the Appalachian Mountains (average of 391 m/yr along Track 1, 283 m/yr along Track 2) and between 10,000 yr B.P. and 8000 yr B.P. along the Atlantic Seaboard (average of 312 m/yr along Track 5) (Fig. 5.24e–h; Fig. 6.7a,b; Table 6.4). Throughout the Holocene, spruce populations became progressively restricted to high elevations within the Appalachian Mountains; there, the continuous southern boundary shifted northward to Pennsylvania by the late Holocene (Fig. 6.7a). Within the last several thousand years, the southern range limit of spruce reexpanded southward in the western Great Lakes region and across southern New England.

The mean rate of retreat of the southern continuous range margin of spruce (averaged across all five tracks) was 144 m/yr, roughly comparable to the mean rate of advance of spruce of 141 m/yr. Although the northern boundary of spruce began to move northward 4000 years earlier than the first retreat of the southern boundary, the time of most rapid movement occurred synchronously on both northern and southern range limits, between 10,000 yr B.P. and 8000 yr B.P. The greatest mean rate of advance during that time interval was 242 m/yr (Fig. 6.6b; Table 6.3), and the most rapid retreat was 275 m/yr, averaged across all tracks (Fig. 6.7b; Table 6.4). The time lag observed for the final extinctions of *Picea* along the southern periphery of its full- and late-glacial range may be attributed to persistence of small populations locally around the margins of lakes and bogs because of microclimatic or microhabitat effects.

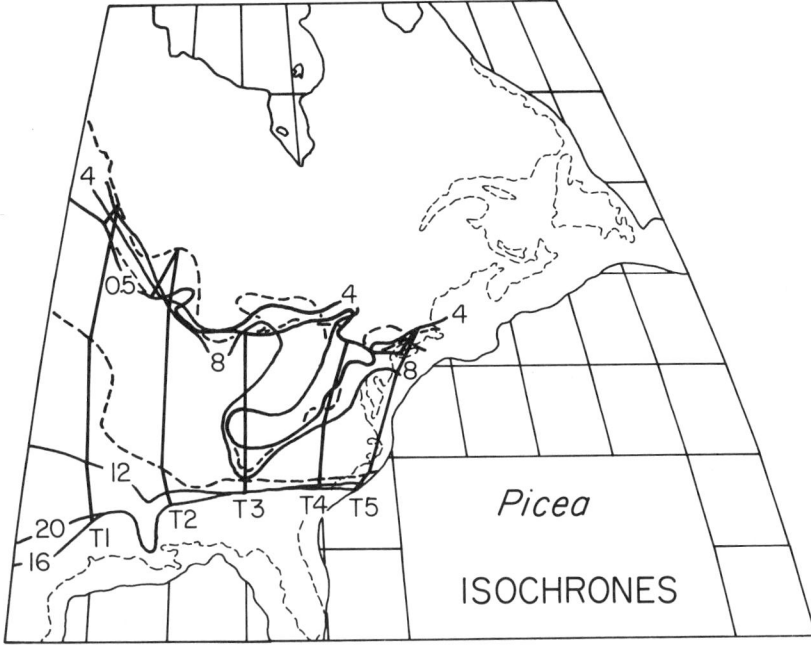

Figure 6.7. a. Isochrone map for *Picea*, showing its southern (retreating) edge of migration (the 0% paleo-dominance contour) at 2000-year intervals from 20,000 yr B.P. to 2000 yr B.P. and at 500 yr B.P.

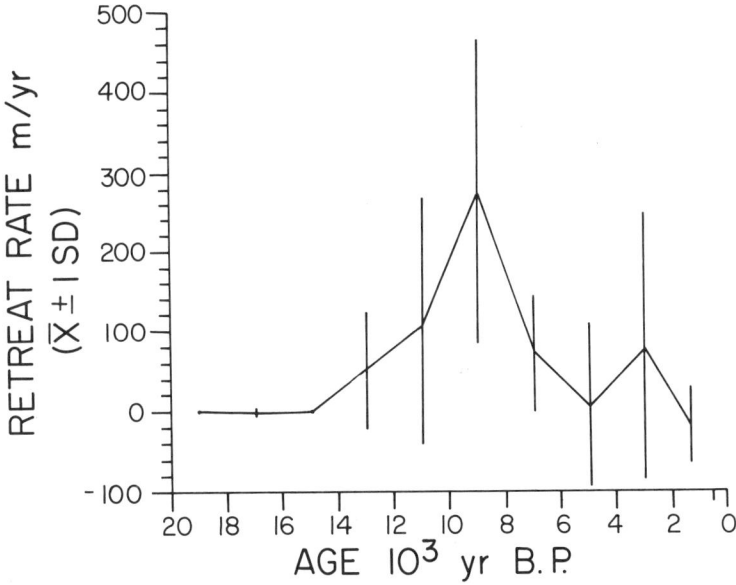

Figure 6.7. b. Rates of retreat of *Picea* from 20,000 yr B.P. to 500 yr B.P., with mean ± 1 SD (vertical line) for migration rates averaged for all five tracks and calculated for successive (2000-year) time intervals.

Table 6.4. *Picea* retreat rates (based on the southern edge of distribution)

Age (yr B.P.)	Time Span (yr)	Track 1		Track 2		Track 3		Track 4		Track 5		Mean Rate ($\bar{X} \pm 1$ SD) (m/yr)
		Distance (km)	Rate (m/yr)	Distance (km)	Rate (m/yr)	Distance (km)	Rate (m/yr)	Distance (km)	Rate (m/yr)	Distance (km)	Rate (m/yr)	
20,000	2000	0	0	0	0	0	0	0	0	0	0	0 ± 0
18,000	2000	-34	-17	0	0	0	0	0	0	0	0	-3 ± 8
16,000	2000	0	0	0	0	0	0	0	0	0	0	0 ± 0
14,000	2000	354	177	72	36	0	0	30	15	35	18	49 ± 73
12,000	2000	776	388	147	74	79	40	19	10	92	46	112 ± 156
10,000	2000	785	393	985	493	15	8	342	171	624	312	275 ± 190
8000	2000	44	22	367	184	19	10	95	48	190	95	72 ± 71
6000	2000	-78	-39	-294	-147	159	80	151	76	122	61	6 ± 99
4000	2000	-62	-31	-53	-27	696	348	283	142	-51	-26	81 ± 166
2000	2000	0	0	0	0	0	0	0	0	-154	-103	-21 ± 46
500	1500											
Net Migration												
Distance		1959 km		1571 km		968 km		920 km		1063 km		
Interval (yr B.P.)		14,000 to 6000		14,000 to 6000		12,000 to 2000		14,000 to 2000		14,000 to 4000		
Total Time		8000 yr		8000 yr		10,000 yr		12,000 yr		10,000 yr		
Overall Rate		245 m/yr		196 m/yr		97 m/yr		77 m/yr		106 m/yr		144 m/yr
Minimum Rate		-39 m/yr		-147 m/yr		0 m/yr		0 m/yr		-103 m/yr		-147 m/yr
Maximum Rate		393 m/yr		493 m/yr		348 m/yr		171 m/yr		312 m/yr		493 m/yr

Examples of K-migration (K_m) strategists

Oak (*Quercus*)

The northern (leading) edge of expansion. The northern continuous limit of oak occurred between 37°N and 38°N through full-glacial times (Fig. 6.8a). The northern distributional limit of oak remained stationary from 20,000 yr B.P. to 16,000 yr B.P. within the central Atlantic Seaboard and the central Appalachian Mountains. However, west of the mountains, oak advanced as much as 100 m/yr between 18,000 yr B.P. and 16,000 yr B.P. across the Lower Midwest. Averaged across all five tracks, the migration of oak displayed the greatest mean rates of 181 m/yr, 146 m/yr, and 172 m/yr for the intervals from 16,000 yr B.P. to 14,000 yr B.P., 14,000 yr B.P. to 12,000 yr B.P., and 12,000 yr B.P. to 10,000 yr B.P., respectively (Fig. 6.8b; Table 6.5). In the west, along Tracks 1 and 2, the most rapid migration rates for oak of 243 m/yr and 290 m/yr, respectively, were recorded from 16,000 yr B.P. to 14,000 yr B.P. In contrast, the most rapid rates of migration of oak along Tracks 3 and 4 were 250 m/yr and 235 m/yr, occurring in the interval from 14,000 yr B.P. to 12,000 yr B.P. In the east, along Track 5, the greatest rate of advance was 421 m/yr, occurring between 12,000 yr B.P. and 10,000 yr B.P. Thus, the response of oak to late-glacial climatic warming occurred earliest west of the Appalachians and was delayed until the early Holocene in the Atlantic Coastal Plain. The Appalachian Mountains served as a partial physical barrier, uncoupling the latitudinal advance of oak populations from 16,000 yr B.P. to 10,000 yr B.P.

Oak populations reached their northernmost continuous limits in eastern Canada and in the central Great Lakes region by 8000 yr B.P. and in the western Great Lakes region by 6000 yr B.P. During the last 6000 years, the northern range limit of oak oscillated about 47°N in the eastern portion of its range. During the late Holocene, oak retreated slightly (from 50°N to 49°N) in the western part of its range.

The map pattern of oak populations resembles the conceptual model of the K_m strategist (Fig. 6.3) through much of the late-Quaternary interval. Between 20,000 yr B.P. and 16,000 yr B.P., the region of > 20% oak dominance occurred at least 5° latitude south of the northern range margin at about 38°N. Between 16,000 yr B.P. and 12,000 yr B.P., oak populations advanced northward at a typical rate of about 164 m/yr and increased to > 40% south of 36°N. By 12,000 yr B.P., the northward extension of the range limit of oak was separated by as much as 10° latitude from the population center, the latitudinal band of > 40% oak between about 33°N and 36°N. By 8,000 yr B.P., oak generally reached its modern northern limit generally along 47°N. Between 35°N and 40°N 10,000 years ago, oak expanded to 20% to 40% of the forest composition, and its populations increased to 40% to > 80% in the Southeast between 25°N and 35°N. In the interval from 10,000 yr B.P. to 8000 yr B.P., oak consolidated its position as a key forest dominant, attaining > 40% of the overall composition in temperate forests ranging from 42°N to 25°N. Oak required 2000 years after reaching its northernmost interglacial limit before oak populations achieved effective canopy capture and thence became dominant across the entire deci-

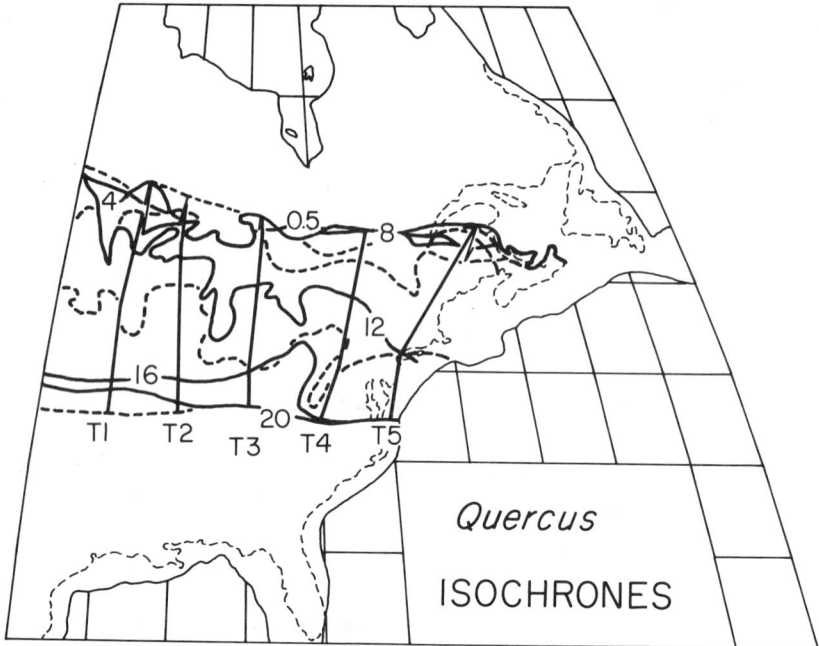

Figure 6.8. a. Isochrone map for *Quercus*, showing its northern (leading) edge of migration (the 0% paleo-dominance contour) at 2000-year intervals from 20,000 yr B.P. to 2000 yr B.P. and at 500 yr B.P.

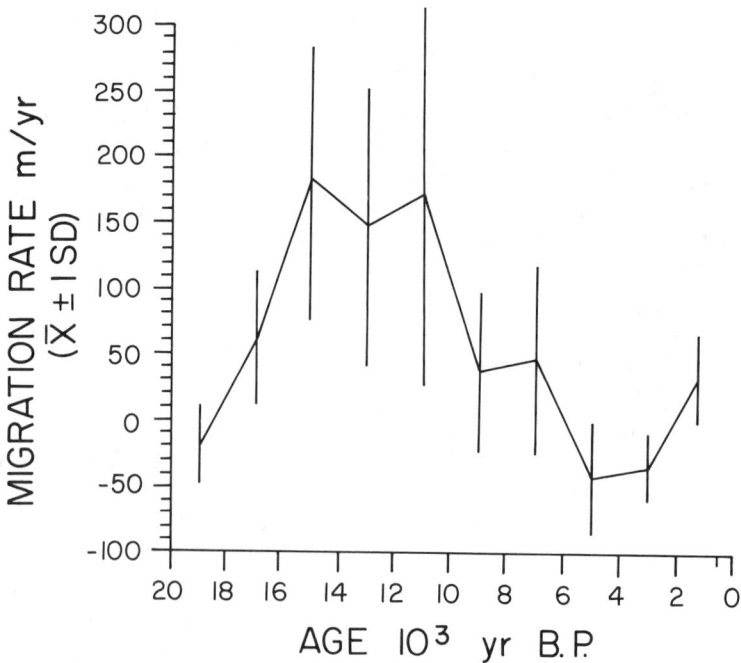

Figure 6.8. b. Rates of migration of *Quercus* from 20,000 yr B.P. to 500 yr B.P., with mean ± 1 SD (vertical line) for migration rates averaged for all five tracks and calculated for successive (2000-year) time intervals.

Table 6.5. *Quercus* migration rates (based on the northern edge of distribution)

Age (yr B.P.)	Time Span (yr)	Track 1 Distance (km)	Track 1 Rate (m/yr)	Track 2 Distance (km)	Track 2 Rate (m/yr)	Track 3 Distance (km)	Track 3 Rate (m/yr)	Track 4 Distance (km)	Track 4 Rate (m/yr)	Track 5 Distance (km)	Track 5 Rate (m/yr)	Mean Rate ($\overline{X} \pm 1$ SD) (m/yr)
20,000	2000	-133	-67	-64	-32	0	0	0	0	0	0	-20 ± 30
18,000	2000	186	93	200	100	208	104	23	12	0	0	62 ± 51
16,000	2000	485	243	579	290	42	21	301	151	401	201	181 ± 103
14,000	2000	323	162	170	85	499	250	469	235	0	0	146 ± 105
12,000	2000	246	123	98	49	303	152	228	114	842	421	172 ± 144
10,000	2000	-132	-66	114	57	117	59	193	97	63	32	36 ± 61
8000	2000	321	161	153	77	0	0	0	0	0	0	48 ± 72
6000	2000	-36	-18	-193	-97	-31	-16	0	0	-148	-74	-41 ± 42
4000	2000	-99	-50	16	8	-54	-27	-112	-56	-83	-42	-33 ± 26
2000	2000	0	0	41	27	0	0	99	66	103	69	32 ± 34
500	1500											

Net Migration	Track 1	Track 2	Track 3	Track 4	Track 5	Mean Rate
Distance	1429 km	1314 km	1169 km	1214 km	1306 km	
Interval (yr B.P.)	18,000 to 6000	18,000 to 6000	18,000 to 8000	18,000 to 8000	16,000 to 8000	
Total Time	12,000 yr	12,000 yr	10,000 yr	10,000 yr	8000 yr	
Overall Rate	119 m/yr	110 m/yr	117 m/yr	121 m/yr	163 m/yr	126 m/yr
Minimum Rate	-67 m/yr	-97 m/yr	-27 m/yr	-56 m/yr	-74 m/yr	-97 m/yr
Maximum Rate	243 m/yr	290 m/yr	250 m/yr	235 m/yr	421 m/yr	421 m/yr

duous forest region. During the early- and mid-Holocene intervals, oak populations were stressed along their western boundary as prairie invaded eastward. Oak did not display the K_m strategy after 6000 yr B.P.; however, late-Holocene replacement of oak by southern pines diminished oak's importance in forests across the southeastern coastal plains.

The southern (retreating) edge of distribution. The late-Quaternary displacement of *Quercus* along its continuous southern margin illustrates a pattern representative of many temperate forest taxa. Averaged across five tracks, the late-Quaternary shift in oak's southern border displayed a mean rate of 13 m/yr, an order of magnitude less than its typical advance of 126 m/yr on its northern boundary. From the late-Pleistocene through the mid-Holocene intervals, the southern and southeastern margins for oak were progressively curtailed by the landward shifting of the ocean shorelines (Fig. 5.30). This late-glacial and postglacial transgression of the sea across the coastal plain was a result of the rising sea levels produced by the return of glacial meltwater to the oceans. The early-Holocene rise in regional groundwater tables in karst terrain favored the relatively minor advance of forests across central and southern Florida between 12,000 yr B.P. and 10,000 yr B.P. (Fig. 5.30f,g). Oak's southwestern periphery retracted with the Holocene spread of prairies through the Southern High Plains of Texas and Oklahoma.

Hickory (*Carya*)

Map patterns for hickory indicate a K_m strategy through the late-glacial to mid-Holocene interval. The northward migration of hickory was initiated between 16,000 yr B.P. and 14,000 yr B.P. (Fig. 6.9a), advancing from 34°N to 37°N across the southern Appalachians Mountains. West of the Appalachians, the rates of advance of hickory were as high as 354 m/yr (Fig. 6.9b; Table 6.6) from 14,000 yr B.P. to 12,000 yr B.P. as it spread to 41°N. The northernmost interglacial limit of hickory west of the Appalachian Mountains was reached at 45°N by 8000 yr B.P. In contrast, east of the Appalachians in North Carolina (along Track 5) hickory populations advanced more slowly, at about 20 m/yr, from 16,000 yr B.P. to 12,000 yr B.P. Along the central Atlantic Seaboard from Virginia to Delaware, hickory advanced at about 93 m/yr from 12,000 yr B.P. to 8000 yr B.P., and it then dispersed across the southern New England region at 235 m/yr from 8000 yr B.P. to 6000 yr B.P. When averaged across all five tracks, hickory's average migration rate was 119 m/yr, comparable with the mean migration rate of 126 m/yr for oak. The fastest mean rates of migration for hickory, from 96 m/yr to 193 m/yr, occurred west of the Appalachians along Tracks 1, 2, and 3. However, hickory colonized new habitats more slowly along the Appalachian Mountains (mean rate of 76 m/yr along Track 4) as well as along the Atlantic Coastal Plain (74 m/yr along Track 5).

From 20,000 yr B.P. to 14,000 yr B.P., the mapped dominance pattern for *Carya* (Fig. 5.8b–e) shows the population center of $> 10\%$ located south of 33°N, with a relatively steep gradient to its northern distributional limit, located at 35°N. Between 14,000 yr B.P. and 12,000 yr B.P. (Fig. 5.8f), hickory pop-

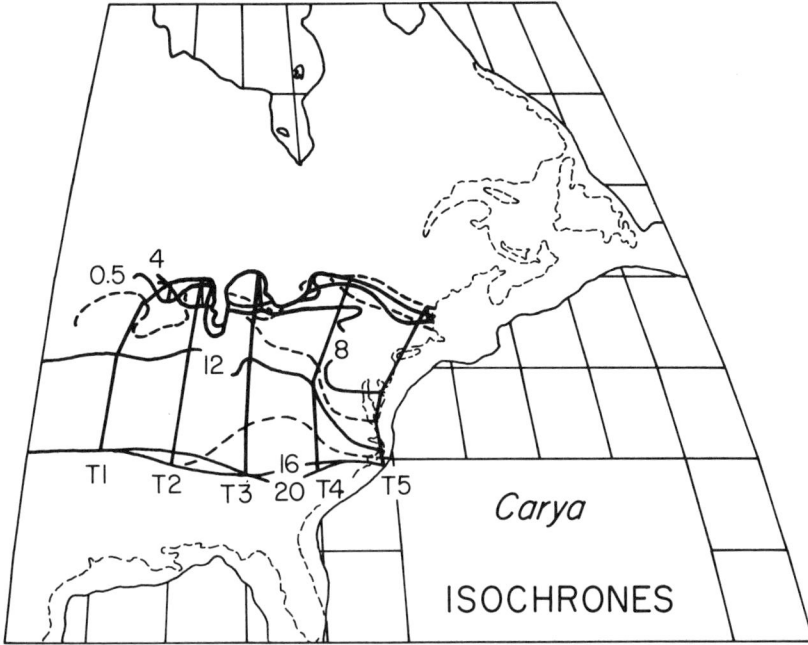

Figure 6.9. a. Isochrone map for *Carya*, showing its northern (leading) edge of migration (the 0% paleo-dominance contour) at 2000-year intervals from 20,000 yr B.P. to 2000 yr B.P. and at 500 yr B.P.

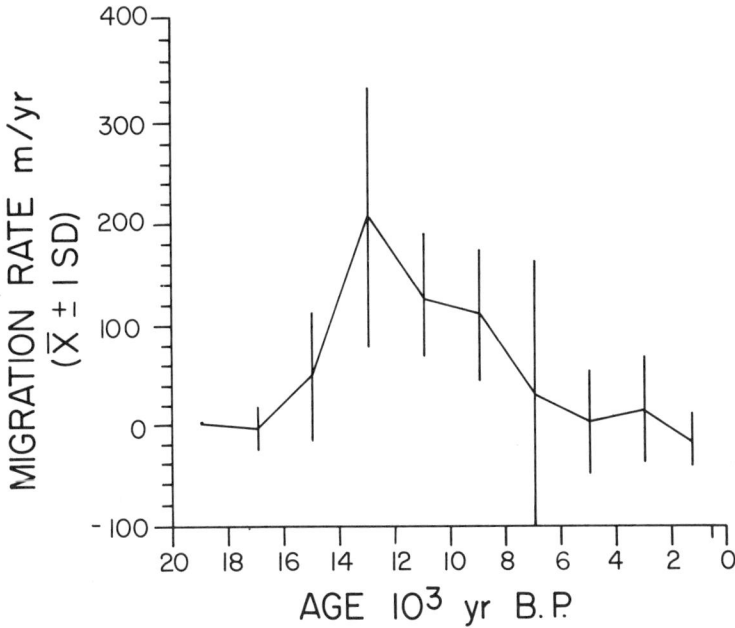

Figure 6.9. b. Rates of migration of *Carya* from 20,000 yr B.P. to 500 yr B.P., with mean ± 1 SD (vertical line) for migration rates averaged for all five tracks and calculated for successive (2000-year) time intervals.

Table 6.6. *Carya* migration rates (based on the northern edge of distribution)

Age (yr B.P.)	Time Span (yr)	Track 1 Distance (km)	Track 1 Rate (m/yr)	Track 2 Distance (km)	Track 2 Rate (m/yr)	Track 3 Distance (km)	Track 3 Rate (m/yr)	Track 4 Distance (km)	Track 4 Rate (m/yr)	Track 5 Distance (km)	Track 5 Rate (m/yr)	Mean Rate ($\overline{X} \pm 1$ SD) (m/yr)
20,000	2000	0	0	0	0	0	0	0	0	0	0	0 ± 0
18,000	2000	0	0	−79	−40	0	0	38	19	0	0	−4 ± 22
16,000	2000	0	0	0	0	299	150	163	82	39	20	50 ± 65
14,000	2000	604	302	707	354	402	201	321	161	42	21	208 ± 130
12,000	2000	354	177	406	203	228	114	114	57	185	93	129 ± 60
10,000	2000	198	99	39	20	299	150	378	189	188	94	110 ± 64
8000	2000	0	0	0	0	−272	−136	107	54	470	235	31 ± 134
6000	2000	0	0	−174	−87	79	40	78	39	59	30	4 ± 54
4000	2000	−114	−57	178	89	0	0	20	10	58	29	14 ± 53
2000	2000	0	0	0	0	0	0	−34	−23	−91	−61	−17 ± 27
500	1500											

	Track 1	Track 2	Track 3	Track 4	Track 5
Net Migration Distance	1156 km	1156 km	1228 km	1219 km	1041 km
Interval (yr B.P.)	14,000 to 8000	14,000 to 2000	16,000 to 8000	18,000 to 2000	16,000 to 2000
Total Time	6000 yr	12,000 yr	8000 yr	16,000 yr	14,000 yr
Overall Rate	193 m/yr	96 m/yr	154 m/yr	76 m/yr	74 m/yr
Minimum Rate	−57 m/yr	−87 m/yr	−136 m/yr	−23 m/yr	−61 m/yr
Maximum Rate	302 m/yr	354 m/yr	201 m/yr	189 m/yr	235 m/yr

ulations increased to $> 10\%$ in the region south of 35°N, and the northern limit of distribution extended to about 41°N west of 83°W. By 12,000 yr B.P., the sigmoid-shaped pattern of *Carya* dominance depicting a migrational front and plateau, consistent with the interpretation of a K_m strategy, was evident across the region of hickory distribution. A gentle slope from the 0% to 5% contours occurred on its northern leading edge of migration, followed by a steepened gradient with closely spaced contours of 5% to 20%, reaching a plateau of 25% across the southern Atlantic and Gulf Coastal Plains.

Between 12,000 yr B.P. and 10,000 yr B.P., hickory populations were split into two major centers, separated by the physical barrier of the Appalachian Mountain chain. The early-Holocene environments at middle and high elevations were characterized by intensified periglacial activity and geomorphic disturbance that maintained a mosaic of open ground and disturbance-favored, cold hardy tree taxa (Delcourt and Delcourt 1986; Larabee 1986). Consequently, under these relatively stressful conditions, temperate, late-successional populations of trees such as hickory were limited to lower elevations in surrounding areas.

At 10,000 yr B.P. (Fig. 5.8g), northernmost populations of hickory extended to about 45°N across the central and western Great Lakes region. The steep gradient of 5% to 20% dominance of hickory was located behind this leading edge of distribution between 38°N and 41°N, reaching a plateau of between 20% and 30% south of 38°N. Along the central and northern portion of the Atlantic Seaboard, the combination of rainshadow from the Appalachians and persistence of the cold Labrador Current (Delcourt and Delcourt 1984) provided cold and dry climatic conditions that limited the northward advance and increase in dominance of hickory. Along the Atlantic Coastal Plain, only 1° latitude separated the northern distributional limit of hickory from the steepened gradient of 5% to 20% and the plateau of $> 20\%$ located in South Carolina. Hickory populations declined below 20% in Georgia and across Florida.

With continued northward migration of hickory from 10,000 yr B.P. to 8000 yr B.P. (Fig. 5.8h), the K_m map pattern persisted both west and east of the Appalachian Mountains. In the last 6000 years, however, populations of hickory have diminished on the coastal plains, but prominent population centers persisted in the deciduous forest region from the central Appalachian Mountains west to the Ozark Plateaus (Fig. 5.8i–l).

Examples of fugitive-migration (f_m) strategists

Tamarack (*Larix*)

The genus *Larix*, today represented by only one species in eastern North America, *Larix laricina*, exhibited late-Quaternary migrational dynamics characteristic of a fugitive taxon (Fig. 6.4). Reconstructed populations of tamarack during the full-glacial and late-glacial intervals were located in the Lower Mississippi Alluvial Valley from 20,000 yr B.P. to 14,000 yr B.P. (Fig. 5.20b–e). Between 16,000 yr B.P. and 14,000 yr B.P., tamarack invaded deglaciated terrains from approximately 80°W to at least 95°W. Locally, tamarack populations were up to 11% of the total forest dominance at 16,000 yr B.P.

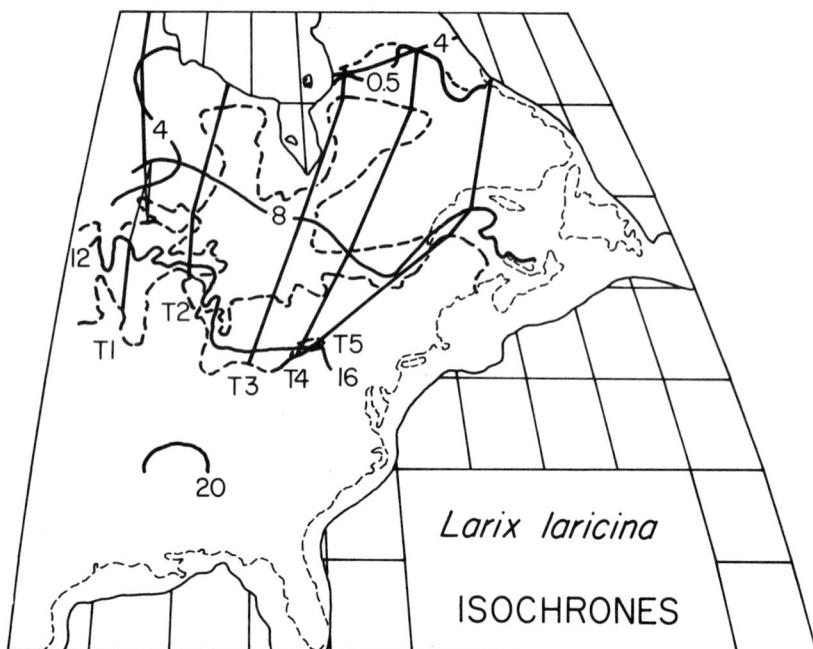

Figure 6.10. a. Isochrone map for *Larix laricina*, showing its northern (leading) edge of migration (the 0% paleo-dominance contour) at 2000-year intervals from 20,000 yr B.P. to 2000 yr B.P. and at 500 yr B.P.

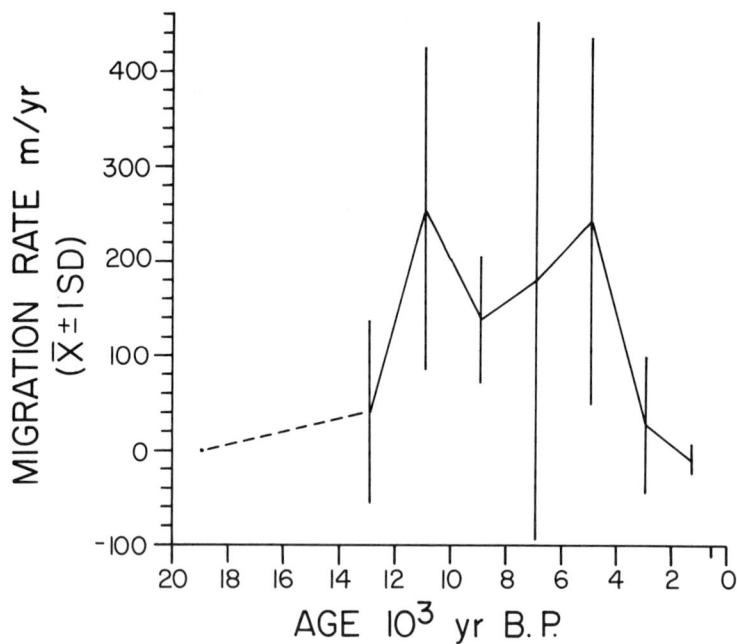

Figure 6.10. b. Rates of migration of *Larix laricina* from 20,000 yr B.P. to 500 yr B.P., with mean ± 1 SD (vertical line) for migration rates averaged for all five tracks and calculated for successive (2000-year) time intervals.

Table 6.7. *Larix* migration rates (based on the northern edge of distribution)

Age (yr B.P.)	Time Span (yr)	Track 1 Distance (km)	Track 1 Rate (m/yr)	Track 2 Distance (km)	Track 2 Rate (m/yr)	Track 3 Distance (km)	Track 3 Rate (m/yr)	Track 4 Distance (km)	Track 4 Rate (m/yr)	Track 5 Distance (km)	Track 5 Rate (m/yr)	Mean Rate ($\overline{X} \pm 1$ SD) (m/yr)
20,000*	2000	0	0	0	0	0	0	0	0	0	0	0 ± 0
18,000*	2000	—	—	—	—	—	—	—	—	—	—	0 ± 0
16,000*	2000	—	—	—	—	—	—	—	—	—	—	0 ± 0
14,000*	2000	408	204	46	23	86	43	−36	−18	−91	−46	41 ± 97
12,000	2000	279	140	311	156	365	183	504	252	1101	551	256 ± 170
10,000	2000	240	120	267	134	501	251	154	77	211	106	138 ± 67
8000	2000	−360	−180	416	208	789	395	954	477	0	0	180 ± 273
6000	2000	969	485	152	76	121	61	364	182	826	413	243 ± 195
4000	2000	301	151	0	0	49	25	0	0	−71	−36	28 ± 72
2000	2000	0	0	0	0	−53	−35	0	0	0	0	−7 ± 16
500	1500											

Net Migration		Track 1		Track 2		Track 3		Track 4		Track 5		
Distance		1837 km		1192 km		1911 km		1976 km		2138 km		
Interval (yr B.P.)		14,000 to 2000		14,000 to 4000		14,000 to 2000		12,000 to 4000		12,000 to 4000		
Total Time		12,000 yr		10,000 yr		12,000 yr		8000 yr		8000 yr		
Overall Rate		153 m/yr		119 m/yr		159 m/yr		247 m/yr		267 m/yr		
Minimum Rate		−180 m/yr		0 m/yr		−35 m/yr		−18 m/yr		−46 m/yr		
Maximum Rate		485 m/yr		208 m/yr		395 m/yr		477 m/yr		551 m/yr		

Larix populations were reconstructed in western Tennessee for 20,000 yr B.P. and 18,000 yr B.P., but no populations are documented from the Lower Mississippi Valley for 16,000 yr B.P. Migration rates are quantified for the interval from 14,000 yr B.P. to 500 yr B.P.

From 14,000 yr B.P. to 12,000 yr B.P., tamarack invaded northward preferentially west of the Appalachian Mountains, advancing by as much as 204 m/yr along Track 1 (Figs. 6.10a,b; Table 6.7). Between 12,000 yr B.P. and 10,000 yr B.P., colonies of tamarack extended north and northeastward at a mean rate of 256 m/yr, averaged across all tracks. During this early-Holocene interval, tamarack spread from the eastern Great Lakes across the northern Appalachians and through the northern New England region. This early-Holocene rate of advance averaged 551 m/yr along Track 5, the highest mean rate of spread for any 2000-year interval in the late-Quaternary history of *Larix*.

In the interval from 10,000 yr B.P. to 8000 yr B.P., tamarack typically occupied the latitudinal band of boreal forest within 5° of the southern Laurentide Ice Sheet margin (Fig. 5.20f–h). Between 8000 yr B.P. and 6000 yr B.P. (Fig. 5.20i), tamarack invaded northward at rates up to 477 m/yr along poorly drained sites along the retreating ice sheet to the vicinity of Hudson Bay (see Tracks 3 and 4; Fig. 6.10a). Rather than expanding as a wave front behind the leading edge of migration, or increasing to a plateau of relatively high values far south of the leading edge, however, populations of tamarack were generally < 5% throughout its widespread Holocene distributional range. Tamarack became abundant only locally within specialized, poorly drained, generally acidic habitats including riverine swamps, muskegs, and bogs, reaching values in such sites as high as 55%, *e.g.*, along Tamarack Creek in southwestern Wisconsin at 500 yr B.P. (Fig. 5.20l).

Walnut (*Juglans*)

The genus *Juglans* has two representatives native to eastern North America, butternut (*J. cinerea*) and black walnut (*J. nigra*). Both species characteristically are found as infrequent, overstory trees in rich, lowland woods and along mesic, well-drained stream terraces (Fernald 1970). During the late Quaternary, *Juglans* exhibited the fugitive strategy of migration (Fig. 6.4).

From 20,000 yr B.P. to 14,000 yr B.P., populations of walnut have been reconstructed for only a few localities, for example, in north-central Louisiana and southwestern Tennessee (Fig. 5.18b,c; Fig. 6.11a). During the full-glacial interval, walnut was present in those refugia at < 2% of the total forest composition. At Nonconnah Creek, Tennessee (Delcourt *et al.* 1980), full-glacial populations of black walnut are represented by both pollen grains and a nutshell that was dated at 17,200 yr B.P.

Between 14,000 yr B.P. and 12,000 yr B.P., walnut dispersed northeastward, establishing populations in geographically separated areas extending from west-central Mississippi to West Virginia (Fig. 6.11a). Between 12,000 yr B.P. and 10,000 yr B.P., walnut became widely distributed, invading suitable habitats extending from the northern Appalachian Mountains west across the Great Lakes region to the prairie–forest border. During this 2000-yr interval, the mean rate of spread for walnut was 390 m/yr, averaged across all five measured tracks (Fig. 6.11a,b; Table 6.8). The migration rates were greatest in the west, reaching 620 m/yr along Track 1 and 630 m/yr along Track 2. Intermediate migration rates were recorded through the central Great Lakes region, 341 m/yr along

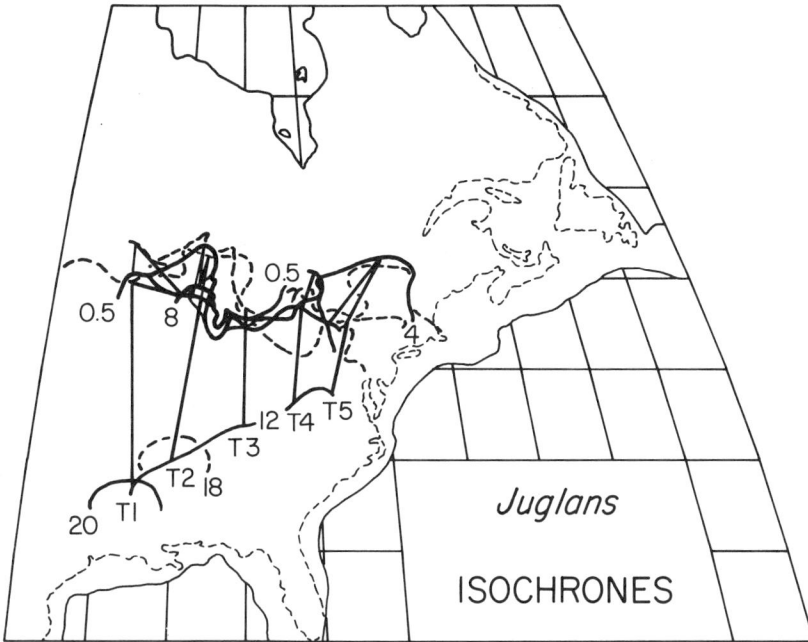

Figure 6.11. a. Isochrone map for *Juglans*, showing its northern (leading) edge of migration (the 0% paleo-dominance contour) at 2000-year intervals from 20,000 yr B.P. to 2000 yr B.P. and at 500 yr B.P.

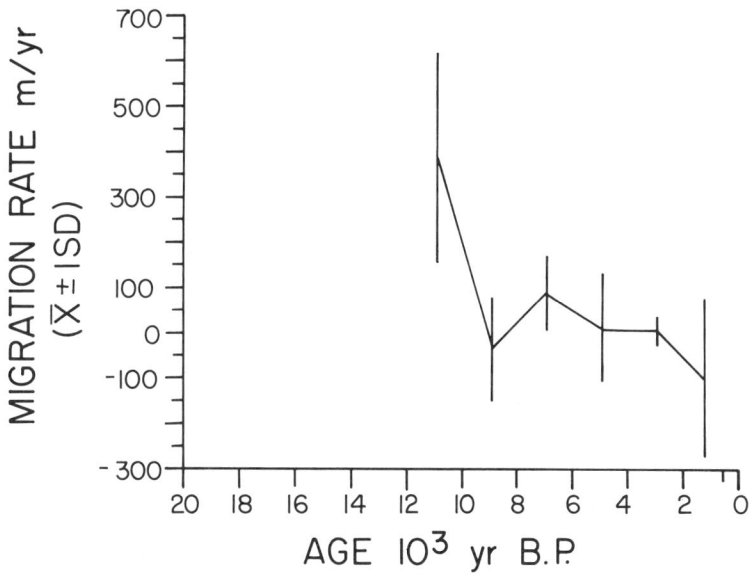

Figure 6.11. b. Rates of migration of *Juglans* from 20,000 yr B.P. to 500 yr B.P., with mean ± 1 SD (vertical line) for migration rates averaged for all five tracks and calculated for successive (2000-year) time intervals.

Table 6.8. *Juglans* migration rates (based on the northern edge of distribution)

Age (yr B.P.)	Time Span (yr)	Track 1 Distance (km)	Track 1 Rate (m/yr)	Track 2 Distance (km)	Track 2 Rate (m/yr)	Track 3 Distance (km)	Track 3 Rate (m/yr)	Track 4 Distance (km)	Track 4 Rate (m/yr)	Track 5 Distance (km)	Track 5 Rate (m/yr)	Mean Rate ($\overline{X} \pm 1$ SD) (m/yr)
20,000*	2000	—	—	—	—	—	—	—	—	—	—	—
18,000*	2000	—	—	—	—	—	—	—	—	—	—	—
16,000*	2000	—	—	—	—	—	—	—	—	—	—	—
14,000*	2000	—	—	—	—	—	—	—	—	—	—	—
12,000*	2000	—	—	—	—	—	—	—	—	—	—	—
10,000	2000	1240	620	1265	633	681	341	282	141	426	213	390 ± 228
8000	2000	−321	−161	−190	−95	−76	−38	293	147	−41	−21	−34 ± 115
6000	2000	418	209	252	126	0	0	139	70	100	50	91 ± 80
4000	2000	−219	−110	0	0	0	0	−91	−46	428	214	12 ± 122
2000	2000	43	22	0	0	102	51	0	0	−67	−34	8 ± 31
500	1500	−44	−29	−322	−215	−53	−35	177	118	−489	−326	−97 ± 174
Net Migration Distance		1337 km		1327 km		707 km		800 km		913 km		
Interval (yr B.P.)		12,000 to 6000		12,000 to 6000		12,000 to 2000		12,000 to 500		12,000 to 4000		
Total Time		6000 yr		6000 yr		10,000 yr		11,500 yr		8000 yr		
Overall Rate		223 m/yr		221 m/yr		71 m/yr		70 m/yr		114 m/yr		
Minimum Rate		−161 m/yr		−215 m/yr		−38 m/yr		−46 m/yr		−326 m/yr		
Maximum Rate		620 m/yr		633 m/yr		341 m/yr		147 m/yr		214 m/yr		

*Migration rates are quantified for the interval from 12,000 yr B.P. to 500 yr B.P.

Track 3. The migration of walnut between 12,000 yr B.P. and 10,000 yr B.P. was slowest along Tracks 4 and 5, at 141 m/yr and 213 m/yr, respectively. Through the remaining 10,000 years of the Holocene, the northern continuous limit for walnut's distribution has oscillated across several hundred kilometers distance (Fig. 6.11a).

During the early-Holocene migrational phase, no wave front or plateau in dominance was observed that would indicate an r_m or K_m strategy. During the Holocene, the geographic center for *Juglans* distribution has remained within the region of the Great Lakes and Lower Midwest. The outer distributional limits for walnut, however, have shifted sporadically along all margins, often with ephemeral outlier populations established as much as 200 km beyond the continuous distributional limits. In the Holocene, the overall distribution of walnut has been characterized by widespread occurrence throughout the deciduous forest region, but as a generally infrequent tree with very low dominance values. Throughout the past 20,000 years, walnut has been a fugitive taxon, with the pattern of northward spread following climatic amelioration occurring as a patchwork.

Geographic and temporal variation in late-Quaternary migration rates

Mean migration rates for 20 taxa for which information could be summarized in this study (Table 6.9) ranged from 45 m/yr to 287 m/yr. Previously published estimates of migration rates of comparable magnitude have been calculated based upon pollen data (Davis 1976b, 1981a; Davis and Jacobson 1985) rather than reconstructed forest composition. The key factors that have influenced the mean migration rates for the major tree taxa of eastern North America are (1) the primary vector (*e.g.*, wind, water, bird, mammal, or gravity) and rate of dispersal of propagules; (2) the environmental tolerances of the tree taxa; (3) the extent of habitat specificity or requirements of the taxa; and (4) the response of tree populations to prevailing disturbance regimes. In the context of the overall migration of a taxon, the timing of the fastest migration depends not only on the location and dominance level of the populations at the start of migration, but upon the limitations imposed by environmental and/or biological resistance encountered along the path of migration.

Mechanisms of dispersal include several different vectors for many of the tree taxa. Wind and water dispersal tend to be most effective for those taxa with winged seeds (*e.g.*, *Picea* and *Pinus*) or with buoyant, downy seed coats (*e.g.*, *Salix*). For the 15 taxa for which wind and water represent primary dispersal vectors, mean migration rates encompassed the entire observed range of values (Table 6.9). Bird dispersal of propagules was important for six taxa which have fruits that are berry-like, drupes, or nuts (*e.g.*, *Fagus, Juniperus, Nyssa,* and *Tilia*). These taxa had mean migration rates ranging from 45 m/yr to 209 m/yr. Mammals and gravity were also partly responsible for the transport of heavy nuts or large fruits characteristic of five of the taxa (including *Quercus, Carya, Juglans, Fagus,* and *Tilia*); and those taxa typically displayed intermediate rates of spread between 119 m/yr and 209 m/yr.

Table 6.9. Migration rates (expressed in m/yr) for the leading (northern) edge of distribution. Because of the "jump dispersal" character of *Celtis*, its migration rates are not included

Taxon	Dispersal Mode*	Overall Rate Averaged For 5 Tracks (m/yr)	Minimum Overall Track Rate in m/yr (Trk #)	Maximum Overall Track Rate in m/yr (Trk #)	Maximum Individual Rate on any track (m/yr)	Overall Migration Interval (yr B.P.)	Time Interval (yr B.P.) with Fastest Mean Rate Averaged for all tracks	Corresponding Mean Rate in m/yr ($\bar{X} \pm 1$ SD)
Salix	Wi, Wa*	287	182 (1)	416 (5)	726	16,000 to 10,000	16,000 to 14,000	230 ± 285
Populus	Wi, Wa	263	167 (1)	350 (5)	637	18,000 to 4000	14,000 to 12,000	544 ± 86
Betula	Wi, Wa	212	195 (4)	232 (5)	639	16,000 to 4000	16,000 to 14,000	244 ± 244
Tilia	B, M, G	209	116 (3)	357 (5)	753	14,000 to 500	14,000 to 12,000	494 ± 294
Tsuga	Wi	202	113 (3)	278 (5)	1044	16,000 to 500	16,000 to 14,000	496 ± 313
Larix	Wi, Wa	189	119 (2)	267 (5)	551	14,000 to 2000	12,000 to 10,000	256 ± 170
Fagus	B, M, G	169	95 (1)	214 (5)	581	16,000 to 2000	12,000 to 10,000	257 ± 297
Abies	Wi	159	110 (1)	235 (5)	401	20,000 to 4000	8000 to 6000	273 ± 108
Picea	Wi	141	104 (3)	221 (5)	376	18,000 to 2000	10,000 to 8000	242 ± 91
Juglans	M, G	140	70 (4)	223 (1)	633	12,000 to 500	12,000 to 10,000	390 ± 228
Northern Cupressaceae (*Thuja & Juniperus*)	Wi, Wa, B	138	101 (5)	205 (4)	428	18,000 to 2000	14,000 to 12,000	213 ± 92
Pinus (Northern group)	Wi	135	105 (4)	170 (5)	613	18,000 to 2000	12,000 to 10,000	330 ± 212
Ulmus	Wi, Wa	134	123 (1 & 2)	162 (3)	462	18,000 to 2000	14,000 to 12,000	251 ± 148
Acer	Wi, Wa	126	80 (2)	172 (5)	405	19,000 to 500	14,000 to 12,000	225 ± 121
Quercus	B, M, G	126	110 (2)	163 (5)	421	18,000 to 6000	16,000 to 14,000	181 ± 103
Fraxinus	Wi, Wa	123	94 (1)	177 (4)	487	20,000 to 2000	14,000 to 12,000	209 ± 115
Carya	M, G	119	74 (5)	193 (1)	354	18,000 to 2000	14,000 to 12,000	208 ± 130
Pinus (Southern group)	Wi	81	22 (3)	174 (4)	423	14,000 to 500	10,000 to 8000	151 ± 243
Nyssa	Wa, B	70	42 (2)	115 (5)	338	20,000 to 2000	10,000 to 8000	152 ± 183
Southern Cupressaceae & Taxodiaceae (*Chamaecyparis, Juniperus, & Taxodium*)	Wi, Wa, B	45	15 (3)	87 (5)	424	18,000 to 500	2000 to 500	144 ± 174

*Dispersal vectors: Wi, Wind; Wa, Water; B, Bird; M, Mammal; G, Gravity.

The predominant drainage patterns in unglaciated eastern North America flow southward west of the Appalachian Mountains or eastward through the Atlantic Coastal Plain to the Atlantic Ocean. Only after migrating tree populations invaded deglaciated terrain was northward transport of propagules possible along meltwater streams or proglacial lake margins. The northward postglacial migrations of certain taxa may therefore have been facilitated by water transport of propagules, but only after they crossed the glacial boundary.

The three tree taxa with the greatest mean migration rates (*Salix*, *Populus*, and *Betula*) are characterized primarily by wind dispersal, their populations are broadly tolerant of climatic conditions, and they are typically able to invade disturbed habitats such as found along the retreating margin of a continental ice sheet. Boreal and cool-temperate taxa that display intermediate migration rates, from 119 m/yr to 209 m/yr, are typically those that invade after initial colonization by pioneer taxa. In part, the migration rates for these taxa may be limited by soil development and accumulation of essential nutrients such as nitrogen and phosphorus, establishment of appropriate mycorrhizae within the soil litter, competiton with previously established tree taxa, or inability to establish within the prevailing disturbance regime.

The slowest group of tree taxa had mean migration rates of 45 m/yr to 81 m/yr. This group included taxa characteristic of the southeastern evergreen forest region, including southern *Pinus*, *Nyssa*, and the group of southern Cupressaceae and Taxodiaceae (*Chamaecyparis*, *Juniperus*, and *Taxodium*). The slower migration rates of this group presumably reflect their more limited climatic tolerances and cold hardiness. Their migration "lag" may as well reflect the limited late-glacial to early-Holocene availability of suitable swamp habitats along the coastal plains, habitats which expanded substantially in areal extent only in the mid- and late-Holocene intervals.

On the whole, during the late Quaternary, the rate of dispersal has apparently not been the overriding factor limiting the rate of migration for any of the taxa. The ratio of maximum documented rate of advance along any measured migration track to the mean migration rate averaged for all tracks typically varied between 2:1 and 9:1 for all tree taxa studied. Thus, for all taxa, the typical rate of advance has been far less than their inherent dispersal capability.

We examined the question of geographic patterning in the rates of migration for the 20 taxa for which measurements were obtained on the five migration tracks (Table 6.9). If the rate of dispersal were the only factor involved in determining the rates of spread of the taxa, then a diffusion model in which propagules were disseminated uniformly across all available habitats might account for the observed migration of a taxon. Such a migration would be observed to proceed generally from south to north across the landscape with no preferential migration occurring along any particular track. Comparison of several taxa would indicate that, although their rates of spread might differ among themselves, nevertheless geographic patterning would not be evident for any of them. If, on the other hand, dispersal were not a limiting factor, but instead climatic change or other environmental conditions strongly controlled the rates and directions of species migrations, then groups of species might migrate fastest along certain tracks and more slowly along others. A third possibility is that each species would exhibit individualistic behavior with respect to both the direction and rate of migration, thus making both the direction and timing of the fastest rate of spread unpredictable.

For 14 of the 20 eastern North American tree taxa (Fig. 6.12a, Table 6.9), the fastest measured overall rate of migration occurred along Track 5, located along the Atlantic Coastal Plain. Secondarily, for three taxa, Track 4 along the Appalachian Mountains was the fastest migrational pathway, two taxa migrated with greatest speed along Track 1, and one taxon spread northward most rapidly along Track 3. Although the start of the migrational event was delayed along the Atlantic Seaboard until about 12,000 yr B.P., four thousand years after

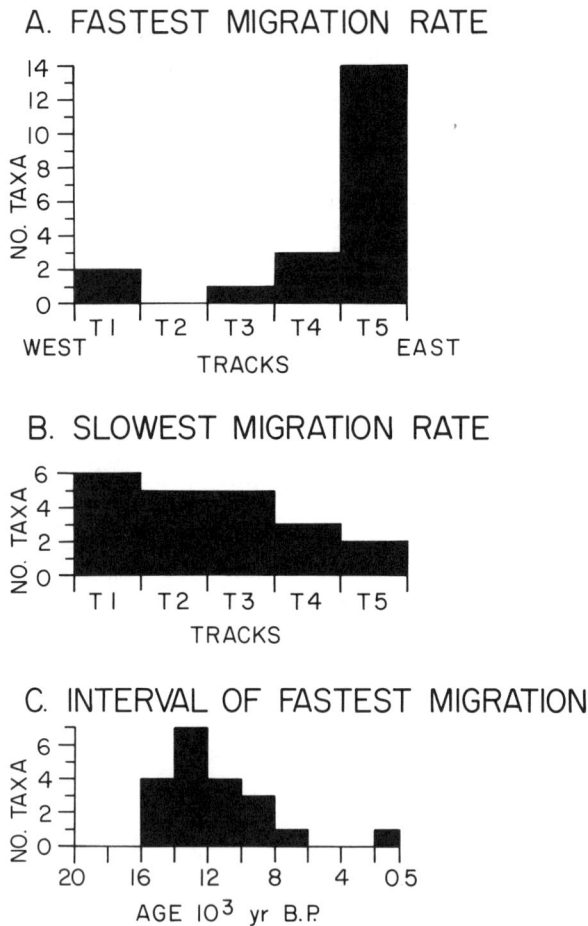

A. FASTEST MIGRATION RATE

B. SLOWEST MIGRATION RATE

C. INTERVAL OF FASTEST MIGRATION

Figure 6.12. a. Geographic patterns with overall migration rates along fastest tracks. b. Geographic patterns with overall migration rates along slowest tracks. c. Temporal pattern for fastest mean migration rates averaged across all five tracks for each 2000-year interval for the time from 20,000 yr B.P. to the present.

northward spread of trees began in the continental interior, once migrations began on the Atlantic Coastal Plain, they proceeded at a very rapid rate. These results indicate that environmental conditions, along with the individualistic capabilities of the taxa, were very important in determining the geographic patterning as well as the the rates of migration of tree taxa during the Pleistocene–Holocene transition.

The complementary analysis of the slowest overall rate of migration documented for each taxon along the five tracks (Fig. 6.12b) shows a distinctive longitudinal trend from west to east. Six of the 20 taxa had their slowest overall migration rates along Track 1 in the continental interior. In each of Tracks 2 and 3, five taxa displayed their slowest rates of migration. In Tracks 4 and 5, the lowest number of taxa (three and two, respectively) had slowest overall migration rates of any track measured across eastern North America. Although for many of the taxa, the migrational advances occurred earliest (and continued for a longer span of time) to the west of the Appalachian Mountains, those rates of advance were characteristically slower than those to the east.

The time interval of active migration for the 20 tree taxa spanned from as short as 6000 to as long as 18,000 years (Table 6.9). Temporal patterns in mean rates of migration were summarized for 2000-year intervals, with means and standard deviations calculated from the rates measured over all five tracks. For each taxon, the time interval with the fastest mean rate of migration was indicated on a time scale covering the past 20,000 years (Fig. 6.12c). For four taxa (*Quercus, Tsuga, Betula,* and *Salix*), the interval of fastest mean rate of migration was between 16,000 yr B.P. and 14,000 yr B.P. From 14,000 yr B.P. to 12,000 yr B.P., seven taxa (*Carya, Fraxinus, Acer, Ulmus,* northern Cupressaceae, *Tilia,* and *Populus*) advanced northward most rapidly. Four taxa (northern *Pinus, Larix, Juglans,* and *Fagus*) expanded northward most rapidly from 12,000 yr B.P. to 10,000 yr B.P., with three taxa (*Nyssa,* southern *Pinus,* and *Picea*) expanding most rapidly between 10,000 yr B.P. and 8000 yr B.P. Only one taxon (*Abies*) spread fastest between 8000 yr B.P. and 6000 yr B.P., and one (southern Cupressaceae plus Taxodiaceae) moved most rapidly from 2000 yr B.P. to 500 yr B.P.

With climatic amelioration starting in the early late-glacial interval, boreal taxa including *Picea* and northern *Pinus* maintained major population centers near the margin of the Laurentide Ice Sheet; their migrations were delayed, however, until major retreat of the ice sheet created opportunities for population spread around the northeastern and northwestern margins of the ice. Thus, four boreal taxa had peak rates of expansion between 12,000 yr B.P. and 6000 yr B.P. in response to the disappearance of the Laurentide Ice Sheet and the creation of new migration routes northward. In contrast, 10 of the 20 taxa, representing cool-temperate forest constituents, advanced most rapidly northward between 16,000 yr B.P. and 12,000 yr B.P., much earlier than many boreal taxa. These cool-temperate tree taxa shifted from southerly refugia in relatively low latitudes to as far north as the glacial-ice margin during this time interval. Warm-temperate taxa, including *Nyssa* and southern *Pinus,* responded most rapidly during postglacial warming between 10,000 yr B.P. to 8000 yr B.P.

Conclusions

1. Migrational rates and routes are influenced by the intensity of both environmental and biotic resistance the populations encounter. Despite the overprint of environmental control on directions and rates of migrations, however, individual taxa exhibit mapped patterns of dominance that are interpretable in terms of "migration strategies" that are analogous to the r-, K-, and fugitive life-history strategies. The migration patterns observed are a function of the inherent biology of the taxa as conditioned by the changing opportunities for colonization and persistence afforded by the environment.
2. Both the spatial and temporal patterns observed for greatest mean rates of migration should be viewed in the context of the overall migration history of each taxon. These patterns are conditioned by the starting positions of the taxa in space and time as well as the initial dominance levels of the populations along their northern distributional limits.
3. Where inherent dispersal capability does not limit the spread of a taxon, the relative importance of environmental versus biological interactions such as competition must be quantified in order to understand long-term community dynamics.

7. The Role of Competition in Long-Term Forest Dynamics

The ghost of competition past

Species interactions, their dispersal, and the physical environment are generally considered to be three primary factors responsible for structuring communities (Diamond and Case 1986; Roughgarden and Diamond 1986). In recent years, however, traditional community theory has been called into question by the results of investigations that find little or no evidence of density dependence or interspecific competition (Strong 1984). The typical interpretation of niche theory (Hutchinson 1957) has been that, where two species do not today occupy similar niches, their differences may be attributed to competitive interactions sometime in the past that led to divergence (Giller 1984). This explanation of the cause of present-day community structure has been termed "the ghost of competition past" (Connell 1980; Rosenzweig 1979a,b, 1981). Because of the difficulty in obtaining empirical evidence to demonstrate the importance of past competition to present-day niche separation or resource partitioning between species (Tilman 1982; Diamond 1986), the validity of this assumption has been challenged (Rosenzweig 1981; Simberloff 1981, 1984; Strong 1984).

The paleoecological record provides an important source of empirical data with which to test directly hypotheses concerning the relative importance of competition, species migrations, and climatic change in the development of forest communities during the late Quaternary (Chesson and Case 1986). Climatic change is the most widely recognized forcing function for changes in community composition and species migrations in temperate forest regions during the time since the last full-glacial interval, 18,000 years ago (Davis 1978, 1981a, 1983; Delcourt et al. 1982; Webb 1980; Wright 1968, 1984). Vegetational patterns and processes resulting from competitive interactions within and among species populations during times of climatic change have been studied only indirectly through application of succession models that simulate gap-phase dynamics of individual forest stands (Solomon et al. 1980, 1981; Solomon and Webb 1985; Shugart 1984; Davis and Botkin 1985). In these studies, late-Quaternary climatic change is simulated by varying the magnitude and duration of changes in growing-degree days. Observed community responses result from plant succession, where temperature, shade, and soil-moisture tolerance constitute the niche defined for each available species (Botkin et al. 1972a).

The empirical paleoecological record offers the opportunity to examine the dynamics of species populations on the spatial scale of watershed and the temporal scale of hundreds to thousands of years (Watts 1973a; Delcourt et al. 1982). Study of competitive interactions requires developing quantitative data sets from which we can infer absolute changes in population size for taxa that are of direct interest. The potential for use of **pollen accumulation rates** (referred to as **PAR**, typically expressed as the number of pollen grains that accumulate on a square centimeter of sediment surface per year, or $gr \cdot cm^{-2} \cdot yr^{-1}$) as a measure of population size has been long recognized (Davis 1969; Davis et al. 1973). In the past two decades, major research efforts have been made to understand the factors that complicate the PAR record, including imprecision in radiocarbon dating (Davis 1969), changing sedimentation patterns through time within lakes (Lehman 1975; Davis et al. 1984), and deterioration of pollen within sediments (Delcourt and Delcourt 1980; Hall 1981; Havinga 1984). Despite these potential complicating factors, net measures of PAR have been used to evaluate the time of first arrival of immigrating species (Davis 1976a, 1981a) as well as intrinsic rates of increase and population doubling times of tree taxa during times of invasion (Tsukada 1982a,b; Bennett 1983; Chen 1986; Walker and Chen 1987).

To date, the use of PAR in estimating logistic growth rates of tree populations has dealt only with the expansions of individual taxa. The importance of intraspecific and interspecific competition in late-Quaternary vegetational dynamics can be evaluated, however, using models both of logistic growth of populations of individual tree taxa and of competition among two or more taxa (Harper 1977; Begon and Mortimer 1982; Pimm 1984; Shugart 1984).

Use of pollen accumulation rates (PAR) for estimating changes in tree populations

Davis (1969; Davis et al. 1973) used the record of PAR from Rogers Lake, Connecticut, to show convincingly the sequential nature of immigrations of first coniferous, then deciduous trees into New England during the late-glacial and Holocene intervals. For example, sharp rises in PAR of *Pinus* between about 9000 and 8500 yr B.P. were interpreted to represent a rapid increase in population size of jack pine and/or red pine to levels similar to those in pine forests and pine–oak forests of central Lower Michigan (Davis *et al.* 1973). *Pinus* declined rapidly at Rogers Lake after about 8000 yr B.P., but the PAR record showed that a series of cool-temperate hardwood trees established between 9000 and 6000 yr B.P., followed by warm-temperate species of hickory and chestnut (*Castanea dentata*) after 4000 yr B.P. A key to this interpretation was the establishment of an absolute time scale for the Rogers Lake site based upon over 30 radiocarbon dates. Although it was noted that the rates of increase of PAR were different for taxa such as pine and beech during the phase of immigration, no further statistical evaluation was made to quantify differences in terms of the expansion times for these tree populations (Davis *et al.* 1973).

In several recent studies, Tsukada (1982a,b), Bennett (1983), Chen (1986), and Walker and Chen (1987) have used exponential and logistic growth models to evaluate the rates of expansion of populations of tree species during late-glacial and early-Holocene times as reflected in PAR data. Tsukada (1982a) found that the rate of expansion of Japanese beech (*Fagus*) populations following initial establishment on the watershed of Lake Nojiri approximated a **logistic or sigmoid growth** curve. Because the rates of increase differed at different sites, Tsukada suggested that the rate of population growth of invading species was conditioned both by environmental conditions and by the presence of competing species.

Bennett (1983; Fig. 7.1) studied a pollen sequence from Hockham Mere, Norfolk, U.K. To examine the question of rates of population expansion, he made pollen counts at close intervals through the early-Holocene portion of the record; the chronology was based on 23 radiocarbon dates. Under the assumption that the pollen record reflected primarily the changes in vegetation on the local watershed, the estimates of PAR obtained from the site were considered to reflect changes in population size of trees in the nearby vegetation mosaic. The pollen record thus represented changes in the overall population size of each tree taxon on the watershed over hundreds to thousands of years. On this scale, population models are appropriate to examine the invasions and expansions of relatively long-lived taxa such as trees (Bennett 1983). Bennett calculated the **exponential or unrestricted rate of increase** for each species during the time interval of invasion (Fig. 7.1). Logistic growth curves gave a more realistic approximation for several taxa, *e.g., Corylus avellana* (Fig. 7.1), since the logistic curves incorporate the effects of intraspecific competition, allowing population expansion to increase asymptotically to a finite carrying capacity (represented at the site by the highest value of PAR reached for each species after its phase of invasion and population growth). Bennett found that even simple population-growth models gave highly statistically significant results when applied to his empirical data, and he was able to calculate realistic population doubling times from the equations.

Chen (1986) and Walker and Chen (1987) used both exponential and logistic models in conjunction with PAR data in order to examine early- and mid-Holocene immigration of tropical rainforest taxa into the northeastern portion of

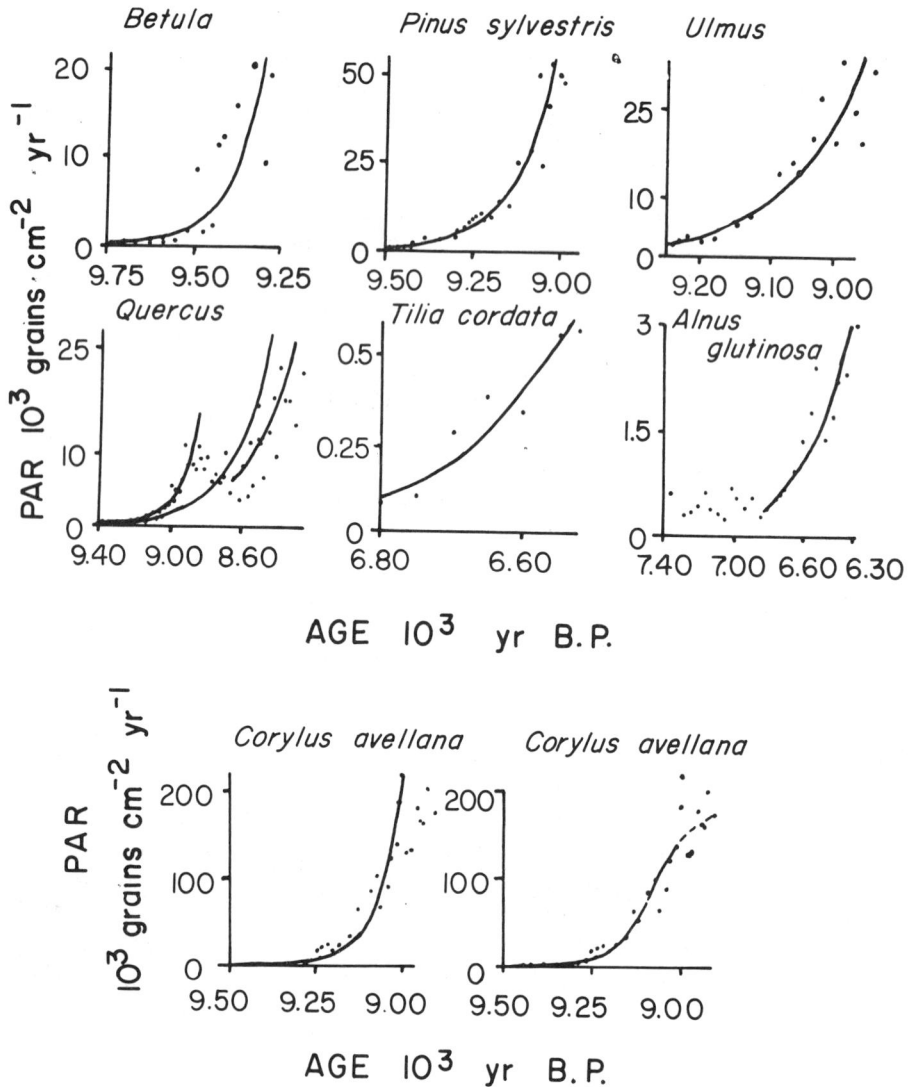

Figure 7.1. Pollen accumulation rates (PAR) for selected tree taxa based upon the fossil-pollen record from Hockham Mere, U.K. Exponential population growth curves are shown for all taxa; in addition, the second plot for *Corylus avellana* shows a logistic population growth curve (modified from Bennett 1983; reprinted by permission from *Nature*, Vol. 303, p. 166 © Macmillan Journals Ltd).

Queensland in Australia. Fine-grained temporal resolution (samples at 34-year intervals for the past 10,000 years) was available for PAR data from a sediment core from Lake Barrine. Using these paleoecological data, population doubling times were documented for the Holocene establishment of Southern Hemisphere tree taxa such as *Agathis*, *Eugenia*, *Podocarpus*, and *Rapanea* as rainforest replaced sclerophyll communities between approximately 6800 and 6100 yr B.P.

(Chen 1986; Walker and Chen 1987). Thus, the examination of PAR data using both exponential and logistic population models has been applied productively to evaluate the invasions and establishment of tree taxa within both temperate and tropical forest systems.

Black ash and hornbeam: A case study in late-glacial and early-Holocene competition

During the last full-glacial interval (approximately 20,000 to 16,500 yr B.P.), the Laurentide Ice Sheet extended into central Minnesota and Wisconsin and nearly as far south as the Ohio River Valley in Indiana and Ohio. Tundra was present locally south of the ice sheet in Minnesota (Birks 1976) and Illinois (King 1981). Boreal coniferous forest was extensive across the region from the Ozarks of Missouri to northern Georgia and the Carolinas (Watts 1970, 1980a; Delcourt 1979; Delcourt and Delcourt 1979, 1981; Delcourt *et al.* 1980; Smith 1984; Wilkins 1985).

Beginning about 16,500 yr B.P., deciduous trees increased in representation in sites in Missouri (King 1981; Smith 1984), Tennessee (Delcourt 1979; Delcourt *et al.* 1980), and South Carolina (Watts 1980a). Black ash (represented by the *Fraxinus nigra* pollen type) was among the first of deciduous tree taxa to increase in abundance in the late-glacial period, reaching peak values in southeastern Missouri by 13,000 yr B.P. (Smith 1984), then expanding into Illinois and Iowa by 12,000 yr B.P. (Van Zant 1979; King 1981). During the late-glacial to early-Holocene transition (approximately over the interval from 13,000 to 9000 yr B.P., hornbeam (represented by the *Ostrya/Carpinus* pollen type) became widespread and abundant throughout the Lower Midwest region, reaching values of 25% to 30% of the arboreal pollen (AP) assemblages in sites from Middle Tennessee (Delcourt 1979), Kentucky (Wilkins 1985), Missouri (Smith 1984), and Iowa (Van Zant 1979). Elm (*Ulmus*) was also present during this time interval, and increased to values of 10% to 20% of the AP in sites in Iowa and southern Minnesota.

In central Kentucky (Wilkins 1985), boreal coniferous woodland of spruce (*Picea*) and jack pine (northern Diploxylon *Pinus*) was replaced by deciduous woodland after 11,000 yr B.P. Following initial early-Holocene peaks in black ash and hornbeam, warm-temperate deciduous forest taxa established, including species of oak (*Quercus*), hickory (*Carya*), and sweetgum (*Liquidambar styraciflua*). Mesic cool-temperate hardwoods such as sugar maple (*Acer saccharum*), beech (*Fagus grandifolia*), hemlock (*Tsuga canadensis*), basswood (*Tilia americana*), and tree birch (*Betula alleghaniensis* and *B. papyrifera*), today characteristic of mixed conifer–northern hardwoods communities of the northern Great Lakes region, were present in only minor amounts in late-glacial forests of the Lower Midwest. Primary migrational routes for these taxa followed the Allegheny Plateau and lower elevations of the Ridge and Valley and Blue Ridge provinces of the Appalachian Mountains (Maxwell and Davis 1972; Davis 1976b, 1981a; Spear and Miller 1976; Kapp 1977a; Watts 1979; Delcourt and Delcourt 1985b, 1986). Beech and hemlock migrated into southern Lower Michigan from the east between 8000 and 6000 yr B.P. (Kapp 1977b; Bailey and Ahearn 1981; Davis *et al.* 1986). Most of the tree species today characteristic of the mixed conifer–northern hardwoods forests of the Great Lakes region (Maycock and Curtis 1960; Beschel *et al.* 1962; Marks 1974; Curtis 1959; Bourdo 1956, 1983) arrived in the region and reached their general northern distributional limits during the early- to mid-Holocene interval (Davis 1981a, 1983; Webb *et al.* 1983a).

Two paleoecological sites with continuous pollen records from full-glacial time to the present have been studied recently that offer new insights concerning late-glacial interactions of tree species. At Cupola Pond, Missouri (36°48'N, 91°05'W, 244 m elevation), an 11.9-m core of lacustrine sediments represents continuous deposition over the past 17,100 years, with a chronology based upon 10 radiocarbon dates (Smith 1984). An organic-rich 6.2-m sediment core from Jackson Pond, Kentucky (37°27'N, 85°43'W, 289 m elevation), continuously

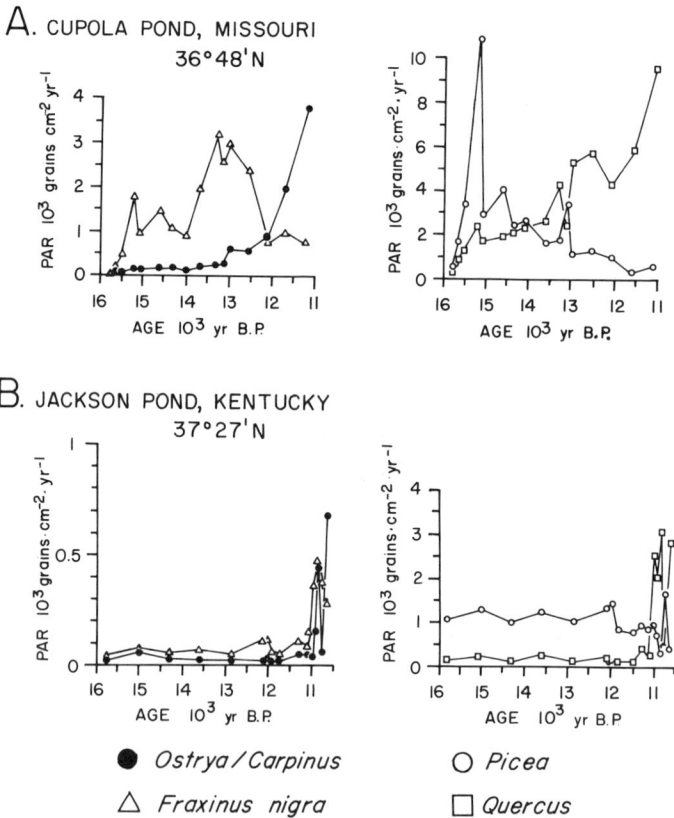

Figure 7.2. Pollen accumulation rates (PAR) for selected tree taxa for the late-glacial and early-Holocene intervals from 16,000 yr B.P. to about 11,000 yr B.P.: a, Cupola Pond, Missouri; b, Jackson Pond, Kentucky.

spans the past 20,500 years, with a chronology based upon six radiocarbon dates (Wilkins 1985).

At Cupola Pond, the first climatic amelioration in the late-glacial interval (Delcourt and Delcourt 1984) was reflected by changes in forest composition as early as 15,500 yr B.P. Boreal coniferous forests in which *Picea* was important were replaced by temperate deciduous forest including *Fraxinus nigra, Ostrya/ Carpinus,* and *Quercus* over approximately a 3000-year interval. These changes in community composition are reflected in the PAR data from the site (Fig. 7.2a).

At Jackson Pond, located farther north and only 190 km from the maximum limit of the Laurentide Ice Sheet, the shift from boreal coniferous forest to temperate hardwoods forest was delayed until substantial climatic amelioration occurred after 11,500 yr B.P. Between 11,500 and 10,000 yr B.P., *Fraxinus nigra, Ostrya/Carpinus,* and *Quercus* all established on the watershed of Jackson Pond. This is demonstrated by simultaneous increases in their PAR (Fig. 7.2b) that are independent of changes in accumulation rates of mineral sediments (Wilkins 1985). These two situations thus contrasted in terms of proximity to full-glacial refuge areas for temperate deciduous trees, but they were similar in the composition of deciduous woodlands that established in the early post-glacial period.

One-taxon models of exponential and logistic growth

At Jackson Pond, the intervals of establishment and expansion of populations of *Ostrya/Carpinus* (from 11,500 to 10,650 yr B.P., Fig. 7.2b) and *Fraxinus nigra* (from 11,500 to 10,830 yr B.P., Fig. 7.2b) were marked by pronounced increases in PAR. Using exponential and logistic models for population growth (Bennett 1983), we calculated least-squares linear regressions of ln PAR on age and of $\ln[(K - N)/N]$ on age in yr B.P. for both *Ostrya/Carpinus* and *Fraxinus nigra* at Jackson Pond (Figs. 7.3 and 7.4). The equations for these models follow Bennett (1983):

$$\textbf{Exponential increase: } (\delta N/\delta t) = rN \qquad (1)$$

where $(\delta N/\delta t)$ is the change in population size (δN) over a period of time (δt);

N is the total population of individuals present, represented in the pollen data by PAR [PAR (expressed as pollen grains $\cdot$ cm^{-2} $\cdot$ yr^{-1}) = pollen concentration (in grains/cm^3 sediment) times the sediment accumulation rate (in cm/yr)]; and

r is the unrestricted rate of increase per individual.

The integrated form of the equation may be written:

$$\ln N = rt + a \qquad (2)$$

where a is the constant of integration.

Formula (2) gives a linear relationship between ln N and t, with a slope of r.

Ostrya/Carpinus at JACKSON POND, KY.

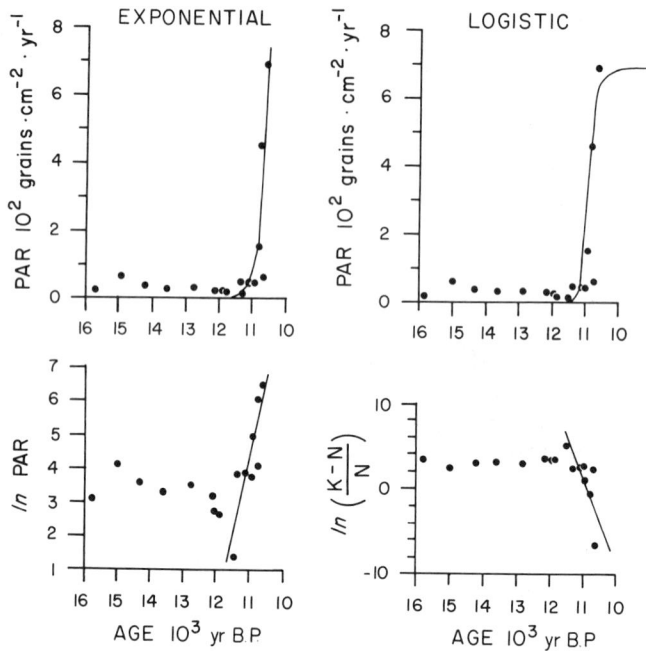

Figure 7.3. Exponential and logistic population growth curves for *Ostrya/Carpinus* at Jackson Pond, Kentucky for the time interval of increasing PAR. Asterisks following correlation coefficients indicate statistical significance as follows: *, $\alpha = 0.05$; **, $\alpha = 0.01$; ***, $\alpha = 0.001$. Statistics for the exponential model: $Y = 2.06 + (4.77 \cdot 10^{-3}X)$; correlation coefficient = 0.85**; n = 8. Statistics for the logistic model: $Y = 5.63 - (9.29 \cdot 10^{-3}X)$; K = 690; correlation coefficient = 0.76*; n = 8.

Logistic growth curve: $(\delta N/\delta t) = [rN(K - N)]/K$ (3)

or

$$\ln [(K - N)/N] = a - r t \qquad (4)$$

where K is an upper asymptote, or carrying capacity, determined for each taxon by the largest PAR value reached after its period of increase at a given paleoecological site.

At Jackson Pond, the results were statistically significant at the 99% and 95% levels of confidence for *Ostrya/Carpinus* and at the 95% and 90% levels for *Fraxinus nigra*. Estimated population-doubling times [calculated as $(\ln 2)/r$] during the interval of expansion were 145 years for *Ostrya/Carpinus* and 215 years for *Fraxinus nigra*. These estimates are of the same order of magnitude obtained by Bennett for early-successional hardwoods invading southeastern England in the early postglacial interval (Bennett 1983). At Jackson Pond, the expansions of *Ostrya/Carpinus* and *Fraxinus nigra* may approximate their intrinsic rates of population increase.

The observed rate of population increase for *Ostrya/Carpinus* was much

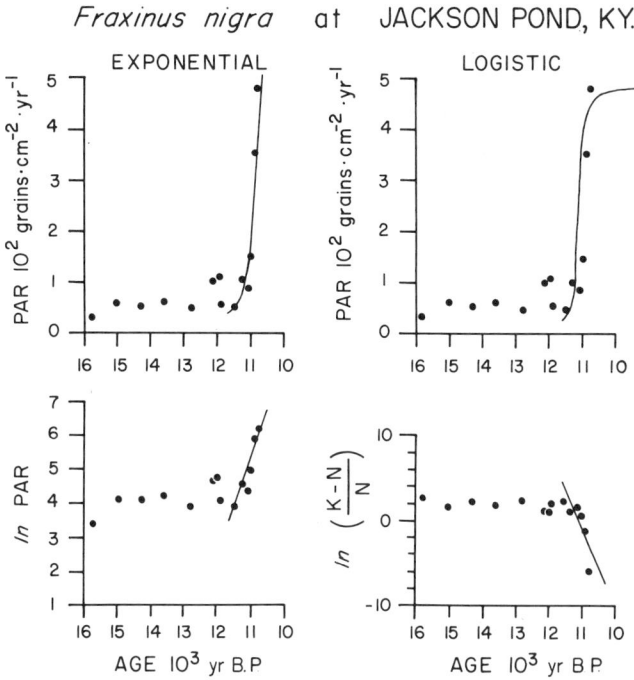

Figure 7.4. Exponential and logistic population growth curves for *Fraxinus nigra* at Jackson Pond, Kentucky for the time interval of increasing PAR. Statistics for the exponential model: Y = 3.79 + (3.22 · 10⁻³X); correlation coefficient = 0.90*; n = 6. Statistics for the logistic model: Y = 3.38 − (9.68 · 10⁻³X); K = 473; correlation coefficient = −0.77*; n = 6.

slower at Cupola Pond (Fig. 7.5) than at Jackson Pond; the population expansion occurred over 3100 years (from 14,330 to 11,180 yr B.P.) with a calculated population-doubling time of 650 years. *Fraxinus nigra* was established on the watershed of Cupola Pond for about 1000 years prior to the beginning of the expansion of *Ostrya/Carpinus* (Fig. 7.2). Values of PAR for *Fraxinus nigra* began to decrease with the initial expansion of *Ostrya/Carpinus*, however, and were reduced to low levels after 12,000 yr B.P. (Fig. 7.1a).

Two-taxa models of interspecific competition

In order to test the hypothesis that the slow increase in *Ostrya/Carpinus* at Cupola Pond resulted at least in part from competitive interaction with *Fraxinus nigra*, we used a two-taxa population-growth model (Begon and Mortimer 1982; Pimm 1984). This model incorporated both intraspecific and interspecific competition, as follows:

$$\dot{X}_1 = X_1 (r_1 + a_{11}X_1 + a_{12}X_2) \tag{5}$$

$$\dot{X}_2 = X_2 (r_2 + a_{22}X_2 + a_{21}X_1) \tag{6}$$

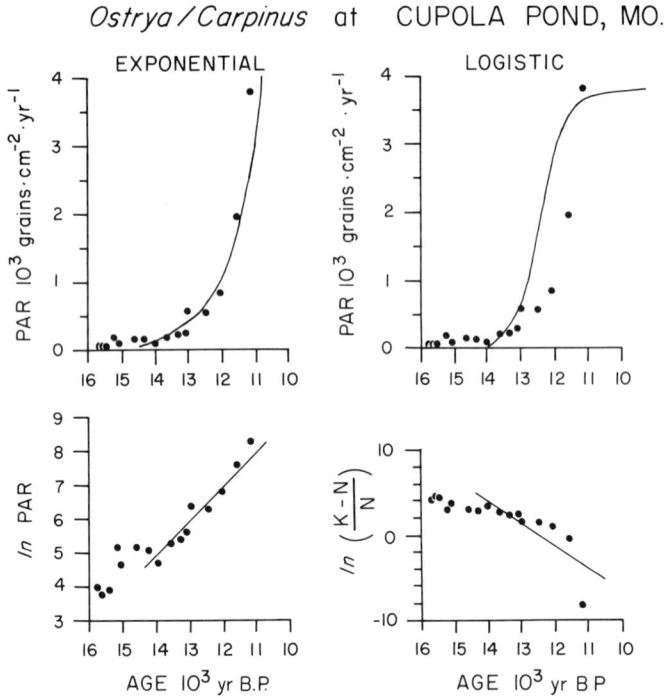

Figure 7.5. Exponential and logistic population growth curves for *Ostrya/Carpinus* at Cupola Pond, Missouri for the time interval of increasing PAR. Statistics for the exponential model: $Y = 4.58 + (1.07 \cdot 10^{-3}X)$; correlation coefficient $= 0.97^{***}$; n $= 10$. Statistics for the logistic model: $Y = 4.99 - (2.65 \cdot 10^{-3}X)$; $K = 3798$; correlation coefficient $= -0.79^{**}$; n $= 10$.

where $\dot{X}_1$ and $\dot{X}_2$ represent the incremental rate of change in population size or PAR for taxa 1 and 2, respectively;

X_1 and X_2 are initial values of population size or PAR for taxa 1 and 2;

r_1 and r_2 are the intrinsic rates of population increase for taxa 1 and 2, calculated from the single-taxon model of logistic expansion at the site with the fastest observed r;

a_{11} and a_{22} are negative values that represent intraspecific competition coefficients for taxa 1 and 2 ($a_{11} = -r_1/K_1$ and $a_{22} = -r_2/K_2$, where K_1 and K_2, represented by the highest observed PAR, are site-specific for each taxon because of differences in the mosaic of habitat types, soil fertility, and watershed size);

a_{12} is a negative value that represents the coefficient of competition of taxon 2 on taxon 1; and

a_{21} is a negative value that represents the coefficient of competition of taxon 1 on taxon 2.

Depending on the values selected for a_{12} and a_{21}, both taxa can coexist indefinitely, or one may replace the other. The number of iterations required to reach stable values or for competitive exclusion to occur vary with the intensity of interspecific competition and can be scaled to the realistic generation time of the taxon modeled (here constrained by the empirical values for r_1, r_2, K_1, K_2, a_{11}, and a_{22}. The shapes of the curves for expansion and decline vary as the interspecific competition coefficients are varied, but within a finite number of possibilities.

We used this two-taxa model to simulate interactions between *Ostrya/Carpinus* and *Fraxinus nigra*, first with the assumption of no interspecific competition between the two taxa, and then with varying degrees of competition. We used the intrinsic rates of increase (r) calculated from the Jackson Pond data. However, in order to test model results against the empirical record of late-glacial and early-Holocene vegetational dynamics at Cupola Pond, we determined the carrying capacity (K) for each taxon from the highest values of PAR attained by *Ostrya/Carpinus* and by *Fraxinus nigra* at Cupola Pond. In this way, we standardized the procedure for the Cupola Pond watershed, as the size and shape of both the watershed and the site of pollen deposition influence the magnitude of PAR values (Davis *et al.* 1973, 1984).

Using only the calculated self-damping effects of intraspecific competition for each species, and beginning at about 13,300 yr B.P. with initially high PAR

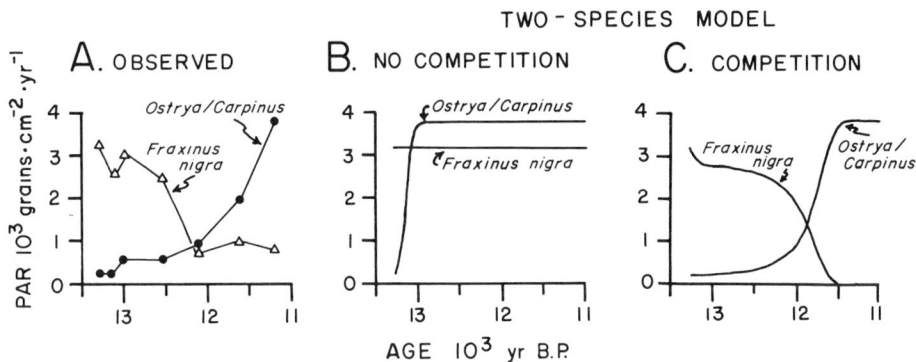

Figure 7.6. a. The empirical PAR record from Cupola Pond, Missouri. b. Results of two-taxa population–growth model of the interactions of *Ostrya/Carpinus* and *Fraxinus nigra* without interspecific competition. c. Results of two-taxa population–growth model with competitive pressure of *Ostrya/Carpinus* on *Fraxinus nigra* of $-5.00 \cdot 10^{-6}$ and of *Fraxinus nigra* on *Ostrya/Carpinus* of $-3.00 \cdot 10^{-6}$. In both models tests, we used intraspecific competition coefficients calculated from the logistic growth relationships at Jackson Pond, as follows: *Ostrya/Carpinus*, $-2.45 \cdot 10^{-6}$; *Fraxinus nigra*, $-3.03 \cdot 10^{-6}$. Initial PAR values were: *Ostrya/Carpinus*, 228; *Fraxinus nigra*, 3194. K values used were the highest observed values of PAR for each taxon at Cupola Pond: *Ostrya/Carpinus*, 3798; *Fraxinus nigra*, 3194.

values for *Fraxinus nigra* and low values for *Ostrya/Carpinus* (as derived from the empirical pollen record; Fig. 7.6a), *Fraxinus nigra* remained at constant, high values indefinitely, and *Ostrya/Carpinus* independently and rapidly increased to its carrying capacity (Fig. 7.6b). In a second case, including possible interaction of interspecific competition (Fig. 7.6c), the model closely simulated the changes observed in PAR for the two taxa. In this particular example (Fig. 7.6c), we applied a competitive pressure of *Ostrya/Carpinus* against *Fraxinus nigra* that was nearly twice as high as the intraspecific damping for either taxon alone. The simulated time interval over which *Ostrya/Carpinus* replaced *Fraxinus nigra* was about 1800 years. The model results indicate that, for relatively long-lived species such as trees, even during times of climatic change, the process of competitive exclusion of populations on a watershed can occur on a time scale as long as millennia.

Potential versus realized niches

Whether the niche breadths of tree taxa have remained constant, or whether these taxa have become progressively restricted through time to narrower portions of key environmental gradients such as the soil-moisture gradient is important to the understanding of how forest communities are structured through time (Hutchinson 1957; Begon and Mortimer 1982; Giller 1984). One example of possible changes in niche breadth is that of the postglacial history of changing abundance of *Fraxinus nigra* in the vegetation of the western Great Lakes region. Pollen percentages of *Fraxinus nigra* type greater than those known today from the region occurred in sites from southeastern Missouri to Iowa and southern Minnesota in the late-glacial and early-Holocene intervals (Fig. 10.10 in Webb *et al.* 1983a). Late-glacial forests of the Midwest region were relatively poor in species compared with forests in the late Holocene (Davis 1983), and *Fraxinus nigra* was among the first to invade northward with climatic amelioration. During the late-glacial interval, the realized niche of *Fraxinus nigra* may have approximated a larger portion of its potential niche, possibly occupying both hydric and mesic habitats across a broad portion of the soil-moisture gradient. With subsequent invasions of deciduous taxa, such as *Ostrya/Carpinus*, intensified interspecific competition may have constituted a major factor in diminishing the importance of *Fraxinus nigra* populations, particularly on mesic sites. During the late-glacial and Holocene intervals, increasing competitive interactions among additional invading tree species may have resulted in the progressive restriction of tree populations to more narrow realized niches as they sorted out along more limited portions of key environmental gradients.

Conclusions

1. The experiments with population-growth and competition models, using PAR records from Cupola and Jackson Ponds, are the first examples to incorporate

interspecific competition between two taxa as tested against empirical data from fossil pollen records. This first attempt, however, should be considered a preliminary study that demonstrates the feasibility and potential of this population-model approach in the field of Quaternary paleoecology.

2. Major factors that structure forest communities include (1) environmental changes, (2) species dispersal and invasions, and (3) intraspecific and interspecific competition.

8. Gradient Analysis, Ecotones, and Forest Communities

Direct gradient analysis

The analysis of changes in distributions and relative dominances of individual tree taxa emphasizes their autecological responses to environmental changes. Analyses of pollen accumulation rates (PAR) indicate that in addition to inherent life-history strategies of individual taxa, changes in dominance may be related in part to competition during times of invasion and population expansion. When the late-Quaternary vegetational record is examined in this light, the impression gained is one of ecological interactions predominantly on the population level. However, the paleovegetation data (the array of dominance of taxa by site and by time) are also pertinent to quantitative plant community analysis. In this chapter we ask questions relevant to determining whether temperate forest communities have been stable through time, in the sense of long-term persistence as intact assemblages. We also examine changes in the position and steepness of ecotones between major forest types through the past 20,000 years.

In his classic study of the vegetation of the Great Smoky Mountains, Whittaker (1956) demonstrated that populations of trees tend to exhibit broadly overlapping distributions along environmental gradients such as elevation, soil moisture, and slope aspect. In combination, the total distributions of species ranges along environmental gradients, together with the changes in their dominance values throughout their distributions, constitute an **ecocline** (Gauch 1982).

Where a number of taxa with similar physiological tolerances tend to reach common limits of distribution, the **ecotone**, or compositional transition zone, between definable plant communities is relatively steep (Chabot and Mooney 1985). If, however, all species are very broadly overlapping in their dominance distributions along an environmental gradient, the rate of turnover of species may be much lower, and consequently ecotones will be more diffuse. Whereas the species number or richness within a community sample is a measure of **alpha diversity**, the rate of turnover of species populations along environmental gradients is a measure of the steepness of ecotones that can be quantified and expressed as an index of **beta diversity** (Whittaker 1975; Gauch 1982; Wilson and Shmida 1984).

Direct gradient analysis has been a mainstay of quantitative analyses in vegetation science for over 30 years (Kessel 1979; McIntosh 1985). However, no quantitative paleovegetational data sets have been previously examined by direct gradient analysis in order to evaluate the question of shifts in location and steepness of ecotones or turnover rates of populations of plant taxa over a glacial–interglacial cycle (Birks and Gordon 1985). Detrended Correspondence Analysis (DCA) is an appropriate technique with which to examine paleo-ecological data, whether in the form of raw pollen percentages or as calibrated paleovegetational estimates (Birks and Gordon 1985). Time series of paleo-ecological data represent changing ecoclines on the landscape surrounding a single site (Delcourt *et al.* 1983b) and can be recognized as vegetational continua through time that are analogous to those across space (Davis and Jacobson 1985). DCA provides a quantitative technique with which to evaluate rates of change in plant communities either at a site through time (Jacobson and Grimm 1986) or across environmental gradients in space and time.

Interpolation of dominance values for individual taxa along a south-to-north transect at 85°W

We approached this problem using the south-to-north transect at 85°W across eastern North America for which we previously developed three hypotheses of vegetational dynamics (Delcourt and Delcourt 1983). This quantitative analysis therefore represents a direct test of hypotheses dealing with vegetational change presented in Chapter 1. We interpolated dominance values for all calibrated tree taxa at 1°-latitude intervals from 28°N (the Gulf of Mexico) to 56°N (Hudson Bay), based upon a subset of paleo-dominance maps (presented in Chapter 5) selected at 20,000 yr B.P., 16,000 yr B.P., 12,000 yr B.P., 8000 yr B.P., 4000 yr B.P., and 500 yr B.P. The interpolated dominance values are smoothed percentages constrained by the geographic array of paleovegetation data used to produce the contour intervals on the paleo-dominance maps. The length of the physical environmental gradient corresponds to the distance (in degrees of latitude) along 85°W that occurs within forest. This gradient is typically bounded by the Gulf of Mexico to the south and to the north by tundra, glacial ice, proglacial lakes, or Hudson Bay.

Changing abundances of individual taxa through the late Quaternary along changing environmental gradients

The changes in population dominance through the past 20,000 years along this south-to-north transect are illustrated by three taxa, *Picea* (Fig. 8.1), *Quercus* (Fig. 8.2), and southern *Pinus* (Fig. 8.3).

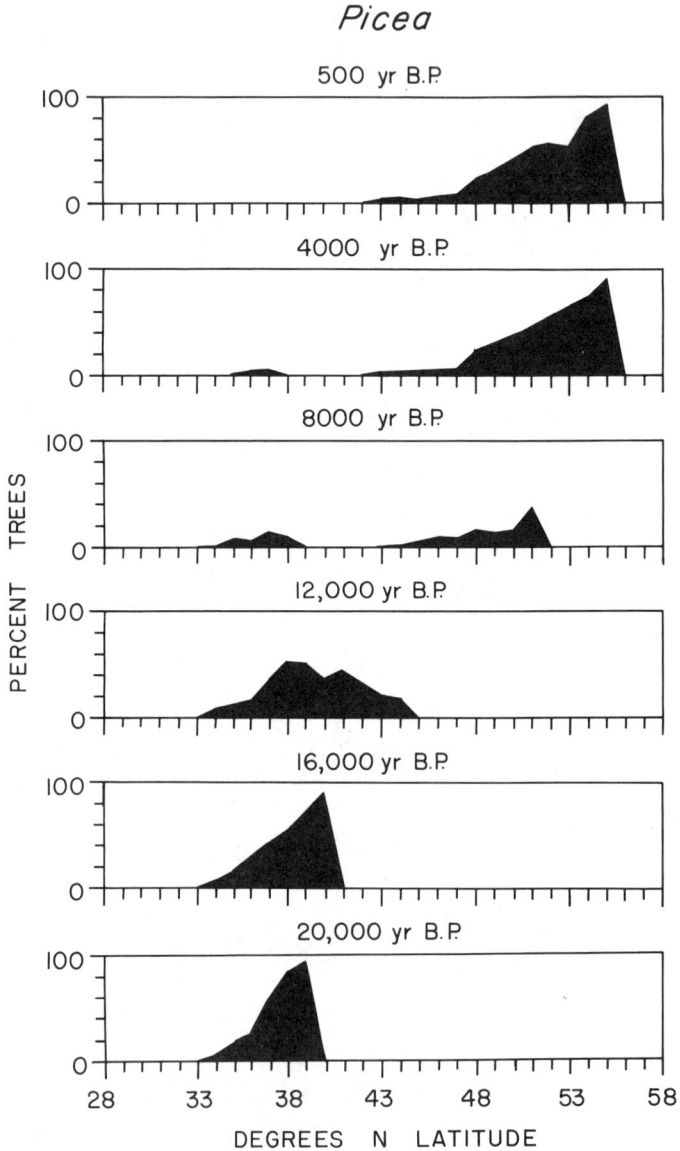

Figure 8.1. Paleo-dominance gradients for *Picea* along 85°W, at 20,000, 16,000, 12,000, 8000, 4000, and 500 yr B.P.

Quercus

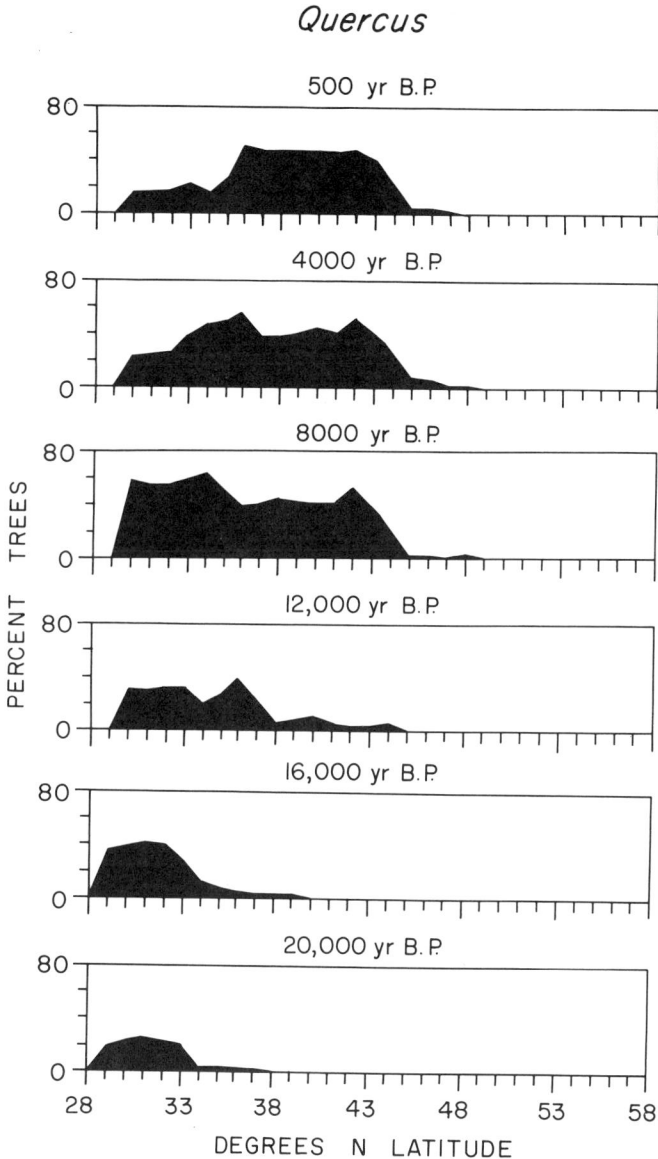

Figure 8.2. Paleo-dominance gradients for *Quercus* along 85°W, at 20,000, 16,000, 12,000, 8000, 4000, and 500 yr B.P.

During the full-glacial interval, between 20,000 yr B.P. and 16,000 yr B.P., spruce was confined to between 33°N and 41°N, with peak dominance $\geqslant 90\%$ of the total reconstructed forest composition near its northern limit of distribution and with relative dominance declining gradually southward (Fig. 8.1). By 12,000 yr B.P., although extending north to 45°N, peak values of about 50% occurred at 38°N, with relative dominance declining both to the north and to

Pinus (SOUTHERN GROUP)

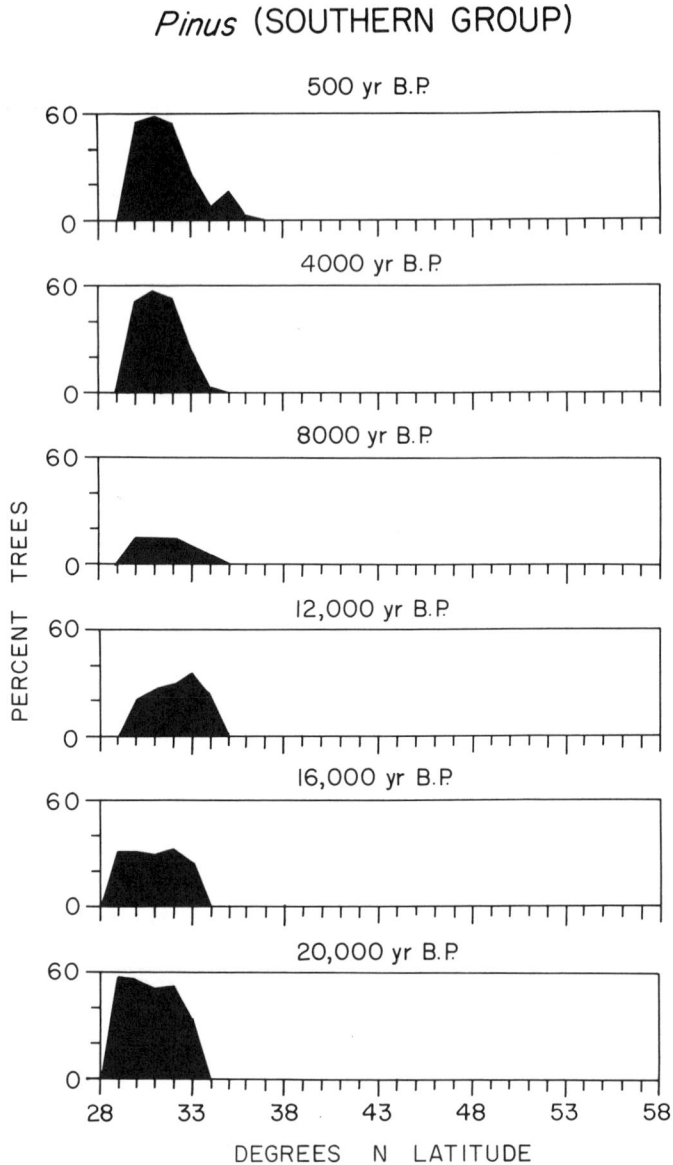

Figure 8.3. Paleo-dominance gradients for southern *Pinus* along 85°W, at 20,000, 16,000, 12,000, 8000, 4000, and 500 yr B.P.

the south. However, the bulk of the spruce population was located between 36°N and 44°N. Between 12,000 yr B.P. and 8000 yr B.P., populations of spruce declined to < 20%, persisting between 33°N and 39°N as the southern populations became separated from those that were migrating northward. Northern populations were also greatly reduced in dominance, reaching up to 40% only at 51°N near their northern limit of distribution. By 4000 yr B.P., southern populations of *Picea* had dwindled to < 10%. Northern spruce populations, however, had regained dominance values > 50% of total forest composition north of about 51°N, reaching nearly 100% dominance at 55°N.

In contrast with *Picea, Quercus* reached maximal full-glacial values along the transect at 85°W between 29°N and 33°N, increasing from about 25% of forest dominance to 40% between 20,000 yr B.P. and 16,000 yr B.P. The northern distribution of oak overlapped with the southern distribution of spruce between about 33°N and 38°N at 20,000 yr B.P., and between 33°N and 40°N at 16,000 yr B.P. By 12,000 yr B.P., two peaks in oak dominance were evident (Fig. 8.2), one at about 30% between 30°N and 33°N, and the other up to 40% at 36°N. The leading (northern) edge of distribution of oak was located at 45°N, with values generally between 5% and 10% dominance from 38°N to 45°N. By 8000 yr B.P., oak increased to between 40% and 60% of the forest composition throughout the latitudinal range from 30°N to 43°N, with only a few percent oak extending northward to 49°N. Oak dominance values remained stable throughout the mid- and late-Holocene intervals in the central and northern portion of its range. However, oak declined to < 40% south of 33°N by 4000 yr B.P., and south of 36°N by 500 yr B.P., producing a more symmetrical distribution along the transect.

Southern *Pinus* exhibits a third, distinctively different pattern along the latitudinal gradient (Fig. 8.3). At 20,000 yr B.P., southern pines were restricted south of 34°N but were 50% to 60% of the forest composition of the Gulf Coastal Plain. The dominance of southern pines diminished during the late-glacial and early-Holocene intervals to between generally 25% and 30% from 16,000 yr B.P. to 12,000 yr B.P., and down to < 15% by 8000 yr B.P. After 8000 yr B.P., southern *Pinus* regained dominance of > 50% between 29°N and 32°N and also began to extend northward to 35°N. By 500 yr B.P., southern pines reached 15% at 35°N and also peaked at 60% at 31°N. The late-Holocene increase in southern pine was responsible for the concomitant decline in oak dominance (Fig. 8.2).

DECORANA ordination of late-Quaternary ecoclines

Ordination is an ecological technique for determining the structure of vegetation along ecoclines. **Detrended Correspondence Analysis (DCA)** provides a method of ordination that is considered superior to alternatives such as reciprocal averaging, multidimensional scaling, and principal components analysis (Hill 1979). DCA is recommended for use for complex plant-ecological data sets with species composition associated with samples and for which the data arrays are heterogeneous (Gauch 1982).

For DCA analysis of the paleoecological data set along the transect at 85°W, we used the FORTRAN program **DECORANA** (Hill 1979) as compiled in MS-FORTRAN for use on the IBM-PC (Clampitt 1985). The DECORANA results included quantitative values for species scores as well as for sample scores (which are weighted mean species scores). Eigenvalues calculated for each ordination axis provided quantitative measures for variability analyzed within the Quaternary plant-ecological data set. We used untransformed data with no downweighting of rare taxa. The standard number of 26 segments was used in order to detrend the sample scores such that no systematic relationship occurred among the ordination Axes. Four iterations were performed for rescaling of ordination Axes; this removed any systematic relationship between within-sample deviation and position along the ordination Axes. We checked the results using an octave transformation of the data (Hill 1979; Gauch 1982). For the purposes of this study, we present numerical results based upon the untransformed data, as the eigenvalues were consistently greater for the first ordination axis than were those calculated by the octave transformation.

Using DECORANA for ordination, the samples are rescaled in order to generate an even turnover of species along the ecocline gradient. The unit measure of length (or distance) along the gradient represents the measure of beta diversity expressed as 1 "standard deviation" (1.0 sd; usage follows Hill 1979) (Whittaker 1960). More specifically, the root-mean-square standard deviation for the dominance curve of the taxon along the gradient is generally equal to the value of 1.0 sd. The gradient length of 1.0 sd is approximately ¼ of a species turnover. That is, the length of 4.0 sd typically reflects the first occurrence of a taxon along the ecocline gradient, then its increase in dominance along a bell-shaped curve to its peak, and subsequent decline to its last occurrence. Thus, when the gradient exceeds 4.0 sd, the forest composition represents completely different species assemblages at the extreme positions of the ecocline (Hill 1979; Gauch 1982).

We applied DECORANA to analyze ecocline data along the 85°W transect separately for the times at 20,000 yr B.P. (Fig. 8.4; Fig. 8.5a), 16,000 yr B.P. (Fig. 8.5b), 12,000 yr B.P. (Fig. 8.5c), 8000 yr B.P. (Fig. 8.5d), 4000 yr B.P. (Fig. 8.5e), and 500 yr B.P. (Fig. 8.5f). In addition, we combined all data for all taxa along the transect at all selected times into a larger data set for ordination analysis through space and time (Fig. 8.6).

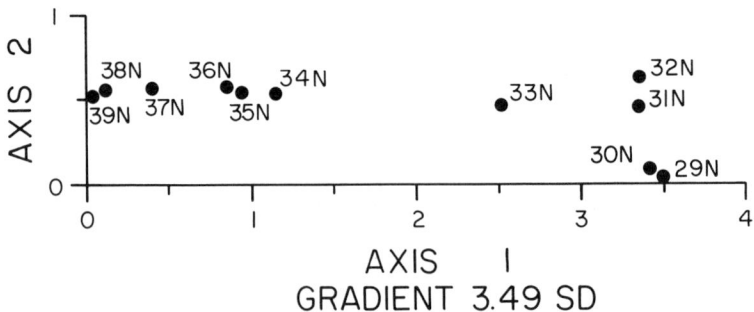

Figure 8.4. DECORANA sample scores at 20,000 yr B.P. for samples distributed at a spacing of 1° latitude along the 85°W transect.

Ordination of sample scores

Beta diversity was calculated between adjacent samples of forest composition along the latitudinal gradient. Values of beta diversity ≥ 1.0 sd were considered to identify prominent and steep ecotones within the vegetation. Values of beta diversity ≥ 0.5 and < 1.0 sd indicated minor or diffuse transition zones in the vegetation.

Ordination of sample scores for specific times

For 20,000 yr B.P. along the latitudinal transect, the dominance curves for all the taxa are illustrated along the full-glacial ecocline (Fig. 8.5a). The DECOR-ANA ordination shows that the species scores represented an eigenvalue of 0.873 along the ordination Axis 1. Together, the sample scores represented a total length of 3.49 sd for the beta diversity gradient. For example, maximum compositional differences occurred between the vegetation samples of 32°N and 33°N (a beta diversity measurement of 0.8 sd), 33°N and 34°N (1.4 sd), and 36°N and 37°N (0.5 sd). The positions in ordination space of the samples arrayed along Axes 1 and 2 are shown for 20,000 yr B.P. in Fig. 8.4. Samples from 29°N to 32°N clustered within a distance of 0.3 sd to the right side of the graph, and samples from 34°N to 39°N were distributed on the left side of the graph across a span of 1.2 sd. The position of the major full-glacial ecotone thus was situated between 32°N and 34°N, as indicated by the wide gap between the corresponding samples (Fig. 8.4). The steepened gradients for beta diversity (portrayed as black bands in Fig. 8.5) illustrate the geographic positions of prominent ecotones between different forest communities, where the rate of species turnover equals or exceeds 0.5 sd over a physical distance of 1° latitude.

Ordination of sample scores for the remaining times (Fig. 8.5b–f) indicates that both the positions and the steepness of ecotones have shifted and that the beta diversity gradient has lengthened through the past 20,000 years. The ecocline at 16,000 yr B.P. (Fig. 8.5b) shows the primary ecotone located between 32°N and 34°N; a beta diversity measurement of 0.6 sd characterized the difference in the composition of forests between 32°N and 33°N, and 1.3 sd represents the accentuated rate of species turnover between 33°N and 34°N. The 12,000 yr B.P. ecocline (Fig. 8.5c) exhibited two ecotonal zones. One vegetational transition was positioned between 34°N and 37°N; the second was located between 40°N and 41°N (0.6 sd). The more southerly ecotone included a measure of beta diversity of 0.9 sd between 34°N and 35°N and 0.5 sd between 36°N and 37°N. The measure of beta diversity characterizing this ecotone decreased from 16,000 yr B.P. to 12,000 yr B.P. as the ecotone expanded across a broader latitudinal zone.

The ecocline at 8000 yr B.P. (Fig. 8.5d) displays two prominent ecotones, one situated between 34°N and 35°N (0.6 sd) and the other located between 42°N and 45°N. This latter ecotone included beta diversity measures of 0.5 sd, 0.6 sd, and 0.5 sd at 1°-latitudinal intervals from south to north, respectively.

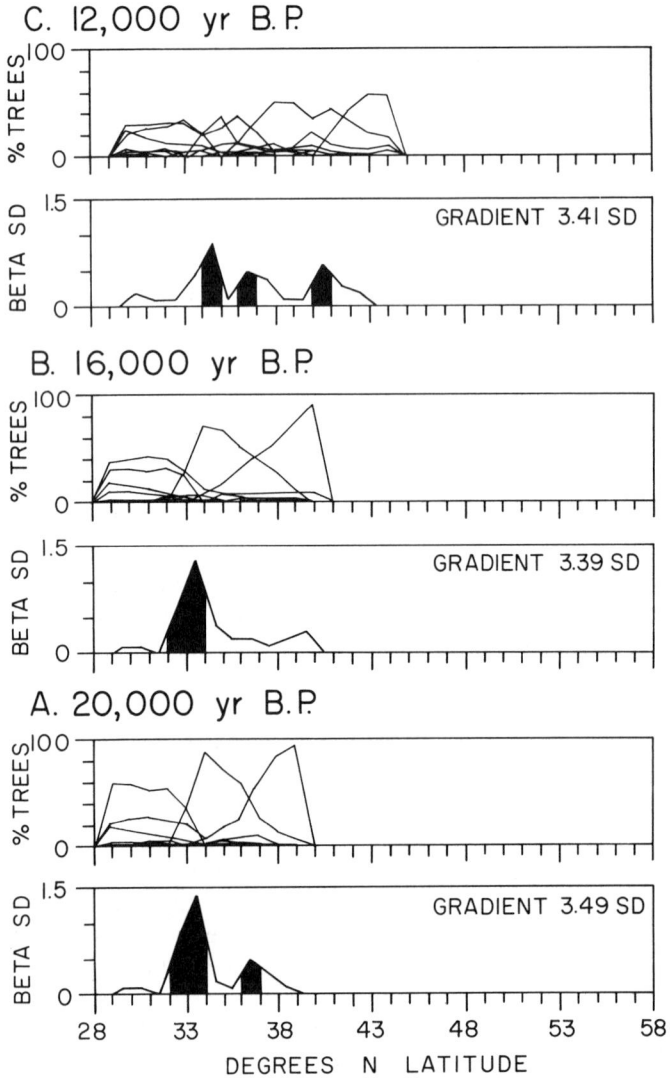

Figure 8.5. Late-Quaternary ecoclines (upper panel with taxa dominance curves) and latitudinal gradients in beta diversity (lower panel with ecotones shown as black segments), at a, 20,000 yr B.P.; b, 16,000 yr B.P.; c, 12,000 yr B.P.

At 4000 yr B.P. (Fig. 8.5e), the southern ecotone had shifted to between 35°N and 36°N, with a corresponding measure of 0.7 sd. The more northerly ecotone had broadened to encompass the region between 44°N and 48°N; rates of species turnover were 0.5 sd from 44°N to 47°N and 0.7 sd between 47°N and 48°N.

F. 500 yr B.P.

GRADIENT 5.18 SD

E. 4000 yr B.P.

GRADIENT 5.06 SD

D. 8000 yr B.P.

GRADIENT 3.83 SD

DEGREES N LATITUDE

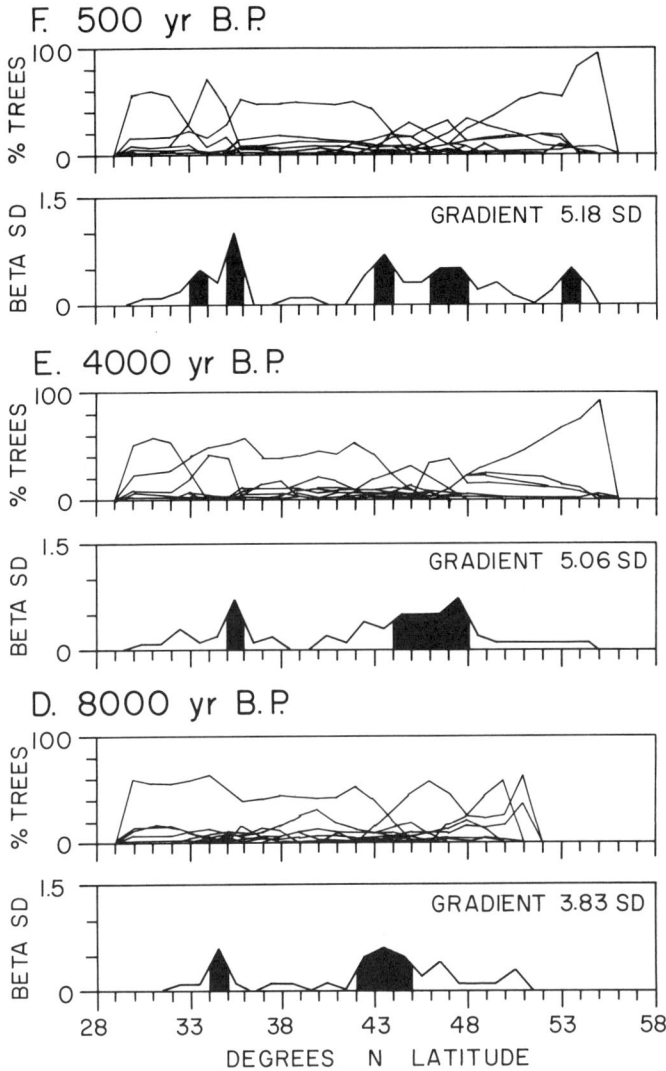

Figure 8.5. *continued.* Late-Quaternary ecoclines (upper panel with taxa dominance curves) and latitudinal gradients in beta diversity (lower panel with ecotones shown as black segments), at d, 8000 yr B.P.; e, 4000 yr B.P.; and f, 500 yr B.P.

The ecocline for 500 yr B.P. (Fig. 8.5f) portrays the persistence of two prominent ecotones, between 33°N and 36°N and between 43°N and 48°N. The southerly ecotone reached a peak beta diversity measure of 1.0 sd between 35°N and 36°N, and the second ecotone attained a value of 0.7 sd between 43°N and

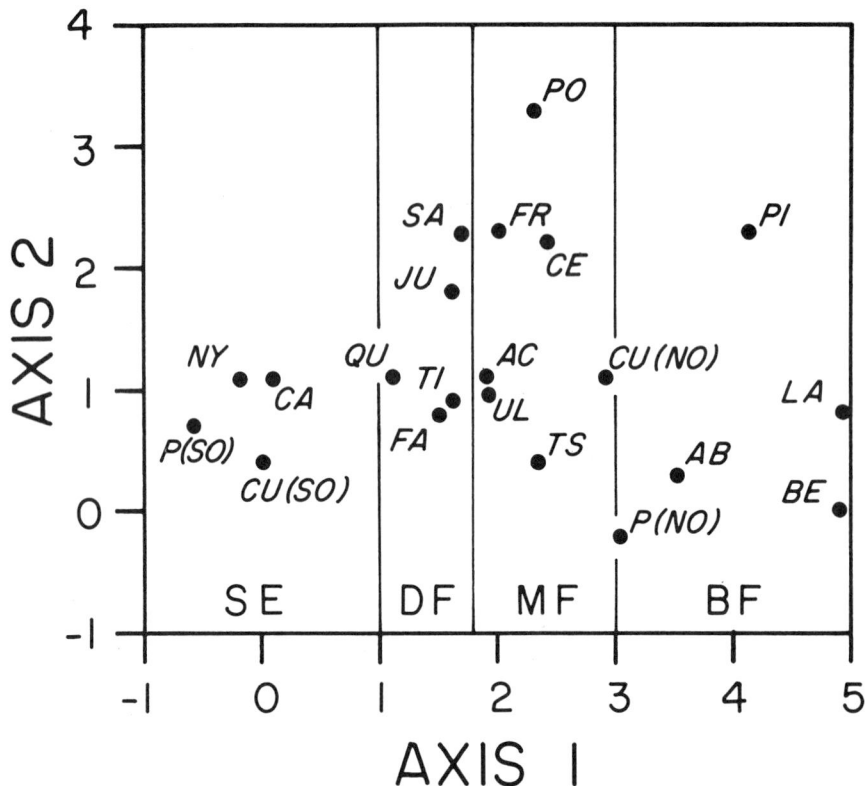

Figure 8.6. DECORANA taxon scores for the untransformed data set including all latitudinal samples from 20,000 yr B.P. to 500 yr B.P. along the transect at 85°W. Southeastern evergreen forest taxa (SE on DCA Axis 1) are indicated by the following letter codes: P(SO), southern *Pinus*; NY, *Nyssa*; CA, *Carya*; and CU(SO), southern group of Cupressaceae and Taxodiaceae. Deciduous forest taxa (DF on DCA Axis 1) have the following letter codes: QU, *Quercus*; FA, *Fagus*; TI, *Tilia*; JU, *Juglans*; and SA, *Salix*. Tree taxa of the mixed conifer–northern hardwoods forest (MF on DCA Axis 1) are coded as follows: TS, *Tsuga*; CU(NO), northern group of Cupressaceae; UL, *Ulmus*; AC, *Acer*; CE, *Celtis*; FR, *Fraxinus*; and PO, *Populus*. Boreal forest taxa (BF on DCA Axis 1) have letter codes as follows: PI, *Picea*; LA, *Larix*; AB, *Abies*; P(NO), northern *Pinus*; and BE, *Betula*.

44°N. A third, minor ecotone developed by presettlement times between 53°N and 54°N, with a beta diversity of 0.5 sd.

Latitudinal gradients in beta diversity. The total length of the beta diversity gradient has changed along the transect at 85°W over the past 20,000 years (Fig. 8.5). At 20,000 yr B.P. (Fig. 8.5a), the length of the beta diversity gradient was 3.49 sd but dropped to 3.39 sd at 16,000 yr B.P. (Fig. 8.5b). At 12,000 yr B.P. and 8000 yr B.P., the beta diversity gradient measured 3.41 sd and 3.83 sd, respectively (Fig. 8.5c,d). The late-Holocene ecoclines at 4000 yr B.P. and 500 yr B.P. (Figs. 8.5e,f) had total lengths of beta diversity gradients of 5.06 sd and 5.18 sd, respectively. Over the past 20,000 years, the total length of the physical environmental gradient has increased from 12° of latitude at 20,000 yr B.P. to 13° at 16,000 yr B.P., 16° at 12,000 yr B.P., 25° at 8000 yr B.P., and to 27° of latitude at 4000 yr B.P. and 500 yr B.P. The absolute bounds of the environmental gradient were dictated from 20,000 yr B.P. to 6000 yr B.P. by the northern position of the Gulf of Mexico and the southern margin of the Laurentide Ice Sheet. During the last 6000 years, the northern transect boundary of the environmental gradient has been determined by the shoreline position of Hudson Bay.

Interpretation of changes in location and breadth of ecotones and beta diversity through time. The changes in beta diversity of the forest-compositional gradient have not been a direct reflection of the physical length of the environmental gradient available for occupation. Between 20,000 yr B.P. and 8000 yr B.P., the physical length of the environmental gradient more than doubled, expanding from 12° to 25° of latitude along the transect at 85°W. However, the total values of beta diversity decreased from full-glacial through late-glacial and early-Holocene times, with the lowest recorded values between 16,000 yr B.P. (a total of 3.39 sd) and 12,000 yr B.P. (3.41 sd), increasing only slightly by 8000 yr B.P. (3.83 sd). The generally low values for beta diversity during the late-glacial and early-Holocene intervals may be attributed to northward migrations of tree taxa across the mid-latitudinal zone that resulted in broad overlaps in their dominance distributions during that time interval. These results are consistent with the space–time domain for which we postulated vegetational disequilibrium (Delcourt and Delcourt 1983; Chapter 1) and in which we have found only poor modern analogs for forest composition (Delcourt and Delcourt 1985a; Chapter 3).

Between 8000 yr B.P. and 4000 yr B.P., the length of the environmental gradient increased by only 2° of latitude, but the beta diversity gradient increased markedly from 3.83 sd to 5.06 sd. During the last 4000 years, the physical gradient did not change in total length, but beta diversity increased to an overall gradient length of 5.18 sd. During the mid- to late-Holocene interval, it is evident that with an increased rate of turnover for tree taxa along the ecocline gradient, there was less compositional overlap between adjacent forest assemblages along the latitudinal transect than had occurred previously.

The increased number of ecotones and the relatively high beta diversity values that developed in the mid- and late-Holocene intervals may be explained by one of two kinds of processes: (1) interspecific competition; or (2) a hysteretic effect of climatic cooling influencing forest composition. The migrations of most tree taxa were completed before the late Holocene, and therefore invasion of late-Holocene forests by new tree species was not a major factor after about 4000 yr B.P. Rather, the complement of tree taxa that coexisted during the late Holocene adjusted in relative dominance along the ecocline. Increased interspecific competition and niche partitioning would have increased the packing of tree taxa along the ecocline, resulting in an increase in the number of discernable ecotones and in the rate of turnover of populations of the taxa, therefore increasing the beta diversity gradient. Alternatively, minor climatic cooling may have also influenced the competitive success of tree taxa near their distributional limits during the late-Holocene interval (Ritchie 1986), contributing to the increase in beta diversity. The relatively low beta diversity gradients of the late-glacial and early-Holocene intervals were associated with biotic response to climatic warming. This situation contrasted markedly with the biotic responses during the last 4000 years associated with late-Holocene cooling. This differential competitive success was probably associated with the differential effectiveness of seedling establishment during changing climatic conditions (Shugart *et al.* 1980). Thus, the differential responses of taxa may have been reflected in the adjustments of their range limits and relative population dominances during this time of climatic cooling. As a consequence, the progressive sharpening of ecotones observed in the late Holocene may in part result from such a hysteretic effect (Shugart *et al.* 1980) observed over the last cooling stages of the present interglacial cycle.

Ordination of all taxon and sample scores from 20,000 yr B.P. to 500 yr B.P.

In order to examine gradients in beta diversity and vegetational composition along the transect at 85°W through the time span of the last 20,000 years, we used DECORANA to ordinate all the taxon and sample scores from all of the six time planes evaluated. For any given time, this analysis permits us to look at compositional change of forests across the entire environmental gradient. In addition, it permits us to examine the nature of vegetational change at any given site through time in terms of the rate of change in beta diversity. Third, it allows us to evaluate the clustering of samples in space and time in order to objectively identify (1) forests that have been relatively consistent in composition throughout the last 20,000 years and therefore have modern analogues; and (2) forest types that were ephemeral and that lack good modern analogues. In addition, the shifting position of ecotones can be tracked through space and time by measuring the distance (both in terms of degrees latitude and in sd units of beta diversity) by which the clusters of latitudinal samples are separated from each other.

Evaluation of clustering of taxa through time. When plotted in ordination space along DCA Axes 1 and 2, the taxon scores illustrate the broad spectrum from warm-temperate taxa characterized by values < 1.0 along Axis 1, to cool-temperate taxa between 1.0 and 3.0 along Axis 1, and boreal taxa with scores > 3.0 along Axis 1 (Fig. 8.6). Four taxa cluster tightly with values < 1.0 on Axis 1, within what we interpret as the southeastern evergreen forest. These taxa include southern *Pinus* [with the letter code P(SO)], *Nyssa* (NY), *Carya* (CA), and the southern group of Cupressaceae and Taxodiaceae [CU(SO)]. The group of taxa characterizing the deciduous forest includes *Quercus* (QU), *Fagus grandifolia* (FA), *Tilia* (TI), *Juglans* (JU), and *Salix* (SA); these taxa are arrayed along ordination Axis 1 between values of 1.0 and 1.8. The third forest type, the mixed conifer–northern hardwoods forest, occurs between 1.8 and 3.0 along ordination Axis 1. Taxa characterizing this vegetation type can be segregated into two subgroups within ordination space (Fig. 8.6). The first group, with values of < 1.4 on Axis 2, includes *Tsuga* (TS), northern Cupressaceae [CU(NO)], *Ulmus* (UL), and *Acer* (AC); the second group, with values > 1.4 on Axis 2, includes *Celtis* (CE), *Fraxinus* (FR), and *Populus* (PO). In the case of the boreal forest, with taxon values > 3.0 along Axis 1, two subgroups can be identified. A species-rich boreal forest with values < 1.4 on Axis 2 includes the coniferous taxa of northern *Pinus* [P(NO)], *Abies* (AB), and *Larix laricina* (LA), as well as the broadleaved deciduous taxon *Betula* (BE). The second subgroup, positioned at > 1.4 on Axis 2, represents species-depauperate boreal forest dominated by *Picea* (PI), as occur today, for example, along the northern periphery of the boreal forest region (Rowe 1972; Delcourt *et al.* 1984).

Evaluation of clustering of samples through time. The plot of sample scores (Fig. 8.7) reveals new insights concerning the changes in composition of late-Quaternary boreal and temperate forest communities. Ordination Axis 1 has an eigenvalue of 0.769, accounting for the majority of compositional variation within the data set. Ordination Axis 2 (Fig. 8.7) has an eigenvalue of 0.265. Axis 1 represents the gradient from warm-temperate to boreal forests. Several

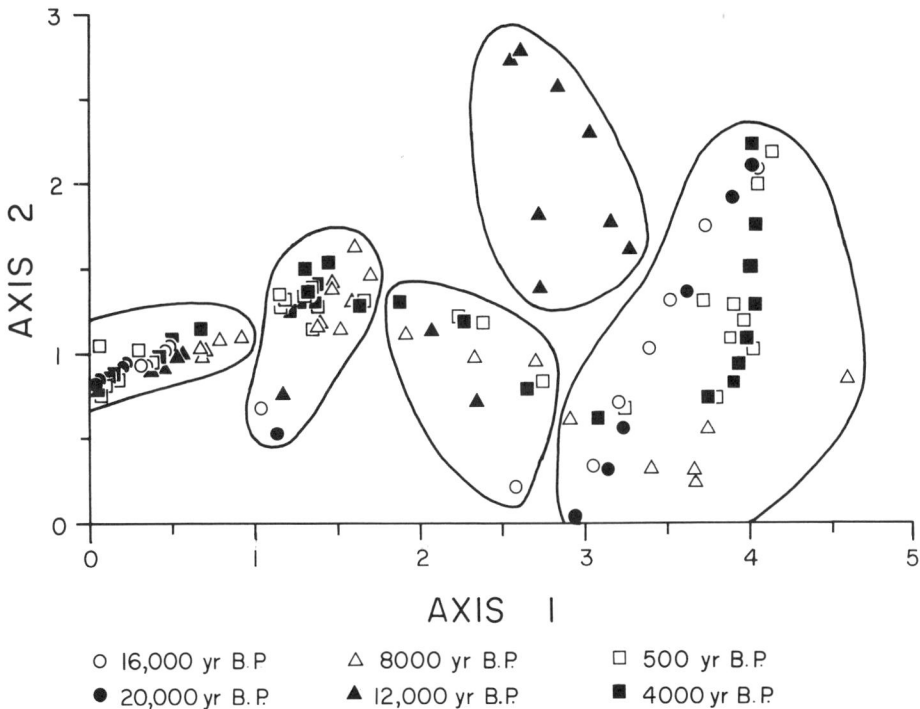

Figure 8.7. DECORANA sample scores for the untransformed data including all latitudinal samples from 20,000 yr B.P. to 500 yr B.P. along the transect at 85°W.

discrete clusters include, from left to right along Axis 1, southeastern evergreen forest (< 1.0 sd), deciduous forest (1.0 sd to 1.8 sd), and then a broad overlap within each of the mixed conifer–northern hardwoods (1.8 sd to 3.0 sd) and of the boreal forest communities (> 3.0 sd). Along Axis 2, both the southeastern evergreen forest and the deciduous forest cluster tightly, but, within the remaining several clusters, the samples are arrayed across a much larger span of ordination space. This indicates much greater compositional variability among samples within both the mixed conifer–northern hardwoods forest and the boreal forest. On the plots for Figs. 8.6 and 8.7, the ordination values used to distinguish species clusters and forest types along the ordination Axes 1 and 2 are comparable although not necessarily exactly the same on the two graphs (Gauch 1982).

In the southeastern evergreen forest cluster, the samples include those generally in latitudes south of 35°N for all times examined over the past 20,000 years (Fig. 8.7). This cluster is displaced by a beta diversity measure of about 0.1 sd along Axis 1 from the cluster representing deciduous forest. The southeastern evergreen forest thus appears to have persisted through the late Quaternary as an intact forest type, with minimal compositional change, at least along this transect through the Gulf Coastal Plain.

Within the deciduous forest cluster, the samples typically represent the forest composition at stations from 36°N to 43°N from 8000 yr B.P. to 500 yr B.P. Forests of similar composition to those in this cluster were not represented along the 85°W transect prior to 8000 yr B.P. Development of this forest community may be explained as either (1) the Holocene establishment of a new forest type that became areally dominant (in what is today known as the central hardwood region of the eastern deciduous forest); or (2) a continuation of deciduous forest communities that spread from another region during the early Holocene and crossed into the area of this 85°W transect.

The boreal forest samples span approximately 2.1 sd along Axis 1 and about 2.5 sd along Axis 2. The boreal forest communities reconstructed for 20,000 yr B.P., 16,000 yr B.P., 8000 yr B.P., and 4000 yr B.P. cluster within the range of presettlement boreal forests reconstructed for 500 yr B.P. However, forest samples from 38°N to 44°N (solid triangles on Fig. 8.7 located at > 1.4 sd on Axis 2) for 12,000 yr B.P. are in a separate cluster from that containing boreal forest samples from 500 yr B.P. As identified in ordination space, full-glacial and mid- to late-Holocene boreal forest communities along the transect are comparable in vegetational composition.

Between the cluster for the deciduous forest and those for the boreal forest (generally from 1.8 sd to 3.0 sd on Axis 1, Fig. 8.7), substantial scatter of samples and wide spacing of ecotonal samples characterize the mixed conifer–northern hardwoods forest communities. Although the ecotone separating temperate from boreal forests expanded from about 32°N to 34°N at 20,000 yr B.P. to between 43°N and 48°N by 500 yr B.P. (Fig. 8.6), the range of compositional variability within the mixed conifer–northern hardwoods forest remained over approximately 2.2 sd along ordination Axis 1. For all times except for 12,000 yr B.P., the cluster of samples of this forest type remained within the broad range of its compositional variability exhibited at 500 yr B.P., < 1.4 sd along Axis 2. Samples from 12,000 yr B.P. (during the transition from Pleistocene to Holocene conditions) represent forest communities dominated by *Populus*, with *Fraxinus* and *Picea*. These mixed forest communities were substantially different in composition from those at 500 yr B.P., with values of 1.4 sd to 2.3 sd along Axis 2 (Fig. 8.7). This cluster of values represents an example of ephemeral late-glacial to early-Holocene communities with poor modern analogues that developed as many tree taxa migrated north of 37°N and invaded deglaciated terrain as far north as 40°N.

Rates of species turnover through the past 20,000 years

The DECORANA ordination of sample scores for all sites and times included along the 85°W transect provides quantitative measures by which the rates of species turnover can be determined through time. This constitutes a quantitative test of the rate of change in forest composition hypothesized for the Southeast, the middle latitudes, and deglaciated eastern North America presented in Delcourt and Delcourt (1983).

Our hypothesis concerning late-Quaternary vegetational dynamics across the Gulf Coastal Plain was presented as a scenario of minimal environmental change and of dynamic vegetational equilibrium maintained through glacial–interglacial cycles. We postulated that this mode of vegetational dynamics was one in which the complement of arboreal taxa remained constant, with no local extinctions or species invasions. Further, the hypothesized dynamic equilibrium implies that, although the flora may have remained constant through time, the populations of tree taxa may have shifted in relative dominance along local gradients in moisture availability and in disturbance regimes such as fire. To test this model explicitly, we examined the measures of beta diversity at one latitudinal position along the transect (32°N, near Goshen Springs, Alabama). At this sample location, the total value of beta diversity was only 0.6 sd for the full time interval from 20,000 yr B.P. to 500 yr B.P. (Fig. 8.7), indicating that both the rate and magnitude of compositional change within southeastern evergreen forests were minimal. For example, individual beta diversity measures from 20,000 yr B.P. to 16,000 yr B.P. and from 16,000 yr B.P. to 12,000 yr B.P. were 0.3 sd and 0.1 sd, respectively. The Holocene measures of beta diversity were 0.2 sd from 12,000 yr B.P. to 8000 yr B.P., 0.6 sd from 8000 yr B.P. to 4000 yr B.P., and 0.0 sd from 4000 yr B.P. to 500 yr B.P. Although the greatest beta diversity value (only 0.6 sd) was associated with the mid-Holocene changeover from oak to pine dominance, that value was still small in comparison with those measured at higher latitudes. This result is consistent with a view of coastal-plain vegetation in dynamic equilibrium, within a relatively coherent climatic regime and with a landscape mosaic of successional stages fluctuating in response to subtle changes in predominant disturbance regimes (*e.g.*, fire and hurricanes).

For the middle latitudes between 33°N and 39°N, we postulated that the vegetational equilibrium established during full-glacial climatic conditions would have been replaced by nonequilibrium vegetation with major changes in forest composition during the time of late-glacial and early-Holocene climatic amelioration between 16,500 yr B.P. and 9000 yr B.P., with the establishment of a new vegetational equilibrium during the mid- and late-Holocene intervals (Delcourt and Delcourt 1983). Late-glacial and early-Holocene instability in the vegetation would have been triggered by major climatic change that exceeded the tolerance limits for species that dominated during the full-glacial interval, causing their local extinctions. With climatic warming, additional species immigrated into the region from refugial areas located generally to the south. We proposed that the full-glacial and interglacial climatic regimes represented two very different sets of boundary conditions in the middle latitudes toward which the vegetation would have tended to equilibrate (Delcourt and Delcourt 1983). The test of this hypothesis is provided by the beta diversity measures from 36°N along the 85°W transect (a location near Anderson Pond, Tennessee). The beta diversity measure of 0.0 sd between 20,000 yr B.P. and 16,000 yr B.P. indicates vegetational equilibrium during the full-glacial interval. At 36°N, the greatest rate of species turnover during the last 20,000 years was 1.1 sd, occurring during the late-glacial interval and the Pleistocene–Holocene transition between 16,000 yr B.P. and 12,000 yr B.P. From 12,000 yr B.P. to 8000 yr B.P., the beta diversity measure decreased to 0.5 sd. During the mid- and late-Holocene intervals, beta diversity decreased progressively to 0.3 sd from 8000 yr B.P. to 4000 yr B.P. and to 0.1 sd between 4000 yr B.P. and 500 yr B.P. This is consistent with an interpretation of a shift from a relatively stable full-glacial vegetation through a time of rapid and major compositional change during the late glacial and early Holocene, and then gradually stabilizing through the middle and late Holocene. From 20,000 yr B.P. to 500 yr B.P., the overall span of change in beta diversity was 2.0 sd, indicating that substantial changes in forest composition occurred over the glacial–interglacial cycle. These results are consistent with those from the analogue comparisons of fossil-pollen assemblages from Anderson Pond with modern pollen samples (Chapter 3; Delcourt and Delcourt 1985a). In that test of the hypothesis, excellent modern analogues (indicated by relatively small coefficients of dissimilarity) were found for full-glacial pollen assemblages within the modern boreal forest region of central and eastern Canada. Relatively poor analogues, from modern mixed conifer–northern hardwoods forests, were found for late-glacial and early-Holocene pollen assemblages from Anderson Pond. Progressively closer modern analogues, from the eastern deciduous forest region, were found for mid- and late-Holocene pollen assemblages from Anderson Pond.

The third mode of vegetational dynamics, within cool-temperate and boreal forest regions, is characterized by successive invasions of tree species through the Holocene across deglaciated landscapes north of 43°N (Delcourt and Delcourt 1983). Within this region, we postulated that forest vegetation was in continual disequilibrium because of changes in climate, long-term nutrient loss accompanying soil leaching, and species migrations during interglacial times. To test this hypothesis concerning vegetational dynamics in deglaciated terrains, we examined the measures of beta diversity at 43°N along 85°W (near Demont

Lake, Michigan; Kapp 1977a). Glacial ice covered that landscape from 20,000 yr B.P. to about 14,000 yr B.P. From 12,000 yr B.P. to 8000 yr B.P., the beta diversity measure was 0.7 sd, and it decreased progressively through the Holocene to 0.3 sd between 8000 yr B.P. and 4000 yr B.P. and to 0.1 sd from 4000 yr B.P. to 500 yr B.P. The total range in beta diversity was 1.0 sd, indicating the persistence of a mixture of populations for both boreal and cool-temperate tree taxa after their arrival in the central Great Lakes region during the last 12,000 years. The last major immigrations of tree taxa occurred by about 7000 yr B.P., and readjustments in relative dominance of taxa resulted in progressive increases in importance of deciduous tree taxa during the last 7000 years. The beta diversity measures indicate that maximum vegetational change, in terms of rate of species turnover, occurred during the early Holocene, with overall forest composition tracking toward equilibrium conditions by late-Holocene times.

Conclusions

1. Over the past 20,000 years, across eastern North America, both the positions and breadth of ecotones have shifted in response to late-Quaternary environmental changes. Examination of ecoclines allows evaluation of the combined influences of physical barriers constraining the area available for occupation and of the dynamic changes in forest community composition due to differential species invasions.

2. The measures of beta diversity permit the quantification of the rates of compositional change of temperate and boreal forests across both space and time. These measures identify the broadening of a latitudinal zone of vegetational disequilibrium across the Pleistocene–Holocene transition, shifting from the middle latitudes northward across deglaciated terrain.

3. Detrended Correspondence Analysis (DCA, performed with the FORTRAN program DECORANA) allows objective evaluation of the degree of persistence of forest communities through the past 20,000 years, as well as identification of ephemeral combinations of tree taxa. In eastern North America, the southeastern evergreen forest, composed predominantly of southern *Pinus*, *Nyssa*, *Taxodium distichum*, *Chamaecyparis thyoides*, and *Carya*, persisted throughout the late Quaternary in the Gulf Coastal Plain. In contrast, the central hardwoods portion of the eastern deciduous forest, characterized by dominance of *Quercus*, with *Juglans*, *Fagus grandifolia*, *Tilia*, and *Salix*, has developed as an areally dominant forest community only in the past 8000 years. The mixed conifer–northern hardwoods forest today characteristic of the Great Lakes region represents a broad vegetational transition between temperate deciduous forest and boreal forest. This transitional mixed forest has included forest types in the past, for example the *Populus–Picea–Fraxinus* forests of 12,000 yr B.P., for which there are no good counterparts today along the transect at 85°W. Full-glacial boreal forests were compositionally similar to mid- and late-Holocene boreal forests.

However, during the late-glacial and early-Holocene intervals, the boreal forest was not a well-defined forest type and had no good counterparts in the modern vegetation.

4. This is the first application of DECORANA to a transect of plant-ecological data along a major late-Quaternary ecocline in space and time. DECORANA's analyses of ecoclines from a series of times has allowed the use of beta diversity measures to quantify rates of compositional change in forest communities through a glacial–interglacial cycle. In a quantitative test of previous hypotheses concerning the nature of vegetational dynamics along the transect at 85°W in eastern North America, the beta diversity measures identify (1) low-latitude forests that have maintained both stability and integrity in a dynamic equilibrium through the past 20,000 years; (2) mid-latitude forests that have shifted from one dynamic equilibrium state during full-glacial times to a new steady state with vegetational equilibrium achieved in the mid- and late-Holocene intervals, separated by vegetational disequilibrium during the late-glacial to early-Holocene transition; and (3) after glacial retreat from higher latitudes, forest communities that were initially in disequilibrium because of successive species invasions and continued climate change have tracked toward a late-Holocene equilibrium in species composition.

9. Quaternary Landscape Ecology

Quaternary landscape ecology represents a holistic approach to the understanding of the long-term patterns and processes operative on landscapes. It is a perspective for understanding the development of terrain, climate, and the biota, including mankind. Contemporary syntheses emphasize that studies of landscape ecology can be undertaken on virtually any geographic scale (Tjallingii and de Veer 1982; Naveh and Lieberman 1984; Forman and Godron 1986). We suggest that additional dimensions of landscape dynamics can be perceived by integrating landscape patterns with processes of landscape development over time scales of hundreds of years to tens of thousands of years.

Integration of pattern and process
in vegetation, climate, and geomorphology

In previous chapters, we have focused upon vegetation processes and patterns apparent on the macro-scale of resolution (the regional to subcontinent spatial scale). In this chapter, we characterize the development of landscapes through the late Quaternary as they are representative of the dynamic interplay of climate, geomorphic regimes, and vegetation characteristic of different physiographic regions. Each physiographic region may be viewed as a coherent landscape unit with a distinctive tectonic (structural) setting and bedrock composition, both of which influence its landforms. Each region has characteristic soils, as well as type of geomorphic, hydrologic, and pedogenic processes. Physiographic regions are also characterized by a particular macroclimate (Bryson and Hare 1974) and by a diagnostic spectrum of plant communities (Braun 1950). Each major landform–climate–vegetation system represents an ecological region or **ecoregion** (Bailey 1978; Delcourt *et al*. 1984). The historical trajectory of landscape development has been influenced by both the nature and the timing of ecosystem responses to the passage of fundamental geomorphic, climatic, and biotic thresholds. Thus, within each physiographic region, the dynamics of landscape interactions are shaped by the consistency, not the constancy, of the environment as it is modified by different kinds, frequencies, and intensities of disturbances (Pickett and White 1985).

The structural components of each landscape are represented by elements or patches that can be characterized by their size, shape, composition, arrangement, density, and temporal span (Forman 1981). Landscape patches can be viewed in the context of a landscape mosaic or matrix, and both physical and biotic processes regulate the flux of energy, nutrients, and species among the patches (Forman 1981; Pickett and White 1985). Rather than isolated from each other, however, patches are to be viewed as integrated within the landscape by patch–matrix interactions across patch boundaries (Forman and Godron 1981). Some patch boundaries are located along environmental discontinuities or steepened portions of environmental gradients. Others represent the geographic limits generated by disturbance events on the landscape. The regionally dominant disturbance regime may produce landscape patches as a function of geomorphic processes operative on landforms (Hack and Goodlett 1960; Hupp 1983), river regimes (Starkel 1982; Swanson and Lienkaemper 1982), fluctuations in groundwater table (Mills and Delcourt 1987), or even changes in sea level (Clark 1986). Other patches may result from climatically influenced storms, fires (Romme 1982; Romme and Knight 1982), human activities, pathogen outbreaks, or grazing by domesticated animals.

Biotic interactions within plant communities lead to plant succession within each patch. Interactions at higher trophic levels reflect the activities of more mobile herbivore and predator populations that move across patch boundaries and are concentrated by corridors through the landscape mosaic. Fundamental changes in biotic interactions in eastern North America during the late Quaternary have been influenced by migrations of taxa during the late-glacial and early-Holocene intervals, arrival of human populations 12,000 years ago, and the selective loss of key components representative of higher trophic levels, primarily during the Pleistocene–Holocene transition. The loss of Pleistocene megafauna in the late-glacial interval may have represented the substantial decline in grazing pressure and extinction of important seed dispersers. Thus, the ultimate cause of the megafaunal extinctions may have been related to changing patch dynamics during a time interval of vegetational disequilibrium throughout much of their former ranges (Graham and Lundelius 1984). Alternatively, it may have been the result of introduction of a new landscape element, man the hunter (Martin and Klein 1984).

In this chapter, we present examples of the unique contributions provided by Quaternary scientists for understanding the long-term interaction and dynamics of landscapes. This approach is relevant to Quaternary landscape ecology, as a time scale of thousands of years may be required to shape modern landscapes. Much of current research in landscape ecology focuses upon cultural overprints (Burgess and Sharpe 1981; Tjallingii and de Veer 1982), where the nature of disturbance regimes and patch dynamics within the Temperate Zone is obscured by the relatively recent, historic activities of humans. However, in order to understand the dynamics of ecosystems in eastern North America prior to anthropogenic modification, paleoecological and geologic data are required. For example, within montane environments, the disturbance regime, shaping the vegetation may take the form of debris avalanches that are relatively infrequent with respect to the life spans of the dominants within the vegetation, occurring at intervals of up to thousands of years (Kochel and Johnson 1984).

Paleohydrology and Quaternary vegetation of the Interior Low Plateaus

The records of sedimentation within late-Quaternary lacustrine sites can be used together with paleoecological data, derived from both fossil-pollen assemblages and plant-macrofossils, to interpret changes in local water level as well as in regional water tables (paleohydrology) through time (Delcourt et al. 1983b). Radiocarbon-dated sediment sequences from Jackson Pond, Kentucky, and Anderson Pond, Tennessee (Fig. 9.1), located within the Interior Low Plateaus region of the east-central United States, display similar rates of sedimentation through the past 19,000 to 20,000 years. These changing sedimentation patterns can be related to changes in climate, vegetation, geomorphic processes, and paleohydrology in the late Pleistocene and Holocene (Fig. 9.2; Delcourt 1985a).

The region of the Interior Low Plateaus, located immediately west of the Cumberland and Allegheny Plateau regions, is predominantly underlain by limestones and dolomites of Ordovician through Mississippian age. The topography is gently rolling, karst terrain, underlain by subterranean caverns such as the Mammoth Cave system in central Kentucky. Permanent ponds located in the Interior Low Plateaus region contain lacustrine sediments that date as old as 25,000 yr B.P. (Delcourt 1979). During the late Pleistocene, this physiographic region was located south of the glacial margin but north of the Polar Frontal Zone (Delcourt and Delcourt 1984; Fig. 4.5a). The full-glacial vegetation was boreal coniferous forest, dominated by jack pine (*Pinus banksiana*) to the south (Anderson Pond; Delcourt 1979), and by spruce (*Picea*) to the north (Jackson Pond; Wilkins 1985). During the full-glacial interval, water levels in lakes within the region were deep and permanent, as evidenced by macrofossils of submersed aquatic plants such as pondweed (*Potamogeton spirillus*) and mermaid weed (*Najas flexilis* and *N. gracillima*).

The record of paleomagnetic intensity from sediments of Anderson Pond (Lund 1981) indicates that, during the full-glacial and late-glacial intervals, the source area for mineral sediments included a portion of the Cumberland Plateau (with nearest slopes within 1 km of the site). Year-round dominance of the

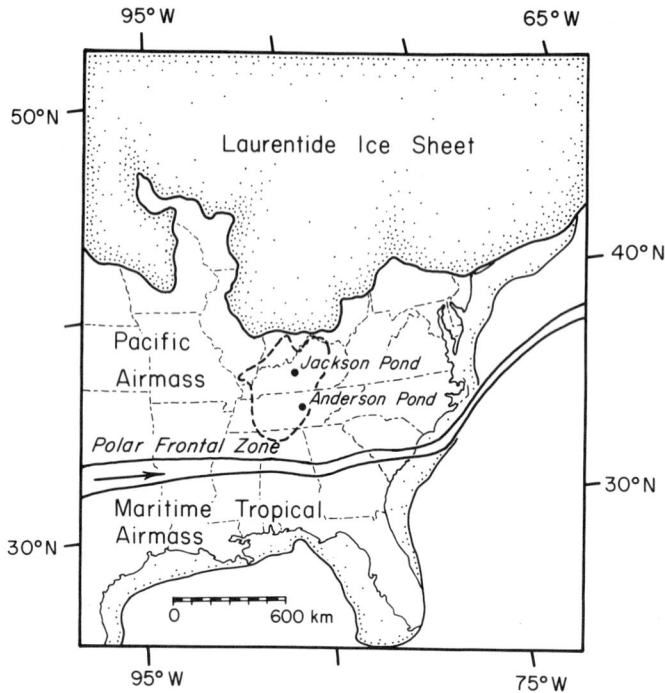

Figure 9.1. Location map for Jackson Pond, Kentucky, and Anderson Pond, Tennessee, in the Interior Low Plateaus physiographic province (region indicated by dashes). The full-glacial positions are shown for the southern limit of the Laurentide Ice Sheet, the Polar Frontal Zone separating the northern Pacific Airmass from the southern Maritime Tropical Airmass, and the marine shoreline.

Polar Frontal Zone would have concentrated moisture from storms within the Interior Low Plateau region. Cool and moist conditions would have been characterized by low rates of evapotranspiration, maintaining high regional ground-water tables, as well as high moisture content within soils and limited water percolation. Surface runoff would have met relatively little resistance from shallow humus in soils beneath the coniferous boreal forest. Under prevailing cool climatic conditions, the process of freezing and thawing of water-saturated soils would have readily transported sediment downslope as colluvium that would have been deposited rapidly within the lake basins (Delcourt 1985a).

With late-glacial climatic amelioration, warmer mean annual temperatures, heightened seasonal contrast in temperature and in moisture availability, and establishment of closed, deciduous forest across the region resulted in major changes in regional hydrology and geomorphic processes. With filling-in of the lake basins, and with both a general lowering and an accentuated seasonal fluctuation of lake levels, emergent and floating-leaved aquatic vascular plants became dominant in the plant-macrofossil records (Delcourt 1979; Wilkins 1985). The establishment of regional deciduous forest and diminishment in number

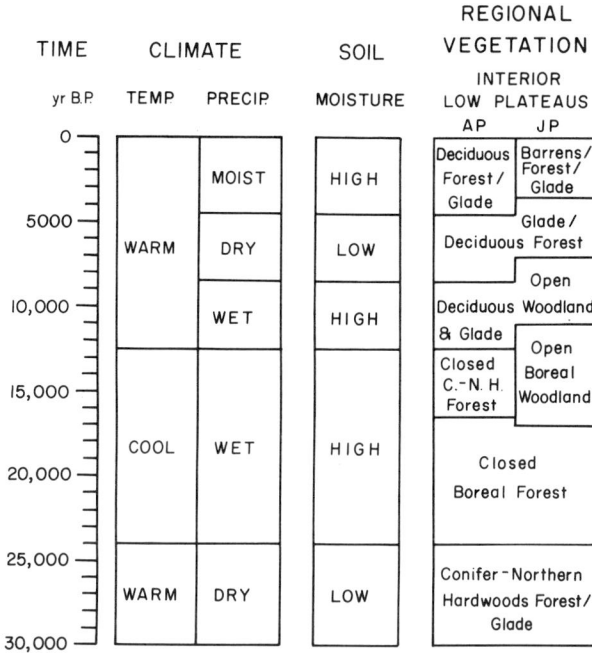

Figure 9.2. Late-Quaternary climatic, environmental, and vegetational change in the Interior Low Plateaus; Anderson Pond, AP; Jackson Pond, JP (modified from Delcourt *et al.* 1986).

and intensity of freeze–thaw cycles resulted in stabilization of upland soils. For example, at Anderson Pond, the concomitant decrease in source area for mineral sediments was reflected in a change in composition of heavy minerals, paleomagnetic intensity, and rate of deposition of mineral sediments (Lund 1981; Delcourt 1985a). The Holocene drop in regional groundwater table fostered the population expansions of temperate, xeric plant taxa. Oak–hickory forests dominated areas of thick soil, and cedar glades and prairie barrens expanded across areas of thin or no soil overlying carbonate bedrock (Delcourt *et al.* 1986b).

Quaternary landscape evolution in the southern Appalachian Mountains

The Appalachian Mountain chain represents a zone of intense periglacial climate, unstable slopes, and colluvial geomorphic activity during the Pleistocene (Pewe 1983). Pleistocene landscapes contrasted markedly with the predominantly temperate climate, stabilized slopes, and alluvial regimes that have characterized Holocene environments in the Appalachians (Hack and Goodlett 1960). In this region of high vertical relief, interactions between changing climate and slope processes have had a profound effect on landscape evolution over the past 20,000 years (Figs. 9.3 and 9.4).

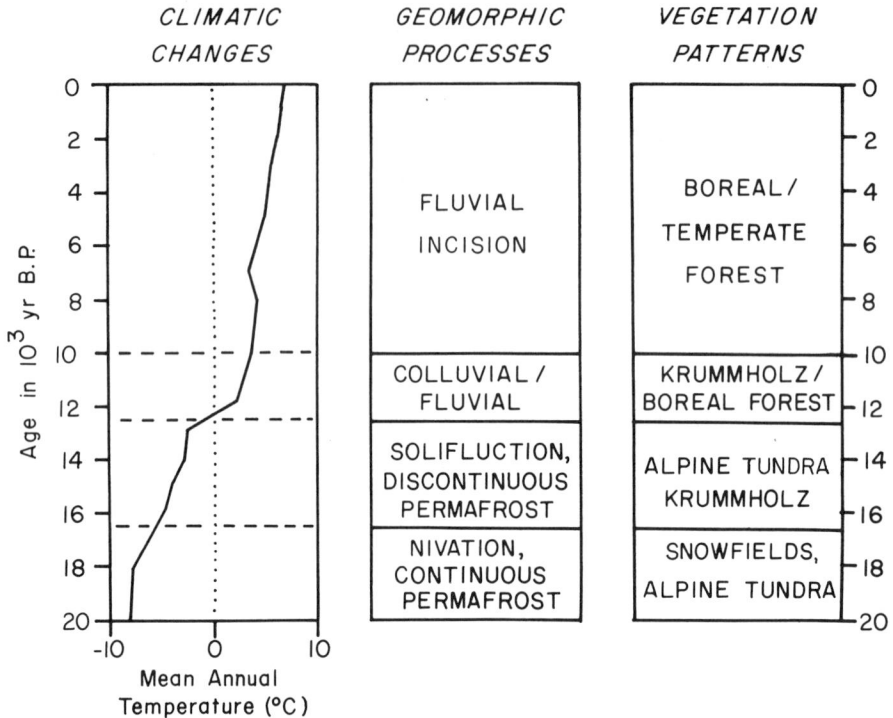

Figure 9.3. Late-Quaternary landscape model for interactions of climate, geomorphology, and vegetation at elevations above 1500 m in the southern Appalachian Mountains (from Delcourt and Delcourt 1985c).

At high elevations in the southern Appalachians, from 20,000 yr B.P. to 16,500 yr B.P., environments were characterized by mean annual temperatures well below 0°C (as much as −8°C; Shafer 1984). Under such conditions, continuous permafrost with nivation and permanent snow packs develop; intense freeze–thaw churning of the soil is a predominant geomorphic disturbance regime that keeps the landscape open and prevents establishment of trees. The mean annual temperatures were low enough during the last full-glacial interval throughout the higher elevations of the central and southern (unglaciated) Appalachian Mountains to have been suitable environments for tundra vegetation (Maxwell and Davis 1972; Watts 1979; Shafer 1984; Larabee 1986; Delcourt and Delcourt 1985c, 1986). Boreal forests covered the landscape only at low elevations both west and east of the Appalachian Mountain chain (Delcourt and Delcourt 1981).

Between 16,500 yr B.P. and 12,500 yr B.P. (Figs. 9.3 and 9.4), permafrost became discontinuous as mean annual temperatures rose to near 0°C. Solifluction became the dominant geomorphic process, and extensive patterned-ground features including stone stripes and nets (Clark 1968), as well as boulder fields (Michalek 1968), developed through active freeze–thaw heaving of surface

D. TODAY

BOREAL SPRUCE-FIR FOREST

HEATH BALD

COVE HDWDS

TEMPERATE FORESTS:

MIXED OAK

PINE

TULIPTREE

N

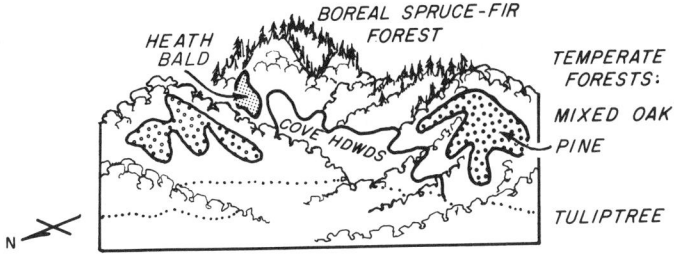

C. 12,500 to 10,000 yr B.P.

KRUMMHOLZ

BOREAL FOREST

TEMPERATE DECIDUOUS FOREST

B. 16,500 to 12,500 yr B.P.

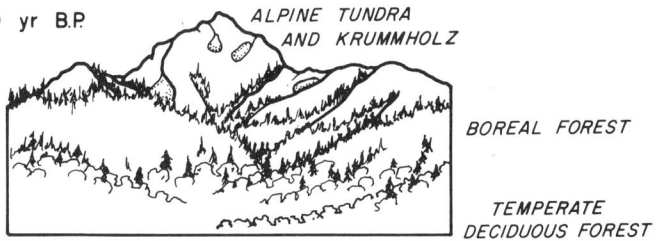

ALPINE TUNDRA AND KRUMMHOLZ

BOREAL FOREST

TEMPERATE DECIDUOUS FOREST

A. 20,000 to 16,500 yr B.P.

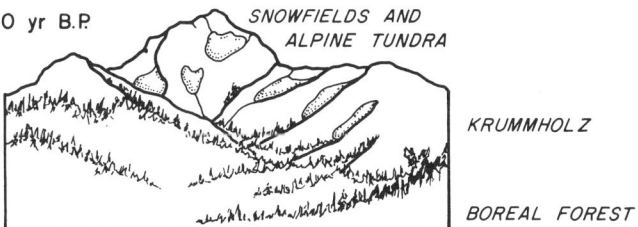

SNOWFIELDS AND ALPINE TUNDRA

KRUMMHOLZ

BOREAL FOREST

Figure 9.4. Schematic illustration of vegetational changes on Mt. LeConte, Great Smoky Mountains National Park: a, full-glacial interval, 20,000 to 16,500 yr B.P.; b, late-glacial interval, 16,500 to 12,500 yr B.P.; c, early-Holocene interval, 12,500 to 10,000 yr B.P.; d, today (from Delcourt and Delcourt 1985c).

sediments on upper slopes in the southern Appalachians (Mills and Delcourt 1987). A patchwork of tundra and krummholz was maintained within this disturbance regime, with boreal and cool-temperate deciduous forests becoming established only at lower elevations in intermontane valleys such as the Saltville Valley of southwestern Virginia (Delcourt and Delcourt 1985c, 1986).

After 12,500 yr B.P., the climate at high elevations in the Appalachians passed the geomorphic threshold between that of predominantly periglacial, colluvial processes and that of temperate, fluvial processes. With stabilization of steep mountain slopes, coniferous and deciduous forests established, occupying the upper slopes and mountain summits by 10,000 yr B.P. (Figs. 9.3 and 9.4). At low elevations, temperate deciduous forests predominated after 10,000 yr B.P. The combination of extensive tree roots, as well as thick soil-litter development and presence of coarse woody debris such as fallen logs ("tree dams"), would have increased the proportion of precipitation infiltrating below the ground surface (Harmon et al. 1985). The effectiveness of sheetwash and stream action in eroding mountain slopes would have been diminished with the development of biological impediments to all but catastrophic events of overland flow of runoff (Keller and Swanson 1979).

On the highest montane summits, intervals of accentuated storm activity during the Holocene, accompanied by high amounts of precipitation, have reactivated water-saturated solifluction flow within periglacial blockfields, causing limited downslope movement of some blockfields until as recently as 7000 yr B.P. (Shafer 1984). Such major storm events have also triggered catastrophic episodes of debris avalanche with recurrence intervals on the order of up to 2600 years (Hupp 1983; Kochel and Johnson 1984). Many relict boreal and endemic plant species persist along chutes of recurring landslide activity and along talus slopes and cliff faces, where these plants are particularly concentrated within areas of periodic geomorphic disturbance (White et al. 1984). In the southern Appalachian Mountains, the relict Pleistocene landscape of colluvial periglacial deposits has generally been incised only to a limited extent by contemporary fluvial processes (Shafer 1984; Mills and Delcourt 1987).

Appalachian forest dynamics during the Holocene have been influenced not only by recurrent disturbance events (Delcourt 1985b), but also by changing competitive interactions as forest dominants have been removed by pathogens [such as the hemlock decline at about 4800 yr B.P. (Davis 1981b) and the introduction of the chestnut blight beginning in the 1920s (Anderson 1974; Shafer 1985)]. In the late Holocene, the Appalachian forest landscape was maintained in a mosaic of different stages of forest regeneration through gap-phase dynamics (Lorimer 1980; Oliver 1981).

Landscape transformation in the deglaciated Great Lakes region

During the majority of the Quaternary, the Great Lakes region was buried beneath a major continental glacier, the Laurentide Ice Sheet. Glacial ice retreated from this region only during relatively brief episodes (ca. 10,000 years) of warming during interstadials or interglacials such as the Holocene. During the last transition from glacial to interglacial conditions, the Laurentide Ice Sheet abandoned the northernmost Great Lakes between 10,000 and 9000 yr B.P.; however, stagnant blocks of glacial ice continued to melt out from beneath a glacially deposited mantle of rock debris, for example, with glacial-ice blocks persisting for as much as 8000 years following ice retreat in Minnesota (Wright 1972). Irregular, hummocky upland surfaces were produced as underlying ice lenses and blocks melted out to form kettle-shaped lake basins (Fig. 9.5).

Late-glacial and early-Holocene landscapes in deglaciated terrain were characterized by dynamic geomorphic change. Buried glacial ice in recessional moraines and development of perennially frozen ground (permafrost) adjacent to the retreating ice margin prolonged the persistence of cooler microhabitats. In addition, the permafrost provided moisture seasonally for an active thaw layer in the upper soil, promoting solifluction activity and development of water-saturated, poorly drained muskegs. Instability of the landscape was accentuated through differential melting of ice, and topographic reversals of the landscape occurred at topographic ridges (such as ice-cored moraines) subsided locally to form the bottom of kettle-shaped lake basins (Fig. 9.5) (Florin and Wright 1969; Wright 1980).

The deglaciated terrain, particularly in the Great Lakes region, was characteristically underlain by "fresh", unleached rock debris in the form of glacial

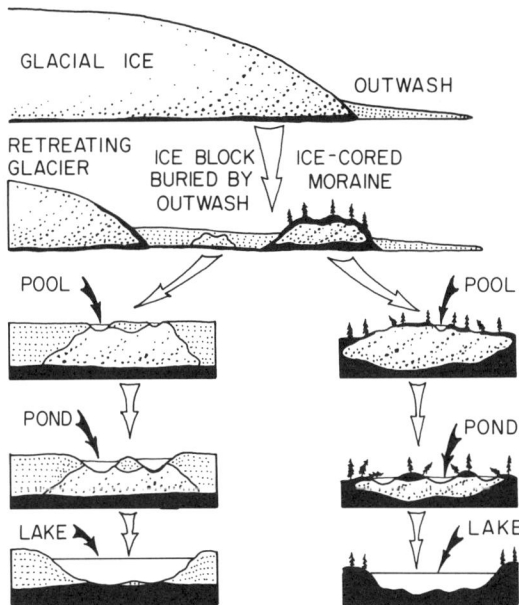

Figure 9.5. Landscape stabilization following glacial retreat in the Great Lakes region (modified from Florin and Wright 1969). The melting of stagnant ice blocks results in irregular upland surfaces with kettle-shaped lake basins.

till or drift, which was initially often calcareous in mineralogical composition (Wright 1972). This provided a substrate of newly deposited, unaltered sediment, and a fresh supply of mineral nutrients available for newly establishing plants immigrating into the region. The invasion of plants into the region began with primary succession of light-demanding, shade-intolerant pioneers. As r_m-strategists, these plant taxa saturated the open landscape with propagules and established populations. In part, the frost churning of the substrate in the periglacial zone bordering the glacier represented a geomorphic disturbance regime that continually exposed mineral sediment that was subsequently available for colonization. The herbaceous and shrubby vascular plants and bryophytes that established initially were primarily broadly tolerant taxa that responded positively to the geomorphic disturbance regime and to the areally extensive calcareous substrate (Miller 1973a, 1980). *Alnus* shrubs and legumes were among the first colonists, and they fixed nitrogen within their roots, supplementing the supply of nutrients otherwise limiting to subsequent invading plant species. Plant succession from tundra through forest communities progressed with continued climatic amelioration and arrival of arboreal r_m-strategists including aspen and spruce (Kapp 1977b).

With extensive development of boreal forests, biological productivity promoted the gradual buildup of a soil-litter horizon, downward percolation of humic acids, and solution of the carbonate mineral particles from the nutrient-rich upper soil. The shift in soil pH from alkaline to neutral produced an environmental change that selected against calciphilous plants and reduced their abundances and distributional ranges within early-Holocene plant communities (Miller 1980). With subsequent postglacial arrival of cool-temperate deciduous and coniferous trees into the Great Lakes region, shade-tolerant and more effective competitors displaced earlier arriving boreal taxa from many of the habitats that they had formerly occupied. The early-interglacial conditions of increasing warmth and maximum seasonal contrast in temperature favored the establishment of more species-rich mixed conifer–northern hardwoods forests. The forest communities produced acidic soil litter which continued to change the soil pH to more acidic conditions, leached nutrients from the soil, and resulted in a long-term loss of nutrients from upland forest ecosystems. Transport of water-soluble nutrients through the groundwater enriched the overall available nutrient supply and the productivity of aquatic communities within lake and stream environments (Whitehead 1979; Davis *et al.* 1984).

In the late stages of the present interglacial cycle (Kapp 1977b), the combination of three environmental stresses on plant populations, including climatic cooling during the last 4000 years; the progressive loss of nutrients within leached, podzolic soils; and the long-term sediment infilling of kettle-lake depressions, expanding the areal extent of poorly drained bog and swamp habitats, have all favored the shift in overall forest composition toward increased populations of evergreen trees and shrubs. Late-Holocene population expansions of spruce, fir, and heath shrubs such as blueberries have occurred in response to colder climates, soil acidification, limited nutrient availability, and the expansion of suitable habitats in poorly drained wetlands (Kapp 1977b).

The interglacial replacement of r_m-strategists and then K_m-strategists by stress-tolerant species (*sensu* Grime 1979) reflects, in part, the progressive leaching of carbonates from soils. Long-term podzolization of soils has resulted in acidification of both upland and aquatic ecosystems over the course of the Holocene (Davis 1987).

Within deglaciated terrains in eastern North America, interglacial patterns of forest development are predictable at the level of vegetation formation (Kapp 1977b); however, within the context of boreal forest, mixed conifer–northern hardwoods forest, or temperate deciduous forest, forest dynamics reflect differences in the migrational timing of arrival and subsequent population expansions of arboreal taxa, based upon chance circumstances, location of full-glacial refugia, and the rates and routes of their late-glacial and postglacial migrations (Davis 1981a, 1983).

Conclusions

1. Geomorphic, climatic, and biological thresholds determine the suite of possible ecosystem interactions on the landscape. These thresholds provide boundary conditions that constrain the dynamic interactions of landscape elements through long periods of time.
2. Fundamental changes in glacial and interglacial forest dynamics are tied to both rate and magnitude of environmental change and their subsequent impact upon environmental thresholds and disturbance regimes.
3. The nature of forest dynamics in part reflects biological responses to the predominant disturbance regime; it is the consistency in kind, frequency, and intensity of physical or biological disturbance, rather than the constancy of the environment, that determines the scale and heterogeneity of interactions across patches in the landscape.
4. The nature of the establishment and development of interglacial forest communities in the northern Temperate Zone is anomalous in the context of the predominant geomorphic and climatic environmental regimes of the Quaternary. An understanding of long-term trajectories of vegetational development is required in order to interpret pattern and process in vegetation on the modern landscape.

10. Long-Term Forest Dynamics of the Temperate Zone

In a recent summary volume, Huntley and Birks (1983) reviewed the late-Quaternary paleoecological literature from the region of continental Europe situated between the Mediterranean Sea and Fennoscandia, and including the British Isles. This region spans from approximately 10°W to 25°E longitude and from 35°N to 70°N latitude, a geographic sector roughly equal to that in eastern North America for which we have synthesized the literature concerning vegetational changes in the past 20,000 years (Chapters 4 and 5). Together, the summaries available from Europe (including Birks 1986) and from eastern North America cover a substantial portion of the northern Temperate Zone and provide a useful comparison of long-term dynamics of temperate forests.

Late-Quaternary vegetational and climatic history of Europe

Huntley and Birks (1983) compiled fossil-pollen data from 843 paleoecological sites throughout Europe. They mapped pollen percentages of key taxa and used Principal Components Analysis to reconstruct changing vegetation through the past 13,000 years. The following discussion of the paleogeography of European ice sheets, changes in vegetation types, locations of Pleistocene refuge areas for the principal tree taxa, as well as routes and rates of migration for tree taxa during the Holocene follow the treatment of Huntley and Birks (1983) unless additional literature is cited.

Environmental conditions from 25,000 yr B.P. to about 13,000 yr B.P. were generally representative of full-glacial environments throughout most of the European continent (Fig. 10.1). Extensive ice caps were located across Fennoscandia (covering most of Norway, Sweden, and Finland), Iceland, and portions of Scotland and Ireland. Alpine glaciers were most extensive at high elevations across the Pyrenees of northern Spain, the Massif Central of France, the west-to-east trending French, Italian, and Swiss Alps, and the Carpathian Mountains. During the late Pleistocene, forests were displaced from most of Europe. The vegetation was tundra to the north of the major mountain chains and xeric, predominantly treeless steppe to the south. Spruce forests and mixed

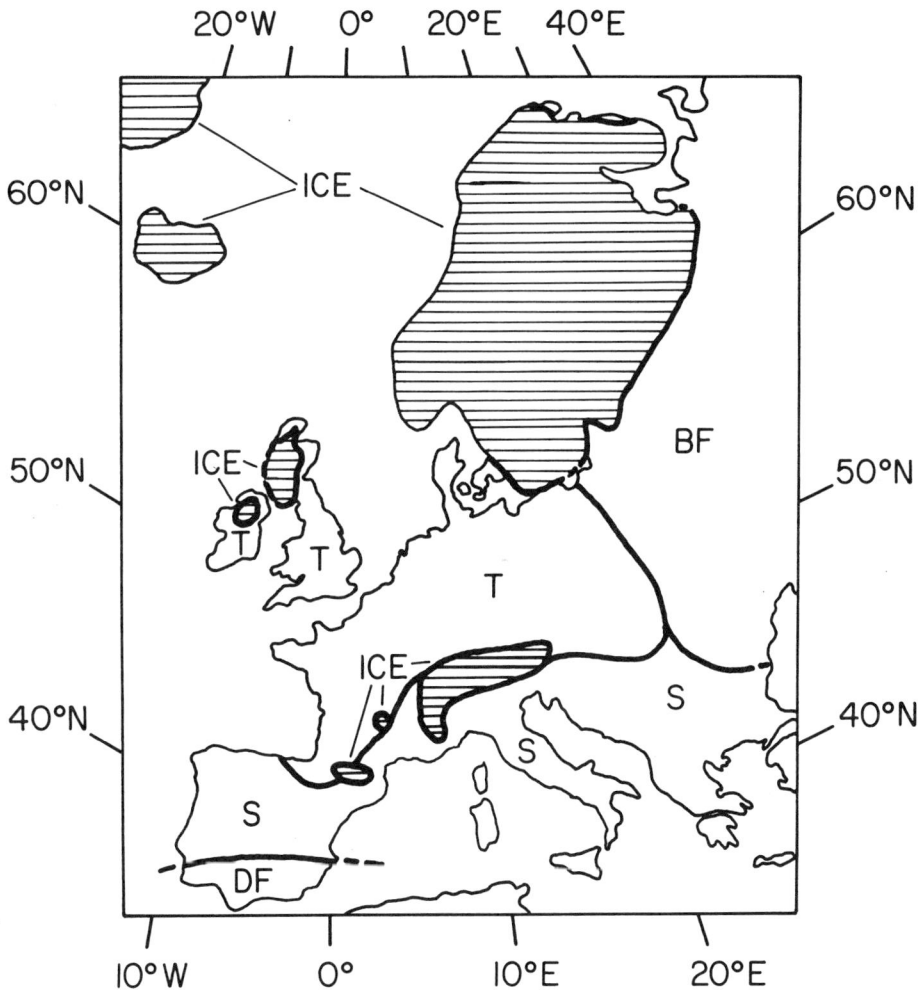

Figure 10.1. European vegetation reconstructed for 13,000 yr B.P.: T, tundra; BF, boreal forest; S, steppe; and DF, deciduous forest (modified from Huntley and Birks 1983, © Cambridge University Press, reprinted with permission).

forests of birch and conifers occurred east of about 20°E and north of 45°N, and mixed deciduous forest was extensive only in southern Spain. These vegetational patterns reflected a cold, continental climate with predominant surface-air flow from the continent toward the oceans. The full-glacial displacement of the marine Polar Frontal Zone southward to Spain (Ruddiman and McIntyre 1981a; Van Campo 1984) contributed to the maintenance of frigid conditions over much of Europe and a steep climatic gradient separating arctic and boreal vegetation from more temperate vegetation at about 35°N to 40°N latitude.

Between approximately 13,000 yr B.P. and 11,000 yr B.P., pronounced climatic warming during the late-glacial interstadial (the general time span represented by the Bolling interstadial, Older Dryas stadial, and Allerod interstadial; Lowe and Gray 1980) occurred across western and northwestern Europe. This initial warming was particularly prominent in the European areas adjacent to the Atlantic Ocean (Watts 1980c). Independent paleo-oceanographic evidence documents that this warm interval corresponded to the northward shift in the marine Polar Frontal Zone and the onset of deglacial warming of surface waters in the northeastern Atlantic Ocean between 13,300 yr B.P. and 11,000 yr B.P. (Duplessy *et al.* 1981).

By 12,000 yr B.P., the ice caps disappeared from Scotland and Ireland but remained throughout Iceland and over most of Fennoscandia. Alpine glaciers were greatly reduced in areal extent. Treeless steppe dominated over much of southern Europe from Greece to Italy and Spain. Tundra persisted throughout much of northwest Europe, from western France to Denmark and across the British Isles. Spruce forest became extensive in the western portion of the U.S.S.R., while mixed forests of birch, pine, larch (*Larix*), and spruce extended westward into the Alps and across central Europe. A refugial pocket of temperate, mixed deciduous forest apparently persisted within Bulgaria in southeastern Europe.

By 11,000 yr B.P., remnant ice sheets still remained across Fennoscandia and Iceland. Tundra was maintained in periglacial landscapes across the British Isles and in northwestern France. The area occupied by treeless steppe was greatly diminished, situated primarily within the Balkan Peninsula and through Italy. Mixed conifer–northern hardwoods forest spread throughout much of eastern and central Europe, as mixed deciduous forest became more extensive in southern Spain. In the late Quaternary, sclerophyllous vegetation, today characteristic of the Mediterranean region, was not known north of the Mediterranean Sea until 11,000 yr B.P., when it first was recorded in the pollen record from southern Greece.

The time interval of most rapid climatic and vegetational change over the European continent was during the glacial–interglacial transition, between 11,000 yr B.P. and 9000 yr B.P. During the interval from 11,000 yr B.P. to 10,000 yr B.P. (the Younger Dryas stadial; Lowe and Gray 1980), the marine Polar Frontal Zone once again shifted from a position north of the British Isles south to offshore Spain, bringing polar waters and major ocean cooling. By 10,400 yr B.P., sea-surface temperatures were as cold as during the last glacial maximum. By 10,000 years ago, the Polar Frontal zone once again tracked northward across the northeast Atlantic Ocean, and the Gulf Stream brought warm water northeast to the British Isles and northwest coasts of Europe. The late-glacial to early-Holocene transition at 10,000 yr B.P. marked the establishment of interglacial climatic conditions across much of the European continent (Duplessy *et al.* 1981). The rapid vegetational changes occurring in the early Holocene, including changes in both distribution and composition of forests, represented biotic responses to major postglacial warming. Two to three thousand years of vegetational disequilibrium accompanied northward and westward migrations of plant species; the early-postglacial vegetation had only poor modern analogues.

By 10,000 yr B.P., although an ice cap still covered most of Fennoscandia, birch and mixed deciduous forest began to invade into southern Great Britain. Tundra became progressively restricted in Scotland, Ireland, and then finally to the coastal zones of Iceland, Norway, and Greenland. Birch–conifer forest was established throughout central Europe, and mixed deciduous forest became more widespread in western Europe, specifically from southern England to Spain and in northern Italy. Steppe was restricted to the southeastern region of Europe, and Mediterranean-type vegetation remained in southern Greece.

By 9000 yr B.P., however, much of Fennoscandia was deglaciated and subsequently invaded by boreal birch forest and mixed birch–conifer forest. Mixed conifer–northern hardwoods forest occupied east-central Europe, and mixed deciduous forest spread throughout western Europe from the British Isles to Spain and eastward across much of France, Italy, and the Balkans. Steppe was eliminated from Europe, and Mediterranean-type vegetation still persisted only in southern Greece.

Between 9000 and 8000 yr B.P., the last remnants of glacial ice disappeared from Iceland and Fennoscandia. Tundra vegetation was mapped only in northernmost coastal Iceland and coastal Greenland. Birch forest and mixed birch–conifer forest extended throughout much of Fennoscandia. Mixed deciduous forest advanced into southern Sweden and Norway as well as throughout much of the region of central Europe formerly occupied by mixed conifer–northern hardwoods forest. Montane mixed conifer–deciduous forest spread throughout the Alps and Carpathians, and plant species diagnostic of Mediterranean-type vegetation colonized the region from southern Italy to Spain.

By 6000 yr B.P. (Fig. 10.2), an extensive west-to-east trending strip of montane mixed conifer–hardwoods forest had developed across the mountains of central Europe. By 4000 yr B.P., this forest type extended westward to the Pyrenees in northeastern Spain. Mediterranean-type vegetation continued to spread westward across Spain by 2000 yr B.P.

From the qualitative identification of good modern analogues for fossil-pollen assemblages, Huntley and Birks (1983) inferred that the major vegetation types present today across Europe established in the mid- and late-Holocene intervals (Fig. 10.2). They concluded that the mid-Holocene vegetation had attained an interglacial condition of stability and equilibrium. However, the development of modern European vegetation during the last 6000 years reflects continuing adjustments in forest distribution and composition to a new, anthropogenic disturbance regime. In the mid- and late-Holocene intervals, human populations

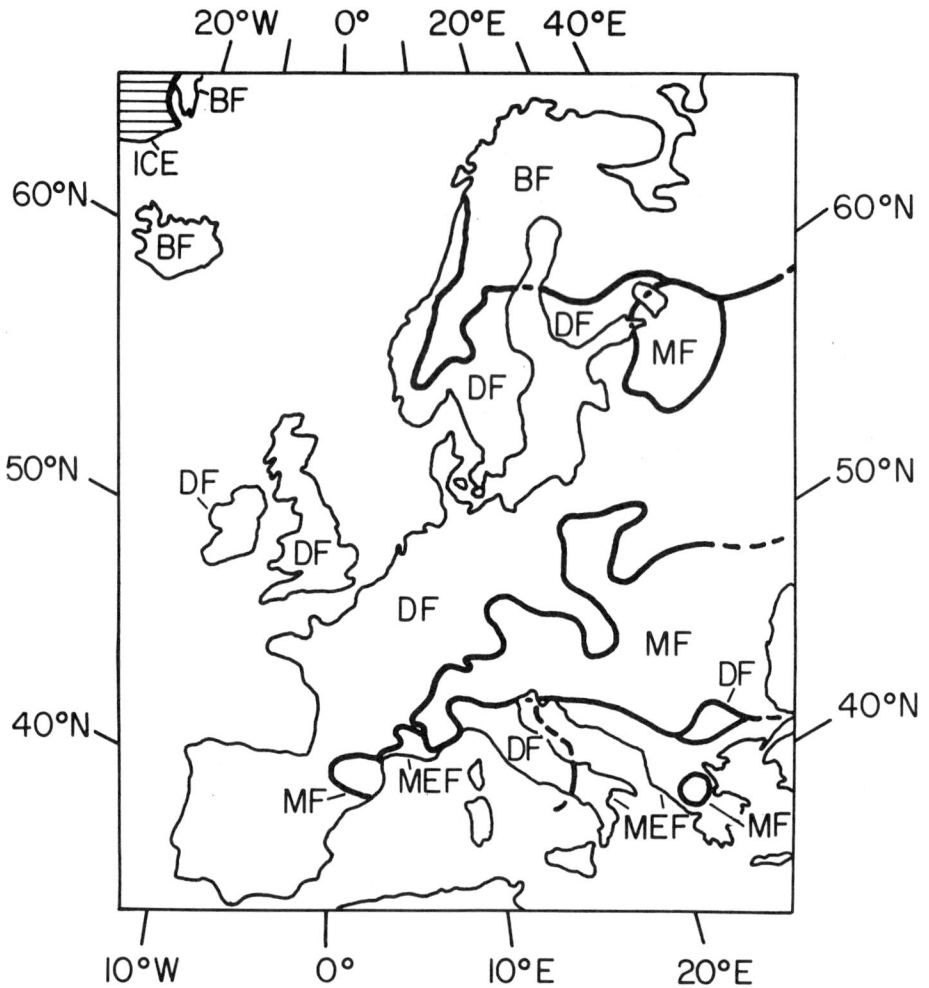

Figure 10.2. European vegetation reconstructed for 6000 yr B.P.: BF, boreal forest; MF, mixed conifer–northern hardwoods forest; DF, deciduous forest; MEF, Mediterranean forest (modified from Huntley and Birks 1983, © Cambridge University Press, reprinted with permission).

were involved in forest clearance and cultivation (*e.g.*, the initial episode of burning and cutting of forests during the "landnam" or "land occupation phase", Iversen 1941, 1960), cutting of tree sprouts and limbs for fodder (Troels-Smith 1960), intensifying the territorial concentration of grazing effects of herds of domesticated animals, and dispersing arboreal taxa such as *Juglans* and *Olea* (olive) as commercially desirable plant resources. During the initial forest clearance between 6000 and 4000 yr B.P., opening of the forest canopy created increased opportunities for invasion by early-successional forest taxa. Long-term fragmentation of European forests is the result of Neolithic and post-Neo-

lithic deforestation associated with cultivation (Barker 1985); agricultural practices involved the deliberate introduction of crop plants and the inadvertent
but rapid spread of weeds and ruderals (West 1977, Behre 1981). Particularly
during the last 2000 years, since the Roman Period, tree populations in Europe
have been cultivated and managed for their wood and food products (Barker
1985).

Locations of refugia for European temperate tree taxa

Because relatively few radiocarbon-dated sites studied from southern Europe
contain evidence of vegetation composition prior to 13,000 yr B.P., Huntley
and Birks (1983) considered that sites in which temperate tree taxa first occurred
after 13,000 yr B.P. represented their primary refugial areas during the full-
glacial interval. Of 17 refuge areas identified in southern Europe, the regions
with highest species richness of tree taxa were concentrated in the southeast,
with a pronounced gradient westward as well as northward from the Mediterranean Sea to the southern Alps. As in southeastern North America, European
refuge areas for temperate (and boreal) trees were small and patchy in distribution. The locations of forest refugia were determined by local diversity in
topography, soils, and the ameliorated climatic conditions that existed near the
Balkan Peninsula relative to the remainder of the European continent (Huntley
and Birks 1983).

Because of the extreme cold and dry climatic conditions that prevailed across
Europe during glacial times, and in part because of the east-to-west trending
barrier of the major mountain chains, the arboreal flora of Europe has become
progressively more depauperate in taxa through the Quaternary (Van der Hammen *et al.* 1971; West 1977; Nilsson 1983). The arboreal taxa that survive today
in Europe tend to be generalists with broad ecological amplitudes; populations
of most of these taxa survived the last major episode of continental glaciation
in a number of different refuge areas relatively isolated from each other. This
may have increased the genetic diversity of the populations and hence increased
the resilience of the tree taxa when released during interglacial periods of
warming and increased precipitation (Huntley and Birks 1983).

Comparison of northward migration routes and rates between European and eastern North American tree taxa

In eastern North America, the effects of late-glacial climatic amelioration were
reflected in northward migrations of temperate tree taxa within and west of the
Appalachian Mountains as early as 16,500 yr B.P. (along migration Tracks 1
to 4). However, northward tree migrations were delayed along the Atlantic
Coastal Plain of North America until generally after 12,000 yr B.P. Thus, the
"delayed" timing of the migrations of tree species along the Atlantic Seaboard
of eastern North America (Track 5) is comparable with the European situation
along the coast of the northeast Atlantic Ocean.

Migration rates for European tree taxa ranged from a minimum value of
25 m/yr to a maximum recorded rate of 2000 m/yr. Slowest advances were

recorded for taxa such as ash (*Fraxinus ornus*) that were climatically restricted in distribution during the Holocene and moved only several hundred kilometers from the locations of their full-glacial refugia. The fastest migration rates were displayed by early-successional, riparian taxa such as alder (*Alnus*), with broad climatic tolerances and extraordinary circumstances of dispersal that included long-distance transport of their seeds by northward flowing rivers. The majority of taxa had typical maximum rates of migration between 100 m/yr and 500 m/yr, values generally of the same order of magnitude as those calculated for eastern North American trees (see Table 6.9 in Chapter 6, as well as consistent with preliminary migration rates calculated from pollen data by Davis 1976, 1981a, and by Davis and Jacobson 1985).

Maximum rates of migration greater than 1000 m/yr were calculated by Huntley and Birks (1983) for six European tree taxa (*Acer, Alnus, Carpinus betulus, Corylus, Pinus,* and *Ulmus*). These rates of advance are considerably higher than the maximum rates calculated in this study (Table 6.9) for eastern North American tree taxa. However, the measurements of migration rates for European trees (Table 6.13 in Huntley and Birks 1983) are based on intuitive interpretations of the threshold of percent pollen in pollen spectra that indicate the first presumed establishment of a taxon's population, rather than on quantitative reconstructions based upon calibrations of pollen percentages with representation in vegetation assemblages. The migration rates calculated for European tree taxa therefore may be overestimates in certain cases, related more to the distances that arboreal pollen grains were dispersed into vegetational regions with very low levels of pollen production (*e.g.*, into tundra or across a broad tundra–forest ecotone).

Nevertheless, the paleogeographic settings of the two subcontinents during the time intervals of most rapid migrations in the late Pleistocene and early Holocene were quite different from each other and help to explain some of the apparent discrepancies in migration rates. For example, the major river systems of central and western Europe drain to the north and west, providing a means of rapid transit to newly deglaciated regions for seeds of early-successional, cold hardy riparian taxa such as *Alnus* and *Corylus* (Huntley and Birks 1983). In contrast, during the late-glacial interval, the meltwater-fed streams of eastern North America either drained south and southeast from the glacial margin (*e.g.*, from the confluence of the Ohio and Mississippi Rivers) or east from the Appalachian Mountains across the Atlantic Coastal Plain. Transport of seeds and fruits by flowing water therefore did not favor rapid dispersal for eastern North American tree taxa during times of predominantly northward migrations of forest populations following climatic amelioration.

With only a few mountain passes below 2000 m elevation, the generally west-to-east trending mountain chains of central and western Europe were a formidable barrier to migrations of temperate trees. Few routes were accessible for northward dispersal and establishment of tree populations from refugial areas in Bulgaria and the Balkan Peninsula. In contrast, the southwest-to-northeast trend of the Appalachian Mountains, as well as their generally lower topographic relief (with few peaks exceeding 2000 m elevation), allowed for widespread establishment of boreal and temperate trees in the early Holocene. These tree taxa, for example, hickory, were able to migrate northward both to the west and to the east of the Appalachian Mountains.

Late-Quaternary distributional histories of European tree genera

In the following sections, the late-glacial and Holocene histories of changes in distribution and pollen percentages of European trees (Huntley and Birks 1983) are reviewed for taxa with taxonomic (although not necessarily ecologic) equivalents in eastern North America. This discussion is restricted to those trees that are taxonomic counterparts of eastern North American taxa for which we have presented paleo-dominance maps based upon quantitative calibrations. The tree genera are discussed in alphabetical sequence, based upon their scientific names.

Fir (*Abies*)

The genus *Abies* includes five principal species in Europe. One species (silver fir, *Abies alba*) is widespread throughout central Europe, where it characteristically occurs in mid-elevation, mixed deciduous–coniferous montane forests codominated by beech (*Fagus sylvatica*). Silver fir is restricted in distribution to the south by its limited drought tolerance. It is replaced in the mountains of the Mediterranean region by several endemic species of fir, including *A. pinsapo* in southern Spain, *A. nebrodensis* in Sicily, and *A. borisii-regis* and *A. cephalonica* in Greece.

During the late-glacial interval, *Abies* was locally present in Greece, Italy, and possibly the Iberian Peninsula of Spain, with principal refuge areas located in the Balkans and throughout Italy. By 10,000 yr B.P., fir expanded into the southern Alps of northernmost Italy at a rate of up to 50 m/yr and reached the Pyrenees by 8000 yr B.P. Between 7500 yr B.P. and 7000 yr B.P., fir expanded in abundance to > 10% (based on the total sum of tree and shrub pollen) throughout the southern European Mountains, including the Pyrenees, the western and central Alps, the Apennine Mountains, as well as across Yugoslavia. Between 6000 and 5000 yr B.P., a second expansion phase with rates of migration up to 300 m/yr resulted in colonization of fir populations throughout the montane (beech–fir) forests of the Massif Central and Vosges of France, the southern German mountains, and the Romanian Carpathians. Subsequently, fir populations became fragmented and *Abies alba* became geographically separated in distributional range from the Mediterranean species. The mid-Holocene expansion of fir in central Europe may have resulted from climatic cooling or from (perhaps inadvertent) introduction by humans; the late-Holocene reduction in fir populations after 4000 yr B.P. is attributed largely to forest clearance and other anthropogenic activities (Huntley and Birks 1983).

Central and western European species of *Abies* are more temperate in overall climatic tolerances than are the eastern North American species of fir (note, however, that the boreal *Abies sibirica* is distributed throughout much of the U.S.S.R. but does not generally extend into western Europe). The northward migration of *Abies alba* occurred generally after 9000 yr B.P., was restricted to montane regions of both southern and central Europe, and may have been facilitated by human activities. In contrast, in North America, *Abies balsamea* spread northward along with spruce in a broad latitudinal band following the retreating Laurentide Ice Sheet. Where fir (both *A. balsamea* and *A. Fraseri*) persists today in the Appalachian Mountains, it forms the upper tree line or highest forest zone rather than occupying an intermediate position between spruce forest and deciduous forest. Fir populations achieved maximum migration rates in the middle Holocene in both North America (fastest mean rate from 8000 to 6000 yr B.P.) and in Europe (from 6000 to 5000 yr B.P.). Although a progressive reduction in fir populations occurred in Europe during the past 4000 years, it is only in the past several hundred years that anthropogenic activities (EuroAmerican logging, then accidental introduction of the balsam wooly adelgid parasite on infected stock of *Abies alba* trees; White 1984) have resulted in historic fragmentation of fir populations in eastern North America.

Maple (*Acer*)

Thirteen native species of *Acer* occur within the portion of Europe mapped by Huntley and Birks (1983). European species of maple grow primarily within mesic, mixed deciduous forests. The four most widespread species of maple are *Acer campestre*, occurring across the Mediterranean region from France east to Bulgaria and north to the North Sea; *A. pseudoplatanus*, extensive in central Europe but absent both from Fennoscandia and from the western Mediterranean region; *A. platanoides*, growing from central Europe north to beyond 63°N in Sweden and introduced by humans into the British Isles; and *A. monspessulanum*, widespread in southern Europe.

Glacial-age refugia for *Acer* were located throughout southern Europe and the southern U.S.S.R. (Grichuk 1984). By 10,000 yr B.P., the distributional range of maple was continuous throughout southeastern Europe. Rapid northward expansion at migration rates from 500 m/yr to 1000 m/yr occurred during the interval from 10,000 yr B.P. to 9000 yr B.P., and *Acer* was widespread throughout the European mountains by 9000 yr B.P. Maple retracted its range from southeastern Europe by 8000 yr B.P., but it continued to expand northward, becoming extensive throughout eastern and central Europe by 4000 yr B.P. The mid-Holocene expansion was probably a combination of populations of *Acer* species moving northward from southern Europe and the immigration into northeast Europe by *Acer platanoides* from the European Plain of the U.S.S.R. A general decline in both areal extent and abundance of maple populations occurred throughout western Europe after 4000 yr B.P. because of anthropogenic forest clearance.

Throughout the Holocene, European *Acer* was represented by 5% or less of the total pollen of trees and shrubs. In contrast, percentages of reconstructed paleo-dominance values reached 25% to 50% in portions of eastern North America during the late Holocene. Eastern North American species of maple migrated northward beginning with first climatic amelioration in the late-glacial interval, achieving their most rapid mean migration rates between 14,000 and 12,000 yr B.P., but they became forest dominants only in the late-Holocene interval within the Great Lakes region. In Europe, the fastest rates of migration for maple occurred between 10,000 and 9000 yr B.P.; maple populations reached their maximum areal extent at 4000 yr B.P. On both continents, maples tend to be mesic species, codominating on nutrient-rich sites with other mesic deciduous tree taxa such as *Fagus*, *Tilia*, *Fraxinus*, and *Ulmus*.

Birch (*Betula*)

Of the four species of *Betula* native in Europe, two (*B. humulis* and *B. nana*) are shrubs, one (*B. pubescens*) attains the stature of a low tree, and only one species (*B. pendula*) has individuals that can grow into tall forest trees. These taxa are all broadly tolerant, disturbance-favored, early-successional species with distributional ranges extending generally from central Europe northward to the forest limit in Fennoscandia. Tree birches decline in height and are replaced by shrub birches in northern Norway, Sweden, and Finland.

In the late-glacial interval, *Betula* (with values reaching > 25% of the pollen rain) formed widespread birch woodlands in central and northeastern Europe (Huntley and Birks 1983) and grew in the forest steppe of the western U.S.S.R. (Grichuk 1984). Because of the proximity of source populations and the ability to invade open, newly deglaciated terrain, the rate of migration of birch into Fennoscandia was apparently limited only by the rate of retreat of the ice sheet. The contoured isopoll maps for 13,000 yr B.P. and 12,000 yr B.P. (Huntley and Birks 1983) clearly delineated a prominent migration wave front, reaching a ridge of high pollen values along the west and northwest advancing border of pioneer birch populations invading tundra. A minor diminishment of birch populations occurred between 11,000 and 10,000 yr B.P. in northwestern Europe because of pronounced climatic cooling (associated with the southward shift of the marine Polar Front during the Younger Dryas stadial). By 10,000 yr B.P., however, renewed expansion in tree-birch dominance resulted in a continuous zone of birch woodland extending from northwestern France and the British Isles along the North Atlantic Coast and Baltic Sea to the northwestern U.S.S.R. Early-Holocene birch values > 50% of the total tree and shrub pollen occurred as far north as about 70°N in northernmost Norway. Substantial birch populations spread to Greenland, Iceland, and northernmost Fennoscandia by 9000 yr B.P. Between 8000 and 5000 yr B.P., birch was excluded as other tree taxa such as *Pinus* invaded and replaced birch in Finland, Sweden, and Norway. Birch was only sparsely represented in the mountains of central Europe and became progressively more restricted in distribution as temperate tree taxa migrated northward in the early- to mid-Holocene intervals. After 5000 yr B.P., however, birch increased in importance, with a northern and a southern region of dominance developing by 3000 yr B.P. The late-Holocene increase in birch to the south in central Europe and the British Isles probably resulted from its ability to opportunistically recolonize areas deforested by humans. To the north, particularly in northern Norway and Sweden, birch populations increased after

5000 yr B.P., possibly responding to a late-interglacial trend toward climatic cooling.

In contrast with Europe, tree birches of eastern North American were restricted in distribution to small, isolated refugia during the full-glacial and late-glacial intervals. *Betula* became widespread south of the Laurentide Ice Sheet after 14,000 yr B.P. but reached dominance values > 40% only in the mid- to late-Holocene intervals within the modern boreal forest region of central and eastern Canada. Temperate species of birch persisted in the Appalachian Mountains throughout the Holocene. In both Europe and North America, the temporal pattern for birch migration was tied to the availability of landscapes suitable for colonization; for both continents, birch invasion was rate-limited during the late-glacial and early-Holocene intervals by physical barriers such as the retreating glacial margin, proglacial lakes, and marine embayments.

Beech (*Fagus*)

The two native European species of beech are *Fagus sylvatica* and *F. orientalis*. *Fagus sylvatica* is a codominant or dominant tree within closed-canopy, mesic deciduous forests; its native distributional range extends throughout much of Europe except for southern portions of Greece, Italy, and Spain, the northern and western areas in the British Isles, and the northern two-thirds of Scandinavia. *Fagus orientalis* occurs as a native tree species in southeastern Bulgaria, southern Yugoslavia, and northern Greece. Both species are widely naturalized across much of the European continent. *Fagus* often grows today within mixed deciduous and coniferous forests with *Taxus*, *Tilia*, *Acer*, *Ostrya*, and *Abies*.

During the late-glacial interval, *Fagus* pollen reached as much as 5% of the sum of total terrestrial pollen only in the Calabrian region of southernmost Italy. Beech was present in scattered localities in southern and southeastern Europe until 9000 yr B.P., when it increased in population size and areal extent across the southern Balkans. During the interval from 9000 to 7000 yr B.P., beech populations dispersed northward across Bulgaria and Yugoslavia, into the Carpathian Mountains in Romania, and northwestward into northern Italy. Westward migration of beech resulted in its initial establishment in the French Massif Central by 7500 yr B.P. and its subsequent occupation of southernmost France by 6500 yr B.P. Following the Danube River Valley, beech migrated from 6500 to 5000 yr B.P. around the northern Alpine barrier. During the mid-Holocene interval, beech expanded at rates up to 300 m/yr; by 5000 yr B.P., beech populations occupied a continuous distributional area from the Carpathian Mountains, along the Alps, to the Massif Central of France. In the next 1000 years, *Fagus* spread northward into central Germany and northern Poland, westward into the northern Pyrenees of northeastern Spain, and it reexpanded southeastward within the Balkans of Greece and Bulgaria. After 4000 yr B.P., beech reached and maintained values of > 50% of the total sum for tree and shrub pollen in the mountains of central and eastern Europe. By 3000 yr B.P., it reached from northern Germany west to the Atlantic Coast and dispersed across the English Channel to southeastern England; by 2000 yr B.P., beech populations were established in Denmark and southern Sweden.

Huntley and Birks (1983) speculate that the complex behavior of *Fagus* in the Holocene may be related to the emergence of *Fagus sylvatica* as a new species since the last (Eemian) interglacial interval. Fossil evidence from deposits of previous interglacial intervals indicates that beech was generally restricted to the southeastern quadrant of continental Europe. *Fagus sylvatica* is more cold-tolerant than *F. orientalis* and thus was able to spread farther to the north during the Holocene. Although found in rich, mixed deciduous forests, *F. sylvatica* is relatively shade-intolerant, only reproducing today in disturbed areas or light gaps within otherwise closed forest canopies (Watt 1947). Thus, it expanded northward into open-ground environments rapidly in the early-Holocene interval following climatic warming. However, beech did not successfully invade lowlands of central and western Europe during the mid-Holocene interval, because the landscapes there were already occupied by closed canopies of deciduous forest dominated by late-successional taxa such as elm (*Ulmus*). The late-Holocene expansion of beech into central and western Europe followed the decline of elm, as beech successfully invaded into forests fragmented by human activities.

The Holocene history of European beech contrasts markedly with that of eastern North American beech (*Fagus grandifolia*). Although also restricted to few, small, scattered refugia during the late Pleistocene, American beech populations expanded progressively through the early Holocene, following a K_m strategy of colonization and population increase. American beech is a shade-tolerant, late-successional forest tree (Fowells 1965) that has continued to invade closed-canopy deciduous forests as its range has expanded throughout the eastern Great Lakes region and New England during the mid-Holocene and late-Holocene intervals. The most rapid migrational advances of American beech occurred during the early Holocene (fastest mean rate of 257 m/yr between 12,000 and 10,000 yr B.P.; Table 6.9). In Europe, sustained migration of beech was later, with two stages of areal expansion from 6000 to 5000 yr B.P. and from about 4000 to 2000 yr B.P.

Ash (*Fraxinus*)

Two groups of ash are recognized in the European pollen record: *Fraxinus excelsior*-type (including pollen from the species *F. excelsior*, *F. angustifolia*, and *F. pallisiae*, all with individuals potentially growing into tall trees) and *Fraxinus ornus*. Although rarely a forest-canopy dominant, *Fraxinus excelsior* is widespread today, growing within mixed deciduous woodlands in alluvial habitats (Gordon 1964). *Fraxinus angustifolia* and *F. pallisiae* are characteristic of floodplain woodlands in southern Europe. *Fraxinus ornus* is an understory tree characteristic of scrubby woodlands in dry rocky habitats. It is a sub-Mediterranean species that extends as far north as the southern Alps and the Carpathian Mountains.

Glacial-age refugia for *F. excelsior*-type were located south of the Alps in northern Italy and in Austria, as well as probably in southeastern Europe. After 10,000 yr B.P., *F. excelsior*-type expanded northward at rates of from 200 m/yr up to 500 m/yr, first along the southern mountain chains, and then along the riverine corridors of the Danube and Rhine river valleys, across the northern European plains, reaching the Baltic Coast, southern Scandinavia, and the British Isles by 6000 yr B.P. During the early- and mid-Holocene intervals, values for pollen percentages of *F. excelsior*-type generally did not exceed 5%, except in the Swiss Alps and in eastern Bulgaria. After 5000 yr B.P., this ash type experienced an areal decrease and fragmentation of its range, especially in central and eastern Europe. A mid-Holocene decline in ash resulted from forest clearance and/or competitive exclusion of ash populations following the immigration of taxa such as *Carpinus betulus* (hornbeam) that became more abundant as ash decreased. In the late Holocene, populations of ash producing the pollen of *F. excelsior*-type increased only in the British Isles. This localized late-Holocene increase in ash in Great Britain is attributed by Huntley and Birks (1983) to the increased occurrence of secondary-successional woodlands, limited competition from restricted populations of beech and hornbeam, and a maritime climatic trend toward increased precipitation.

Fraxinus ornus is known from early-Holocene pollen sequences in southern Greece and the Adriatic Coast. During the Holocene, it expanded progressively in southeastern Europe at rates of 25 m/yr to 200 m/yr, reaching up to 5% of the sum of tree and shrub pollen in Yugoslavia by 5000 yr B.P. This species of ash advanced along its northern distributional limit as far as the Italian and Austrian Alps by 1000 yr B.P. Within the central Mediterranean area, *Fraxinus ornus* was cultivated for medicinal purposes, but its populations were apparently neither greatly curtailed nor favored by anthropogenic disturbances.

In contrast with the European ashes, in eastern North America, *Fraxinus* was widespread throughout temperate forests during the full-glacial and late-glacial intervals, and it extended into the late-Pleistocene boreal forest region. Ash reached its greatest abundance of from 20% to 50% paleo-dominance over a broad area of the Lower Midwest at about 12,000 yr B.P. Its populations subsequently continued to expand in range over much of eastern North America generally north to about 50°N, although diminishing in dominance to typically < 5% of the forest composition after 10,000 yr B.P. At the genus level, *Fraxinus* achieved its fastest migration rates earlier in the eastern United States (with the highest mean rate of 209 m/yr averaged for all measured tracks between 14,000 and 12,000 yr B.P.) than in Europe (between 10,000 and 6000 yr B.P.).

Walnut (*Juglans*)

Juglans regia is the sole native European species of walnut. Although its contemporary distribution is southern European, it is probably only native to the Balkans, Italy, and the southern Alps. In the Balkan Peninsula, walnut occurs in mixed forest stands with *Castanea sativa*, *Pinus nigra*, and *Platanus orientalis*.

Walnut is not recorded in the pollen record in Europe in the late-glacial interval. Its earliest appearance was in the Balkans and Italy at 5000 yr B.P. By 2000 yr B.P., *Juglans* occurred throughout the western Balkans and westward into southern France. By 1000 yr B.P., its range extended throughout France and into northwest Italy. The nut of this tree is considered highly desirable as a food resource, and the Holocene range extension of walnut may in part be attributable to human introductions (particularly during the times of the Greek and Roman Empires) beyond its native range (Barker 1985). Huntley and Birks (1983) consider that the late-Pleistocene refugia for walnut were located in the Balkans, the southern Alps, and Italy, as well as possibly farther east in Turkey and Iran. Its range probably increased because of disturbance following Neolithic land clearance, but it was probably not widely cultivated until Roman times about 2000 yr B.P. *Juglans* expanded northward slowly from 5000 yr B.P. to 2000 yr B.P., but it subsequently spread more rapidly, at rates up to 400 m/yr, possibly reflecting its introduction by humans into western Europe. *Juglans* was widely cultivated by 1000 yr B.P.

As in eastern North America, walnut exhibited a fugitive (f_m) strategy for survival during the late-glacial and early- to mid-Holocene intervals. The fastest mean migration rate for North American *Juglans* (390 m/yr) occurred between 12,000 and 10,000 yr B.P. However, in Europe, walnut experienced its major range expansion during the late Holocene, probably a result of human transport of its fruits and deliberate planting of walnut trees. This type of human assistance is probably the reason for accelerated late-Holocene migrations recorded for other cultivated species of trees, such as olive (*Olea europaea*) and pistachio (*Pistachia*).

Larch (*Larix*)

Larix decidua is distributed today primarily along the Alpine Mountains, with secondary populations located along the Tatra Mountains and the Carpathians in southern Poland, eastern Czechoslovakia, and Romania. *Larix decidua* is characteristic of subalpine forests of the central European mountains, where it may codominate with *Pinus cembra*. A second species of larch, native to northeastern Europe, is *Larix russica*, closely related to the Siberian larch (*L. sibirica*) which occurs in boreal forests of the northwestern U.S.S.R. (Huntley and Birks 1983).

In the late-glacial interval, larch was present in Europe in the southern Alps and in the Polish Carpathians, as well as in the forest steppe of the western U.S.S.R. (Grichuk 1984). Larch populations expanded to the north and west across Poland between 12,000 and 10,000 yr B.P., but they were virtually eliminated there during the transition from late-glacial to Holocene conditions. In the early Holocene, larch populations increased first in importance in the Swiss, Austrian, and Italian Alps. *Larix* persisted in those alpine environments near timberline through the middle and late Holocene. Timberline has dropped in elevation in the Alps during the last 4000 years, probably responding to a combination of climatic cooling, anthropogenic forest clearance (Lang and Tobolski

1985), and increased browsing pressure by domesticated cattle and sheep. However, larch has increased in population abundance in subalpine forests within the last 2000 years.

Larix laricina, the eastern North American tamarack, is not an ecological equivalent to *L. decidua*, as the former is characteristic of lowland bog environments rather than subalpine forests. The distributional range of *L. laricina* is much more widespread than that of the western European *Larix* species, occupying much of the boreal forest region of central and eastern Canada. Tamarack migrated most rapidly (with a mean rate of 256 m/yr averaged for all measured migrational tracks) between 12,000 and 10,000 yr B.P. through the spruce-dominated forests in the deglaciated landscapes between 42°N and 48°N in the Great Lakes and New England regions of eastern North America. Similarly, European larch invaded most rapidly into spruce-dominated forests of Poland during the same time interval.

Spruce (*Picea*)

Two species of spruce, *Picea abies* and *P. omorika*, are considered native to Europe. *Picea omorika* is restricted to central Yugoslavia. *Picea abies* is widespread within coniferous montane and boreal forests in continental Europe and Fennoscandia, where it codominates with *Pinus* species, as well as within mixed conifer–northern hardwoods forests that occur across the European Plain of the western U.S.S.R.

During the late-glacial interval, two refugial areas were important for spruce: (1) the coniferous forest refugia west of the Ural Mountains in the U.S.S.R. (Grichuk 1984); and (2) eastern portions of the Austrian Alps, the mountains of northern Yugoslavia, and the Polish and Romanian Carpathians (Huntley and Birks 1983). During the early Holocene, the northern spruce populations expanded in both abundance and areal extent within taiga forests west of the Urals in the U.S.S.R. (Khotinskiy 1984). By 7000 yr B.P., spruce consolidated a major northern population center (indicated by > 25% of the total sum for tree and shrub pollen) across the European U.S.S.R. and expanded into Finland by 5000 yr B.P. The northern group of spruce populations continued its westward dispersal at rates of up to 500 m/yr, occupying Sweden and most of Norway by 500 yr B.P. (Moe 1970). Southern populations of spruce migrated westward from their southern refugia beginning at about 10,000 yr B.P., extending along the montane axes of the Alps and Carpathians at rates of 80 m/yr to 240 m/yr, and slowing in their advance only after 4000 yr B.P. In the late-Holocene interval, spruce populations declined in the central European mountains; they reexpanded in the foothills and lowlands north of the Alps only during the last 2000 years. *Picea abies* is extensively planted in that region today.

The contrasting histories of the two European population centers for spruce are interpreted in terms of changing climates and human disturbances during the Holocene. In the central European mountains, spruce spread westward during early-Holocene climatic warming. The spread of spruce into the western Alps after 5000 yr B.P. may have been a consequence of human disturbance or climatic cooling; its decline after 4000 yr B.P. could reflect the displacement of spruce by beech populations, by clearance of forests, or by both factors. Either deliberate planting of spruce, or its secondary succession after anthropogenic forest clearance, could have resulted in the increase reconstructed for spruce populations in the Alpine foothills after 2000 yr B.P. The northern populations of spruce expanded initially during episodes of climatic warming such as occurred during the late-glacial and early-Holocene intervals, then reexpanded in a major migrational pulse during the late-Holocene interval. This may reflect the response of spruce to late-interglacial climate cooling, its invasion of areas disturbed by humans (Moe 1970), or it may represent the long-term result of trampling or grazing by domesticated animals that selected against arboreal competitors of spruce.

The late-Quaternary history of the northern population of *Picea abies* is strikingly different in geographic direction of migration from that of eastern North American *Picea*, but the two taxa are similar in their histories of changing population abundances during the late-glacial and Holocene intervals. With late-glacial and early-Holocene climatic amelioration, eastern North American *Picea* initially migrated rapidly northward following the retreating Laurentide Ice Sheet, but its populations were diminished in the early- to mid-Holocene intervals as unfavorable warm climates became prevalent throughout much of its distributional range. Subsequently, after their respective early- to mid-Holocene population collapses, both North American and northern European groups of spruce species reexpanded in area. By the late Holocene, they occupied extensive regions of deglaciated terrain in relatively high latitudes, environments suitable for colonization by boreal conifers.

Pine (Diploxylon *Pinus*)

Of the ten species of Diploxylon *Pinus* in Europe, seven are narrowly restricted to southern Europe and the Mediterranean region (*Pinus brutia*, *P. halepensis*, *P. heldreichii*, *P. leucodermis*, *P. nigra*, *P. pinaster*, and *P. pinea*). Two additional pines are characteristic of montane environments of central Europe; these include *P. mugo*, a large shrub characteristic of alpine environments at tree line in the Alps, the Carpathians, and the mountains of Yugoslavia, and *P. uncinata*, a tree that grows in the Pyrenees, the Alps, and the northern Carpathians. Only one pine species (*P. sylvestris*) is a forest dominant distributed widely throughout Europe, excluding portions of the Mediterranean, the Arctic, and northwest Europe. These pines are typically broadly tolerant, early-successional taxa that can grow in nutrient-poor substrates; they are favored by natural disturbances such as fire and successfully invade abandoned cultivated fields. Pines produce large quantities of pollen grains that are carried by wind long distances from their source trees, especially when transported across open, treeless landscapes. On the modern European isopoll maps, Diploxylon *Pinus* is commonly represented by values from 25% to 50% of the total sum for tree and shrub pollen, even when pine is not a regional dominant in the vegetation. High values of 50% to more than 75% pollen of *Pinus* indicate presence of regionally extensive pine-dominated forests (Huntley and Birks 1983).

In the late-glacial interval, high (> 50%) pollen values of *Pinus* indicate the locations of pine refugia primarily occurring in the Alps of northern Italy and the Romanian Carpathians. Although pine pollen was represented by moderate values (25% to 50%) in western Iberia as well as throughout central and eastern Europe, this was probably a result of long-distance dispersal of pollen grains into steppe or tundra environments (Huntley and Birks 1983). Additional glacial-age refugia for pine included forest steppe communities of larch, pine, and birch in the European Plain of the western U.S.S.R. (Grichuk 1984). Between 13,000 and 11,000 yr B.P., pine migrated west along the Alps to northern Spain, and both north of the Carpathians and west from the U.S.S.R. into northeastern Europe. The cold stadial interval from 11,000 to 10,000 yr B.P. resulted in a temporary decline in pine abundance in the mountains as well as in the lowlands of northeastern Europe. At 10,000 yr B.P., pine became extensive throughout Europe except for Fennoscandia and parts of the Mediterranean region. Between 10,000 yr B.P. and 9000 yr B.P., pine declined in the central lowlands but remained most important in the central European mountains. Pine invaded eastern and central portions of Fennoscandia by 8000 yr B.P. and continued to advance to 70°N latitude in northernmost Fennoscandia by 7000 yr B.P. Thereafter, pine began to decline, first in southern Sweden by 6000 yr B.P., then in Ireland and Scotland, as well as in the Alps and southern Europe by 4000 yr B.P. Pine continued to decline generally through the late Holocene except for reexpansion in southern Europe. By 1000 yr B.P., fragmentation of pine forests occurred in northern Fennoscandia.

In the late-glacial and Holocene intervals, only *Pinus sylvestris* migrated substantial distances from southerly refugial areas. The most rapid rates of advance of pine occurred during the Pleistocene–Holocene transition, between 10,500 yr B.P. and 10,000 yr B.P., with rates up to 1500 m/yr as it advanced through northeastern Europe and into Fennoscandia. In western Europe the migration of pine was halted as deciduous taxa including *Quercus* invaded from the south and west. Diploxylon *Pinus sylvestris* was able to persist throughout the Holocene only on sites with sandy or dry, nutrient-poor soil that were unfavorable for the establishment of deciduous trees. In the late Holocene, pine populations expanded locally in areas of abandoned cultivation. The montane species of Diploxylon *Pinus* persisted at high elevations with minor changes in distribution through the Holocene. Mediterranean species of Diploxylon *Pinus* became prominent only in the late Holocene, probably as a result of colonization of anthropogenically disturbed sites.

In eastern North America, both northern and southern groups of Diploxylon *Pinus* were relatively widespread throughout the full-glacial and late-glacial intervals. Northern Diploxylon pines spread northward most rapidly between 12,000 yr B.P. and 10,000 yr B.P., following interglacial climatic warming; during the Holocene, populations of these northern pines were most important on nutrient-poor, dry sites prone to recurrent fires, in regions such as central Canada. During the Holocene, southern Diploxylon pines in both the southeastern United States and in the Mediterranean region did not migrate far to the north of their late-Pleistocene refugial areas. The importance of pine in the vegetation of the southeastern United States increased during the mid- and late-Holocene intervals as a result of increased seasonal precipitation and fire frequency, the latter possibly influenced by activities of prehistoric Native Americans (Delcourt 1980). This parallels the late-Holocene "pine rise" or expansion of pine populations attributed to human activities in Europe, particularly in Spain, southern France, and Greece (Huntley and Birks 1983).

Pine (Haploxylon *Pinus*)

Two species of Haploxylon *Pinus*, *P. cembra* and *P. peuce*, were represented in the isopoll maps for Europe (Huntley and Birks 1983). A third species, *P. sibirica*, is widespread through much of the U.S.S.R. but is generally not important in the vegetation of continental Europe. *Pinus cembra* occurs at tree line on the high peaks of the Alps, as well as the Tatra Mountains of Poland and Czechoslovakia, and extends into the Romanian Carpathians. *Pinus peuce* is more geographically restricted, occurring at high elevations of mountain regions within southeastern Europe, in southern Yugoslavia, northern Greece, Albania, and Bulgaria.

Late-glacial refugia for *Pinus cembra* were located in the central Alps; *P. cembra* became locally important there by 11,500 yr B.P., as well as developing a population center in the Polish Carpathians. These populations declined during the stadial interval from 11,000 yr B.P. to 10,000 yr B.P. and then increased again briefly in the Alps before declining in the early Holocene in both regions. From 7000 yr B.P. to 4000 yr B.P., both *P. cembra* and *P. peuce* increased in population size, in the Alps and in Bulgaria, respectively. A general decline in their abundance occurred in the late-Holocene interval after 4000 yr B.P. *Pinus cembra* was progressively restricted to tree line as it was displaced from lower elevations of the Carpathians with subsequent Holocene arrival of additional tree species. Late-Holocene declines in Haploxylon pines also occurred on high-elevation sites where forests were cleared and converted to alpine meadows.

Eastern white pine (*Pinus strobus*) is distributed throughout the Great Lakes region and along the axis of the Appalachian Mountains today. It does not form a tree line in the mountains. Rather, it is characteristic of middle elevation montane sites, where it grows in mixed conifer–northern hardwoods stands. In the Great Lakes region, eastern white pine occupied extensive sandy outwash plains along with northern Diploxylon pines (*P. banksiana* and *P. resinosa*) prior to EuroAmerican settlement and land clearance (Flader 1983). In the last 125 years, Great Lakes populations of eastern white pine have collapsed because, unlike the European white pines, its wood was used extensively and its seed source has been virtually eliminated (Flader 1983). The late-glacial to early-Holocene migration of eastern white pine was primarily along the axis of the Appalachian Mountains, then west across the Great Lakes region, reaching Minnesota by about 7000 yr B.P. (Jacobson 1979). *Pinus strobus* extended to its northernmost limit in east-central Canada by about 5000 yr B.P. (Terasmae and Anderson 1970) and then retreated, acquiring its present distributional range by about 4000 yr B.P.

Aspen (*Populus*)

Four species of *Populus* are native to Europe. Of these, three (*P. alba, P. canescens,* and *P. nigra*) are riparian, occurring along floodplains of the major river systems of central and southern Europe. One species, *P. tremula,* occupies a variety of sites and is widespread throughout much of the region. It is today an important forest tree in the zone of mixed conifer–northern hardwoods forest situated between boreal forests of Fennoscandia and temperate deciduous forests of mainland Europe.

Full- and late-glacial refugia for aspen may have been located near the tundra–forest ecotone in eastern Europe. By 10,000 yr B.P., however, aspen occupied the southern British Isles and southern Scandinavia south of the remaining ice sheets, the northern lowlands of continental Europe, and was possibly also located in montane sites in the Alps. Aspen spread throughout Fennoscandia immediately following deglaciation, and it became important in the Alps by 9000 yr B.P. During the mid-Holocene interval, the southern range limit of aspen moved northward across the Baltic region, presumably a result of increasing competition from additional immigrating tree taxa (Huntley and Birks 1983). By 4000 yr B.P., the distributional area of aspen was constricted to an interglacial minimum as both the northern boundary of aspen contracted southward (possibly because of climatic cooling) and its southern boundary continued to retreat northward. After 4000 yr B.P., the main range of aspen shifted southward across the northern European lowlands. During the last several thousand years, aspen populations increased locally within the Alps, probably reflecting increased opportunities for establishment in areas disturbed by anthropogenic forest clearance.

In eastern North America and in Europe, aspen tracked the retreating margin of ice sheets during the late-glacial and early-Holocene intervals, rapidly colonizing newly deglaciated landscapes. For North America, the paleo-dominance maps of reconstructed forest composition reveal the distinctive r_m strategy of migration as aspen populations responded to availability of new habitats in the deglaciated landscape. During the mid- and late-Holocene intervals, the range of eastern North American aspen, as well as its overall population dominance, diminished throughout eastern Canada. As in Europe, this interglacial constriction in distributional area of aspen was, in large part, probably the result of decreased openness of forests in the landscape and increased interspecific competition from boreal conifers in the northern part of aspen's range and from temperate tree taxa at its southern range limit.

Oak (*Quercus*)

Twenty-five species of oak are native to Europe. Of these, 16 produce individuals that attain large tree stature, with the remainder of the oak species typically small trees to shrubs. Two different pollen types (deciduous and evergreen oak types) are typically distinguished in European Quaternary deposits. However, the paleoecological interpretation of isopoll maps for these two types is complicated, as one evergreen and seven semievergreen oak species are included within the "deciduous" pollen type (Huntley and Birks 1983). The European oaks occupy a great diversity of xeric to mesic habitats. Two oak species with prominent populations in northern and northwestern Europe are *Quercus robur*, characteristically growing in habitats with loamy or clayey soils, and *Q. petraea*, occurring on sites with sandy soils. Three oak species characteristic of mesic deciduous forests of central and southern Europe include *Q. pubescens, Q. pyrenaica*, and *Q. cerris*. In the Mediterranean region, the endemic oak species tend to be restricted to xeric, rocky habitats, and include evergreen species such as *Q. ilex* and *Q. coccifera*. Several species are important economically, including *Q. suber*, grown for its cork bark. In Britain, oak-dominated woodlands are common today because oak is grown for commercial timber (and, previously, for charcoal used in smelting of iron).

Several widely separated refugial areas for oak occurred across the northern Mediterranean region during the late Pleistocene. Oak populations also probably survived the last maximum in continental glaciation in the latitudinal band of broad-leaved montane forests, situated between the Black and Caspian Seas (Grichuk 1984). "Deciduous" oak pollen reached > 10% of the total pollen grain and spore sum of terrestrial vascular plants at 13,000 yr B.P. in southern Spain, the Calabrian region of southernmost Italy, and in southern Greece. Along the Mediterranean coast of southern Spain, deciduous-type oak pollen was > 25% of the total pollen sum at 13,000 yr B.P. By 12,000 yr B.P., this value decreased to < 10% in southern Spain but increased to > 25% in southern Greece. Between 12,000 yr B.P. and 11,000 yr B.P., the northern range limit of deciduous oaks advanced across northern Spain and along the western coast of France. During this late-glacial interstadial of minor warming, the relative abundance of deciduous-type oak pollen increased to > 10% throughout western and southern Spain as well as Italy, and remained > 25% in southern Greece. Subsequent stadial cooling (related to the southward shifting in the position of the oceanic Polar Frontal Zone) affected boreal tree populations in northern Europe and Fennoscandia but did not have a pronounced effect on oak distribution or abundance in southern Europe. By 10,000 yr B.P., oak expanded its range across France, nearly reaching the British Isles, and spread across the southern Alps and the Carpathians. Deciduous-type oak pollen reached values (based on the sum of pollen of trees and shrubs) > 50% generally south of 41°N throughout central and southern Spain, portions of northeastern Italy, and throughout the Balkan Peninsula and Greece. Northward expansion in the range limit of oak continued until 8000 yr B.P. at rates up to 500 m/yr as it invaded the British Isles, Germany, northern Poland, and southern Sweden. The migration of oak then slowed to a maximum rate of 200 m/yr between 8000 and 5000 yr B.P., advancing at low population levels (< 10% of the pollen sum of trees and shrubs) into southern Norway and Finland. Throughout the interval from 10,000 yr B.P. to 6000 yr B.P., oak population centers (indicated by pollen values > 50%) were generally maintained in central to southern Spain, southern Italy, and southern Greece. The southern (Mediterranean) population centers of oak became fragmented after 5000 yr B.P., and oak diminished in importance throughout much of its range throughout the late Holocene. Populations of deciduous oaks experienced a minor increase in the British Isles in the last 2000 years. Evergreen-type oak pollen primarily represented species endemic to the Mediterranean region and remained minor in its contribution to the pollen rain (and, presumably, to forest composition) throughout the Holocene.

During the late Pleistocene, European oak was generally restricted in distribution south of about 40°N latitude, the approximate position of the oceanic Polar Frontal Zone offshore in the Atlantic Ocean (Duplessy *et al.* 1981). The early- to mid-Holocene expansion of oak across Europe clearly reflects a K_m strategy of migration. The first limited northward range extensions of oak occurred at relatively low latitudes as the Polar Frontal Zone fluctuated across the North Atlantic Ocean and European continent during the late-glacial interval. Major northward movement of the Polar Front and subsequent influence of the marine Gulf Stream ameliorated the climatic conditions along the Atlantic Coast of western Europe. The shifts in these atmospheric and oceanographic conditions promoted rapid northward expansion of oak through France to the British Isles in the early Holocene. Oak populations increased to dominate Holocene forests across southern Europe from Spain to Bulgaria. Only with anthropogenic forest clearance in the middle to late Holocene did the populations of oak diminish, particularly in eastern Europe.

Representation of oak in eastern North America is also considered to be primarily of deciduous species, with the pollen of the few evergreen and semi-evergreen species (*e.g.*, *Quercus virginiana*) indistinguishable from that of other species within the genus (Solomon 1983a,b). In the late-glacial interval in the central and eastern United States, *Quercus* had its most rapid migrational advance (mean value of 181 m/yr, averaged across all five measured migrational tracks) between 16,000 yr B.P. and 14,000 yr B.P. Its northern distributional limit advanced far beyond its primary areas of high population dominance. The patterns of the expansion in population dominance of oak in Europe during the early Holocene are similar to those in eastern North America, where the primary population center of oak ($> 40\%$ of forest composition) was restricted generally south of about 35°N until about 12,500 yr B.P., but extended northward rapidly after the northward passage of the Polar Frontal Zone. Eastern North American oak species also progressively increased their populations to become dominant in the southern and western portions of their ranges through the Holocene. Only with the mid- to late-Holocene rise in dominance of southern pine and the eastward extension of the Prairie Peninsula did oak become partially fragmented in distribution and dominance.

Willow (*Salix*)

Most of the 69 species of *Salix* native to Europe are shrubs with montane or boreal distributions, with only a few species of trees (*e.g.*, *Salix fragilis* and *S. alba*). The latter occur primarily in riparian habitats in southern Europe and tend to have sporadic representation in the fossil-pollen record.

Shrub willow was important across much of the tundra and steppe region of western and central Europe during the late-glacial and early-Holocene intervals. The history of tree *Salix* is less clear, although Pleistocene refugia for its populations probably occurred in northeastern Spain, southwestern France, and the Balkan Peninsula. On the isopoll maps of Huntley and Birks (1983), northern (shrub) and southern (tree and shrub) willow populations become disjunct at about 8000 yr B.P., with fossil-pollen values of 2% to 10% of the sum of total terrestrial pollen. During the early Holocene, willows (capable of tall-shrub and tree stature) became geographically isolated in southwestern France, the eastern European mountains, and Bulgaria, and they persisted there through the remainder of the Holocene.

As in eastern North America, the European record of tree willow is influenced by the location of sediment-coring sites. *Salix* pollen tends to be deposited locally in poorly drained habitats where the willows grow. Thus, the generalized distribution of tree willow can be reconstructed, but detailed interpretation of its abundance pattern or rates of migration is lacking in Europe.

Linden (*Tilia*)

Four species of *Tilia* are native to Europe; of these, two (*T. rubra* and *T. tomentosa*) are restricted in distribution to southeastern Europe within the Balkan Peninsula. The other tree species (*Tilia platyphyllos* and *T. cordata*) are widespread, shade-tolerant, late-successional trees occurring on well-developed, nutrient-rich soils. Linden trees grow in mixed deciduous forests dominated by *Quercus* and *Ulmus* and including *Fagus, Acer, Ostrya, Carpinus,* and *Castanea*. Linden is not tolerant of extreme drought, however, and is generally excluded in its modern distribution from the western and central Mediterranean region. *Tilia cordata* is limited in distribution along its northern limit (in the northern British Isles and southern Scandinavia) by spring temperatures below the critical threshold for successful fertilization; once established in the forest canopy, however, it can persist vegetatively in clones for hundreds to thousands of years (Huntley and Birks 1983).

At 13,000 yr B.P., *Tilia* was present in limited populations east of 10°E in central Italy and the Balkan Peninsula, with scattered occurrences to the northeast across Poland. By 11,500 yr B.P., it had spread throughout southeastern Europe and into the southern Alps of northern Italy, west into northwest France, and northeast from Poland into the European Plain of the western U.S.S.R. During the early Holocene, after 10,000 yr B.P., *Tilia* extended south into Greece and southwest across southern France, and it also increased in abundance within the southern Alps. By 9000 yr B.P., linden was widespread throughout southern Europe as far west as the Pyrenees, and across the northern European lowlands from France to Poland. At 9000 yr B.P., pollen values of > 5% of the total trees and shrubs were mapped by Huntley and Birks (1983) across the region from the eastern Alps to the Balkans and into Bulgaria. *Tilia* continued to consolidate its range northward and westward across continental Europe, with migration rates up to 500 m/yr until 8000 yr B.P. Between 8000 yr B.P. and 5000 yr B.P., linden migrated into England, southern Norway, southern Sweden, and Finland, slowing to between 50 m/yr and 130 m/yr as it presumably approached its climatically controlled limits of distribution. By 6000 yr B.P., linden attained pollen percentages (of total pollen sum for trees and shrubs) that exceeded 10%, indicating that *Tilia* dominated forests in regions of western Europe including central France, southern England, southwestern Sweden, and mountain areas of southeastern Europe. During the mid-Holocene interval, *Tilia* became an important forest tree throughout much of Europe. *Tilia* reached its maximum interglacial limits along its northern and western distributional margin at 5000 yr B.P., reaching the central British Isles as well as the southern half of Fennoscandia. In the late Holocene, linden retracted southward from its northern limit at rates of 100 m/yr to 200 m/yr, probably retreating in response to climatic cooling; the distribution of linden also retracted along its southern and eastern borders, and *Tilia* generally declined in relative abundance throughout the remainder of its range. This late-Holocene population decline, occurring during the last 4000 years, has been attributed to widespread forest clearance. By 2000 yr B.P., *Tilia* persisted only in fragmented portions of its former interglacial range, principally within eastern Europe.

Eastern North American *Tilia* also exhibited primarily a continental distribution, expanding during the late-glacial interval from initially low dominance values of restricted full-glacial refugial populations. By the late Holocene, *Tilia*'s distribution covered much of the region of eastern deciduous forest. Unlike European linden, American basswood became prominent only in the mid- and late-Holocene intervals as it attained dominance within the maple–basswood forests of the western Great Lakes region. In eastern North America, basswood migrated most effectively in the late-glacial interval from 14,000 yr B.P. to 12,000 yr B.P.; the postglacial response of European linden resulted in its maximal migration rates between 10,000 yr B.P. and 8000 yr B.P.

Elm (*Ulmus*)

Five species of elm trees are considered native to the region of Europe mapped by Huntley and Birks (1983). *Ulmus glabra* is widespread throughout the region, extending to the northern and western limits of mixed deciduous forest and as far east as the western U.S.S.R., and south to the Mediterranean Sea. The other elm species are more restricted in distribution, occurring primarily in central to southeastern Europe. *Ulmus glabra* occurs on mesic, nutrient-rich soils, except in the northwestern part of its range, where it may occupy rocky cliff edges in areas bordering the North and Baltic Seas. Elms tend to grow in mesic habitats with an admixture of deciduous trees including *Fraxinus, Quercus,* and *Acer*.

Elm was restricted to small, scattered refugia during the late Pleistocene. *Ulmus* occurred primarily in southeastern Europe by 11,000 yr B.P. as well as in the southern Alps and the northern Carpathians. Between 11,000 yr B.P. and 10,000 yr B.P., however, elm had migrated at rates up to 1000 m/yr north to the eastern Baltic region and west to the Pyrenees. During the early Holocene, elm spread rapidly across northern Europe, into the southern portion of the British Isles and Fennoscandia. From 8500 yr B.P. to 6000 yr B.P., the migration of elm slowed to between 100 m/yr and 200 m/yr, and the southern limit of its distribution in the Mediterranean area receded to the north. Pollen percentages for elm diminished throughout southeastern Europe between 7000 yr B.P. and 6500 yr B.P., and across central Europe during the interval from 6500 yr B.P. to 5500 yr B.P. A major decline in elm pollen occurred in northern and northwestern Europe between 5500 yr B.P. and about 5000 yr B.P. During the late Holocene, elm continued to diminish in importance and to fragment in range.

European elm responded rapidly to climatic warming in the late-glacial and early-Holocene intervals. Elm began to decline in importance between 7000 yr B.P. and 6500 yr B.P., possibly in response to mid-Holocene climatic changes and to limited outbreaks of a fungal pathogen detrimental to elm (possibly responsible for the "Dutch elm disease" observed in modern elm forests). Subsequently, *Fagus* and *Carpinus* replaced elm in the deciduous forests of southeastern Europe. A more pronounced decline occurred between 6500 yr B.P. and 4500 yr B.P. in *Ulmus* populations located in central and northern Europe, where they were replaced largely by early-successional *Betula* and *Corylus*. This latter decline in elm populations is attributed to a combination of the activities of Neolithic farmers (including cutting of branches for fodder for cattle, as well as land clearance for shifting agriculture), and to continued outbreaks of fungal pathogen (Iversen 1941, 1960; Watts 1961; Smith 1970; Sims 1973; Huntley and Birks 1983).

In detailed late-Quaternary mapping of *Ulmus* pollen values, Ralska-Jasiewiczowa (1983) used fossil-pollen data from 64 sites from Poland to demonstrate convincingly that the geographic areas of "*Ulmus*-fall" were concentrated in regions of fertile loess–chernozem (black-earth) soils. She noted the "striking correspondence" in timing and location between the geographic areas of elm decline and archaeological evidence for the distribution of human settlements from middle Neolithic cultures radiocarbon-dated from 5800 yr B.P. to 4500 yr B.P. Ralska-Jasiewiczowa (1983) speculated that the palynological evidence for the elm decline may reflect selective exploitation of elm populations by mid-Holocene human populations.

The late-Quaternary history of elm in eastern North America differs markedly from that of the European elm. The late-glacial and early-Holocene intervals were times characterized by increases in both the area of distribution and the dominance of American elm species. Only in the 20th Century A.D., with the introduction of the Dutch elm disease, has elm decreased substantially in importance throughout its range in eastern North America.

Conclusions

1. Ninety percent of the time within the Quaternary Period has been dominated by glacial climatic regimes and environments, and only 10% of the past 2 million years has corresponded with warm interglacial conditions such as today. Thus, long-term vegetational dynamics within the Temperate Zone have been primarily shaped by conditions very different from those of today.

2. Full-glacial conditions in eastern North America permitted the development of extensive temperate and boreal forests situated south of the Laurentide Ice limit between approximately 28°N and 43°N. In contrast, the majority of the European continent was covered by tundra or steppe vegetation with geographically restricted forests situated in refugial areas along the Mediterranean region generally south of 40°N. In both situations, temperate forests were restricted south of the climatic gradient imposed by the Polar Frontal Zone.

3. The timing and nature of the late-glacial to early-Holocene transition differed between Europe and North America. Eastern North American tree taxa initiated their migrations as early as 16,000 yr B.P. to 14,000 yr B.P., whereas the northward spread of European tree taxa was delayed until approximately 12,000 yr B.P. to 10,000 yr B.P.

4. In both Europe and North America, where populations of early-successional taxa had refugial areas relatively near the ice sheets, their rates of migration were limited not by their intrinsic rates of seed dispersal but rather by the barriers imposed by slowly retreating glacial ice. However, early-successional tree and shrub taxa with refuge areas in southern Europe had rates of post-glacial migration that exceeded those of their North American counterparts. In these cases, advance populations of these taxa were dispersed over long distances, in part aided in their dispersal of seeds by the northward flowing river systems, and rapidly colonized extensive areas of northern and western Europe formerly occupied by tundra.

5. On both continents, temperate and boreal taxa achieved their maximum interglacial limits in northern distribution typically between 6000 yr B.P. and 4000 yr B.P. The northern range limits of many arboreal taxa have retracted southward during the last 4000 years in response to late-Holocene global climatic cooling. In Europe, human activities including forest clearance, cultivation, and land management have greatly modified both the distributions and the abundances of tree species throughout the mid- and late-Holocene intervals. In eastern North America, humans had relatively little impact on native vegetation until the last 1000 years, with cultivation focused within major river valleys of the southeastern United States. Extensive fragmentation of North American forests has occurred only in the last few hundred years with EuroAmerican settlement, land clearance, and introduction of pathogens.

References

Ahearn P J, Bailey R E (1980) Pollen record from Chippewa Bog, Lapeer County, Michigan. Michigan Acad 12:297–308

Ahlgren C E, Ahlgren I F (1983) The human impact on northern forest ecosystems. In: Flader S L (ed) The Great Lakes forest: an environmental and social history. University of Minnesota Press, Minneapolis, pp 33–51

Albert L E, Wyckoff D G (1981) Ferndale Bog and Natural Lake: five thousand years of environmental change in southeastern Oklahoma. Okla Archaeol Surv, Stud Okla Past 7:1–125

Allen T H F, Starr T B (1982) Hierarchy: perspectives for ecological complexity. University of Chicago Press, Chicago

Allison T D, Moeller R E, Davis M B (1986) Pollen in laminated sediments provides evidence for a mid-Holocene forest pathogen outbreak. Ecology 67:1101–1105

Andersen S T (1970) The relative pollen productivity of North European trees, and correction factors for tree pollen spectra. Danm Geol Unders 2:1–99

Andersen S T (1973) The differential pollen productivity of trees and its significance for the interpretation of a pollen diagram from a forested region. In: Birks H J B, West R G (eds) Quaternary plant ecology, Blackwell, Oxford, pp 109–115

Andersen S T (1980a) Influence of climatic variation on pollen season severity in wind-pollinated trees and herbs. Grana 19:47–52

Andersen S T (1980b) The relative pollen productivity of the common forest trees in the early Holocene in Denmark. Danm Geol Unders Arbog 1979:5–19

Andersen S T (1984) Forests at Lovenholm, Djursland, Denmark, at present and in the past. Det Kongelige Danske Videnskabernes Selskab Biologiske Skrifter 24:1–208

Anderson T W (1974) The chestnut pollen decline as a time horizon in lake sediments in eastern North America. Can J Earth Sci 11:678–685

Anderson T W (1980) Holocene vegetation and climatic history of Prince Edward Island, Canada. Can J Earth Sci 17:1152–1165

Anderson T W (1983) Preliminary evidence for Late Wisconsinan climatic fluctuations from pollen stratigraphy in Burin Peninsula, Newfoundland. Current Res, Part B, Geol Surv Can, Pap 83-1B:185–188

Andrews J T (1982) On the reconstruction of Pleistocene ice sheets: a review. Quat Sci Rev 1:1–30

Andrews J T, Mahaffy M A W (1976) Growth rate of the Laurentide Ice Sheet and sea level lowering (with emphasis on the 115,000 BP sea level low). Quat Res 6:167–183

Armentano T V, Ralston C W (1980) The role of Temperate Zone forests in the global carbon cycle. Can J For Res 10:53–60

Asch D L, Asch N B (1985) Prehistoric plant cultivation in west-central Illinois. In: Ford R I (ed) Prehistoric food production in North America. Museum of Anthropology, University of Michigan, Anthropological Papers No. 75, pp 149–203

Axelrod D I (1983) Biogeography of oaks in the Arcto-Tertiary province. Ann Mo Bot Gard 70:629–657

Bailey R E (1972) Late- and postglacial environmental changes in northwestern Indiana. PhD Dissertation, Indiana University, Bloomington

Bailey R E, Ahearn P J (1981) A late and postglacial pollen record from Chippewa Bog, Lapeer Co, MI: further examination of white pine and beech immigration into the central Great Lakes region. In: Romans R (ed) Geobotany II. Plenum Press, New York, pp 53–74

Bailey R G (1978) Description of the ecoregions of the United States. United States Department of Agriculture, Forest Service, Odgen, Utah

Baker R G (1970) A radiocarbon-dated pollen chronology for Wisconsin: Disterhaft Farm Bog revisited. Geol Soc Am Abstr 2:488

Baker R G, Van Zant K L, Dulian J J (1980) Three late-glacial pollen and plant macrofossil assemblages from Iowa. Palynology 4:197–203

Balsam W (1981) Late Quaternary sedimentation in the western North Atlantic: stratigraphy and paleoceanography. Palaeogeogr Palaeoclimatol Palaeoecol 35:215–240

Barclay F H (1957) The natural vegetation of Johnson County, Tennessee, past and present. PhD Dissertation, University of Tennessee, Knoxville

Bareis C J, Porter J W (1984) American Bottom Archaeology. University of Illinois Press, Urbana

Barker G (1985) Prehistoric farming in Europe. Cambridge University Press, Cambridge

Barry R G (1983) Late-Pleistocene climatology. In: Porter S C (ed) Late-Quaternary environments of the United States, Volume 1, The Late Pleistocene. University of Minnesota Press, Minneapolis, pp 390–407

Barry R G, Elliott D L, Crane R G (1981) The palaeoclimatic interpretation of exotic pollen peaks in Holocene records from the eastern Canadian Arctic: a discussion. Rev Palaeobot Palynol 33:153–167

Bartlein P J, Webb III T, Fleri E (1984) Holocene climatic change in the northern Midwest: pollen-derived estimates. Quat Res 22:361–374

Bartlein P J, Prentice I C, Webb III T (1986) Climatic response surfaces from pollen data for some eastern North American taxa. J Biogeogr 13:35–57

Bassett I J, Terasmae J (1962) Ragweeds, *Ambrosia* species, in Canada and their history in postglacial time. Can J Bot 40:145–150

Begon M, Mortimer M (1982) Population ecology. Blackwell Scientific, Oxford

Behre K E (1981) The interpretation of anthropogenic indicators in pollen diagrams. Pollen et Spores 23:225–245

Belding H F, Holland W C (1970) Bathymetric maps, eastern continental margin, U.S.A. Sheets 1,2, and 3, map scale 1:1,000,000. Am Assoc Pet Geol, Tulsa, Oklahoma.

Bender M M, Bryson R A, Baerreis D A (1971) University of Wisconsin radiocarbon dates IX. Radiocarbon 13:475–486

Bender M M, Bryson R A, Baerreis D A (1975) University of Wisconsin radiocarbon dates XII. Radiocarbon 17:121–134

Bennett K D (1983) Postglacial population expansion of forest trees in Norfolk, UK. Nature 303:164–167

Bennett K D (1985) The spread of *Fagus grandifolia* across eastern North America during the last 18,000 years. J Biogeogr 12:147–164

Berger A L (1978) Long-term variations of caloric insolation resulting from the Earth's orbital elements. Quat Res 9:139–167

Berggren W A (1982) Role of ocean gateways in climatic change. In: Geophysics Study Committee (eds) Climate in Earth history. National Academy Press, Washington, DC pp 118–125

Berggren W A, Hollister C D (1974) Paleogeography, paleobiogeography and the history of circulation in the Atlantic Ocean. In: Hay W W (ed) Studies in paleooceanography. Soc Econ Paleontol Mineral Spec Publ 20:126–186

Bernabo J C (1981) Quantitative estimates of temperature changes over the last 2700 years in Michigan based on pollen data. Quat Res 15:143–159

Bernabo J C, Webb T III (1977) Changing patterns in the Holocene pollen record of northeastern North America: a mapped summary. Quat Res 8:64–96

Beschel R E, Webber P J, Tippett R (1962) Woodland transects of the Frontenac Axis Region, Ontario. Ecology 43:386–396

Billings W D, Mooney H A (1968) The ecology of arctic and alpine plants. Biol Rev 43:481–529

Birks H J B (1973) Modern pollen rain studies in some arctic and alpine environments. In: Birks H J B, West R G (eds) Quaternary plant ecology. Blackwell, Oxford, pp 143–170

Birks H J B (1976) Late-Wisconsinan vegetational history at Wolf Creek, central Minnesota. Ecol Monogr 46:395–429

Birks H J B (1981a) Late Wisconsin vegetational and climatic history at Kylen Lake, northeastern Minnesota. Quat Res 16:322–355

Birks H J B (1981b) The use of pollen analysis in the reconstruction of past climates: a review. In: Wigley T M L, Ingram M J, Farmer G (eds) Climate and history. Cambridge University Press, Cambridge, pp 111–138

Birks H J B (1986) Late-Quaternary biotic changes in terrestrial and lacustrine environments, with particular reference to north-west Europe. In: Berglund B E (ed) Handbook of Holocene palaeoecology and palaeohydrology. John Wiley & Sons, New York, pp 3–65

Birks H J B, Birks H H (1980) Quaternary palaeoecology. University Park Press, Baltimore

Birks H J B, Gordon A D (1985) Numerical methods in Quaternary pollen analysis. Academic Press, New York

Birks H J B, Webb III T, Berti A A (1975) Numerical analysis of pollen samples from central Canada: a comparison of methods. Rev Palaeobot Palynol 20:133–169

Bloom A L (1977) Atlas of sea-level curves, IGCP Project 61. Int. Union Quat Res

Bloom A L (1983) Sea level and coastal morphology of the United States through the Late Wisconsin glacial maximum. In: Porter S C (ed) Late-Quaternary environments of the United States, Volume 1, The Late Pleistocene. University of Minnesota Press, Minneapolis, pp 215–229

Boellstorff J (1978) North American Pleistocene stages reconsidered in light of probable Pliocene-Pleistocene continental glaciation. Science 202:305–307

Bonny A P (1978) The effect of pollen recruitment processes on pollen distribution over the sediment surface of a small lake in Cumbria. J Ecol 66:385–416

Bormann F H, Likens G E (1979) Pattern and process in a forested ecosystem: disturbance, development and the steady state based on the Hubbard Brook Ecosystem Study. Springer-Verlag, New York

Botkin D B (1973) Life and death in a forest: the computer as an aid to understanding. In: Hall C A S, Day J W Jr. (eds) Ecosystem modeling in theory and practice: an introduction with case histories. John Wiley & Sons, New York, pp 213–233

Botkin D B, Sobel M J (1975) Stability in time-varying ecosystems. Am Nat 109:625–646

Botkin D B, Janak J F, Wallis J R (1972a) Rationale, limitations, and assumptions of a northeastern forest growth simulator. IBM J Res Devel 16:101–116

Botkin D B, Janak J G, Wallis J R (1972b) Some ecological consequences of a computer model of forest growth. J Ecol 60:849–873

Bourdo E A Jr (1956) A review of the General Land Office Survey and of its use in quantitative studies of former forests. Ecology 37:754–768

Bourdo E A Jr (1983) The forest the settlers saw. In: Flader S L (ed) The Great Lakes forest, an environmental and social history. University of Minnesota Press, Minneapolis, pp 3–16

Bowen D Q (1978) Quaternary geology, a stratigraphic framework for multidisciplinary work. Pergamon, Oxford

Bradshaw R H W (1981a) Quantitative reconstruction of local woodland vegetation using pollen analysis from two small basins in Norfolk, England. J Ecol 69:941–955

Bradshaw R H W (1981b) Modern pollen-representation factors for woods in south-east England. J Ecol 69:45–70

Bradshaw R H W, Webb T III (1985) Relationships between contemporary pollen and vegetation data from Wisconsin and Michigan, U.S.A. Ecology 66:721–737

Bradstreet T E, Davis R B (1975) Mid-postglacial environments in New England with emphasis on Maine. Arctic Anthropol 12-2:7–22

Braun E L (1950, reprinted in 1974) Deciduous forests of eastern North America. Hafner, New York

Broecker W A, van Donk J (1970) Insolation changes, ice volumes, and O^{18} record in deep sea cores. Rev Geophys 8:169–198

Brown J G (1981) Palynologic and petrographic analyses of bayhead hammock and marsh peats at Little Salt Spring archeological site (8So18), Florida. MS Thesis, University of South Carolina, Columbia

Brown J G, Cohen A D (1985) Palynologic and petrographic analyses of peat deposits, Little Salt Spring. Nat Geogr Res 1:21–31

Brubaker L B (1975) Postglacial forest patterns associated with till and outwash in northcentral Upper Michigan. Quat Res 5:499–527

Brunner C A (1982) Paleoceanography of surface waters in the Gulf of Mexico during the late Quaternary. Quat Res 17:105–119

Brush G S (1967) Pollen analyses of late-glacial and post-glacial sediment in Iowa. In: Cushing E J, Wright H E Jr (eds) Quaternary paleoecology. Yale University Press, New Haven, pp 99–105

Brush G S, Brush L M Jr (1972) Transport of pollen in a sediment-laden channel: a laboratory study. Am J Sci 272:359–381

Bryant V M Jr (1977) A 16,000 year pollen record of vegetational change in central Texas. Palynology 1:143–156

Bryant V M Jr, Holloway R G (1985a) Pollen records of late-Quaternary North American sediments. American Association of Stratigraphic Palynologists Foundation, Dallas

Bryant V M Jr, Holloway R G (1985b) A late-Quaternary paleoenvironmental record of Texas: an overview of the pollen evidence. In: Bryant V M Jr, Holloway R G (eds) Pollen records of late-Quaternary North American sediments. American Association of Stratigraphic Palynologists Foundation, Dallas, pp 39–70

Bryson R A (1966) Air masses, streamlines, and the boreal forest. Geogr Bull 8:228–269

Bryson R A, Hare F K (1974) The climates of North America. In: Bryson R A, Hare F K (eds) Climates of North America, World Survey of Climatology 11. Elsevier, New York, pp 1–47

Bryson R A, Wendland W M (1967) Tentative climatic patterns for some late glacial and post-glacial episodes in central North America. In: Mayer-Oakes W J (ed) Life, land, and water, proceedings of the 1966 Conference on Environmental Studies of the Glacial Lake Agassiz Region. University of Manitoba Press, Winnepeg, pp 271–298

Budyko M I (1977) Climatic changes. American Geophysical Union, Washington, DC

Burgess R L, Sharpe D M (1981) Forest island dynamics in man-dominated landscapes. Springer-Verlag, New York

Burke M J, George M F, Bryant R G (1975) Water in plant tissues and frost hardiness. In: Duckworth R B (ed) Water relations of foods. Academic Press, New York, pp 111–135

Chabot B F, Mooney H A (eds) (1985) Physiological ecology of North American plant communities. Chapman and Hall, New York

Chapin F S III, Shaver G R (1985) Arctic. In: Chabot B F, Mooney H A (eds) Physiological ecology of North American plant communities. Chapman and Hall, New York, pp 16–40

Chapman J (1985) Tellico archaeology: 12,000 years of Native American history. University of Tennessee Press, Knoxville

Chapman J, Shea A B (1981) The archaeobotanical record: Early Archaic period to contact in the Lower Little Tennessee River Valley. Tenn Anthropol 6:61–84

Chen Y (1986) Early Holocene vegetation dynamics of Lake Barrine basin, northeast Queensland, Australia. PhD Dissertation, Australian National University, Canberra

Chesson P L, Case T J (1986) Overview: nonequilibrium community theories: chance, variability, history, and coexistence. In: Diamond J, Case T J (eds) Community ecology. Harper & Row, New York pp 229–239

Clampitt C (1985) DECORANA for IBM-PCs. BioScience 35:738.

Clark G M (1968) Sorted patterned ground: new Appalachian localities south of the glacial border. Science 161:355–356

Clark J A, Farrell W E, Peltier W R (1978) Global changes in postglacial sea level: a numerical calculation. Quat Res 9:265–287

Clark J S (1986) Coastal forest tree populations in a changing environment, southeastern Long Island, New York. Ecol Monogr 56:259–277

Clayton L, Moran S R (1982) Chronology of late Wisconsinan glaciation in middle North America. Quat Sci Rev 1:55–82

CLIMAP (1976) The surface of the Ice-Age Earth. Science 191:1131–1137

CLIMAP (1981) Seasonal reconstructions of the Earth's surface at the last glacial maximum. Geol Soc Am Map Chart Ser MC-36

CLIMAP (1984) The last interglacial ocean. Quat Res 21:123–224

Cloud P (1972) A working model of the primitive earth. Am J Sci 272:537–548

Cohen A D, Casagrande D J, Andrejko M J, Best G R (1984) The Okefenokee Swamp: its natural history, geology, and geochemistry. Wetland Surveys, Los Alamos

Comanor P L (1968) Forest vegetation and the pollen spectrum: an examination of the usefulness of the R value. Bull NJ Acad Sci 13:7–19

Connell J H (1980) Diversity and the coevolution of competitors, or the ghost of competition past. Oikos 35:131–138

Cotter J F P, Crowl G H (1981) The paleolimnology of Rose Lake, Potter Co., Pennsylvania, a comparison of palynologic and paleo-pigment studies. In: Romans R (ed) Geobotany II. Plenum, New York, pp 91–122

Cox D D, Lewis D M (1965) Pollen studies in the Crusoe Lake area of prehistoric Indian occupation. NY State Mus Sci Serv Bull No. 397:1–29

Craig A J (1969) Vegetational history of the Shenandoah Valley, Virginia. Geol Soc Am Spec Pap 123:283–296

Craig A J (1972) Pollen influx to laminated sediments: a pollen diagram from northeastern Minnesota. Ecology 53:46–57

Cronin T M, Szabo, B J, Ager, T A, Hazel J E, Owens J P (1981) Quaternary climates and sea levels of the U.S. Atlantic Coastal Plain. Science 211:233–240

Crowder A A, Cuddy D G (1973) Pollen in a small river basin: Wilton Creek, Ontario. In: Birks H J B, West R G (eds) Quaternary plant ecology. Blackwell, Oxford, pp 61–77

Crowell J C (1982) Continental glaciation through geologic times. In: Geophysics Study Committee (eds) Climate in Earth history. National Academy Press, Washington, DC, pp 77–82

Currier P J, Kapp R O (1974) Local and regional pollen components at Davis Lake, Montcalm County, Michigan. Mich Acad 7:211–225

Curtis J T (1959) The vegetation of Wisconsin: an ordination of plant communities. University of Wisconsin Press, Madison

Cwynar L C, Ritchie J C (1980) Arctic steppe–tundra: a Yukon perspective. Science 208:1375–1377

Davidson J L (1983) Paleoecological analysis of Holocene vegetation, Lake in the Woods, Cades Cove, Great Smoky Mountains National Park. MS Thesis, University of Tennessee, Knoxville

Davis A M (1977) The prairie–deciduous forest ecotone in the Upper Middle West. Ann Assoc Am Geogr 67:204–213

Davis A M (1979) Wetland succession, fire, and the pollen record: a midwestern example. Am Midl Nat 102:86–94

Davis M B (1963) On the theory of pollen analysis. Am J Sci 261:897–912

Davis M B (1969) Climatic changes in southern Connecticut recorded by pollen deposition at Rogers Lake. Ecology 50:409–422

Davis M B (1973) Redeposition of pollen grains in lake sediment. Limnol Oceanogr 18:44–52

Davis M B (1976a) Erosion rates and land-use history in southern Michigan. Environ Conserv 3:139–148

Davis M B (1976b) Pleistocene biogeography of temperate deciduous forests. Geosci Man 13:13–26

Davis M B (1978) Climatic interpretation of pollen in Quaternary sediments. In: Walker D, Guppy J C (eds) Biology and Quaternary environments. Australian Academy of Science, Canberra, A.C.T., pp 35–51

Davis M B (1981a) Quaternary history and the stability of forest communities. In: West D C, Shugart H H, Botkin D B (eds) Forest succession, concepts and application. Springer-Verlag, New York, pp 132–153

Davis M B (1981b) Outbreaks of forest pathogens in Quaternary history. Proc IV Int Palynol Conf Lucknow (1976-1977) 3:216–227

Davis M B (1983) Quaternary history of deciduous forests of eastern North America and Europe. Ann Mo Bot Gard 70:550–563

Davis M B, Botkin D B (1985) Sensitivity of cool-temperate forests and their fossil pollen record to rapid temperature change. Quat Res 23:327–340

Davis M B, Brubaker L B (1973) Differential sedimentation of pollen grains in lakes. Limnol Oceanogr 18:635–646

Davis M B, Deevey E S Jr (1964) Pollen accumulation rates: estimates from late-glacial sediment of Rogers Lake. Science 145:1293–1295

Davis M B, Goodlett J C (1960) Comparison of the present vegetation with pollen spectra in surface samples from Brownington Pond, Vermont. Ecology 41:346–357

Davis M B, Brubaker L B, Beiswenger J M (1971) Pollen grains in lake sediments: pollen percentages in surface sediments from southern Michigan. Quat Res 1:450–467

Davis M B, Brubaker L B, Webb T III (1973) Calibration of absolute pollen influx. In: Birks H J B, West R G (eds) Quaternary plant ecology. Wiley, New York, pp 9–25

Davis M B, Moeller R E, Ford J (1984) Sediment focusing and pollen influx. In: Haworth E Y, Lund J W G (eds) Lake sediments and environmental history: studies in palaeolimnology and palaeoecology in honour of Winifred Tutin. University of Minnesota Press, Minneapolis, pp 261–293

Davis M B, Woods K, Webb S, Futyma R P (1986) Dispersal versus climate: expansion of *Fagus* and *Tsuga* into the Upper Great Lakes region. Vegetatio 67: 93–103

Davis O K (1984) Pollen frequencies reflect vegetation patterns in a Great Basin (U.S.A.) mountain range. Rev Palaeobot Palynol 40:295–315

Davis R B (1967) Pollen studies of near-surface sediments in Maine lakes. In: Cushing E J, Wright H E Jr (eds) Quaternary paleoecology. Yale University Press, New Haven, Connecticut, pp 143–173

Davis R B (1987) Paleolimnological diatom studies of acidification of lakes by acid rain: an application of Quaternary science. Quat Sci Rev 6:(in press)

Davis R B, Webb T III (1975) The contemporary distribution of pollen in eastern North America: a comparison with the vegetation. Quat Res 5:395–434

Davis R B, Jacobson G L Jr. (1985) Late glacial and early Holocene landscapes in northern New England and adjacent areas of Canada. Quat Res 23:341–368

Davis R B, Bradstreet T E, Stuckenrath R Jr, Borns H W Jr (1975) Vegetation and associated environments during the past 14,000 years near Moulton Pond, Maine. Quat Res 5: 435–465

Deevey E S Jr (1949) Biogeography of the Pleistocene, Part I: Europe and North America. Bull Geol Soc Am: 60:1315–1416.

Delcourt H R (1979) Late-Quaternary vegetation history of the eastern Highland Rim and adjacent Cumberland Plateau of Tennessee. Ecol Monogr 49:255–280

Delcourt H R (1985b) Holocene vegetational changes in the southern Appalachian Mountains, U.S.A. Ecol Mediterr 11:9–16

Delcourt H R (1987) The impact of prehistoric agriculture and land occupation on natural vegetation. Trends Ecol Evol 2:39–44

Delcourt H R, Delcourt P A (1975) The Blufflands: Pleistocene pathway into the Tunica Hills. Am Midl Nat 94:385–400

Delcourt H R, Delcourt P A (1977a) Presettlement Magnolia–Beech Climax of the Gulf Coastal Plain: quantitative evidence from the Apalachicola River Bluffs, north-central Florida. Ecology 58:1085–1093

Delcourt H R, Delcourt P A (1985a) Comparison of taxon calibrations, modern analogue techniques, and forest-stand simulation models for the quantitative reconstruction of past vegetation. Earth Surf Proc Landforms 10:293–304

Delcourt H R, Delcourt P A (1985b) Quaternary palynology and vegetational history of the southeastern United States. In: Bryant V M Jr, Holloway R G (eds) Pollen records of late-Quaternary North American sediments. American Association of Stratigraphic Palynologists Foundation, pp 1–37

Delcourt H R, Delcourt P A (1986) Late-Quaternary vegetational history of the central Atlantic states. In: McDonald J, Bird S O (eds) The Quaternary of Virginia. Virginia Commonwealth Division of Mineral Resources, Charlottesville, pp 23–35

Delcourt H R, Harris W F (1980) Carbon budget of the southeastern U.S. biota: analysis of historical change in trend from source to sink. Science 210:321–323

Delcourt H R, Pittillo J D (1986) Comparison of contemporary vegetation and pollen assemblages: an altitudinal transect in the Balsam Mountains, Blue Ridge Province, western North Carolina, USA. Grana 25:131–141

Delcourt H R, West D C, Delcourt P A (1981) Forests of the southeastern United States: quantitative maps for aboveground woody biomass, carbon, and dominance of major tree taxa. Ecology 62:879–887

Delcourt H R, Delcourt P A, Webb T III (1982) Dynamic plant ecology: the spectrum of vegetational change in space and time. Quat Sci Rev 1:153–175

Delcourt H R, Delcourt P A, Spiker E C (1983b) A 12,000-year record of forest history from Cahaba Pond, St. Clair County, Alabama. Ecology 64:874–887

Delcourt H R, Delcourt P A, Wilkins G R, Smith E N Jr (1986b) Vegetational history of the cedar glades regions of Tennessee, Kentucky, and Missouri during the past 30,000 years. Assoc Southeastern Biol Bull 33:128–137

Delcourt P A (1980) Goshen Springs: late-Quaternary vegetation record for southern Alabama. Ecology 61:371–386

Delcourt P A (1985a) The influence of late-Quaternary climatic and vegetational change on paleohydrology in unglaciated eastern North America. Ecol Mediterr 11:17–26

Delcourt P A, Delcourt H R (1977b) The Tunica Hills, Louisiana-Mississippi: late glacial locality for spruce and deciduous forest species. Quat Res 7:218–237

Delcourt P A, Delcourt H R (1979) Late Pleistocene and Holocene distributional history of the deciduous forest in the southeastern United States. Veroff Geobot Inst ETH, Stift. Rubel (Zurich) 68:79–107

Delcourt P A, Delcourt H R (1980) Pollen preservation and Quaternary environmental history in the southeastern United States. Palynology 4:215–231

Delcourt P A, Delcourt H R (1981) Vegetation maps for eastern North America: 40,000 yr BP to the present. In: Romans R (ed) Geobotany II. Plenum Press, New York, pp 123–166

Delcourt P A, Delcourt H R (1983) Late-Quaternary vegetational dynamics and community stability reconsidered. Quat Res 19:265–271

Delcourt P A, Delcourt H R (1984) Late-Quaternary paleoclimates and biotic responses across eastern North America and the northwestern Atlantic Ocean. Palaeogeogr Palaeoclimatol Palaeoecol 48:263–284

Delcourt P A, Delcourt H R (1985c) Dynamic Quaternary landscapes of East Tennessee: an integration of paleoecology, geomorphology, and archaeology. University of Tennessee Depart Geol Sci Studies in Geology 9:191–220

Delcourt P A, Delcourt H R, Brister R C, Lackey L E (1980) Quaternary vegetation history of the Mississippi Embayment. Quat Res 13:111–132

Delcourt P A, Delcourt H R, Davidson J L (1983a) Mapping and calibration of modern pollen–vegetation relationships in the southeastern United States. Rev Palaeobot Palynol 39:1–45

Delcourt P A, Delcourt H R, Webb T III (1984) Atlas of mapped distributions of dominance and modern pollen percentages for important tree taxa of eastern North America. Am Assoc Strat Palynol Contrib Ser No 14:1–131

Delcourt P A, Delcourt H R, Cridlebaugh P A, Chapman J. (1986a) Holocene ethnobotanical and paleoecological record of human impact on vegetation in the Little Tennessee River Valley, Tennessee. Quat Res 25:330–349

Denny C S, Owens J P, Sirkin L A, Rubin M (1979) The Parsonburg Sand in the central Delmarva Peninsula, Maryland and Delaware. US Geol Surv Prof Pap 1067B:1–16

Denton G H, Hughes T J (eds) (1981) The last great ice sheets. John Wiley & Sons, New York

Denton G H, Hughes T J (1983) Milankovitch theory of Ice Ages: hypothesis of ice-sheet linkage between regional insolation and global climate. Quat Res 20:125–144

Diamond J (1986) Evolution of ecological segregation in the New Guinea montane avifauna. In: Diamond J, Case T J (eds) Community ecology. Harper & Row, New York, pp 98–125

Diamond J, Case T J (1986) Community ecology. Harper & Row, New York

Dreimanis A (1977) Late Wisconsinan glacial retreat in the Great Lakes region, North America. Ann NY Acad Sci 288:70–89

Durkee L H (1971) A pollen profile from Woden Bog in north-central Iowa. Ecology 52:837–844

Duplessy J C, Delibrias G, Turon J L, Pujol C, Duprat J (1981) Deglacial warming of the northeastern Atlantic Ocean: correlation with the paleoclimatic evolution of the European continent. Palaeogeogr Palaeoclimatol Palaeoecol 35:121–144

Elliott-Fisk D L, Andrews J T, Short S K, Mode W N (1982) Isopoll maps and an analysis of the distribution of the modern pollen rain, eastern and central northern Canada. Geogr Phys Quat 37:91–108

Elson J A (1967) Geology of Glacial Lake Agassiz. In: Mayer-Oakes W J (ed) Life, land, and water, proceedings of the 1966 Conference on Environmental Studies of the Glacial Lake Agassiz Region. University of Manitoba Press, Winnepeg, pp 37–95

Elton C S (1958) The ecology of invasions by animals and plants. Chapman and Hall, New York

Emiliani C (1966) Isotopic paleotemparatures. Science 154:851–856

Eyster-Smith N M (1977) Holocene pollen stratigraphy of Lake St. Croix, Minnesota-Wisconsin, and some aspects of the depositional history. MS Thesis, University of Minnesota, Minneapolis

Faegri K, Iversen J (1975) Textbook of pollen analysis, 3rd ed. Hafner, New York

Faegri K, van der Pijl L (1979) The principles of pollination ecology, 3rd ed. Pergamon, Oxford

Fastook J L, Hughes T (1982) A numerical model for reconstruction and disintegration of the Late Wisconsin Glaciation in the Gulf of Maine. In: Larson G J, Stone B D (eds) Late Wisconsin Glaciation of New England. Kendall/Hunt, Dubuque, Iowa, pp 229–242

Fearn L B (1981) A paleoecological reconstruction of Okefenokee swamp based on pollen stratigraphy and peat petrography. MS Thesis, University of South Carolina, Columbia

Fearn L B, Cohen A D (1984) Palynologic investigations of six sites in the Okefenokee Swamp. In: Cohen A D, Casagrande D J, Andrejko M J, Best G R (eds) The Okefenokee Swamp: its natural history, geology, and geochemistry. Wetland Surveys, Los Alamos, New Mexico, pp 423–443

Fenner M (1985) Seed ecology. Chapman and Hall, London

Fernald M L (1970) Gray's manual of botany. D. Van Nostrand, New York

Field M E, Meisburger E P, Stanley E A, Williams S J (1979) Upper Quaternary peat deposits on the Atlantic inner shelf of the United States. Bull Geol Soc Am 90:618–628

Flader S L (1983) The Great Lakes forest, an environmental and social history. University of Minnesota Press, Minneapolis

Florin M-B, Wright H E Jr (1969) Diatom evidence for the persistence of stagnant glacial ice in Minnesota. Bull Geol Soc Amer 80:695–704

Forman R T T (1981) Interaction among landscape elements: a core of landscape ecology. Proc Int Congr Neth Soc Landscape Ecol, Veldhoven, 1981:35–48

Forman R T T, Godron M (1981) Patches and structural components for a landscape ecology. Bioscience 31:733–740

Forman R T T, Godron M (1986) Landscape ecology. John Wiley & Sons, New York

Fowells H A (1965) Silvics of forest trees of the United States. US Forest Service Agric Handbk 271:1–762

Frakes L A (1979) Climates throughout geologic time. Elsevier Press, Amsterdam

Frey D G (1951) Pollen succession in the sediments of Singletary Lake, North Carolina. Ecology 32:518–533

Frey D G (1953) Regional aspects of the late-glacial and post-glacial pollen succession of southeastern North Carolina. Ecol Monogr 23:289–313

Frey D G (1955) A time revision of the Pleistocene pollen chronology of southeastern North Carolina. Ecology 36:762–763

Fries M (1962) Pollen profiles of late Pleistocene and Recent sediments from Weber Lake, northeastern Minnesota. Ecology 43:295–308

Fulton R J (ed) (1984) Quaternary stratigraphy of Canada, a Canadian contribution to IGCP Project 24. Geol Soc Canada, Ottawa

Gauch H (1982) Multivariate analysis in community ecology. Cambridge University Press, Cambridge

Geiger R (1966) The climate near the ground. Harvard University Press, Cambridge, Massachusetts

Geikie J (1874) The great Ice Age and its relation to the antiquity of man. Isbister, London

Giller P S (1984) Community structure and the niche. Chapman and Hall, London

Glaser P H (1981) Transport and deposition of leaves and seeds on tundra: a late-glacial analog. Arctic Alpine Res 13:173–182

Goldthwaite R P (1965) Great Lakes-Ohio River Valley. Guidebook for Field Conference G, International Association for Quaternary Research, VIIth Congress

Gordon A G (1964) The nutrition and growth of ash, *Fraxinus excelsior*, in natural stands in the English Lake District as related to edaphic site factors. J Ecol 52:169–187

Gordon A G (1985) "Budworm! What about the forest?" US Dept Agric For Serv Gen Tech Rep NE-99:3–29

Graham A (1972) Outline of the origin and historical recognition of floristic affinities between Asia and eastern North America. In: Graham A (ed) Floristics and paleofloristics of Asia and eastern North America. Elsevier Press, Amsterdam, pp 1–16

Graham R W, Lundelius E L Jr (1984) Coevolutionary disequilibrium and Pleistocene extinctions. In: Martin P S, Klein R G (eds) Quaternary extinctions, a prehistoric revolution. University of Arizona Press, Tucson, pp 223–249

Green D G (1981) Time series and postglacial forest ecology. Quat Res 15:265–277

Grichuk V P (1984) Late Pleistocene vegetation history. In: Velichko A A, Wright H E Jr., Barnosky C W (eds) Late Quaternary environments of the Soviet Union. University of Minnesota Press, Minneapolis, pp 155–178

Grime J P (1979) Plant strategies and vegetation processes. John Wiley & Sons, New York

Gruger J (1973) Studies on the late Quaternary vegetation history of northeastern Kansas. Geol Soc Am Bull 84:239–250

Hack J T, Goodlett J C (1960) Geomorphology and forest ecology of a mountain region in the central Appalachians. US Geol Surv Prof Pap 347:1–66

Hadden K A (1975) A pollen diagram for a postglacial peat bog in Hants County, Nova Scotia. Can J Bot 53:39–47

Hall S A (1981) Deteriorated pollen grains and the interpretation of Quaternary pollen diagrams. Rev Palaeobot Palynol 32:193–206

Halliday W E D, Brown A W A (1943) The distribution of some important forest trees in Canada. Ecology 24:353–373

Hamilton W (1983) Cretaceous and Cenozoic history of the northern continents. Ann Mo Bot Gard 70:440–458

Harmon M E, Franklin J F, Swanson F J, Sollins P, Gregory S V, Lattin J D, Anderson N H, Cline S P, Aumen N G, Sedell J R, Lienkaemper G W, Cromack K Jr, Cummins K W (1985) Ecology of coarse woody debris in temperate ecosystems. Adv Ecol Res 15:133–302

Harper J L (1977) The population biology of plants. Academic Press, New York

Harrington J B Jr., Metzger K (1963) Ragweed pollen density. Am J Bot 50:532–539

Harrar E S, Harrar J G (1962) Guide to southern trees. Dover, New York

Harrison W, Malloy R J, Rusnak G A, Terasmae J (1965) Possible late Pleistocene uplift: Chesapeake Bay Entrance. J Geol 73:201–229

Havinga A J (1984) A 20-year experimental investigation into the differential corrosion susceptibility of pollen and spores in various soil types. Pollen et Spores 26:541–558

Hays J D, Imbrie J, Shackleton N J (1976) Variations in the Earth's orbit: pacemaker of the Ice Ages. Science 194:1121–1132

Hebda R J (1985) Museum collections and paleobiology. In: Miller E H (ed) Museum collections: their roles and future in biological research. Brit Columbia Prov Mus Occ Pap 25, pp 93–111

Heide K (1984) Holocene pollen stratigraphy from a lake and small hollow in north-central Wisconsin, USA. Palynology 8:3–20

Heide K, Bradshaw R (1982) The pollen-tree relationship within forests of Wisconsin and upper Michigan, U.S.A. Rev Palaeobot Palynol 36:1–23

Hill M O (1979) DECORANA—a FORTRAN program for Detrended Correspondence Analysis and Reciprocal Averaging. Cornell University, Ithaca, New York

Holloway R G, Valastro S (1983) Palynological investigations along the Yazoo River. In: Thorne R M, Curry H K (eds) Cultural Resources Survey of Items 3 and 4, Upper Yazoo River Projects, Mississippi, with a paleoenvironmental model of the Lower Yazoo Basin. Archaeol Pap, Center Archaeol Res, No 3, University of Mississippi, University, pp 159–257

Hopkins D M, Matthews J V Jr, Schweger C E, Young S B (1982) Paleoecology of Beringia. Academic Press, New York

Howe S, Webb T III (1983) Calibrating pollen data in climatic terms: improving the methods. Quat Sci Rev 2:17–51

Hughes N F (1976) Palaeobiology of Angiosperm origins. Cambridge University Press, Cambridge

Hughes T, Hyland M R, Lowell T V, Kite J S, Fastook J L, Borns H W Jr (1985) Models of glacial reconstruction and deglaciation applied to maritime Canada and New England. In: Borns H W Jr, LaSalle P, Thompson W B (eds) Late Pleistocene history of northeastern New England and adjacent Quebec. Geol Soc Am Spec Pap 197:139–150

Huntley B, Birks H J B (1983) An atlas of past and present pollen maps for Europe: 0-13000 years ago. Cambridge University Press, Cambridge

Hupp C R (1983) Geobotanical evidence of late Quaternary mass wasting in block field areas of Virginia. Earth Surf Proc Landforms 8:439–450

Hutchinson G E (1957) Concluding remarks. Cold Spring Harbor Symp Quant Biol 22:415–427

Hutson W H (1977) Transfer functions under no-analog conditions: experiments with Indian Ocean planktonic Foraminifera. Quat Res 8:355–367

Imbrie J, Imbrie K P (1979) Ice Ages, solving the mystery. Enslow, Hillside, New Jersey

Imbrie J, Kipp N G (1971) A new micropaleontological method for quantitative paleo-climatology: application to a late Pleistocene Caribbean core. In: Turekian K K (ed) The Late Cenozoic glacial ages. Yale University Press, New Haven, pp 71–181

Imbrie J, McIntyre A, Moore T C Jr (1983) The ocean around North America at the last glacial maximum. In: Porter S C (ed) Late-Quaternary environments of the United States, Vol I, the Late Pleistocene. University of Minnesota Press, Minneapolis, pp 230–236

Iversen J (1941) Landnam i Danmarks Stenalder. Danm Geol Unders Ser II 66:1–68

Iversen J (1960) Problems of the early Post-glacial forest development in Denmark. Danm Geol Unders Ser IV 4(3):1–32

Iversen J (1969) Retrogressive development of a forest ecosystem demonstrated by pollen diagrams from fossil mor. Oikos, Suppl 12:35–49

Ives J D, Andrews J T, Barry R G (1975) Growth and decay of the Laurentide Ice Sheet and comparisons with Fenno-Scandinavia. Naturwissenschaften 62:118–125

Jacobson G L Jr. (1979) The palaeoecology of white pine (*Pinus strobus*) in Minnesota. J Ecol 67:697–726

Jacobson G L Jr., Birks H J B (1980) Soil development on recent end moraines of the Klutlan Glacier, Yukon Territory, Canada. Quat Res 14:87–100

Jacobson G L Jr., Bradshaw R H W (1981) The selection of sites for paleovegetational studies. Quat Res 16:80–96

Jacobson G L Jr., Grimm E C (1986) A numerical analysis of Holocene forest and prairie vegetation in central Minnesota. Ecology 67:958–966

Janssen C R (1966) Recent pollen spectra from the deciduous and coniferous–deciduous forests of northwestern Minnesota: a study in pollen dispersal. Ecology 47:804–825

Janssen C R (1967a) Stevens Pond: a postglacial pollen diagram from a small *Typha* swamp in northwestern Minnesota, interpreted from pollen indicators and surface samples. Ecol Monogr 37:145–172

Janssen C R (1967b) A comparison between the recent regional pollen rain and the subrecent vegetation in four major vegetation types in Minnesota (U.S.A.). Rev Palaeobot Palynol 2:331–342

Janssen C R (1968) Myrtle Lake: a late- and post-glacial pollen diagram from northern Minnesota. Can J Bot 46:1397–1408

Janssen C R (1973) Local and regional pollen deposition. In: Birks H J B, West R G (eds) Quaternary plant ecology. Blackwell Scientific Publications, Oxford, England, pp 31–42

Janssen C R (1984) Modern pollen assemblages and vegetation in the Myrtle Lake Peatland, Minnesota. Ecol Monogr 54:213–252

Janzen D H, Martin P S (1982) Neotropical anachronisms: fruits the gomphotheres left behind. Science 215:19–27

Johannessen S (1984) Paleoethnobotany. In: Bareis D J, Porter J W (eds) American Bottom Archaeology. University of Illinois Press, Urbana, pp 197–214

Johnson D L (1977) The late Quaternary climate of coastal California: evidence for an Ice Age refugium. Quat Res 8:154–179

Johnson W C, Adkisson C S (1985) Dispersal of beech nuts by blue jays in fragmented landscapes. Am Midl Nat 113:319–324

Jordan R (1975) Pollen diagrams from Hamilton Inlet, central Labrador, and their environmental implications for the northern maritime Archaic. Arctic Anthropol 12-2:92–116

Kapp R O (1969) How to know pollen and spores. Wm C Brown Company, Dubuque

Kapp R O (1977a) Paleoecology in central Lower Michigan, field trip guide. Paleoecology Section, Ecological Society of America

Kapp R O (1977b) Late Pleistocene and postglacial plant communities of the Great Lakes region. In: Romans R (ed) Geobotany. Plenum, New York, pp 1–27

Kapp R O, Bushouse S, Foster B (1969) A contribution to the geology and forest history of Beaver Island, Michigan. Proc 12th Conf Great Lakes Res:225–236

Karrow P F, Anderson T W, Clarke A H, Delorme L D, Sreenivasa M R (1975) Stratigraphy, paleontology, and age of Lake Algonquin sediments in Southwestern Ontario, Canada. Quat Res 5:49–87

Keller E A, Swanson F J (1979) Effects of large organic material on channel form and fluvial processes. Earth Surf Proc Landforms 4:361–380

Kellogg W W (1983) Impacts of a CO_2-induced climate change. In: Bach W *et al.* (eds) Carbon Dioxide: current views and developments in energy/climate research. D. Reidel Publishing Company, Dordrecht, pp 379–413

Kerfoot W C (1974) New accumulation rates and the history of Cladoceran communities. Ecology 55:51–61

Kessell S R (1979) Gradient modeling: resource and fire management. Springer-Verlag, New York

Khotinskiy N A (1984) Holocene vegetation history. In: Velichko A A, Wright H E Jr., Barnosky C W (eds) Late Quaternary environments of the Soviet Union. University of Minnesota Press, Minneapolis, pp 179–200

King F B (1985) Early cultivated Cucurbits in eastern North America. In: Ford R I (ed) Prehistoric food production in North America. Museum of Anthropology, University of Michigan, Anthropological Papers, No. 75, pp 73–97

King J E (1973) Late Pleistocene palynology and biogeography of the western Missouri Ozarks. Ecol Monogr 43:539–565

King J E (1981) Late Quaternary vegetational history of Illinois. Ecol Monogr 51:43–62

King J E, Allen W H Jr (1977) A Holocene vegetation record from the Mississippi River Valley, southeastern Missouri. Quat Res 8:307–323

Kochel R C, Johnson R A (1984) Geomorphology and sedimentology of humid temperate alluvial fans in central Virginia, U.S.A. In: Koster E, Steel R (eds) Sedimentology of gravels and conglomerates. Can Soc Petrol Geol Mem 10:109–122

Kolb C R, Fredlund G G (1981) Palynological studies, Vacherie and Rayburn's Domes, north Louisiana salt dome basin. Inst Env Studies Top Rept E530-02200-T-2, Louisiana State University, Baton Rouge, pp 1–50

Kurten B, Anderson E (1980) Pleistocene mammals of North America. Columbia University Press

Kutzbach J E (1976) The nature of climate and climatic variations. Quat Res 6:471–480

Kutzbach J E, Wright H E Jr (1985) Simulation of the climate of 18,000 years BP: results for the North American/North Atlantic/European sector and comparison with the geologic record of North America. Quat Sci Rev 4:147–187

Labelle C, Richard P J H (1981) Vegetation tardiglaciaire et postglaciaire au sud-est du Parc des Laurentides, Quebec. Geogr Phys Quat 35:345–359

Lamb H F (1980) Late Quaternary vegetation and glacial history of southeastern Labrador. Arctic Alpine Res 12:117–135

Lamb H F (1984) Modern pollen spectra from Labrador and their use in reconstructing Holocene vegetational history. J Ecol 72:37–59

Lang G, Tobolski K (1985) Hobschensee—late-glacial and Holocene environment of a lake near the timberline. J Cramer Diss Bot 87:209–228

Larabee P A (1986) Late-Quaternary vegetational and geomorphic history of the Allegheny Plateau at Big Run Bog, Tucker County, West Virginia. MS Thesis, University of Tennessee, Knoxville

Larson D A, Bryant V M Jr, Patty T S (1972) Pollen analysis of a central Texas bog. Am Midl Nat 88:358–367

Lawrenz R W (1975) The developmental paleoecology of Green Lake, Antrim County, Michigan. MS Thesis, Central Michigan University, Mount Pleasant, Michigan

Lehman J T (1975) Reconstructing the rate of accumulation of lake sediment: the effect of sediment focusing. Quat Res 5:541–550

Leith H, Whittaker R H (eds) (1975) Primary productivity of the biosphere, Ecological Studies 14. Springer-Verlag, New York

Leventer A, Williams D F, Kennett J P (1982) Dynamics of the Laurentide ice sheet during the last deglaciation: evidence from the Gulf of Mexico. Earth Planet Sci Let 59:11–17

Lieux M H (1980) An atlas of pollen of trees, shrubs, and woody vines of Louisiana and other southeastern states, Part II. Platanaceae to Betulaceae. Pollen et Spores 22:191–243

Likens G E, Davis M B (1975) Post-glacial history of Mirror Lake and its watershed in New Hampshire, U.S.A.: an initial report. Verh Internat Verein Limnol 19:982–993

Little E L Jr (1971) Atlas of United States trees, Volume 1, Conifers and important hardwoods. US Dept Agric For Serv Misc Publ 1146 (8 p and 200 maps)

Little E L Jr (1977) Atlas of United States trees, Volume 4, Minor eastern hardwoods. US Dept Agric For Serv Misc Publ 1342 (17 p and 230 maps)

Liu K B (1982) Postglacial vegetational history of northern Ontario: a palynological study. PhD Dissertation, University of Toronto, Toronto

Liu K B, Lam N S N (1985) Paleovegetational reconstruction based on modern and fossil pollen data: an application of discriminant analysis. Ann Assoc Am Geogr 75:115–130

Livingstone D A (1968) Some interstadial and postglacial pollen diagrams from eastern Canada. Ecol Monogr 38:87–125

Livingstone D A (1969) Communities of the past. In: Greenidge K N H (ed) Essays in plant geography and ecology. Nova Scotia Museum, Halifax, pp 83–104

Lorimer C G (1980) Age structure and disturbance history of a southern Appalachian virgin forest. Ecology 61:1169–1184

Loucks O L (1970) Evolution of diversity, efficiency and community stability. Am Zool 10:17–25

Lowe J J, Gray J M (1980) The stratigraphic subdivision of the lateglacial of NW Europe: a discussion. In: Lowe J J, Gray J M, Robinson J E (eds) Studies in the lateglacial of north-west Europe. Pergamon Press, Oxford, pp 157–175

Lund S P (1981) Late-Quaternary secular variation of the Earth's magnetic field as recorded in the wet sediments of three North American lakes. PhD Dissertation, University of Minnesota, Minneapolis

MacArthur R H, Wilson E O (1967) The theory of island biogeography. Princeton University Press, Princeton

McAndrews J H (1966) Postglacial history of prairie, savanna, and forest in northwestern Minnesota. Mem Torrey Bot Club 22:1–72

McAndrews J H (1970) Fossil pollen and our changing landscape and climate. Rotunda, Royal Ontario Mus Bull 3:30–37

McAndrews J H (1976) Fossil history of man's impact on the Canadian flora: an example from southern Ontario. Can Bot Assoc Bull 9:1–6

McAndrews J H (1981) Late Quaternary climate of Ontario: temperature trends from the fossil pollen record. In: Mahaney W C (ed) Quaternary paleoclimate. Geo Abstracts Ltd, Norwich, England, pp 319–333

McAndrews J H (1982) Holocene environment of a fossil bison from Kenora, Ontario. Ontario Archaeol 37:41–51

McAndrews J H, Power D M (1973) Palynology of the Great Lakes: the surface sediments of Lake Ontario. Can J Earth Sci 10:777–792

McAndrews J H, Riley J L, Davis A M (1982) Vegetation history of the Hudson Bay lowland: a postglacial pollen diagram from the Sutton Ridge. Nat Can 109:597–608

McDonald J E (1962) Collection and washout of airborne pollens and spores by raindrops. Science 135:435–437

McDowell L L, Dole R M Jr, Howard M Jr, Farrington R A (1971) Palynology and radiocarbon chronology of Bugbee wildflower sanctuary and natural area, Caledonia County, Vermont. Pollen et Spores 13:73–92

McIntosh R P (1985) The background of ecology, concept and theory. Cambridge University Press, Cambridge

McIntyre A, Kipp N, Be A W H, Crowley T, Gardner J V, Prell W L, Ruddiman W F (1976) Glacial North Atlantic 18,000 years ago: a CLIMAP reconstruction. In: Kline R M, Hays J D (eds) Investigations of late Quaternary paleoceanography and paleoclimatology. Geol Soc Am Mem 145:43–76

McPherson J B (1982) Postglacial vegetational history of the eastern Avalon Peninsula, Newfoundland, and Holocene climatic change along the eastern Canadian seaboard. Geogr Phys Quat 36:175–196

Manny R A, Wetzel R G, Bailey R E (1978) Paleolimnological sedimentation of organic carbon, nitrogen, phosphorus, fossil pigments, pollen, and diatoms in a hyper-eutrophic, hardwater lake: a case history of eutrophication. Pol Arch Hydrobiol 25:243–267

Markewich H W, Christopher R A (1982) Pleistocene (?) and Holocene fluvial history of Uphapee Creek, Macon County, Alabama. US Geol Surv Bull 1522:1–16

Marks P L (1974) The role of pin cherry (*Prunus pensylvanica* L.) in the maintenance of stability in northern hardwood ecosystems. Ecol Monogr 44:73–88

Marschner F J (1959) Land use and its patterns in the United States. US Dept Agric, Agric Handbk 153

Marshall L G, Webb S D, Sepkoski J J Jr., Raup D M (1982) Mammalian evolution and the great American interchange. Science 215:1351–1357

Martin P S (1958) Pleistocene ecology and biogeography of North America. In: Hubbs C S (ed) Zoogeography. Am Assoc Adv Sci Publ 51, pp 375–420

Martin P S (1973) The discovery of America. Science 179:969–974

Martin P S, Klein R G (1984) Quaternary extinctions, a prehistoric revolution. University of Arizona Press, Tucson

Martin P S, Mehringer P J Jr. (1965) Pleistocene pollen analysis and biogeography of the Southwest. In: Wright H E Jr., Frey D G (eds) The Quaternary of the United States. Princeton University Press, Princeton, New Jersey, pp 433–451

Martin P S, Wright H E Jr (1967) Pleistocene extinctions: the search for a cause. Yale University Press, New Haven

Martin W H (1978) White oak communities in the Great Valley of East Tennessee—a vegetation complex. Central Hardwood For Conf 11:39–61

Maxwell J A, Davis M B (1972) Pollen evidence of Pleistocene and Holocene vegetation on the Allegheny Plateau, Maryland. Quat Res 2:506–530

Maycock P F, Curtis J T (1960) The phytosociology of boreal conifer–hardwood forests of the Great Lakes region. Ecol Monogr 30:1–35

Menard H W (1971) The Late Cenozoic history of the Pacific and Indian Ocean basins. In: Turekian K K (ed) The Late Cenozoic glacial ages. Yale University Press, New Haven, pp 1–14

Michalek D D (1968) Fanlike features and related periglacial phenomena of the southern Blue Ridge. PhD Dissertation, University of North Carolina, Chapel Hill

Mickelson D M, Clayton L, Fullerton D S, Borns H W Jr (1983) The Late Wisconsin glacial record of the Laurentide Ice Sheet in the United States. In: Porter S C (ed) Late-Quaternary environments of the United States, Volume 1, The Late Pleistocene. University of Minnesota Press, Minneapolis, pp 3–37

Milankovitch M M (1941) Canon of insolation and the Ice-Age problem. Royal Serb Acad Spec Publ 133

Miller N G (1973a) Lateglacial plants and plant communities in northwestern New York State. J Arnold Arbor 54:123–159

Miller N G (1973b) Late-glacial and postglacial vegetation change in southwestern New York State. N Y State Mus Sci Serv Bull 420:1–102

Miller N G (1980) Mosses as paleoecological indicators of lateglacial terrestrial environments: some North American studies. Bull Torrey Bot Club 107:373–391

Mills H H, Delcourt P A (1987) Appalachian Highlands and Interior Low Plateaus. In: Morrison R B (ed) Quaternary non-glacial geology of the conterminous United States, Volume K2, Decade of North American Geology. Geological Society of America, Boulder Colorado (in press)

Moe D (1970) The post-glacial immigration of *Picea abies* into Fennoscandia. Botaniska Notiser 123:61–66

Morrison A (1970) Pollen diagrams from interior Labrador. Can J Bot 48:1957–1975

Morse D F, Morse P A (1983) Archaeology of the Central Mississippi Valley. Academic Press, New York

Mott R J (1975a) Postglacial history and environments in southwestern New Brunswick. Proc NS Inst Sci 27:67–82

Mott R J (1975b) Palynological studies of lake sediment profiles from southwestern New Brunswick. Can J Earth Sci 12:273–288

Mott R J (1976) A Holocene pollen profile from the Sept-Iles area, Quebec. Nat Can 103:457–467

Mott R J (1977) Late-Pleistocene and Holocene palynology in southeastern Quebec. Geogr Phys Quat 31:139–149

Mott R J, Farley-Gill L D (1978) A late-Quaternary pollen profile from Woodstock, Ontario. Can J Earth Sci 15:1101–1111

Mott R J, Farley-Gill L D (1981) Two late Quaternary pollen profiles from Gatineau Park, Quebec. Geol Surv Can Pap 80-31:1–9

Muller J (1981) Fossil pollen records of extant angiosperms. Bot Rev 47:1–142

Naveh Z, Lieberman A S (1984) Landscape ecology, theory and application. Springer-Verlag, New York

Nicholas J (1968) Late Pleistocene palynology of southeastern New York and northern New Jersey. PhD Dissertation, New York University

Niklas K J, Tiffney B H, Knoll A H (1985) Patterns in vascular land plant diversification: an analysis at the species level. In: Valentine J W (ed) Phanerozoic diversity patterns: profiles in macroevolution. Princeton University Press, Princeton, pp 97–128

Nilsson T (1983) The Pleistocene, geology and life in the Quaternary Ice Age. D Reidel, Dordrecht

O'Rourke M K (1976) An absolute pollen chronology of Seneca Lake, New York. MS Thesis, University of Arizona, Tucson

O'Sullivan P E (1983) Annually-laminated lake sediments and the study of Quaternary environmental changes—a review. Quat Sci Rev 1:245–313

Oechel W C, Lawrence W T (1985) Taiga. In: Chabot B F, Mooney H A (eds) Physiological ecology of North American plant communities. Chapman and Hall, New York, pp 66–94

Ogden E C, Lewis D M (1960) Airborne pollen and fungus spores of New York State. NY State Mus Sci Serv Bull 378:1–104

Ogden E C, Raynor G S, Hayes J V, Lewis D M, Haines J H (1974) Manual for sampling airborne pollen. Hafner Press, New York

Ogden J G III (1963) The Squibnocket Cliff peat: radiocarbon dates and pollen stratigraphy. Am J Sci 261:344–353

Ogden J G III (1966) Forest history of Ohio, I, radiocarbon dates and pollen stratigraphy of Silver Lake, Logan County, Ohio. Ohio J Sci 66:387–400

Ogden J G III (1969) Correlation of contemporary and Late Pleistocene pollen records in the reconstruction of postglacial environments in northeastern North America. Mitt Int Verein Limnol 17:64–77

Oliver C D (1981) Forest development in North America following major disturbances. For Ecol Manage 3:153–168

Overpeck J T, Webb T III, Prentice I C (1985) Quantitative interpretation of fossil pollen spectra: dissimilarity coefficients and the method of modern analogs. Quat Res 23:87–108

Parsons R W, Prentice I C (1981) Statistical approaches to R-values and the pollen-vegetation relationship. Rev Palaeobot Palynol 32:127–152

Parsons R W, Prentice I C, Saarnisto M (1980) Statistical studies on pollen representation in Finnish lake sediments in relation to forest inventory data. Ann Bot Fennici 17:379–393

Pearsall W H (1959) The ecology of invasion: ecological stability and instability. New Biol 29:95–101

Peck R M (1973) Pollen budget studies in a small Yorkshire catchment. In: Birks H J B, West R G (eds) Quaternary plant ecology. Blackwell, Oxford, pp 43–60

Pewe T L (1983) The periglacial environment in North America during Wisconsin time. In: Porter S C (ed) Late-Quaternary environments of the United States, Volume 1, The Late Pleistocene. University of Minnesota Press, Minneapolis, pp 157–189

Pickett S T A (1976) Succession: an evolutionary interpretation. Am Nat 110:107–119

Pickett S T A, White P S (1985) The ecology of natural disturbance and patch dynamics. Academic Press, New York

Pielou E C (1979) Biogeography. John Wiley & Sons, New York

Pimm S (1984) Food webs. Chapman and Hall, London

Pisias N G, Moore T C Jr. (1981) The evolution of Pleistocene climate: a time series approach. Earth Planet Sci Let 52:450–458

Pohl F (1937) Die Pollenerzeugung der Windbluter. Bot Centralblatt 56A:365–470

Prentice I C (1978) Modern pollen spectra from lake sediments in Finland and Finnmark, north Norway. Boreas 7:131–153

Prentice I C (1980) Multidimensional scaling as a research tool in Quaternary palynology: a review of theory and methods. Rev Palaeobot Palynol 31:71–104

Prentice I C (1982) Calibration of pollen spectra in terms of species abundance. In: Berglund B E (ed) Palaeohydrological changes in the Temperate Zone in the last

15,000 years, Volume III: specific methods. Lund University, Lund, Sweden, pp 25–51

Prentice I C (1983) Pollen mapping of regional vegetation patterns in south and central Sweden. J Biogeogr 10:441–454

Prentice I C (1985) Pollen representation, source area, and basin size: toward a unified theory of pollen analysis. Quat Res 23:76–86

Prentice I C (1986) Forest-composition calibration of pollen data. In: Berglund B E (ed) Handbook of Holocene palaeoecology and palaeohydrology. John Wiley & Sons, New York, pp 799–816

Prentice I C, Parsons R W (1983) Maximum likelihood linear calibration of pollen spectra in terms of forest composition. Biometrics 39:1051–1057

Prentice I C, Webb T III (1986) Pollen percentages, tree abundances and the Fagerlind effect. J Quat Sci 1:35–43

Prest V K (1969) Retreat of Wisconsin and Recent ice in North America (map scale 1:5,000,000). Geol Surv Canada Map 1257A

Prest V K (1970) Chapter 12, Quaternary geology. In: Douglas, R J W (ed) Geology and economic minerals of Canada. Geol Surv Canada Econ Geol Rept No. 1:675–765

Quarterman E, Keever C (1962) Southern Mixed Hardwood Forest: climax in the south-eastern Coastal Plain, U.S.A. Ecol Monogr 32:167–185

Ralska-Jasiewiczowa M (1983) Isopollen maps for Poland: 0–11,000 years B.P. New Phytol 94:133–175

Rapoport E H (1982) Areography: geographical strategies of species. Pergamon Press, Oxford

Raynor G S, Hayes J V, Ogden E C (1974a) Particulate dispersion into and within a forest. Boundary-Layer Meteorol 7:429–456

Raynor G S, Hayes J V, Ogden E C (1974b) Mesoscale transport and dispersion of airborne pollens. J Appl Meteorol 13:87–95

Regal P J (1982) Pollination by wind and animals: ecology of geographic patterns. Ann Rev Ecol Sys 13:497–524

Richard P J H (1973a) Histoire postglaciaire comparee de la vegetation dans deux localites au sud de la ville de Quebec. Nat Can 100:591–603

Richard P J H (1973b) Histoire postglaciaire comparee de la vegetation dans deux localites au nord du Parc des Laurentides, Quebec. Nat Can 100:577–590

Richard P J H (1975) Contribution a l'histoire postglaciaire de la vegetation dans les Canton-de-l'est: etude des sites de Weedon et Albion. Cahiers de Geogr de Quebec 19:267–284

Richard P J H (1977) Histoire post-wisconsinienne de la vegetation du Quebec meridional par l'analyse pollinique. Serv Recherche, Dir Gen Forets, Min Terres Forets Quebec, Tome 1, 2

Richard P J H (1978) Histoire tardiglaciaire et postglaciaire de la vegetation au Mont Shefford, Quebec. Geogr Phys Quat 32:81–93

Richard P J H (1979) Contribution a l'histoire postglaciaire de la vegetation au nord-est de la Jamesie, nouveau-Quebec. Geogr Phys Quat 33:93–112

Richard P J H (1981) Paleophytogeographie postglaciaire en Ungava par l'analyse pollinique, Collection Paleo-Quebec. University du Quebec a Montreal.

Richard P J H (1985) Couvert vegetal et paleoenvironnements du Quebec entre 12000 et 8000 ans BP. Res Amerindiennes Quebec 15:39–56

Richard P J H, Larouche A, Bouchard M A (1982) Age de la deglaciation finale et histoire postglaciaire de la vegetation dans la partie centrale du nouveau-Quebec. Geogr Phys Quat 36:63–90

Richard P J H, Poulin P (1976) La diagramme pollinique au Mont des Eboulements, region de Charlevoix, Quebec. Can J Earth Sci 13:145–156

Riegel W L (1965) Palynology of environments of peat formation in southwestern Florida. PhD Dissertation, Pennsylvania State University

Ritchie J C (1964) Contributions to the Holocene paleoecology of West-central Canada. I. The Riding Mountain area. Can J Bot 42:181–197

Ritchie J C (1969) Absolute pollen frequencies and Carbon-14 age of a section of Holocene lake sediment from the Riding Mountain area of Manitoba. Can J Bot 47:1345–1349

Ritchie J C (1976) The late-Quaternary vegetational history of the western interior of Canada. Can J Bot 54:1793–1818

Ritchie J C (1984) Past and present vegetation of the far northwest of Canada. University of Toronto Press, Toronto

Ritchie J C (1985) Late-Quaternary climatic and vegetational change in the lower Mackenzie Basin, northwest Canada. Ecology 66:612–621

Ritchie J C (1986) Climate change and vegetation response. Vegetatio 67:65–74

Ritchie J C, Hadden K A (1975) Pollen stratigraphy of Holocene sediments from the Grand Rapids area, Manitoba, Canada. Rev Palaeobot Palynol 19:193–202

Ritchie J C, Lichti-Federovich S (1967) Pollen dispersal phenomena in arctic–subarctic Canada. Rev Paleobot Palynol 3:255–266

Ritchie J C, Lichti-Federovich S (1968) Holocene pollen assemblages from the Tiger Hills, Manitoba. Can J Earth Sci 5: 873–880

Ritchie J C, Yarranton G A (1978) The late-Quaternary history of the boreal forest of central Canada, based on standard pollen stratigraphy and principal components analysis. J Ecol 66:199–212

Robinson A H, Sale R D (1969) Elements of cartography. John Wiley & Sons, New York

Romme W H (1982) Fire and landscape diversity in subalpine forests of Yellowstone National Park. Ecol Monogr 52:199–221

Romme W H, Knight D H (1982) Landscape diversity: the concept applied to Yellowstone Park. BioScience 32:644–670

Rosenzweig M L (1979a) Three probable evolutionary causes for habitat selection. In: Patil G P, Rosenzweig M L (eds) Quantitative ecology and related econometrics. International Co-operative Publishing House, Fairland, Maryland, pp 49–60

Rosenzweig M L (1979b) Optimal habitat selection in two-species competitive systems. Fortschr Zool 25:283–293

Rosenzweig M L (1981) A theory of habitat selection. Ecology 62:327–335

Roughgarden J, Diamond J (1986) Overview: the role of species interactions in community ecology. In: Diamond J, Case T J (eds) Community ecology. Harper & Row, New York, pp 333–343

Rowe J S (1972) Forest regions of Canada. Canadian Forestry Service Publication No. 1300

Ruddiman W F, McIntyre A (1981a) The mode and mechanism of the last deglaciation: oceanic evidence. Quat Res 16:125–134

Ruddiman W F, McIntyre A (1981b) Oceanic mechanisms for amplification of the 23,000-year ice-volume cycle. Science 212:617–627

Runkle J R (1985) Disturbance regimes in temperate forests. In: Pickett S T A, White P S (eds) The ecology of natural disturbance and patch dynamics. Academic Press, New York, pp 17–33

Saarnisto M (1974) The deglaciation history of the Lake Superior region and its climatic implications. Quat Res 4:316–339

Saarnisto M (1975) Stratigraphical studies on the shoreline displacement of Lake Superior. Can J Earth Sci 12:300–319

Saarnisto M (1979) Studies of annually laminated lake sediments. In: Berglund B E (ed) Volume II, Project Guide, Palaeohydrological changes in the Temperate Zone in the last 15000 years, Subproject B: Lake and mire environments. Lund University, Lund, Sweden, pp 61–80

Saucier R T (1974) Quaternary geology of the Lower Mississippi Valley. Arkansas Archeol Surv, Publ Archeol, Res Ser No. 6:1–26

Saucier R T (1978) Sand dunes and related eolian features of the Lower Mississippi River Alluvial Valley. Geosci Man 19:23–40

Savoie L, Richard P J H (1979) Paleophytogeographie de l'episode de Saint-Narcisse dans la region de Sainte-Agathe, Quebec. Geogr Phys Quat 33:175–188

Shackleton N J, Kennett J P (1975) Paleotemperature history of the Cenozoic and the initiation of Antarctic glaciation: oxygen and carbon isotope analyses in DSDP sites 277, 279, and 281. Init Rep Deep Sea Drill Proj 29:743–755

Shafer D S (1984) Late-Quaternary paleoecologic, geomorphic, and paleoclimatic history of Flat Laurel Gap, Blue Ridge Mountains, North Carolina. MS Thesis, University of Tennessee, Knoxville

Shafer D S (1985) Flat Laurel Gap Bog, Pisgah Ridge, North Carolina: late-Holocene development of a high-elevation heath bald. Castanea 51:1–10

Shane L C K (1975) Palynology and radiocarbon chronology of Battaglia Bog, Portage County, Ohio. Ohio J Sci 75:96–102

Shane L C K (1976) Late-glacial and postglacial palynology and chronology of Darke County, west-central Ohio. PhD Dissertation, Kent State University

Sheldon J M, Hewson E W (1960) Atmospheric pollution by aeroallergens. University of Michigan Res Inst Prog Rept 4:1–191

Short S K (1978) Palynology: a Holocene environmental perspective for archaeology in Labrador–Ungava. Arctic Anthropol 15-2:9–35

Short S K, Nichols H (1977) Holocene pollen diagrams from subarctic Labrador–Ungava: vegetational history and climatic change. Arctic Alpine Res 9:265–290

Shugart H H Jr (1984) A theory of forest dynamics: the ecological implications of forest succession models. Springer-Verlag Press, New York

Shugart H H Jr, West D C (1977) Development of an Appalachian deciduous forest succession model and its application to assessment of the impact of the chestnut blight. J Env Manage 5:161–179

Shugart H H Jr, West D C (1979) Size and pattern of simulated forest stands. For Sci 25:120–122

Shugart H H Jr, West D C, Emanuel W R (1981) Patterns and dynamics of forests: an application of simulation models. In: West D C, Shugart H H Jr, Botkin D B (eds) Forest succession: concepts and application. Springer-Verlag, New York, pp 74–94

Shugart H H Jr, Emanuel W R, West D C, DeAngelis D L (1980) Environmental gradients in a simulation model of a beech–yellow-poplar stand. Math Biosci 50:163–170

Shumard C B (1974) Palynology of a lacustrine sinkhole facies, and the geologic history of a (late Pleistocene?) basin in Clark County, southwestern Kansas. MS Thesis, Wichita State University

Simberloff D (1981) Community effects of introduced species. In: Nitecki M H (ed) Biotic crises in ecological and evolutionary time. Academic Press, New York, pp 53–83

Simberloff D (1984) Properties of coexisting bird species in two archipelagoes. In: Strong D R Jr, Simberloff D, Abele L G, Thistle A B (eds) Ecological communities, conceptual issues and the evidence. Princeton University Press, Princeton, New Jersey, pp 234–253

Simola H, Tolonen K (1981) Diurnal laminations in the varved sediment of Lake Lovojarvi, South Finland. Boreas 10:19–26

Sims R E (1973) The anthropogenic factor in East Anglian vegetational history: an approach using A.P.F. techniques. In: Birks H J B, West R G (eds) Quaternary plant ecology. Blackwell Scientific, Oxford, pp 223–236

Sirkin L A, Owens J P, Minard J P, Rubin M (1970) Palynology of some upper Quaternary peat samples from the New Jersey coastal plain. US Geol Surv Prof Pap 700-D:D77–D87

Sirkin L A, Denny C S, Rubin M (1977) Late Pleistocene environment of the central Delmarva Peninsula. Geol Soc Am Bull 88:139–142

Smith A G (1965) Problems of inertia and threshold related to post-Glacial habitat changes. Proc Roy Soc London 161:331–342

Smith A G (1970) The influence of Mesolithic and Neolithic man on British vegetation: a discussion. In: Walker D, West R G (eds) Studies in the vegetational history of the British Isles. Cambridge University Press, London, pp 81–96

Smith B D (1978) Mississippian settlement patterns. Academic Press, New York

Smith B D (1986) The archaeology of the southeastern United States: from Dalton to de Soto, 10,500–500 B.P. Adv World Archaeol 5:1–92

Smith E N Jr (1984) Late-Quaternary vegetational history at Cupola Pond, Ozark National Scenic Riverways, southeastern Missouri. MS Thesis, University of Tennessee, Knoxville

Solomon A M (1979) Pollen. In: Edmonds R L (ed) Aerobiology: the ecological systems approach. Dowden, Hutchinson & Ross, Stroudsburg, Pennsylvania, pp 41–84

Solomon A M (1983a) Pollen morphology and plant taxonomy of white oaks in eastern North America. Am J Bot 70:481–494

Solomon A M (1983b) Pollen morphology and plant taxonomy of red oaks in eastern North America. Am J Bot 70:495–507

Solomon A M, Delcourt H R, West D C, Blasing T J (1980) Testing a simulation model for reconstruction of prehistoric forest-stand dynamics. Quat Res 14:275–293

Solomon A M, Webb T III (1985) Computer-aided reconstruction of late-Quaternary landscape dynamics. Ann Rev Ecol Syst 16:63–84

Solomon A M, West D C, Solomon J A (1981) Simulating the role of climate change and species immigration in forest succession. In: West D C, Shugart H H, Botkin D B (eds) Forest succession: concepts and application. Springer-Verlag, New York, pp 154–177

Spackman W, Dolsen C P, Riegel W (1966) Phytogenic organic sediments and sedimentary environments in the Everglades-mangrove complex. Part I: evidence of a transgressing sea and its effects on environments of the Shark River area of southwestern Florida. Sonderdruck aus Palaeontogr Beitr zur Natur der Vorzeit 117(B):135–152

Spear R W, Miller N G (1976) A radiocarbon dated pollen diagram from the Allegheny Plateau of New York State. J Arnold Arbor 57:369–403

Sprugel D G, Bormann F H (1981) Natural disturbance and the steady state in high-altitude balsam fir forests. Science 211:390–393

Starkel L (1982) Evolution of the Vistula River Valley during the last 15,000 years. Polish Acad Sci, Inst Geogr Spat Organ, Geogr Stud Spec Issue No. 1:1–169

Statistical Analysis System (1981) Tech Rept P-115, SAS Institute Inc, Raleigh, NC

Statistical Graphics Corporation (1985) STATGRAPHICS statistical graphics system. STSC, Rockville, Maryland

Steyermark J A (1963) Flora of Missouri. Iowa State University Press, Ames

Stockmarr J (1975) Retrogressive forest development, as reflected in a mor pollen diagram from Mantingerbos, Brenthe, The Netherlands. Palaeohistoria 17:38–51

Strong D R Jr (1984) Exorcising the ghost of competition past: phytophagous insects. In: Strong D R Jr, Simberloff D, Abele L G, Thistle A B (eds) Ecological communities, conceptual issues and the evidence. Princeton University Press, Princeton, New Jersey, pp 28–41

Stuiver M (1971) Evidence for the variation of atmospheric C^{14} content in the late Quaternary. In: Turekian K K (ed) The Late Cenozoic glacial ages. Yale University Press, New Haven, pp 57–70

Stuiver M, Deevey E S Jr, Rouse I (1963) Yale natural radiocarbon measurements VIII. Radiocarbon 5:312–341

Swanson F J, Lienkaemper G W (1982) Interactions among fluvial processes, forest vegetation, and aquatic ecosystems, South Fork Hoh River, Olympic National Park. In: Franklin J F, Starkey E E, Matthews J W (eds) Ecological Research in National Parks of the Pacific Northwest. Oregon State University, Forest Research Laboratory, Corvallis, pp 30–34

Tauber H (1965) Differential pollen dispersion and the interpretation of pollen diagrams. Geol Surv Denmark II Ser No 89:1–69

Tauber H (1967a) Investigations of the mode of pollen transfer in forested areas. Rev Palaeobot Palynol 3:277–286

Tauber H (1967b) Differential pollen dispersion and filtration. In: Cushing E J, Wright H E Jr (eds) Quaternary paleoecology. Yale University Press, New Haven, pp 131–141

Taylor T N (1981) Paleobotany, an introduction to fossil plant biology. McGraw-Hill, New York

Teller J T (1985) Glacial Lake Agassiz and its influence on the Great Lakes. In: Karrow P F, Calkin P E (eds) Quaternary evolution of the Great Lakes. Geol Assoc Canada Spec Paper 30, pp 1–16

Teller J T, Clayton (eds) (1983) Glacial Lake Agassiz. Geol Assoc Canada Spec Paper 26

Teller J T, Thorleifson L H, Dredge L A, Hobbs, H C, Schreiner B T (1983) Maximum extent and major features of Lake Agassiz. In: Teller J T, Clayton L (eds) Glacial Lake Agassiz. Geol Ass Can Spec Pap 26:43–45

Terasmae J (1967) Postglacial chronology and forest history in the northern Lake Huron and Lake Superior regions. In: Cushing E J, Wright H E Jr (eds) Quaternary paleoecology. Yale University Press, New Haven, pp 45–58

Terasmae J (1969) A discussion of deglaciation and the boreal forest history in the northern Great Lakes region. Proc Entomol Soc Ontario 99:31–43

Terasmae J (1973) Notes on Late Wisconsin and early Holocene history of vegetation in Canada. Arctic Alpine Res 5:201–222

Terasmae J (1976) In search of a palynological tundra. Geosci Man 15:77–82

Terasmae J, Anderson T W (1970) Hypsithermal range extension of white pine (*Pinus strobus* L.) in Quebec, Canada. Can J Earth Sci 7:406–413

Thom B G (1967) Humate and coastal geomorphology. Louisiana State University Coastal Studies Bull 1:15–17

Tiffney B H (1985) Perspectives on the origin of the floristic similarity between eastern Asia and eastern North America. J Arnold Arbor 66:73–94

Tilman D (1982) Resource competition and community structure. Monogr Population Biol 17, Princeton University Press, Princeton, New Jersey

Tjallingii S P, de Veer A A (eds) (1982) Perspectives in landscape ecology: contributions to research, planning and management of our environment. Proc Internat Congr Neth Soc Landscape Ecol, Veldhoven 1981

Trautman M A, Walton A (1962) Isotopes, Inc. radiocarbon measurements II. Radiocarbon 4:35–42

Traverse A (1982) Response of world vegetation to Neogene tectonic and climatic events. Alcheringa 6:197–209

Troels-Smith J (1960) Ivy, mistletoe, and elm: climatic indicator-fodder plants. Danm Geol Unders Ser IV 4(4):1–32

Tsukada M (1982a) Late-Quaternary development of the *Fagus* forest in the Japanese archipelago. Jap J Ecol 32:113–118

Tsukada M (1982b) *Cryptomeria japonica*: glacial refugia and late-glacial and postglacial migration. Ecology 63:1091–1105

Tsukada M (1985) Map of vegetation during the last glacial maximum in Japan. Quat Res 23:369–381

Ugolini F C (1968) Soil development and alder invasion in a recently deglaciated area of Glacier Bay, Alaska. In: Trappe G M *et al.* (eds) Biology of alder. US Forest Service Pacific NW Forest and Range Experiment Station, Portland, Oregon, pp 115–140

US Forest Service (1967) Forest survey handbook. US Dept Agric For Serv, Washington, DC

Van Campo M (1984) Relations entre la vegetation de l'Europe et les temperatures de surface oceaniques apres le dernier maximum glaciaire. Pollen et Spores 26:497–518

Van der Hammen T, Wijmstra T A, Zagwijn W H (1971) The floral record of the Late Cenozoic of Europe. In: Turekian K K (ed) The Late Cenozoic glacial ages. Yale University Press, New Haven, pp 391–424

Van der Pijl L (1969) Principles of dispersal in higher plants. Springer-Verlag, Berlin

van Donk J (1976) O¹⁸ record of the Atlantic Ocean for the entire Pleistocene Epoch. In: Cline R M, Hays J D (eds) CLIMAP: investigation of late Quaternary paleoceanography and paleoclimatology. Geol Soc Am Mem 145:147–163

Van Zant K L (1979) Late glacial and postglacial pollen and plant macrofossils from Lake West Okoboji, northwestern Iowa. Quat Res 12:358–380

Vincent J S (1973) A palynological study for the Little Clay Belt, northwestern Quebec. Nat Can 100:59–70

von Post L (1916, translated by Davis M B, Faegri K, 1967) Forest tree pollen in south Swedish peat bog deposits. Pollen et Spores 9:378–401

Waddington J C B (1969) A stratigraphic record of the pollen influx to a lake in the Big Woods of Minnesota. Geol Soc Am Spec Pap 123:263–282

Walker D (1982) Vegetation's fourth dimension. New Phytol 90:419–429

Walker D, Chen Y (1987) Palynological light on rainforest dynamics. Quat Sci Rev 6:(in press)

Walker P H, Brush G S (1964) Observations on bog and pollen stratigraphy of the Des Moines glacial lobe, Iowa. Iowa Acad Sci Proc 70:253–260

Walker P H, Hartman R T (1960) The forest sequence of the Hartstown Bog area in western Pennsylvania. Ecology 41:461–474

Walter H (1979) Vegetation of the Earth and ecological systems of the geo-biosphere. Springer-Verlag, New York

Warner B G, Hebda R J, Hann R J (1984) Postglacial paleoecological history of a cedar swamp, Manitoulin Island, Ontario, Canada. Palaeogeogr Palaeoclimatol Palaeoecol 45:301–345

Watt A S (1947) Pattern and process in the plant community. J Ecol 35:1–22

Watts W A (1961) Post-Atlantic forests in Ireland. Proc Linn Soc Lond 172:33–38

Watts W A (1969) A pollen diagram from Mud Lake, Marion County, north-central Florida. Geol Soc Am Bull 80:631–642

Watts W A (1970) The full-glacial vegetation of northwestern Georgia. Ecology 51:17–33

Watts W A (1971) Postglacial and interglacial vegetation history of southern Georgia and central Florida. Ecology 52:676–690

Watts W A (1973a) Rates of change and stability in vegetation in the perspective of long periods of time. In: Birks H J B, West R G (eds) Quaternary plant ecology. Blackwell Scientific Publications, Oxford, England, pp 195–206

Watts W A (1973b) The vegetation record of a Mid-Wisconsin Interstadial in northwest Georgia. Quat Res 3:257–268

Watts W A (1975) A late Quaternary record of vegetation from Lake Annie, south-central Florida. Geology 3:344–346

Watts W A (1979) Late Quaternary vegetation of central Appalachia and the New Jersey coastal plain. Ecol Monogr 49:427–469

Watts W A (1980a) The late Quaternary vegetation history of southeastern United States. Ann Rev Ecol Syst 11:387–409

Watts W A (1980b) Late-Quaternary vegetation history at White Pond on the Inner Coastal Plain of South Carolina. Quat Res 13:187–199

Watts W A (1980c) Regional variation in the response of vegetation to lateglacial climatic events in Europe. In: Lowe J J, Gray J M, Robinson J E (eds) Studies in the lateglacial of north-west Europe. Pergamon Press, Oxford, pp 1–21

Watts W A (1983) Vegetational history of the eastern United States 25,000 to 10,000 years ago. In: Porter S C (ed) Late-Quaternary environments of the United States, Volume I, The Late Pleistocene. University of Minnesota Press, Minneapolis, pp 294–310

Watts W A, Bright R C (1968) Pollen, seed, and mollusk analysis of a sediment core from Pickerel Lake, northeastern South Dakota. Geol Soc Am Bull 79:855–876

Watts W A, Stuiver M (1980) Late Wisconsin climate of northern Florida and the origin of species-rich deciduous forest. Science 210:325–327

Watts W A, Winter T C (1966) Plant macrofossils from Kirchner Marsh, Minnesota— a paleoecological study. Geol Soc Am Bull 77:1339–1360

Webb T III (1974a) Corresponding patterns of pollen and vegetation in Lower Michigan: a comparison of quantitative data. Ecology 55:17–28

Webb T III (1974b) A vegetational history from northern Wisconsin: evidence from modern and fossil pollen. Am Midl Nat 92:12–34

Webb T III (1980) The reconstruction of climatic sequences from botanical data. J Interdiscipl Hist 10:749–772

Webb T III, Bryson R A (1972) Late and postglacial climatic change in the northern Midwest, U.S.A.: quantitative estimates derived from fossil pollen spectra by multivariate statistical analysis. Quat Res 2:70–115

Webb T III, McAndrews J H (1976) Corresponding patterns of contemporary pollen and vegetation in Central North America. Geol Soc Am Mem 145:267–297

Webb T III, Howe S, Bradshaw R H W, Heide K (1981) Estimating plant abundances from pollen percentages: the use of regression analysis. Rev Palaeobot Palynol 34:269–300

Webb T III, Laseski R A, Bernabo J C (1978) Sensing vegetational patterns with pollen data: choosing the data. Ecology 59:1151–1163

Webb T III, Cushing E J, Wright H E Jr (1983a) Holocene changes in the vegetation of the Midwest. In: Wright H E Jr (ed) Late-Quaternary environments of the United States, Volume 2, The Holocene. University of Minnesota Press, Minneapolis, pp 142–165

Webb T III, Richard P J H, Mott R J (1983b) A mapped history of Holocene vegetation in southern Quebec. In: Harington C R (ed) Climatic Change in Canada 3. Syllogeus 49, pp 273–336

Wendland W M (1977) Tropical storm frequencies related to sea surface temperatures. J Appl Meteorol 16:477–481

Werfft R (1951) Uber die Lebensdauer der Pollenkorner in der freien Atmosphare. Biol Zentralbl 70:354–367

West F H (1983) The antiquity of man in America. In: Porter S C (ed) Late-Quaternary environments of the United States, Volume 1, The Late Pleistocene. University of Minnesota Press, Minneapolis. pp 364–382

West R G (1961) Late- and postglacial vegetational history in Wisconsin, particularly changes associated with the Valders readvance. Am J Sci 259:766–783

West R G (1970) Pleistocene history of the British flora. In: Walker D, West R G (eds) Studies in the vegetational history of the British Isles. Cambridge University Press, Cambridge, pp 1–11

West R G (1977) Pleistocene geology and biology, with especial reference to the British Isles. Longmans, London

White J (1985) Studies in plant demography. Academic Press, New York

White P S (1979) Pattern, process, and natural disturbance in vegetation. Bot Rev 45:229–299

White P S (1983) Eastern Asian-eastern North American floristic relations: the plant community level. Ann Mo Bot Gard 70:734–747

White P S (ed) (1984) The southern Appalachian spruce–fir ecosystem: its biology and threats. US National Park Serv Res Res Manage Rept Ser-71:1–268

White P S, MacKenzie M D, Busing R T (1985) A critique on overstory/understory comparisons based on transition probability analysis of an old growth spruce–fir stand in the Appalachians. Vegetatio 64:37–45

White P S, Miller R I, Ramseur G S (1984) The species–area relationship of the southern Appalachian high peaks: vascular plant richness and rare plant distributions. Castanea 49:47–61

Whitehead D R (1964) Fossil pine pollen and full-glacial vegetation in southeastern North Carolina. Ecology 45:767–777

Whitehead D R (1967) Studies of full-glacial vegetation and climate in southeastern United States. In: Cushing E J, Wright H E Jr (eds) Quaternary paleoecology. Yale University Press, New Haven, pp 237–248

Whitehead D R (1969) Wind pollination in the Angiosperms: evolutionary and environmental considerations. Evolution 23:28–35

Whitehead D R (1972) Development and environmental history of the Dismal Swamp. Ecol Monogr 42:301–315

Whitehead D R (1973) Late-Wisconsin vegetational changes in unglaciated eastern North America. Quat Res 3:621–631

Whitehead D R (1979) Late-glacial and post-glacial vegetational history of the Berkshires, western Massachusetts. Quat Res 12:333–357

Whitehead D R (1981) Late-Pleistocene vegetational changes in northeastern North Carolina. Ecol Monogr 51:451–471

Whitehead D R (1985) Wind pollination: some ecological and evolutionary perspectives. In: Real L (ed) Pollination biology. Academic Press, New York, pp 97–108

Whitehead D R, Sheehan M C (1985) Holocene vegetational changes in the Tombigbee River Valley, eastern Mississippi. Am Midl Nat 113:122–137

Whitehead D R, Jackson S T, Sheehan M C, Leyden B W (1982) Late-glacial vegetation associated with caribou and mastodon in central Indiana. Quat Res 17:241–257

Whittaker R H (1956) Vegetation of the Great Smoky Mountains. Ecol Monogr 26:1–80

Whittaker R H (1960) Vegetation of the Siskiyou Mountains, Oregon and California. Ecol Monogr 30:279–338

Whittaker R H (1975) Communities and ecosystems. MacMillan, New York

Wilkins G R (1985) Late-Quaternary vegetational history at Jackson Pond, Larue County, Kentucky. MS Thesis, University of Tennessee, Knoxville

Williams A S (1974) Late-glacial—postglacial vegetational history of the Pretty Lake region, northeastern Indiana. Geol Surv Prof Pap 686-B:1–23

Williams J, Barry R G, Washington W M (1974) Simulation of the atmospheric circulation using the NCAR global circulation model with ice age boundary conditions. J Appl Meteorol 13:305–317

Williams L D (1978) Ice-sheet initiation and climatic influences of expanded snow cover in Arctic Canada. Quat Res 10:141–149

Williams S (1982) The Vacant Quarter hypothesis: a discussion symposium. Paper presented at the Annual Meeting, Southeastern Archaeological Conference, Memphis, Tennessee, October 29, 1982

Wilson L R (1966) Domebo. Contrib Mus Great Plains 1:44–50

Wilson M V, Shmida A (1984) Measuring beta diversity with presence–absence data. J Ecol 72:1055–1064

Wright H E Jr (1968) History of the Prairie Peninsula. In: Bergstrom R E (ed) The Quaternary of Illinois. Spec Rept 14, College of Agriculture, University of Illinois, Urbana, pp 78–88

Wright H E Jr (1972) Quaternary history of Minnesota. In: Sims P K, Morey G B (eds) Geology of Minnesota: a centennial volume. Minn Geol Surv, St. Paul, pp 515–547

Wright H E Jr (1980) Surge moraines of the Klutlan Glacier, Yukon Territory, Canada: origin, wastage, vegetation succession, lake development, and application to the late-glacial of Minnesota. Quat Res 14:2–18

Wright H E Jr (1981) Vegetation east of the Rocky Mountains 18,000 years ago. Quat Res 15:113–125

Wright H E Jr (1984) Sensitivity and response time of natural systems to climatic change in the late Quaternary. Quat Sci Rev 3:91–131

Wright H E Jr, Patten H L (1963) The pollen sum. Pollen et Spores 5:445–450

Wright H E Jr, Watts W A (1969) Glacial and vegetational history of northeastern Minnesota. Minnesota Geol Surv Spec Pub 11:1–59

Wright H E Jr, Winter T C, Patten H L (1963) Two pollen diagrams from southeastern Minnesota: problems in the regional late-glacial and postglacial vegetational history. Geol Soc Am Bull 74:1371–1396

Appendix

List of late-Quaternary sites used in the FORMAP Project

This appendix includes the locations, chronology, and citation for each of the 162 paleoecological sites used in the construction and quantitative analysis of late-Quaternary paleo-dominance maps for eastern North America. The time range represents the specific times expressed in 10^3 yr B.P. mapped for each site. The letter codes for vegetation correspond to the mapped occurrence of forest (F), tundra (T), or prairie (P).

Site Name	Location	Time Range & Vegetation	Analyst	Publication
Albion. Quebec	45.7°N, 71.3°W	11T, 10–0F	P. Richard	Richard (1975)
Alexander Lake, Labrador	53.2°N, 60.3°W	6–0F	R.H. Jordan	Jordan (1975)
Alfies Lake, Ontario	47.9°N, 84.9°W	10–0F	M. Saarnisto	Saarnisto (1974. 1975)
Aliuk Pond, Labrador	54.4°N, 57.2°W	9–7T, 6–3F, 0T	R.H. Jordan	Jordan (1975)
Anderson Pond, Tennessee	36.0°N, 85.5°W	20–0F	H.R. Delcourt	Delcourt (1979)
Attawapisket Lake, Ontario	53.0°N, 85.2°W	5.5–0F	J. Terasmae	Terasmae (1969)
Baie St. Paul (Lac a l'Ange), Quebec	47.5°N, 70.7°W	12–10T, 9–0F	C. Labelle	Labelle and Richard (1981)
Barney Lake, Michigan	45.7°N, 85.5°W	8–0F	B. Foster, R.O. Kapp	Kapp et al. (1969)
Basswood Road Lake, New Brunswick	45.3°N, 67.3°W	14–13T, 12–0F	R.J. Mott	Mott (1975a,b)
Battaglia Bog. Ohio	41.1°N, 81.2°W	16–10F	L.C.K. Shane	Shane (1975)
Belmont Bog. New York	42.3°N, 77.9°W	14–13T, 12–0F	R.W. Spear	Spear and Miller (1976)
Bereziuk Bog, Quebec	54.0°N, 76.1°W	7T. 6–0F	P. Richard	Richard (1979)
Berry Pond. Massachusetts	42.5°N, 73.3°W	12T, 11–0F	D.R. Whitehead	Whitehead (1979)
Big Basin, Kansas	37.3°N, 100.0°W	1.5–0P	C.B. Shumard	Shumard (1974)
Big Pond. Pennsylvania	39.8°N, 78.6°W	11–3F, 1–0F	W.A. Watts	Watts (1979)
Black Pond. Tennessee	35.6°N, 84.2°W	3–0F	P.A. Cridlebaugh	Delcourt et al. (1986a)
B.L. Bigbee Oxbow, Mississippi	33.6°N, 88.5°W	10–0F	M.C. Sheehan	Whitehead and Sheehan (1985)
Blue Mounds Creek, Wisconsin	43.1°N, 89.9°W	10–0F	A.M. Davis	Davis (1977)
Bog D. Minnesota	47.2°N, 95.2°W	11–9F, 8–4P, 3.5–0F	J.H. McAndrews	McAndrews (1966)
Boney Springs. Missouri	38.1°N, 93.4°W	20–14F	J.E. King	King (1973)
Boriack Bog. Texas	30.3°N, 97.1°W	16–13F, 12–0P	V.M. Bryant. Jr.	Bryant (1977)
Boundary Pond. Maine	45.6°N, 70.7°W	12T, 11–0F	R.J. Mott	Mott (1977)
Buckle's Bog. Maryland	39.6°N, 79.3°W	18–13T, 12–0F	J. Maxwell	Maxwell and Davis (1972)
Bugbee Bog. Vermont	44.4°N, 72.2°W	10–0F	R.M. Dole. Jr.	McDowell et al. (1971)
Cahaba Pond. Alabama	33.6°N, 86.5°W	12–0F	H.R. Delcourt	Delcourt et al. (1983b)
Camp 11 Lake, Michigan	46.7°N, 88.0°W	10–0F	L.B. Brubaker	Brubaker (1975)
Chatsworth Bog. Illinois	40.7°N, 88.3°W	14–13T, 12–9F, 8–1P	J.E. King	King (1981)
Chesapeake Bay Opening. Virginia	37.0°N, 76.1°W	15–11F	J. Terasmae	Harrison et al. (1965)

Site	Coordinates	Code	Investigator	Reference
Chippewa Bog, Michigan	43.1°N, 83.3°W	10–0.5F	P.J. Ahearn	Ahearn and Bailey (1980); Bailey and Ahearn (1981)
Chism 1, Quebec	54.8°N, 76.1°W	6–0F	P. Richard	Richard (1979)
Chism 2, Quebec	53.1°N, 76.3°W	6–0F	P. Richard	Richard (1979)
Christensen Mastodon Site, Indiana	39.9°N, 85.8°W	14–12F	D.R. Whitehead, B.W. Leyden, M.C. Sheehan	Whitehead et al. (1982)
Churchill Falls South, Labrador	53.6°N, 64.3°W	5.5–0F	A. Morrison	Morrison (1970)
Clear Lake, Indiana	41.7°N, 86.5°W	12–0F	R.E. Bailey	Bailey (1972)
Colo Bog, Iowa	42.0°N, 93.3°W	12–9F, 8–0P	G.S. Brush	Walker and Brush (1964)
Cranberry Glades, West Virginia	38.2°N, 80.3°W	20–14T, 13–0F	W.A. Watts	Watts (1979)
Crates Lake, Ontario	49.2°N, 81.3°W	8T, 7–0F	K.B. Liu	Liu (1982)
Crider's Pond, Pennsylvania	40.0°N, 77.6°W	15–0F	W.A. Watts	Watts (1979)
Crusoe Lake, New York	43.1°N, 76.8°W	12–0F	D.D. Cox	Cox and Lewis (1965); Trautman and Walton (1962)
Crystal Lake, Pennsylvania	41.6°N, 80.4°W	12–1F	R.T. Hartman	Walker and Hartman (1960)
Cupola Pond, Missouri	36.8°N, 91.1°W	17–0F	E.N. Smith, Jr.	Smith (1984)
Demont Lake, Michigan	43.5°N, 85.0°W	11–0F	R.O. Kapp	Kapp (1977a)
Dismal Swamp, Virginia	36.4°N, 76.5°W	10–0F	D.R. Whitehead	Whitehead (1972)
Disterhaft Farm Bog, Wisconsin	43.9°N, 89.2°W	15–0.5F	R.G. Baker	West (1961); Baker (1970); Bender et al. (1971, 1975)
Domebo Site, Oklahoma	35.0°N, 98.3°W	11P, 0P	L.R. Wilson	Wilson (1966)
Dosquet, Quebec	46.5°N, 71.5°W	8–0F	P. Richard	Richard (1973a)
Eagle Lake, Labrador	53.2°N, 58.6°W	10–7T, 6–0F	H.F. Lamb	Lamb (1980)
East Baltic Bog, Prince Edward Island	46.4°N, 62.1°W	8–0F	T.W. Anderson	Anderson (1980)
Edward Lake, Ontario	44.4°N, 80.3°W	12–0F	C. Manville, J.H. McAndrews	McAndrews (1981)
Federalsburg, Maryland	38.7°N, 75.8°W	13F	L.A. Sirkin	Sirkin et al. (1977); Denny et al. (1979)
Ferndale Bog, Oklahoma	34.4°N, 95.8°W	12F, 11–2P, 1–0F	L.E. Albert, R.G. Holloway	Albert and Wyckoff (1981); Bryant and Holloway (1985b)
Flat Laurel Gap Bog, North Carolina	35.4°N, 82.8°W	3–0F	D.S. Shafer	Shafer (1985)

(continued)

Site Name	Location	Time Range & Vegetation	Analyst	Publication
Found Lake, Ontario	45.5°N, 78.5°W	11–0F	M. Boyko, J.H. McAndrews	McAndrews (1981)
Frains Lake, Michigan	42.3°N, 83.6°W	12–0F	W.C. Kerfoot	Kerfoot (1974)
FRB (Hayes) Lake, Ontario	49.6°N, 93.8°W	12T, 11–0F	J.H. McAndrews	McAndrews (1982)
Gabriel Lake, Ontario	46.3°N, 73.5°W	9T, 8–0F	P. Richard	Richard (1977)
Glenboro Lake, Manitoba	49.4°N, 99.3°W	13–11F, 10–3P, 2–0F	J.C. Ritchie, S. Lichti-Federovich	Ritchie and Lichti-Federovich (1968); Ritchie and Yarranton (1978)
Goshen Springs, Alabama	31.7°N, 86.1°W	20–0F	P.A. Delcourt	Delcourt (1980)
Grand Falls, New Brunswick	47.2°N, 67.7°W	12–0F	J. Terasmae	Terasmae (1973)
Grand Rapids Lake, Manitoba	53.0°N, 98.3°W	7–6P, 5.5–0F	J.C. Ritchie, K.A. Hadden	Ritchie and Hadden (1975)
Green Lake, Michigan	44.9°N, 85.1°W	13–12T, 11–0F	R.W. Lawrenz	Lawrenz (1975)
Greenbush Swamp, Ontario	45.9°N, 81.9°W	10–0F	B.G. Warner	Warner et al. (1984)
Hack Pond, Virginia	38.0°N, 79.0°W	18–0F	A.J. Craig	Craig (1969)
Harrowsmith Bog, Ontario	44.4°N, 76.7°W	10–0F	J. Terasmae	Terasmae (1969)
Helmetta Bog, New Jersey	40.4°N, 74.4°W	10–0F	W.A. Watts	Watts (1979)
Hell's Bay (Core 59-T1), Florida	25.4°N, 81.2°W	3–0F	W.L. Riegel	Riegel (1965); Spackman et al. (1966)
Hershop Bog, Texas	29.6°N, 97.6°W	11–0P	T.W. Patty	Larson et al. (1972)
Hopedale Pond, Labrador	55.5°N, 60.3°W	6–4.5T, 4–3F, 2–0T	S.K. Short	Short (1978)
Jack Lake, Ontario	47.3°N, 81.8°W	9–0F	K.B. Liu	Liu and Lam (1985)
Jackson Pond, Kentucky	37.4°N, 85.7°W	20–0F	G.R. Wilkins	Wilkins (1985)
Jacobson Lake, Minnesota	46.4°N, 92.7°W	10–0F	W.A. Watts	Wright et al. (1969)
Kenogami Bog, Quebec	48.4°N, 71.6°W	7–0F	P. Richard	Richard (1973b)
Kincardine Bog, Ontario	44.2°N, 81.7°W	12T, 11–0F	T.W. Anderson	Karrow et al. (1975)
Kirchner Marsh, Minnesota	44.8°N, 93.1°W	13–10F, 9–3.5P, 3–0.5F	T.C. Winter	Wright et al. (1963); Watts and Winter (1966)
Kogaluk Plateau Lake, Labrador	56.1°N, 63.8°W	9–5T, 4.5–3F, 2–0T	S.K. Short	Short (1978)
Kylen Lake, Minnesota	47.3°N, 91.8°W	15–13T, 12–9F	H.J.B. Birks	Birks (1981a)

Site	Latitude	Longitude		Investigator	Reference
Lac a St. Germain, Quebec	45.9°N,	74.4°W	11T, 10–0F	L. Savoie	Savoie and Richard (1979)
Lac Colin, Quebec	46.7°N,	70.3°W	11T, 10–0F	R.J. Mott	Mott (1977)
Lac Delorme II, Quebec	54.4°N,	69.9°W	6–3.5F, 3–0T	P. Richard	Richard *et al.* (1982)
Lac Faribault, Quebec	58.9°N,	71.7°W	4–0T	P. Richard	Richard (1981)
Lac Louis, Quebec	47.3°N,	79.1°W	9–0F	J.S. Vincent	Vincent (1973)
Lac Mimi, Quebec	47.5°N,	70.4°W	12–11T, 10–1F	P. Poulin, P. Richard	Richard and Poulin (1976)
Lac Nedlouc, Quebec	57.6°N,	71.6°W	4–2F, 1–0T	P. Richard	Richard (1981)
Lake 27, Michigan	45.1°N,	84.8°W	3–0F	J.C. Bernabo	Bernabo (1981)
Lake Annie, Florida	27.2°N,	81.4°W	13–11P, 10–0F	W.A. Watts	Watts (1975)
Lake of the Clouds, Minnesota	48.2°N,	91.1°W	11T, 10–0.5F	A.J. Craig	Craig (1972); Stuiver (1971)
Lake Louise, Georgia	30.7°N,	83.3°W	8–0F	W.A. Watts	Watts (1971)
Lake Mary, Wisconsin	46.3°N,	89.9°W	10–0.5F	T. Webb III	Webb (1974b)
Lake Rogerine, New Jersey	41.5°N,	74.3°W	13–1F	J. Nicholas	Nicholas (1968)
Lake Site South, Newfoundland	46.9°N,	55.4°W	13–11T	T.W. Anderson	Anderson (1983)
Lake Six, Ontario	48.4°N,	81.3°W	7–0F	K.B. Liu	Liu (1982)
Lake West Okoboji, Iowa	43.4°N,	95.2°W	14–10F, 9–0P	K. Van Zant	Van Zant (1979)
LD Lake (Sept-Iles), Quebec	50.1°N,	67.1°W	7–0F	R.J. Mott	Mott (1976)
Little Salt Spring, Florida	27.1°N,	82.3°W	9–0F	J.G. Brown	Brown (1981); Brown and Cohen (1985)
Longswamp, Pennsylvania	40.5°N,	75.7°W	12–0F	W.A. Watts	Watts (1979)
Maplehurst Lake, Ontario	43.2°N,	80.7°W	12–0F	R.J. Mott	Mott and Farley-Gill (1978)
Mauricie (Sud du Lac Du Noyer), Quebec	46.8°N,	72.8°W	10T, 9–0.5F	P. Richard	Richard (1977)
Mirror Lake, New Hampshire	43.9°N,	71.7°W	14–12T, 11–0F	M.B. Davis	Likens and Davis (1975); Davis (1978); Davis *et al.* (1984)
Mont Shefford, Quebec	45.4°N,	72.6°W	11–0F	P. Richard	Richard (1978)
Moulton Pond, Maine	44.6°N,	68.6°W	14–11T, 10–0F	T.E. Bradstreet	Davis *et al.* (1975); Bradstreet and Davis (1975)
Mud Lake, Florida	29.3°N,	81.9°W	8–0F	W.A. Watts	Watts (1969)
Muscotah Marsh, Kansas	39.5°N,	95.5°W	20–11F, 10–0P	J. Gruger	Gruger (1973)
Myrtle Lake, Minnesota	48.0°N,	93.4°W	11–0F	C.R. Janssen	Janssen (1968)
Nain Pond, Labrador	56.5°N,	61.8°W	10–5T, 4.5–3F, 2–0T	S.K. Short	Short (1978)
Nina Lake, Ontario	46.7°N,	81.5°W	10T, 9–0F	K.B. Liu	Liu (1982)
Ninepin 24, Maryland	38.3°N,	75.3°W	20–17F	L.A. Sirkin	Sirkin *et al.* (1977); Denny *et al.* (1979)

(*continued*)

Site Name	Location	Time Range & Vegetation	Analyst	Publication
Nonconnah Creek, Tennesee	35.1°N, 89.9°W	20–13F	P.A. Delcourt	Delcourt *et al.* (1980)
North Bay Bog, Ontario	46.5°N, 79.5°W	9–0F	J. Terasmae	Terasmae (1969)
Nungesser Lake, Ontario	51.5°N, 93.3°W	8–0F	J. Terasmae	Terasmae (1967, 1973)
Okefenokee Swamp (Territory Prairie), Georgia	30.9°N, 82.2°W	6–0F	L.B. Fearn	Fearn (1981); Fearn and Cohen (1984)
Old Field, Missouri	37.1°N, 89.8°W	9–3F	J.E. King	King and Allen (1977)
Panther Run Pond, Pennsylvania	40.8°N, 77.4°W	13–0F	W.A. Watts	Watts (1979)
Paradise Lake, Labrador	53.1°N, 57.8°W	10–7T, 6–0F	H.F. Lamb	Lamb (1980)
Percy Bluff, Louisiana	30.9°N, 91.5°W	13F, 5–3.5F	H.R. Delcourt	Delcourt and Delcourt (1977b)
Pickerel Lake, South Dakota	45.5°N, 97.3°W	11–10F, 9–0P	W.A. Watts	Watts and Bright (1968)
Pink Lake, Quebec	45.5°N, 75.8°W	11T, 10–0F	R.J. Mott	Mott and Farley-Gill (1981)
Pond Mills Pond, Ontario	42.9°N, 81.3°W	14–12T, 11–0F	J.H. McAndrews	McAndrews (1981)
Portage Bog, Prince Edward Island	46.7°N, 64.1°W	10T, 9–0F	T.W. Anderson	Anderson (1980)
Potts Mountain Pond, Virginia	37.6°N, 80.1°W	11–0F	W.A. Watts	Watts (1979)
Pretty Lake, Indiana	41.6°N, 85.3°W	14T, 13–0F	A.S. Williams	Williams (1974)
Prince Lake, Ontario	46.6°N, 84.6°W	11T, 10–7F	M. Saarnisto	Saarnisto (1974)
Protection Bog, New York	42.6°N, 78.5°W	12–0F	N.B. Miller	Miller (1973b)
Pyramid Hills Lake, Labrador	57.6°N, 65.2°W	7–5T, 4.5–3F, 2–0T	S.K. Short	Short (1978)
Quicksand Pond, Georgia	34.3°N, 84.9°W	20–0F	W.A. Watts	Watts (1970)
Ramsay Lake, Quebec	45.6°N, 76.1°W	10–0F	R.J. Mott	Mott and Farley-Gill (1981)
Rayburns Salt Dome, Louisiana	32.5°N, 93.1°W	20–14F, 5–0F	G.G. Fredlund	Kolb and Fredlund (1981)
Riding Mountain, Manitoba	50.7°N, 99.7°W	11–10F, 9–3P, 2–0F	J.C. Ritchie	Ritchie (1964, 1969)
Riley (R) Lake, Ontario	54.3°N, 84.6°W	8–7T, 6–0F	J.H. McAndrews	McAndrews *et al.* (1982)
Rockyhock Bay, North Carolina	36.2°N, 76.7°W	20–0F	D.R. Whitehead	Whitehead (1981)
Rogers Lake, Connecticut	41.4°N, 72.1°W	14–12T, 11–0F	M.B. Davis	Stuiver *et al.* (1963); Davis (1969); Davis and Deevey (1964)
Rose Lake, Pennsylvania	41.9°N, 77.9°W	14–0F	J.F.P. Cotter	Cotter and Crowl (1981)
Rossburg Bog, Minnesota	46.6°N, 93.6°W	12T, 11–9F, 8–7P, 6–2F	W.A. Watts	Wright and Watts (1969)

Site	Latitude	Longitude		Author	Reference
Rutz Lake, Minnesota	44.9°N,	93.9°W	11–9F, 8–2P, 1–0F	J.C.B. Waddington	Waddington (1969)
Saltville Valley, Virginia	36.9°N,	81.7°W	15–3F	H.R. Delcourt	Delcourt and Delcourt (1986)
Scott Lake, Florida	28.0°N,	82.0°W	4.5–0F	W.A. Watts	Watts (1971)
Shady Valley Bog, Tennessee	36.5°N,	81.9°W	11–0F	F.H. Barclay	Barclay (1957)
Shaws Bog, Nova Scotia	45.0°N,	64.2°W	9–0F	K.A. Hadden	Hadden (1975)
Sheelar Lake, Florida	29.5°N,	82.0°W	20–19F, 14–0F	W.A. Watts	Watts and Stuiver (1980)
Silver Lake, Nova Scotia	44.6°N,	63.7°W	11–10T, 9–0F	D.A. Livingstone	Livingstone (1968)
Silver Lake, Ohio	40.4°N,	83.7°W	14–13T, 12–0F	J.G. Ogden III	Ogden (1966)
Singletary Lake, North Carolina	34.5°N,	78.5°W	20–0F	D.G. Frey, D.R. Whitehead	Frey (1951); Whitehead (1967)
Squibnocket Cliff Peat, Massachusetts	41.4°N,	70.5°W	13–12T, 11–5F	J.G. Ogden III	Ogden (1963)
Stotzel-Leis Site, Ohio	40.2°N,	84.7°W	14–6F, 0F	L.C.K. Shane	Shane (1976)
Sugarloaf Pond, Newfoundland	47.6°N,	52.7°W	9T, 8–0F	J.B. MacPherson	MacPherson (1982)
Tamarack Creek, Wisconsin	44.2°N,	91.5°W	4–0F	A.M. Davis	Davis (1977)
Tannersville Bog, Pennsylvania	41.0°N,	75.3°W	14T, 13–0F	W.A. Watts	Watts (1979)
Torrens Bog, Ohio	40.4°N,	82.5°W	14–0F	L. Gallein, J.G. Ogden III	Goldthwaite (1965)
Track Lake, Labrador	55.8°N,	65.2°W	6–4.5T, 4–2F, 1–0T	S.K. Short	Short (1978)
Ublik Pond, Labrador	57.4°N,	62.1°W	11–4.5T, 4–3F, 2.5–0T	S.K. Short	Short (1978)
Uphapee Creek, Alabama	32.5°N,	85.8°W	7–0.5F	R.A. Christopher	Markewich and Christopher (1982)
Val St. Gilles, Quebec	49.0°N,	79.1°W	6–0.5F	J. Terasmae, T.W. Anderson	Terasmae and Anderson (1970)
Van Nostrand Lake, Ontario	44.0°N,	79.4°W	12–0F	J.H. McAndrews	McAndrews (1970, 1976)
Victoria Road Bog, Ontario	44.6°N,	79.0°W	10–0F	J. Terasmae	Terasmae (1969, 1973)
Volo Bog, Illinois	42.4°N,	88.2°W	11–0F	J.E. King	King (1981)
Weber Lake, Minnesota	47.5°N,	91.7°W	8–0F	M. Fries	Fries (1962)
White Pond, South Carolina	34.2°N,	80.7°W	19–0F	W.A. Watts	Watts (1980b)
Whitney's Gulch, Labrador	51.5°N,	57.3°W	10–8T, 7–0F	H.F. Lamb	Lamb (1980)
Wintergreen Lake, Michigan	42.4°N,	85.4°W	12–0F	R.E. Bailey	Manny et al. (1978)
Woden Bog, Iowa	43.2°N,	93.9°W	11–9F, 8–3.5P	L.H. Durkee	Durkee (1971)
Wolf Creek, Minnesota	46.1°N,	94.1°W	20–14T, 13–10F	H.J.B. Birks	Birks (1976)
Wood Lake, Wisconsin	45.3°N,	90.1°W	13–0F	K.M. Heide	Heide (1984)
Yazoo Basin, Mississippi	33.6°N,	90.1°W	13–0F	R.G. Holloway	Holloway and Valastro (1983)

Index